# The Neocortex

# Strüngmann Forum Reports

Julia R. Lupp, series editor

The Ernst Strüngmann Forum is made possible through the generous support of the Ernst Strüngmann Foundation, inaugurated by Dr. Andreas and Dr. Thomas Strüngmann.

This Forum was supported by the Deutsche Forschungsgemeinschaft

# The Neocortex

*Edited by*

Wolf Singer, Terrence J. Sejnowski, and Pasko Rakic

*Program Advisory Committee:*

David Poeppel, Pasko Rakic, Terrence J. Sejnowski,
Wolf Singer, Peter L. Strick, and Julia R. Lupp

The MIT Press

Cambridge, Massachusetts
London, England

© 2019 Massachusetts Institute of Technology and
the Frankfurt Institute for Advanced Studies

Series Editor: J. R. Lupp
Editorial Assistance: M. Turner, A. Ducey-Gessner, C. Stephen
Photographs: N. Miguletz
Lektorat: BerlinScienceWorks

The book was set in TimesNewRoman and Arial.
Printed and bound in the United States of America.

Library of Congress Cataloging-in-Publication Data

Names: Singer, W. (Wolf), editor.
Title: The neocortex / edited by Wolf Singer, Terrence J. Sejnowski, and
    Pasko Rakic.
Description: Cambridge, MA : The MIT Press, [2019] | Series: Strüngmann
    forum reports | Includes bibliographical references and index.
Identifiers: LCCN 2019019582 | ISBN 9780262043243 (hardcover : alk.
paper)
Subjects:  LCSH: Neocortex.
Classification: LCC QP383.12 .N44 2019 | DDC 612.8/25--dc23 LC record
available at https://lccn.loc.gov/2019019582

10 9 8 7 6 5 4 3 2

# Contents

# Preface

Science is a highly specialized enterprise—one that enables areas of enquiry to be minutely pursued, establishes working paradigms and normative standards, and supports rigor in experimental research. Many "problems" addressed by science, however, do not fall neatly into the scope of any one disciplinary field. For such topics, specialization can actually hinder conceptualization and limit the generation of problem-solving approaches.

The Ernst Strüngmann Forum was established in 2006 to address problems that emerge from ongoing research: topics which transcend classic disciplinary boundaries, where conceptualization may have stagnated and the way forward is uncertain. It facilitates open discourse and encourages divergent perspectives, as both are viewed as integral to the expansion of knowledge. Consensus unto itself is never a goal. Instead, topics are scrutinized from multiple vantage points to expose existing gaps in knowledge; key questions are then formulated and ways to fill such gaps are put forward.

Topic proposals are received from leading scientists who are active in their fields and reviewed by an independent Scientific Advisory Board. Once approved, a steering committee is convened to transform the proposal into a scientific framework that will support the focal meeting, or Forum, which is best imagined as a week-long intellectual retreat. Formal presentations are taboo. Instead, invitees engage in an extended dialogue to maximize intellectual output. In preparation, invited papers introduce key topics in advance to expose issues for discussion at the Forum. Four working groups approach the central topic from different perspectives. Groups work autonomously yet build on interactions with each other. To ensure that emerging insights do not get lost, each group generates a draft report during the week.

This 27th Ernst Strüngmann Forum was proposed by Wolf Singer, Terry Sejnowski, and Pasko Rakic to evaluate how research into cerebral cortex has progressed over the last three decades. The starting point was a Dahlem Workshop on the neurobiology of neocortex (Rakic and Singer 1988), followed by the 5th Ernst Strüngmann Forum on dynamic coordination in the brain (von der Malsburg et al. 2010). What trajectories did research take? Which questions are currently being confronted, and what is needed to address these now?

This volume synthesizes the resulting discourse that took place in Frankfurt, Germany, from April 8–13, 2018, and is comprised of two types of contributions. Specific aspects of the theme are presented in chapters that were originally drafted before the Forum. These "background papers" have since been revised, based on extensive peer-review, and offer an up-to-date assessment on these topics. In Chapters 5, 9, 13, and 17, the working groups provide an overview of their multifaceted discussions. Edited to ensure accessibility, these chapters should not be understood as proceedings or consensus statements.

Their intent is to summarize perspectives, expose diverging opinions and remaining open questions, as well as to highlight areas for future enquiry.

Each Forum creates its own unique dynamics and puts demands on all who participate. Every invitee played an active role throughout this process, and for their efforts, I wish to thank them all. I extend a special word of appreciation to the Program Advisory Committee, to the authors and reviewers of the background papers, as well as to the moderators of the individual working groups (Pasko Rakic, Peter Strick, Jennifer Groh, and David Poeppel). Special recognition goes to the rapporteurs of the working groups (Debra Silver, David Leopold, Kenneth Harris, and Lucia Melloni), for to draft a report during the Forum and finalize it in the months thereafter is never a simple matter. Finally, I extend my sincere appreciation to Wolf Singer, Terry Sejnowski, and Pasko Rakic, whose commitment and enthusiasm for science were essential to the entire endeavor.

To conduct its work, the Ernst Strüngmann Forum relies on institutional stability and an environment that encourages free thought. The generous support of the Ernst Strüngmann Foundation, established by Dr. Andreas and Dr. Thomas Strüngmann in honor of their father, enables the Ernst Strüngmann Forum to serve science and pursue its mandate: "to expand knowledge in basic science and identify directions for future research." In addition, I wish to acknowledge the work of our Scientific Advisory Board, the supplemental backing provided by the German Science Foundation, and the support that we receive from the Frankfurt Institute for Advanced Studies.

Breaking new intellectual ground is never easy, and it can be difficult to set aside long-held views. Yet the will to reexamine the past in the quest to identify future research strategies is a most invigorating activity. On behalf of everyone involved, I hope this volume will be successful in guiding and inspiring further enquiry into the cerebral cortex.

Julia R. Lupp, Director
Ernst Strüngmann Forum
Frankfurt Institute for Advanced Studies (FIAS)
Ruth-Moufang-Str. 1, 60438 Frankfurt am Main, Germany
https://esforum.de/

# List of Contributors

**Bassett, Danielle S.**  Department of Bioengineering, School of Engineering and Applied Sciences, University of Pennsylvania, Philadelphia, PA 19104, U.S.A.

**Bruno, Randy M.**  Columbia University, Department of Neuroscience, Zuckerman Mind Brain Behavior Institute, New York, NY 10027, U.S.A.

**Buffalo, Elizabeth A.**  Physiology and Biophysics, University of Washington, Seattle, WA 98195, U.S.A.

**Coulter, Michael E.**  Division of Genetics and Genomics and Howard Hughes Medical Institute, Boston Children's Hospital, Departments of Pediatrics and Neurology, Harvard Medical School, Boston, MA 02115; and Center for Integrative Neuroscience, University of California San Francisco, San Francisco, CA 94122, U.S.A.

**Cuntz, Hermann**  Ernst Strüngmann Institute for Neuroscience, 60528 Frankfurt am Main, Germany

**Dehaene, Stanislas**  Collège de France/Inserm, 91121 Gif sur Yvette, France

**DiCarlo, James**  Brain and Cognitive Sciences, MIT, Cambridge, MA 02139, U.S.A.

**Fries, Pascal**  Ernst Strüngmann Institute for Neuroscience, 60528 Frankfurt am Main, Germany; and Donders Institute for Brain, Cognition and Behaviour, Radboud University Nijmegen, 6525 EN Nijmegen, The Netherlands

**Friston, Karl J.**  Wellcome Trust Centre for Neuroimaging, London WC1N 3BG, U.K.

**Ghazanfar, Asif A.**  Princeton Neuroscience Institute, Departments of Psychology, Ecology, and Evolutionary Biology, Princeton University, Princeton, NJ 08544, U.S.A.

**Giraud, Anne-Lise**  Department of Basic Neuroscience, University of Geneva/Campus Biotech, 1202 Genève, Switzerland

**Gold, Joshua I.**  Department of Neuroscience, Perelman School of Medicine, University of Pennsylvania, Philadelphia, PA 19104, U.S.A.

**Grafton, Scott T.**  Psychological and Brain Sciences, University of California at Santa Barbara, Ucen Drive, Santa Barbara, CA 93105, U.S.A.

**Groh, Jennifer M.**  Psychology and Neuroscience, Neurobiology, Duke University, Durham, NC 27708, U.S.A.

**Grove, Elizabeth A.**  Department of Neurobiology, University of Chicago, Chicago, IL 60637, U.S.A.

**Haegens, Saskia**  Department of Neurosurgery, Columbia University Medical Center, New York, NY 10032, U.S.A.; and Donders Institute for Brain, Cognition and Behaviour, Radboud University Nijmegen, 6500 HB Nijmegen, The Netherlands

**Harris, Kenneth D.**  Institute of Neurology, University College London, London WC1E 6BT, U.K.

**Harris, Kristen M.**  Institute for Neuroscience, Center for Learning and Memory, The University of Texas at Austin, Austin, TX 78746, U.S.A.

**Hatsopoulos, Nicholas G.**  Department of Organismal Biology and Anatomy, and Committees on Computational Neuroscience and Neurobiology, University of Chicago, IL 60637, U.S.A.

**Haydar, Tarik F.**  Department of Anatomy and Neurobiology, Boston University School of Medicine, Boston, MA 02118, U.S.A.

**Hensch, Takao K.**  Center for Brain Science, Department of Molecular and Cellular Biology, Harvard University, Cambridge, MA 02138, U.S.A.

**Huttner, Wieland B.**  MPI Molecular Cell Biology and Genetics, 01307 Dresden, Germany

**Kaschube, Matthias**  Computational Neuroscience, Goethe University and Frankfurt Institute for Advanced Studies, 60438 Frankfurt am Main, Germany

**Laurent, Gilles**  Neural Systems Department, Max Plank Institute for Brain Research, 60438 Frankfurt am Main, Germany

**Leopold, David A.**  Laboratory of Neuropsychology, National Institute of Mental Health, Bethesda, MD 20892, U.S.A.

**Leugering, Johannes**  Institute of Cognitive Science, Department of Neuroinformatics, Osnabrück University, 49090 Osnabrück, Germany

**Lorente-Galdos, Belen**  Department of Neuroscience and Kavli Institute for Neuroscience, Yale School of Medicine, New Haven, CT 06510, U.S.A.

**MacLean, Jason N.**  Department of Neurobiology, Committees on Computational Neuroscience and Neurobiology, University of Chicago, Chicago, IL 60637, U.S.A.

**McCormick, David A.**  Yale School of Medicine, Yale University, New Haven, CT 06520-8001; and Institute of Neuroscience, University of Oregon, Eugene, OR 97403-1254, U.S.A.

**Melloni, Lucia**  Department of Neuroscience, Max Planck Institute for Empirical Aesthetics, 60322 Frankfurt am Main, Germany

**Mitra, Anish**  Department of Radiology, Washington University School of Medicine, St. Louis, MO 53110, U.S.A.

**Molnár, Zoltán**  Department of Physiology, Anatomy and Genetics, University of Oxford, Oxford OX1 3PT, U.K.

**Muchnik, Sydney K.** Department of Neuroscience and Kavli Institute for Neuroscience, Yale School of Medicine, New Haven, CT 06510, U.S.A.

**Nieters, Pascal** Institute of Cognitive Science, Department of Neuroinformatics, Osnabrück University, 49090 Osnabrück, Germany

**Oberlaender, Marcel** Max Planck Group: In Silico Brain Sciences, Center of Advanced European Studies and Research, CAESAR, 52175 Bonn, Germany

**Pesaran, Bijan** Center for Neural Science, New York University, New York, NY 10003, U.S.A.

**Petkov, Christopher I.** Institute of Neuroscience, Newcastle University Medical School, Newcastle upon Tyne NE2 4HH, U.K.

**Pipa, Gordon** Institute of Cognitive Science, Department of Neuroinformatics, Osnabrück University, 49090 Osnabrück, Germany

**Poeppel, David** Department of Psychology, New York University, New York, NY 10003, U.S.A.; and Max Planck Institute for Empirical Aesthetics, 60322 Frankfurt am Main, Germany

**Raichle, Marcus E.** Department of Radiology and Department of Neurology, Washington University School of Medicine, St. Louis, MO 53110, U.S.A.

**Rakic, Pasko** Department of Neuroscience and Kavli Institute for Neuroscience, Yale School of Medicine, New Haven, CT 06510, U.S.A.

**Reynolds, John H.** Systems Neurobiology Laboratory, Salk Institute for Biological Studies, La Jolla, CA 92037, U.S.A.

**Raut, Ryan V.** Department of Radiology, Washington University School of Medicine, St. Louis, MO 53110, U.S.A.

**Rubenstein, John L.** Nina Ireland Laboratory of Developmental Neurobiology, Department of Psychiatry, Weill Institute for Neurosciences, University of California San Francisco, San Francisco, CA 94158, U.S.A.

**Schwartz, Andrew B.** Neurobiology, University of Pittsburgh, Pittsburgh, PA 15261, U.S.A.

**Sejnowski, Terrence J.** Salk Institute for Biological Studies and Division of Biological Sciences, University of California at San Diego, La Jolla, CA 92093, U.S.A.

**Sestan, Nenad** Department of Neuroscience and Kavli Institute for Neuroscience, Yale School of Medicine, New Haven, CT 06510, U.S.A.

**Silver, Debra L.** Departments of Molecular Genetics and Microbiology, Cell Biology, and Neurobiology, Duke University Medical Center, Durham, NC 27710, U.S.A.

**Singer, Wolf**  Ernst Strüngmann Institute for Neuroscience, Max Planck Institute for Brain Research, and the Frankfurt Institute for Advanced Studies, 60438 Frankfurt am Main, Germany

**Strick, Peter L.**  Department of Neurobiology, University of Pittsburgh, Pittsburgh, PA 15261, U.S.A.

**Stryker, Michael P.**  Department of Physiology, University of California, San Francisco, San Francisco, CA 94143, U.S.A.

**Sur, Mriganka**  Picower Institute for Learning and Memory, Department of Brain and Cognitive Sciences, Massachusetts Institute of Technology, Cambridge, MA 02139, U.S.A.

**Sutherland, Mary Elizabeth**  Nature Human Behaviour, New York, NY 10004, U.S.A.

**Tosches, Maria Antonietta**  Max Planck Institute for Brain Research, 60438 Frankfurt am Main, Germany

**Tyler, William A.**  Department of Anatomy and Neurobiology, Boston University School of Medicine, Boston, MA 02118, U.S.A.

**Vinck, Martin**  Ernst Strüngmann Institute for Neuroscience, 60528 Frankfurt am Main, Germany

**Walsh, Christopher A.**  Howard Hughes Medical Institute, Division of Genetics and Genomics, Boston Children's Hospital, and Departments of Pediatrics and Neurology, Harvard Medical School, Boston, MA 02115, U.S.A.

**Zurn, Perry**  Department of Philosophy, American University, Washington, DC 20016, U.S.A.

# 1

# Introduction

Wolf Singer, Terrence J. Sejnowski, and Pasko Rakic

As science works to address any number of complex problems, a certain measure of humility must accompany its quest. Viewed over time, it is clear that myriad intricacies are often undervalued, as our collective wisdom and collaborative efforts have failed to resolve any number of issues. Although ultimate answers may be rare, this should not undercut the process of discovery or diminish the measurable progress that has been, or is currently being, made. It simply puts into context a truism: Science is an iterative process. As knowledge expands, each step forward requires us to test the concepts and ideas that emerge. To do this may require us to develop new methods or tools, which in turn may lead us to uncover completely new aspects of the problem that had hitherto escaped attention, thus bringing us back to a point where we need to evaluate, again, where things stand.

So it is, and has been, with our quest to understand the cerebral cortex.

Three decades ago, two of us (Pasco Rakic and Wolf Singer) chaired a Dahlem Workshop in Berlin on the neurobiology of neocortex. This gathering brought together forty distinguished neuroscientists from comparative and evolutionary biology, developmental neurobiology, neuroanatomy, neurophysiology, and behavioral neuroscience for an in-depth discussion of the cerebral cortex and an assessment of current research. The motivation behind this Dahlem Workshop was the realization that although research had advanced system by system and yielded an immense amount of data, the underlying rules and principles were defined for, and understood in, separate research areas, thus complicating communication and cross-disciplinary research. What was clearly lacking was an overarching theory of cortical organization—one that could account for general principles within particular areas as well as for cooperative interactions between cortical regions and cross-system generalities. From the numerous peer reviews of the results (Rakic and Singer 1988[1]), this book captured the conceptual understanding of the time and stimulated future research in developmental, cellular, functional, and cognitive neuroscience.

Years later, at an annual meeting of the Society for Neuroscience, we started to reflect on how the field had changed since that Berlin meeting: What seminal

---

[1]    Although out of press, this publication is available freely online at https://esforum.de/publications/Neocortex_1/chaps/01_Neurocortex_Rakic_and_Singer.pdf (accessed 31.3.19).

discoveries had actually been made? Which questions remained unanswered, and what might be needed to address these now? Our discussions led us to explore whether it might be worthwhile to convene another group of experts to assess where things currently stand, in an effort to position research with the conceptual means to move ever forward. Marked by the emergence of completely new disciplines, several key areas demonstrated the extent to which research had expanded dramatically over the past three decades:

- Progress in genetics and molecular biology had revolutionized neuroscientific approaches in virtually all domains, from investigations of development all the way to studies of psychiatric conditions.
- The transfection of neurons and glial cells with genetically encoded marker molecules and the development of transgenic animal models had permitted comprehensive analyses of the brain's connectome, massive parallel recording of neuronal activity at the cellular level, as well as cell-specific interference with neuronal activity.
- The advent of noninvasive imaging technologies and methods to stimulate selected regions of the human brain had boosted the field of cognitive neuroscience.
- The availability of powerful and affordable computational resources now allow us to address the large data sets that were produced through advanced electrophysiological and optical recording methods.
- Last, but not least, the rapidly growing field of computational neuroscience enables us, for the first time, to test the validity of theories and concepts through simulation experiments that are able to cope, although still in a rudimentary way, with the mind-boggling complexity and dynamics of neuronal interactions.

This progress convinced us of the necessity for a new collaboration, yet to do justice to these novel developments, the scope of expertise needed to be broadened. We found a willing partner in Terry Sejnowski, who worked with us to develop a proposal for a forum that would explore the extent to which existing data could be embedded in unifying conceptual frameworks of the neocortex.

As the reader may be aware, major changes in 2006 impacted the Dahlem Workshops, and the institution no longer exists. Its guiding spirit, philosophy, and approach, however, continue to flourish in Frankfurt under the auspices of the Ernst Strüngmann Forum. (For an overview of this transition, see Singer 2016:475–476). Briefly, the Ernst Strüngmann Forum creates an environment that ensures open discourse and encourages divergent ideas. Long-established perspectives are questioned and disciplinary idiosyncrasies exposed. Consensus is never a goal. Instead, topics are examined from multiple perspectives: existing gaps in knowledge are exposed, key questions formulated, and ways of filling such gaps (through future research) are proposed. From April 8–13, 2018, the 27th Ernst Strüngmann Forum was convened in Frankfurt, Germany, to which 48 experts from diverse areas in neuroscience participated.

Even a week-long brainstorming encounter of this kind is unable to do complete justice to the state-of-the-art research that has unfolded over three decades, much let alone provide a comprehensive summary. Far more time and effort would be needed just to review the immense amount of data that has accumulated in virtually every domain of research into the cerebral cortex. What could be perceived as a "shortcoming," however, actually gives way to an important insight: In 1987, at the Dahlem Workshop, participants were by and large aware of the developments in the various disciplines and were able to understand the concepts and terminologies used in these fields. In 2018, at the Ernst Strüngmann Forum, transdisciplinary dialogue proved much more difficult: a plethora of abbreviations characterize the language of geneticists and molecular biologists, and the mathematical descriptions of complex dynamics and the highly differentiated taxonomies used in cognitive psychology posed substantial challenges to everyone.

At Dahlem, theories on cortical processing were still dominated by behaviorist concepts, which viewed the brain primarily as a stimulus-response machine. Accordingly, emphasis was placed on serial processing in feedforward architectures. The assumption was that detailed analysis of single-cell responses across the processing hierarchy, all the way up to executive centers, should ultimately permit comprehensive understanding of the system. Hence, the field was mainly interested in describing the gradual transformation of neuronal response properties from sensory surfaces across the hierarchy of cortical processing levels to executive organs. Common concepts for the investigation of sensory processes were feature-selective receptive fields, filter operations to reduce signal-to-noise ratios and redundancies, columns as functional units, maps for the orderly arrangement of neighborhood relations, representations of cognitive objects by responses of individual neurons, and (on the executive side) motor response fields, command neurons, and population vectors. As all information was assumed to be encoded in the discharge rate of neurons, the gold standard was the single-unit recording. Signals reflecting the temporal coordination of population activity, such as field potentials and EEG, were considered too coarse, and it was felt that they provided scant additional information. With a few notable exceptions (see below), these concepts are implemented in the architecture of perceptrons and Hopfield networks as well as their recent extension in deep learning networks. Because of the astonishing performance of these artificial systems in admittedly restricted domains, and because the architecture of these artificial neuronal networks shares similarities with some of the organizational features of the cerebral cortex, one might assume that we now possess valid and explicit models of brain function and hence are close to understanding how the cortex works.

At the Ernst Strüngmann Forum, it became clear that this optimistic view is not warranted; many of the concepts favored during the Dahlem Workshop needed to be abandoned or substantially modified due to novel insights that had since been gained. Importantly, we realized that we are probably further away from a comprehensive understanding of the functions of the cerebral cortex than we imagined thirty years ago. As always in the empirical sciences, technological advances go hand in hand with conceptual developments. In addition to the still valid approach of feedforward processing, the comprehensive study of connectomics (both at the level of intracortical microcircuitry and inter-areal connections) forced us to consider

- functional implications of recurrent coupling within and between cortical areas,
- the immense density of information exchanged among processing streams,
- flat and often reversed hierarchy of putative interactions, and
- distributedness captured by graph theoretical terms such as rich club or small world networks.

These anatomical features are reflected by functional features that could, in part, have been discovered already by single-cell recordings at the time of the Dahlem Workshop. One of them is the concept of an invariant feature-selective receptive field. When feature-selective neurons were exposed to complex patterns, in particular in awake-performing animals, it became obvious how their responses are strongly sensitive to context, behavioral state, and top-down influences resulting from predictions, expectancies, and attention. It was recognized, however, that neuronal responses were variable and not always canonical, in particular in behaving animals, but this variability was attributed to noise fluctuations. The experimenter averaged over trials to extract the "essential" information, as the brain was supposed to average across a population of similar neurons. The new structural data also challenged the concept of columns as a functional unit. They suggest, at least outside input layer four, that horizontal coupling is reciprocal and continuous, even across boundaries between areas. Finally, the flat hierarchy and dense interconnectivity make it appear highly unlikely that areas operate in isolation and only serve as links in a serial processing stream.

Major arguments for an extension and reinterpretation of classical concepts came from experiments in which researchers recorded from more than one neuron at a time. It soon became clear that the fluctuations of neuronal responsiveness were correlated. Some maintain that these correlations reflect noise, hence the term "noise fluctuations." Others, however, observe that correlated firing contained information as it depended on stimulus configuration and behavioral context. Parallel recordings from electrode arrays have also revealed a puzzling but well-coordinated dynamics of cell populations. It was observed that individual neurons can engage in oscillatory patterning of their responses and that these temporally structured responses could synchronize with amazing precision in the millisecond range, depending on stimulus configurations, central states, and top-down signals. Furthermore, these oscillations are organized as traveling waves across the cortex and are both generated spontaneously and induced by stimuli. After the discovery of these coordinated population dynamics in the cerebral cortex, very similar oscillatory phenomena and traveling waves were observed in another structure sharing essential features of recurrency and connected with the cerebral cortex: the hippocampus (Muller et al. 2018). These observations led to a renaissance of interest in dynamics and in recording methods able to capture spatially and temporally coordinated (synchronized) activity of local cell populations with multiunit activity (MUA), intracortical local field potentials (LFPs), electrocorticography recordings from cortical surface electrodes and, at a still coarser spatial and temporal scale, of EEG, MEG, and fMRI signals, respectively. Together with massive parallel recordings of single-cell activity, these approaches revealed a surprising degree of temporal coordination of distributed neuronal activity, both within and across cortical areas, including the nesting of oscillatory activity across distinct frequency bands. Finally, measurements of coherence allowed identification of stimulus and task-dependent formation and dissolution of widespread functional networks and to track the flexible routing of communication between cortical areas. Although the oscillatory patterning of EEG signals in distinct frequency bands was well established at the time of the Dahlem Workshop, and although it was known that these coarse signals reflect synchronized activity, these dynamic signatures of cortical processes were not considered in a functional context: they were merely taken as a state variable correlated with changes in sleep stages and arousal levels. One likely reason is that in the 1980s, most cortical physiology focused on the visual system, and it was thought that processing of (stationary) visual patterns required no computations in the temporal domain. Since then, however, increased research has been devoted to the auditory system, speech recognition, short-term memory, motor control, and spatial navigation, and interest in dynamic processes has increased. A role of precisely timed neuronal activity has also been recognized when it became clear that mechanisms of use-dependent synaptic plasticity were exquisitely sensitive to precise timing relations between pre- and postsynaptic activity, both during development and adult learning. In parallel, computational models became more dynamic, especially those that analyze the computational potential of recurrently coupled networks.

At this Ernst Strüngmann Forum there appeared to be a broad consensus that neuronal information processing capitalizes on the spatial as well as the temporal dimensions of the brain: not only the frequency but also the timing of discharges are informative. However, we are still at the very beginning of our attempts to explore the puzzling complexity of the dynamics that emerge from delay-coupled neuronal networks and to figure out whether and, if so, how the brain actually uses the exceedingly high-dimensional state space provided by these dynamics for computation and the storage of information. One possibility is that the brain exploits these dynamics to define relations that comply with the time-sensitive learning rules for the processing of temporally structured stimuli (the processing of sequences and language) as well as for the realization of generative functions such as are required in predictive coding. In this context, it was noted as surprising that theories on cortical functions took so long to incorporate concepts of pattern generation and dynamic routing, as these had been present in the fields studying pattern generators in invertebrates, lower vertebrates, and insects.

The new evidence on the structural and functional organization of the cerebral cortex suggests that current concepts have to be considerably extended to do justice to the complexity and power of cortical computations. There was consensus that we have to learn to cope with the high-dimensional, nonlinear dynamics of the unimaginably complex interactions among the neurons of cortical networks, and that we will need new tools (e.g., machine learning) to decipher the information content in high-dimensional activity vectors as well as new mathematical instruments to analyze and interpret the trajectories of network states. Concerns were also expressed with respect to the requirement to provide causal evidence for the relations between neuronal activity and behavior. While new methods such as optogenetics and DREADDS permit cell-specific manipulation of neuronal activity, interference with the activity of nodes in a highly interconnected system may have uncontrollable consequences other than those intended. This may force the field to relax the criteria for the establishment of causal relations and in certain cases be satisfied with correlative evidence.

While the new data on connectomics and dynamics has precipitated a shift in concepts and paradigms, which is currently raising more questions than actual answers, the great advances in genetics and molecular biology have dramatically enhanced the resolution of investigations on developmental processes. The basic concepts involved in phylogenetic and ontogenetic development, formulated at the time of the Dahlem Workshop, seem to have passed the test of time. Still, much more is known now about the genetic and molecular networks that determine the birth, division cycles, migration paths, and differentiation steps of stem cells giving rise to excitatory and inhibitory neurons. Among the numerous new insights in the mechanisms determining the fate of precursor cells were the notions that inhibitory interneurons continue to be integrated into cortical circuitry during early postnatal development, that primates possess special mechanisms to increase neuron numbers in supragranular layers, and that genes have been identified that control the overall volume of the neocortex. Since participants at this Forum conduct research on a variety of animal models, the considerable species-specific differences were evident. For example, although radial glial cells in developing rodents are an excellent model to study some aspects of cortical development, the equivalent cells in primates (including humans) have specific genes and molecules as well as possess certain functional capacities that are absent in all subprimate species analyzed thus far. The difference between primary visual cortex in primates and nonprimates is obvious. Likewise, rodents do not even possess some of the cytoarchitectonic and functional areas (e.g., dorsal prefrontal association cortex, Broca and Wernicke areas), which have different neuronal composition and pattern of connections. Thus, the development, anatomy, and function of some human-specific cortical features can only be studied in humans.

In conclusion, and in keeping with the overall nature of science, this Forum was a sincere attempt to understand the major developments that have taken place in neocortex research over the last thirty years, ever with an eye toward the future. It is clear that a large number of methodological breakthroughs in all disciplines of the life sciences drove progress forward, that the analysis of massive new data (especially the big data on connectomics and molecular diversity) is reliant on powerful computational tools, and that a substantial amount of new data has been acquired only through large cooperative efforts, as opposed to research in small groups characteristic of neuroscientific investigation thirty years ago. Equally, however, it is clear that conceptualization lags behind data accumulation. Thus, we posit that the greatest challenge for future endeavors will be to integrate the plethora of facts generated by the highly diverse fields of research into an overarching comprehensive theory on cortical functions. Whether this is at all possible—whether there is even such a thing as a unifying theory of neocortex—remains an open question. Perhaps accumulated knowledge must remain distributed across the community of specialized experts, similar to how functions of the cerebral cortex are distributed. Just as the brain, as a whole, produces intuitively plausible behavior, distributed knowledge might serve to explain a large number of normal and pathological behaviors, ultimately enabling the development of useful tools without meeting the epistemic challenge of having to fit into a unified theory.

## Acknowledgments

We wish to express our gratitude to Silke Bernhard, the director of the Dahlem Workshops from 1974 until 1989. We know of no better way to pay tribute to her memory than to continue the dialogue that took root in Berlin some thirty years ago. Equally, we wish to acknowledge the efforts of Andreas and Thomas Strüngmann, who have enabled this unique approach to continue in Frankfurt. Their vision and support of the Ernst Strüngmann Forum is invaluable to basic science, for it offers researchers a much-needed yet rare opportunity to reflect, to reanalyze, to correct, and to propose directions for future research to pursue. On behalf of all participants, we thank you sincerely.

# Evolution and Ontogenetic Development of Cortical Structures

# 2

# Cortical Specification and Neuronal Migration

William A. Tyler and Tarik F. Haydar

## Abstract

The extraordinary complexity of the mammalian neocortex is the result of millions of years of evolution. Elucidating the principles underlying its development and function has been a major goal in the neurosciences. How a seemingly uniform group of neuroepithelial stem cells produces the vast array of electrically responsive cell types, and how these resulting cells establish such a rich variety of circuits in the mature neocortex remains, in particular, a key focus of the field. This chapter reviews seminal advances in understanding the production, specification, and migration of neocortical neurons prior to the establishment of mature circuits.

## Introduction

Since the introduction of the basic principles of neocortical development at the Dahlem Workshop in 1987 (Rakic 1988a), such as the *protomap hypothesis* and the *radial unit hypothesis*, advances in molecular biology and imaging techniques have revolutionized our ability to interrogate developmental processes in the brain. These technical capabilities have uncovered many new molecular and cellular pathways that operate during neocortical growth, many of which confirm and extend the above-mentioned theories. Here we review the basic principles of neocortical development and highlight key controversies and emerging areas where additional studies are needed.

Although considerable differences exist across mammalian species in terms of timescale and architecture of cerebral cortical development, several major events occur in all mammals. The first stages of neocortical development are characterized by rapid proliferation that leads to exponential expansion of the neuroepithelial progenitor cells (NEPs) within the pseudostratified neuroepithelium (or the ventricular zone, VZ) lining the lateral ventricles of the prosencephalon. This earliest period of precursor proliferation has been termed the "founder cell expansion phase" and is one of the most important phylogenetic determinants of cortical size between species (Rakic 1988b, 1995) (Figure 2.1). Once the telencephalon emerges, rapid NEP proliferation continues until the molecular specification of apical radial glial cells (aRGCs), a cell type now considered to be neocortical stem cells, as they possess the ability to self-renew as well as to generate the mature cell types of the neocortex. The transition from NEP to aRGC occurs at different gestational times, depending on species (e.g., at E10 in mice and ~E40 in primates), prior to the generation of the first neocortical neurons—an event which occurs quickly thereafter. The duration of the founder cell expansion phase controls the resulting number of individual precursor cells. It is thought that these founder cells represent the primordial cortical units that establish columns of neurons in the neocortex.

Following the establishment of aRGCs in the VZ, the first excitatory neurons are born from asymmetrical divisions that result in a newborn neuron and a self-renewed aRGC (Malatesta et al. 2000; Hartfuss et al. 2001; Miyata et al. 2001; Noctor et al. 2001; Tamamaki et al. 2001; Noctor et al. 2002). These neurons exit the VZ and establish the first postmitotic layer of neocortical neurons. In rodents, this first layer is called the preplate and lies just superficial, or basal, to the VZ. In rodents, subsequently generated neurons migrate along the basal fibers of aRGCs (so-called gliophilic migration) to split the preplate

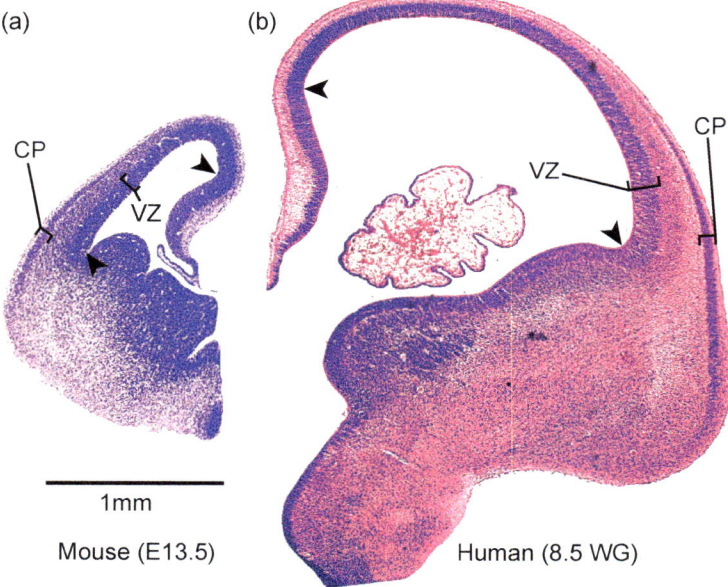

(a)                    (b)

CP

VZ

VZ

CP

1mm

Mouse (E13.5)                    Human (8.5 WG)

**Figure 2.1**    Species-specific effects of founder cell expansion on neocortical size. Coronal sections of E13.5 mouse (a) and 8.5 WG human (b) brains—comparable stages of fetal development—highlight the increase in area of the human ventricular zone (VZ), between arrowheads, due to founder cell expansion. Both brains display the onset of cortical neuron arrival into the cortical plate (CP). Adapted from Tyler and Haydar (2010).

into the deep subplate and the superficial marginal zone (neocortical layer 1). In primates, the firstborn neurons are thought to directly form the cortical plate (Smart et al. 2002). In all mammals, migrating neurons then form the remaining cortical layers (2–6) in an inside-out manner based on their date of birth. This temporal specification results in the later-born neurons migrating past the earlier-generated neurons to form each successive superficial neocortical lamina (Rakic 1975). Once they have arrived at their proper destination, neocortical neurons differentiate molecularly (expressing specific laminar marker genes) as well as cellularly (extending axons to postsynaptic targets and elaborating complex dendritic trees). In addition to producing neurons directly, another property common to all mammals is the ability of aRGCs to generate intermediate precursor cells (IPCs) during the course of neurogenesis. As will be detailed below (see section on "Precursor Heterogeneity"), these IPCs contribute to the expansion of neocortical layers by producing additional neurons. While it is well-established that IPCs come in a variety of different forms, whether and how each cell type uniquely contributes to neocortical formation is only now becoming understood.

Whereas excitatory neurons are produced in the dorsal telencephalon, inhibitory interneurons are generated from precursor cells in the ganglionic eminences of the ventral telencephalon. Once generated, nascent interneurons migrate tangentially into the dorsal telencephalon along axonal fibers (so-called neurophilic migration) and may switch to gliophilic migration as they near their final destination (Polleux et al. 2002). Thus, production of the proper ratio of excitatory to inhibitory neurons is a critical aspect of proper neocortical development (and is altered in some developmental disabilities). Moreover, how the tangentially migrating interneurons coalesce with their radially migrating excitatory cousins in a proper laminar fashion, and with appropriate density, has not yet been fully discovered. It is now known that key, molecularly different germinal fields within the ganglionic eminences produce specific subtypes of interneurons, and the genetic code for this spatiotemporal specification is beginning to be understood (Flames et al. 2007). This causal relationship between differential gene expression in the germinal zones and the fate potential of the daughter cells generated in those areas also exists in the dorsal telencephalon, as we explore further below.

The developmental events outlined above occur in all known mammals, but there are important species-specific differences that are important to consider when arriving at a comprehensive understanding of the developmental mechanisms that govern growth and formation of the neocortex. These differences separate "smooth brain" lissencephalic species (e.g., rodents) from gyrencephalic species (e.g., carnivores and primates). In general, peculiarities in the structure of the gyrencephalic cerebral wall results from precursor cell compartmentalization during fetal development. This organization is thought to play a role in the development of the larger and more complex circuitry found in the gyrencephalic neocortex. One of these specialized architectonic features is the splitting of the subventricular zone (SVZ) into inner and outer compartments by the inner fiber layer (IFL) (Smart et al. 2002). While the inner SVZ resembles the SVZ found in rodent neocortex, the outer SVZ (oSVZ) is greatly expanded in ferret and primates. The predominant precursor cell type in the oSVZ is the basal RGC (bRGC) which has been shown to self-renew and generate neurons via asymmetrical divisions (Fietz et al. 2010; Hansen et al. 2010). The concurrent expansions in the numbers of bRGCs and the size of the oSVZ are thought to underlie the increased radial growth as well as the convoluted surface of the gyrencephalic neocortex. In addition, there are several neuronal groups that appear to be unique to gyrencephalic neocortex, including subpial granular neurons and an expanded population of subplate neurons (Kostovic and Rakic 1990; Meyer et al. 2000). All of these findings suggest that while key mechanisms of neocortical development can, and for many reasons must, still be elucidated in lissencephalic species, the field of neocortical development has arrived at a stage where novel findings of fundamental mechanisms must be confirmed, or at least queried, in gyrencephalic brains as well.

## Control of Mode of Division

The cellular transitions that enable the switch from NEP → aRGC → neuron (direct neurogenesis) or from NEP → aRGC → IPC → neuron (indirect neurogenesis) have a large impact on the eventual size and neuronal complexity of the neocortex. Here we define "mode of division" as the mechanisms that operate within or upon a dividing cell to result in either symmetrical or asymmetrical divisions. Symmetrical divisions occur when the resulting daughter cells share the same fate, whereas asymmetrical divisions lead to daughter cells with different fates. The importance of control of mode of division in neocortical development was first promulgated in the radial unit hypothesis three decades ago (Rakic 1988a, b). Since then, many studies have shown that cell-cycle duration, cleavage plane orientation, diffusible factors (extracellular cues), nascent gene expression, and changes in precursor morphology together control the proper timing and extent of these transitions.

During the founder cell expansion phase prior to the onset of neurogenesis, NEP numbers grow exponentially as the cells symmetrically produce two new NEPs. Following this, NEPs must undergo "consuming" symmetrical divisions since they are rapidly replaced by RGCs, but the factors controlling the transition between these two types of apical precursor cells have not been conclusively identified. Once generated, aRGCs mainly divide asymmetrically, signaling the onset of the neurogenesis period, to produce either neurons or other IPCs which will themselves generate neurons. Many of the extrinsic and intrinsic factors influencing these aRGC divisions have been identified, including Shh, Wnt, BMPs, FGF, IGF, and *FOXG1* (Grove et al. 1998; Lako et al. 1998; Hanashima et al. 2002; Assimacopoulos et al. 2003; Abu-Khalil et al. 2004; Hanashima et al. 2004; Medina et al. 2004; Shimogori et al. 2004; Storm et al. 2006; Clowry et al. 2018). However, as we discuss below (see section on "Precursor Heterogeneity"), all of the IPC cell types known to exist in the mammalian cerebral wall appear to be derived from aRGCs by asymmetrical divisions (i.e., divisions yielding a self-renewed aRGC and an IPC). The factors controlling the genesis of each of these IPC classes have not been identified, and whether this process is stochastic or tightly programmed is as yet unknown. This is a critical knowledge gap, especially since many of these precursor types are simultaneously present during neurogenesis and contemporaneously produce daughter neurons at any given time. In addition, there is evidence that IPC diversity is altered in certain developmental disabilities, such as Fragile X and Down syndrome (Saffary and Xie 2011; Tyler and Haydar 2013). It is also well established that the aRGC cell cycle gradually lengthens during neurogenesis due to increases in S and G1 duration (Takahashi et al. 1995; Turrero Garcia et al. 2016). This increase in cell-cycle length is thought to play a primary role in neuronal production, and recent data also indicate that lengthening the M-phase can lead to precocious neurogenesis at the expense of the precursor pool (Pilaz et al. 2016). Following the asymmetrical division phase of neurogenesis, the aRGC population is largely exhausted by "consuming" symmetrical divisions resulting in two daughter cell neurons. During this stage, it has also been established that aRGCs are direct precursors both to cortical astrocytes and the neural stem cell population that persists in the adult brain, although the factors regulating these developmental pathways remain to be conclusively identified.

## Cleavage Plane Orientation and Segregation of Apical Factors

The angle of the mitotic cleavage plane in relation to the apicobasal polarity of dividing precursors was first identified as critical for specifying mode of division in yeast and in *Drosophila* and *Caenorhabditis elegans* (Skop and White 1998; Theesfeld et al. 1999; O'Connell and Wang 2000; Doe and Bowerman 2001). The overall mechanism at play is that cleavage angle modifications can lead to even or uneven partition of fate-determining molecules to the resulting daughter cells. Numb, Prospero, Pon, and the Par complex (among others) have been identified as key players in this process, and their distribution is affected by vertical or horizontal divisions in many species (Chenn and McConnell 1995; Huttner and Brand 1997). While many groups have shown that mitotic cleavage plane is also important for mode of division in the developing neocortex (especially in the VZ), it is now clear that most VZ divisions occur with relatively little cleavage angle variation (most cleavages occur with vertical cleavage planes). However, even minor deviations from the vertical cleavage plane can result in unequal partitioning of the apical plasma membrane and its associated components, such as cadherin, prominin, and apical junctional complexes (Wang et al. 2009; Kim et al. 2010; Postiglione et al. 2011). Thus, a consensus view has emerged that cleavage plane modification is part of the coordinated process of controlling neurogenesis in the mammalian neocortex. Significant controversy remains, though, in terms of the precise fate of the resulting daughter cells following different orientations (i.e., which daughter differentiates in asymmetrical divisions). This is due primarily to technical challenges in using real-time imaging to follow a cell through mitosis and then track the resulting daughter cells until their fate can be determined. Moreover, because the three-dimensional environment (including cell–cell interactions and gradients of signaling factors) is critical for this process, *in vivo* live imaging is necessary to describe fully how mode of division is controlled during fetal neocortical development.

## Association between Cell Cycle and Mode of Division

In addition to the shift in cleavage plane orientation during the fetal neurogenesis period, the neural precursor cell cycle lengthens during neurogenesis, primarily in G1 phase, and these two physiological changes cooperate to yield the proper numbers of neurons in the overlying cortical plate. Moreover, regional changes in cell-cycle kinetics across the developing neocortical wall fine-tune areal differences in neuronal number and laminar thickness. For example, the neighboring Brodmann areas 17 (primary visual) and 18 (visual association) of primate neocortex differ considerably in the number of supragranular neurons (more in BA17), presumably leading to a discrete functional competence in the visual cortex. Seminal work in 2005 demonstrated significant shortening in cell-cycle duration in the oSVZ of BA17, modulated by differential expression of p27[Kip1] and Cyclin E, enabling supragranular layers in BA17 to expand in comparison to the adjacent layers in BA18 (Lukaszewicz et al. 2005). While this study and others demonstrate some of the cellular mechanisms that lead to final control of neuron number within each area, it remains unclear how specific cell classes within layer 4, such as the spiny stellate neurons, are differentially produced. Regardless, the Lukaszewicz et al. (2005) study illustrates how the primary signals of areal demarcation can be operationalized as the supragranular layers in a given area are generated. In the relevant sections below, we discuss how intrinsic differences, diffusible factors, and various feedback mechanisms within the neocortical wall may provide the initial map and control the implementation of this developmental program.

## Precursor Heterogeneity

It is now known that a variety of neural precursors contribute to the generation of excitatory neurons and the overall expansion of the neocortex. At the center of this process, aRGCs serve as stem cells: they give rise to neurons that form all six layers during the neurogenic interval and subsequently contribute to the production of glial lineages. After the founder cell expansion period ends, aRGCs give rise to an increasing number of IPCs, which in turn amplify neuronal output (Figure 2.2). To date, three distinct classes of IPCs have been characterized in the dorsal germinal compartment of all mammals based on differences in gene expression, morphology, and location of mitoses. These include basal IPCs (bIPCs), apical intermediate progenitors (aIPCs), and bRGCs. The first to be identified were the bIPCs: multipolar cells which form a second proliferative area (the SVZ) that overlays the VZ. These cells express the transcription factor T-box brain protein 2 (Tbr2, or Eomes) protein and undergo primarily symmetric divisions to produce neurons, with increased numbers contributing to the formation of the supragranular layers (Englund et al. 2005; Kowalczyk et al. 2009). Like bIPCs, aIPCs undergo exhaustive symmetric divisions to produce neurons. However, aIPCs reside in the VZ, express paired box 6 (Pax6) but not Tbr2 protein, and divide at the ventricular surface (Gal et al. 2006; Stancik et al. 2010; Tyler and Haydar 2013). Like aRGCs, aIPCs exhibit a radial morphology and maintain contact with the ventricle via an apical process but lack a basal process that extends to the pia mater (Mizutani et al. 2007; Elsen et al. 2013; Nelson et al. 2013; Pilz et al. 2013). While aIPCs constitute a considerable portion of the progenitor pool during mid-neurogenesis in the rodent (Gal et al. 2006; Tyler and Haydar 2013), many details about their numbers and contribution to neuronal production during early and late neurogenesis and across species remain to be elucidated. It should be noted that apical neural precursors without pial-contacting basal processes were recently described in the fetal human brain (Nowakowski et al. 2016). Whether these cells represent the primate version of aIPCs is currently unresolved. A third class of IPCs, bRGCs, undergoes mitoses in the intermediate zone (IZ) and oSVZ, and maintains a long basal fiber extending to the pia. bRGCs were first identified in the developing human brain as an expanded population of neurogenic progenitors in the oSVZ (Fietz et al. 2010; Hansen et al. 2010; Shitamukai et al. 2011; Betizeau et al. 2013). At the molecular level, bRGCs resemble aRGCs in that they express canonical stem cell markers including Pax6 and Sry-box2 (Sox2). Furthermore, bRGCs exhibit a greater propensity to undergo self-renewing or transient-amplifying divisions compared to bIPCs and aIPCs. While these descriptions denote the major classes of neuron-producing progenitors, further complexity has been suggested by evidence which shows that bRGCs and bIPCs are also heterogeneous with respect to their morphology and division parameters. For example, at least five distinct bRGC types with unique lineal attributes have been identified in the primate (Reillo et al. 2011; Pilz et al. 2013; Pfeiffer et al. 2016). Thus, emerging evidence suggests the existence of subclasses of precursors within each of these three IPC types.

The biological significance of neural precursor variation is only now beginning to be understood. Precursor diversity has been postulated to contribute

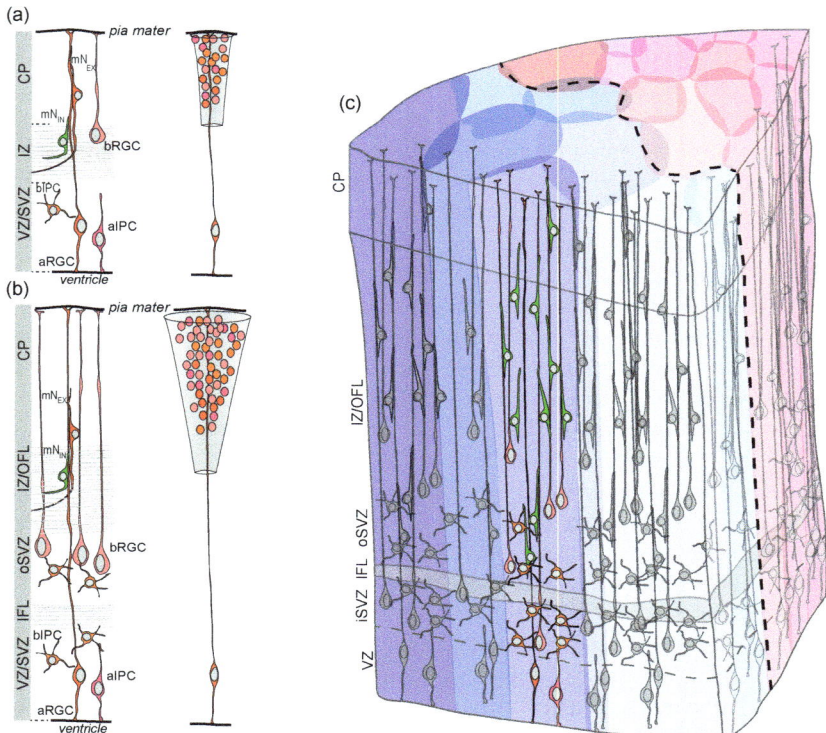

**Figure 2.2** Precursor diversity and clonal output during dorsal cortical neurogenesis. Mouse (a) and primate (b) neocortical progenitors include apical radial glial cells (aRGCs), apical intermediate precursor cells (aIPCs), basal intermediate precursor cells (bIPCs), and basal radial glial cells (bRGCs). Migrating excitatory and inhibitory neurons (mN$_{EX}$ and mN$_{IN}$) climbing to the cortical plate (CP) through the intermediate zone/outer fiber layer (IZ) are also depicted. While both lissencephalic and gyrencephalic neocortices contain similar precursor types, there are comparatively more bRGCs and bIPCs in the gyrencephalic brain neocortical wall, leading to a larger neuron output from each aRGC. Furthermore, these precursors are split into the inner and outer subventricular zones (iSVZ and oSVZ) by the inner fiber layer (IFL). Increased output from the aRGC-derived precursor cells during the late stages of neurogenesis in gyrencephalic species leads to a conical shape of the radial unit due to increased production of supragranular neurons. (c) Cartoon depicting a section of gyrencephalic neocortical wall taken near the boundary of two cortical areas (pink and blue areas separated by dashed line). Each area is populated by the clonal distribution of neurons from aRGCs. Each aRGC generates a cone of neurons through direct and indirect neurogenesis (shaded radial stripes) and the boundaries of these cones of neuronal allocation are thought to slightly overlap.

to two main processes that (a) increase the size of the neocortex, potentially leading to gyrification, and (b) generate neuron diversity. A primary focus of the field for the last decade has been the role of precursor heterogeneity and how it relates to the neocortical growth of higher mammalian species. Several studies have noted that brain size is correlated with an enlargement of both subdivisions of the SVZ and with increased numbers of IPCs. In particular, the gyrification index of a wide array of species has been linked to the size of the bRGC population (Fietz et al. 2010; Hansen et al. 2010). While species with a lissencephalic cortex contain bRGCs, and some (e.g., marmoset) even develop a distinct oSVZ compartment (Garcia-Moreno et al. 2012; Kelava et al. 2012), the numbers of bRGCs within these brains are reduced relative to gyrencephalic species. Nonetheless, how bRGCs contribute on a macro scale to cortical formation is still at a theoretical stage. For instance, the human brain has expanded more in the lateral dimension than in the radial dimension (thickness) and it is unclear how and whether bRGCs lead to a preferential increase surface area rather than thickness. Because of these unanswered questions, a primary mechanism driving species-specific differences in brain growth remains the expansion of the founder cell population prior to the onset of neurogenesis (Figure 2.1). Taken together, differences in brain size across the phylogenetic tree are likely due to the synergistic effects of a larger founder cell population and species-specific differences in the composition of the IPC pool as well as changes in the total neuronal output per precursor type.

A second role for the precursor heterogeneity in brain development may be that it contributes directly to neuronal diversity. The classes of excitatory neurons which comprise the six layers of the neocortex have been characterized by their birth date, molecular expression, electrophysiology, and target-specific projections. Clonal analysis by genetic fate-mapping has shown that individual aRGCs can produce neurons that span all of the neocortical layers, suggesting that aRGCs may be progressively tuned over developmental time to generate different types of excitatory neurons. Transplantation experiments indicate that isolated aRGCs generate neuron types appropriate for their birth date, even when placed in a heterochronic environment (McConnell and Kaznowski 1991), further supporting the notion that aRGCs undergo temporal fate restriction. Another assumption contained within this progressive fate restriction model is that aRGCs of any given developmental age are all identical, but *in vivo* evidence for this assumption is thus far unconvincing. It is also important to note that the molecular mechanisms underlying this temporal fate restriction have not yet been identified. Nevertheless, while this model may explain a primary mechanism for producing the different neuronal cell types across neocortical layers, it does not describe a method for generating different types of neurons within each layer.

Morphological and electrophysiological studies have defined distinct intralaminar populations of excitatory cells and shown that intralaminar diversity differs across areas. At any given time during the neurogenic interval, neurons are contemporaneously produced directly from aRGCs as well as indirectly from multiple classes of IPCs. As shown from birth-dating studies, these neurons are destined for the same neocortical layer, and mouse studies have suggested that neurons born via these parallel routes mature into distinct types of neurons. In particular, we have recently shown in mouse frontal cortex that layer 2/3 neurons generated from Tbr2-expressing progenitors differ from contemporaneously produced non-Tbr2-derived neurons with respect to their electrophysiological properties and the complexity of their apical dendritic arbors (Tyler et al. 2015). These results indicate that neurons are seeded with information from their parental lineage; this information is retained as they migrate to the appropriate layer and manifests a lineage-specific morphological and electrophysiological profile. To test whether this precursor lineage model for intralaminar neuronal diversity is a general rule that applies also to deeper layers and across different areas of the neocortex, we repeated this effort in mouse somatosensory neocortex by fate-mapping the same lineages during production of layer 4 (Guillamon-Vivancos et al. 2018). Results indicate that neurons derived from Tbr2 and non-Tbr2 lineages establish unique settling patterns within the somatosensory barrels and are different in terms of dendrite complexity and firing patterns (as in layer 2/3 of frontal cortex) as well as in their synaptic coupling with thalamocortical afferents.

Collectively, these studies suggest that precursor programs may directly influence how neocortical neurons participate in neocortical microcircuits. However, the lack of understanding of the true scale of precursor heterogeneity is a current roadblock. For example, several different types of aRGCs may exist and there may also be many subtypes of bIPCs and bRGCs. Indeed, there has been a rapid increase in identified neocortical precursor cell types over the past decade, including subapical progenitors (SAPs) (Pilz et al. 2013), quiescent or laminar-fated aRGCs (Franco et al. 2012), and, most recently, truncated RGCs in second trimester human neocortex which resemble aIPCs (Nowakowski et al. 2016). These cells are primarily identified based on morphological criteria and division site, as well as by time-lapse imaging in cortical slices. Despite this type of evidence, newly identified precursor types will remain controversial until they can be molecularly identified. Indeed, one of the rate-limiting steps in this line of research is the development of molecular markers for *in vivo* use of all of the different types of morphologically identified precursor cells. Only once the true scale of precursor heterogeneity is elucidated can comprehensive studies that investigate the role of individual cell types in cortical development and/or evolution be established.

In recent years, single-cell transcriptomics has provided new insight into the molecular signatures of neural precursors in humans, primates, and rodents. Several studies have successfully identified the core gene expression patterns of aRGCs, bIPCs, and to a lesser extent bRGCs, as well as specific differences which may underlie the properties of neuronal progenitors across species. However, the regional differences in aRGC expression predicted by the protomap hypothesis are as yet absent in single-cell RNA sequence analysis. While some studies have presented molecular signatures for potential subpopulations of progenitors, for the most part the number of unique cell types identified has not reflected the full scope of heterogeneity predicted by histological and time-lapse studies. For instance, the five types of bRGCs identified in the primate brain (Betizeau et al. 2013) have not yet revealed themselves as distinct cell classes at the RNA expression level, nor has the genetic fingerprint of aIPCs, SAPs, and truncated RGCs been elucidated. It has already been established that genes like *Trnp1* and *AP2γ* may be expressed in aRGCs producing bIPCs (Pinto et al. 2008, 2009; Stahl et al. 2013), but whether these genes mark divisions yielding other IPCs, such as aIPCs and bRGCs, is not yet clear. Are there molecular differences between these aRGCs (i.e., are there individual subtypes of aRGCs), or is each one of these asymmetrical divisions a stochastic choice? One of the basic assumptions in the field is that any given particular division outcome is reflected by a specific gene expression, yet these molecular signatures remain elusive. The lack of conclusive data in this area raises important questions regarding what defines a cellular "type" versus a distinct cellular "state." One possibility is that the full scope of cellular phenotypes may not be revealed at the RNA level. Instead, translational control may exert an as yet unappreciated influence on progenitor biology. In support of this hypothesis, several studies in progenitors have suggested that certain transcripts may be expressed but not translated into protein, providing a priming mechanism to accelerate the differentiation/commitment of their daughter cells upon cell division (Pinto et al. 2009; Albert et al. 2017).

## Regional Specification of the Neocortical Map

The idea first ramified in the protomap hypothesis—that regional identities first emerge within the neural precursors and that this sets the stage for the mature functional cortical areas—is now widely accepted. Intrinsic characteristics of neocortical precursor cells, such as gene expression, cell-cycle duration, and fate potential, vary across the dorsoventral and rostrocaudal axes of the developing neocortex prior to the influence of subcortical input. In general, the concept that diffusible morphogens circulate across the developing precursor cells and instigate proliferative responses has been proven in multiple CNS areas and in many different species. In rodents and primates, the signaling centers for these various factors, including Shh, Bmps, Wnts, COUP-TFs, and FGFs, develop in multiple and discrete regions of the nascent telencephalon and release their contents during the founder cell expansion phase (Sur and Rubenstein 2005; Clowry et al. 2018). The crossing gradient fields established by these disparate signaling centers lead to regional expression of transcription factor genes in the precursor zones, including *PAX6*, *EMX2*, *COUP-TF1*, *SP8*, *PEA3*, *BHLHB5*, and *OLIG2*. These transcription factors set in motion a cascade of gene expression events that consolidate both the genetic and positional identities of the constituent precursor cells. Recent evidence also suggests that neurogenic factors released into the cerebral spinal fluid (CSF) generate proliferative responses in the neocortical germinal zones. Released into the CSF from the choroid plexus during fetal development, IGF1 and IGF2 can also modulate cortical growth (Lehtinen et al. 2011). There are potentially dozens of molecules in the CSF that can exert similar effects, but whether these factors initiate or merely act upon regional differences in the neocortex remains unclear. The absence of discrete sites of release of these factors and the circulation of fluid within the ventricles makes them an unlikely primary factor for regional diversity, for example, between directly adjacent areas such as BA17 and BA18.

Once the protomap has been established in the precursor cells, part of the intrinsic program includes the expression of cell surface molecules, including cadherins, protocadherins, neurexins, and ephrin receptors that distinguish each area. These molecules are expressed on the cell membranes within each area as well as within the local extracellular matrix. The expression of this "areal marking" mechanism on the axonal membrane of cortical efferents is also thought to be critical for attracting the proper classes of reciprocal thalamic projections as well as ingrowing interneurons that migrate from the basal telencephalon (handshake hypothesis). Once the interneurons have entered the neocortex, cell-adhesion molecules within their growth cones linked to their cytoskeleton mediate the final decision of whether or not to integrate into a particular area.

While the radial migration of cortical pyramidal neurons along the basal fibers of aRGC occurs within each radial unit and does not necessarily rely on areal information, the long-range migration of cortical interneurons and the extension of axonal fibers to disparate targets require mechanisms that confer a navigation system to the developing cortical map. Surprisingly, several pieces of evidence suggest that the spherical telencephalon is organized by a rectilinear map onto which migration routes and axonal pathways are superimposed, much in the way that lines of longitude and latitude organize the surface map of the earth.

### Evidence for a Rectilinear Map

Two major pieces of data indicate that the telencephalon may be organized by an orthogonal grid. First, when tangentially migrating interneurons are imaged near the neocortical pial surface, either in time-lapse imaging studies or by electron microscopy, they are oriented predominantly along perpendicular axes. Intriguingly, the Cajal-Retzius (CR) cells overlying these migrating cells are also oriented along the same axes, indicating that the same cues organizing CR morphology may also direct interneuron migration (Ang et al. 2003). Second, recent diffusion magnetic resonance imaging has shown that cortical fiber pathways in primate brain also conform to a three-dimensional grid, with pathways crossing each other along three main axes (Wedeen et al. 2012). Furthermore, axons labeled with tract tracers turn with near 90-degree precision along these grids and axons from individual neurons labeled with the Golgi impregnation technique (Mortazavi et al. 2017), or with intracellular fluorescent tracers, have long been known to emanate collaterals from 90-degree branch points along the main axonal shaft. More work is needed to identify the molecules that may form such a grid and to determine if the white matter and interneuron migration grids are identical. Such a grid, if present, would constitute a coordinate system laid over the intrinsic determinants of the protomap, enabling a framework for long-range migration and tract formation as the neocortical surface grows and convolutes during development.

## Local Implementation of the Neocortical Map

Once cortical areas have been established in the progenitor zones, several mechanisms ensure that the local program for each area is maintained during radial expansion of the overlying neocortical wall. In all mammals, the fact that regions across the neocortical map develop along different timelines elucidates a general maturational gradient. Within the context of this gradient, changes in mode of division, cell-cycle duration, and precursor subtype control the number and types of neurons across the neocortical laminae within the local area. However, there are many areas of the neocortex that deviate from this gradient-based development, where major differences in gray matter thickness are quite evident between adjacent areas. Moreover, the numbers and types of cells within each lamina are variable across areas and this again is most strikingly observed in the transition between BA17 and BA18 of the primate neocortex. In particular, stellate neurons in the granular layer 4 of BA17 are numerous while they are absent or at least sparsely present in layer 4 in the neighboring BA18. This indicates that areal identity, transmitted initially very early during formation of the telencephalon, is maintained throughout neurogenesis so that specific cellular landscapes develop in each area across the neocortex. These landscapes include markers that recruit ventral interneuron cell types, and this unique cellular and molecular milieu, combined with the ensuing afferent synapses, yield the final architectonic character and function of each cortical area. The fact that some of these area differences are very stark (i.e., not simply a gradation when compared to neighboring areas) indicates that local mechanisms can independently control radial size of the neocortex as well as the intralaminar neuronal diversity within each area. It is important to emphasize that cortical afferents are known to have the ability to modulate area-specific programs of development and to participate in the morphological development of key structural areas in the neocortex including barrels and functional columns. However, it is the combination of the unique marks that identify an area and the precise level of diversity in neuronal number and type that provides the playing field for these afferents.

The transmission of this local developmental program necessitates at least two biological events: the first requirement is a unique molecular "agenda" that is maintained in the precursors during neuronal production so that the proper numbers and types of neurons are produced. This manifests as local control of cell-cycle duration, mode of division, and precursor heterogeneity. Several key studies indicate that gene expression can specify the type of neuron to be produced so that the full panoply of cortical excitatory neurons can develop in the expanding neocortical wall. As an example, we can consider the molecular mechanisms specifying subcortical projection neurons. The gene for the zinc finger protein Fezf2 (or Zfp312) is necessary for subcortical projection neuron morphology and formation of the corticospinal tract. Fezf2 is expressed highly in aRGCs during early phases of neurogenesis when the projection neurons of layers 5 and 6 are generated and is then downregulated during later phases of neurogenesis. As the first-generated neurons migrate to their proper layers, additional transcription factor genes, such as *Sox5* and *Tbr1,* repress *Fezf2* activity in layer 6 neurons, resulting in a layer 5-specific contribution to the corticospinal tract (Chen et al. 2005a, b; Kwan et al. 2008; Han et al. 2011; Guo et al. 2013). While these studies have begun to unpack how gene expression can control neuronal character, the molecular code underlying the specification of the many other types of cortical neurons have not yet been discovered.

Second, this areal identity must be transmitted to neurons as they are produced so that they can support the function-specific aspects of synaptic development that occur only after they achieve their proper positions and differentiation status. While the identity of the molecules imparting the area-specific maps (e.g., cadherins, Protocadherins, Eph) have been partially discovered, new techniques, such as microdissection/bulk RNASeq and single-cell RNASeq, may provide a more comprehensive list of these areal marks in the near future.

Two mechanisms leading to local area development have also been suggested more recently. The first, feedback from the emerging neocortical layers to the precursor zones, presumably operates in each area. The second, local control of precursor heterogeneity, must entail molecular programming that is intrinsic to each area.

## Radial Feedback in Progression of Cortical Growth

In all neocortical areas, neurons of the deepest cortical layers are born first, followed in temporal succession by neurons destined for the more superficial layers. As mentioned above, this temporal specification of neocortical layers could be accomplished by an intrinsic "progressive tuning program" in aRGCs whereby they transition molecularly as they age during neurogenesis. Alternatively, electrical, biochemical, and genetic feedback from the growing laminae could modulate aRGC production parameters to fine-tune the overall extent of neuronal numbers during the neurogenic period. While these two ideas are not mutually exclusive and the system likely operates with both mechanisms, several lines of evidence indicate that coupling of cells within a radial clone could operate a feedback pathway. For example, single-cell fluorescent dye injections are known to label a discrete population of precursor cells surrounding an injected aRGC, and it has been found that the dye-coupled clusters are electrically coupled by gap junction channels (Bittman et al. 1997; Bittman and LoTurco 1999). Indeed, calcium waves initiated near the pial surface spread apically into the VZ and are transmitted for significant distances within the germinal zones (Owens and Kriegstein 1998; Owens et al. 2000). In addition, clonal labeling experiments have shown that neurons born sequentially from a single aRGC and destined for different layers are also gap junction coupled to one another and to their mother aRGC; later, these sister neurons preferentially form chemical synapses (Yu et al. 2009b; Gao et al. 2013; He et al. 2015). Thus, fast intracellular signaling methods have evolved to couple neurons within the developing neocortical layers to their precursors lying below. Moreover, gene expression feedback loops (Toma et al. 2014) as well as gene expression/growth factor loops are another identified mechanism, the latter highlighted in the connection between Sip1 expression by postmitotic neurons and the release of signaling factors to underlying precursor cells (Seuntjens et al. 2009; Parthasarathy et al. 2014). Several other released factors have been implicated in this type of feedback as well, including nitric oxide and ATP.

While regulation of intracellular Notch signaling is thought to be a basic mechanism regulating cellular diversity, including in the neocortical VZ (Rasin et al. 2007; Kopan and Ilagan 2009; Ables et al. 2011; Pierfelice et al. 2011), recent work suggests that there is a Notch-based feedback mechanism between bIPCs and aRGCs during neurogenesis. In particular, Delta 1 and Delta 3 expressing bIPCs contact Notch-expressing aRGCs to modulate their mode of division and repress their differentiation (Mizutani et al. 2007; Kawaguchi et al. 2008; Yoon et al. 2008; Nelson et al. 2013).

Importantly, these feedback circuits enable the germinal zones not only to fine-tune cortical production but also to compensate for developmental or environmental insults by altering division parameters in the VZ and SVZ. For example, potentially in response to reduced neurogenesis and radial growth during early corticogenesis, the bIPC population in the Ts65Dn mouse model of Down syndrome is amplified during the later stages of neurogenesis, largely correcting (in bulk numbers) a severe paucity of early-born neurons with an overproduction of later-born neurons (Chakrabarti et al. 2007). While large insults to the developing system cannot be compensated for and may lead to lasting changes in cortical thickness or surface area (such as microcephaly), there is evidence supporting a level of plasticity in the germinal zones that responds to feedback from cells in the overlying neocortical layers.

## Mechanisms of Gyrification

One of the most remarkable features of the neocortex is the patterned folding of its surface during the evolution of certain mammals. The stereotypical gyri and sulci that develop are thought to be a general mechanism for fitting a neocortex with greater surface area into the volume required to pass through the narrow birth canal and to fit within the confines of the calvarium. Within all gyrencephalic species, the pattern of folding is highly concordant with the neocortical map, such that functional areas routinely lie across the same gyri or sulci in the brains of different individuals. Due to the importance of this event in specifying the size and function of the brain in gyrencephalic species, and the recognition that developmental disorders arising by gene mutations can significantly alter the degree of gyrencephalization, how these folds occur in such a repeated fashion across individuals is a hotly debated topic in the field of cortical development. The primary data supporting each of the theories outlined below come from comparative analysis of lissencephalic and gyrencephalic brains, gene perturbation studies, and complex computer simulations based on imaging studies.

Three overarching theories have been proposed to explain how the neocortex is folded and how this occurs in such a stereotypical pattern within a species. While each theory has a list of studies both supporting and opposing it (an exhaustive list of these corresponding studies is not described here), the most parsimonious explanation is that each of the three concepts partially identify some of the biological events and that they combine to yield cortical folding.

The *axonal tension hypothesis* considers axon connections to be the primary driving force of a folded neocortical sheet (Van Essen 1997; Holland et al. 2015). It proposes that corticocortical fibers primarily connecting two ipsilateral regions can create localized regions of tension, thereby generating a prolonged mechanical force that results in folding of the neocortical surface to produce a gyrus. Correspondingly, neighboring regions that are less well-connected will form the reciprocal event—the formation of a sulcus. The *radial expansion hypothesis* (Richman et al. 1975) suggests that differential production of neurons across the 6 neocortical layers, for example, increased generation of supragranular neurons compared to infragranular neurons, can result in localized tangential spread, or wedging of the neocortical sheet, and that this convexity may later blossom into a gyrus. Conversely, overproduction of deeper layer neurons compared to the superficial layers can result in incipient concavity and formation of a sulcus. Lastly, the *differential tangential expansion hypothesis* (Ronan et al. 2014) suggests that isolated regions of the developing neocortical wall undergo different rates of tangential expansion. This could occur during the founder cell expansion phase or during the period when specific classes of IPCs emerge and begin to divide. The tangential forces generated between neighboring areas then lead to buckling of the neocortical sheet and to formation of gyri and sulci.

In general, these theories can be simplified as mechanisms operating at the level of the precursor cells within each neocortical area and to those influenced by migrating and differentiating neurons and to afferent cortical projections. For example, both of the expansion hypotheses rest on differential cell production, either in the radial domain (i.e., laminar differences) or the tangential domain (i.e., cortical column number). Both, therefore, suggest that isolated programs of neurogenesis along the protomap-specified areas yield the resulting pattern of cortical folding most appreciated after birth. Incidentally, while the axonal tension hypothesis primarily focuses on forces generated by differentiated neurons, it also invokes the area-specific projection and targeting patterns that must be conferred to the neocortical layers upon their generation from the underlying precursor cells. Thus, all theories for gyrification require a significant role for precursor cells in establishing the regions that will eventually undergo convolution.

Recent studies have provided a role for precursor gene expression and its consequences on neurogenesis as a predictor of gyrification. In particular, a key study identified the cell-adhesion molecules FLRT1 and FLRT3 in regulating area-specific formation of gyri and sulci (Del Toro et al. 2017). These two genes are upregulated in the lissencephalic mouse neocortical wall and downregulated in the gyrencephalic ferret and human neocortical wall in regions of incipient gyrus formation. This study shows that perturbations to lower the levels of FLRT1/3 expression lead to faster neuronal migration rates and to clusters of neurons expressing a similar level of these adhesion molecules. The increased radial and tangential pressure caused by these clustered neurons is postulated to lead to localized formation of a gyrus, even in the normally smooth mouse neocortex. Another gene recently identified in the germinal zones that plays a role in neocortical folding is *Trnp1*, which encodes a nuclear protein potentially involved in chromatin state (Stahl et al. 2013). Knockdown of *Trnp1* alters the pattern of cell division in the VZ, causing overproduction of bIPCs and bRGCs. The hypothesis of this study is that increases in the numbers of resulting neurons, and their tangential spread afforded by the increased number of bRGC fibers, result in gyrus formation in the overlying neocortical sheet. A study by de Juan Romero et al. (2015) offers perhaps the most convincing argument for a link between differential gene regulation in the germinal zones, expansion of bRGCs, and formation of overlying cortical convolutions. In this study, regions of the ferret neocortical wall, which later develop either a gyrus or a sulcus, were isolated at a stage in early development prior to the formation of these folds. Upon microarray profiling, many hundreds of genes were differentially expressed between these two areas, and clear expression of these genes in the oSVZ in the future gyrus site was contrasted to the lack of their expression in the oSVZ of the neighboring future sulcus site. In addition, a human RGC-specific gene, *ARGAP11B*, was shown to increase basal precursor proliferation and cortical folding when introduced into the developing mouse germinal zones (Florio et al. 2015).

As described in the precursor heterogeneity section above, individual classes of precursor cells are thought to generate neurons with different dendrite morphologies, electrophysiology, and numbers of synapses with cortical afferents. Thus regions of neocortex with specific numbers and types of precursor cells could differentially produce specific neuron types and yield locally distinct rates of neuropil growth, based on differentiated neuron morphology and synapse capability. This could play a large role in the consolidation or growth of nascent convolutions during later stages of neocortical development. Altogether, several lines of evidence demonstrate that differential neurogenesis may lead to cortical convolutions, by promoting basal precursor production, by modulating neuronal migration rates and adhesive properties, or by providing unique areas for neuropil expansion.

A number of compelling studies also indicate a role in cortical afferent projections in gyrification. First, in enucleation studies, when input from an eye is removed during early cortical development prior to the formation of convolutions, the size and number of resulting gyri and sulci are significantly alterred (Rakic 1988b; Dehay et al. 1996). These studies show that afferent input into the area that will eventually form a gyrus is required for proper development of the convolution pattern. Second, MRI imaging studies from prenatal human brain indicate that gyral patterning is dependent on regional growth heterogeneity as well as axonogenesis and afferentation (Knutsen et al. 2013; Razavi et al. 2017; Wang et al. 2017). Taken together, all of these studies clearly indicate that the pattern and extent of gyrification in certain species is a combined result of programmed events in the germinal zones and influences of cortical afferent systems on synaptogenesis, as well as due to expansion of local regions of the neuropil.

## Summary

Similar to the development of any body organ, formation of the cerebral cortex requires a complex choreography of stem and progenitor cell allocation, cell division, production of the requisite numbers and types of cells, and the eventual differentiation of these cells into mature functional components of the maturing organ. In this chapter, we have discussed several developmental events that pertain especially to the cerebral cortex, including the migration of neurons from their site of birth to their proper location, which can be over several millimeters in the primate brain. In addition, the numbers of excitatory neurons and inhibitory neurons must be tuned within each area and layer, an incredibly intricate process due to the distant proliferative zones from which these cell types are derived. How these cells initiate and consolidate the synapses and circuits necessary for complex function is being elucidated at a rapid pace, as are the genetic and molecular mechanisms which underlie all of these crucial developmental events. Focused effort must be paid in the near future to fully elucidate the molecular controls of cell commitment and allocation, in particular the precise relationships between neocortical precursor cells and their resulting neuron offspring. The exceptional advances over the past several decades described herein will soon culminate in a clear understanding of how cell number and type relates to circuit formation and eventually to behavior and cognitive function. When this is accomplished, a clear roadmap will exist not only for understanding the most complicated biological machine currently known, but also for the design of therapeutic approaches for developmental disorders that affect cognitive and intellectual function.

# 3

# The Evolution of the Human Cerebral Cortex Development

## A Genomic Perspective

Belen Lorente-Galdos, Sydney K. Muchnik, and Nenad Sestan

## Abstract

The extraordinary cognitive abilities that humans possess, such as syntactical-grammatical language, abstract thinking, episodic memory, or complex reasoning, are largely dependent on the brain, and more specifically on its surface, the cerebral cortex, as was initially proposed by Thomas Willis in 1664. Since then, neuroscience has endeavored to decipher what makes the human brain so unique when compared with other species. Early studies were based on comparative brain anatomy between humans and other extant or extinct species, in the latter case based on data compiled from fossil records. More recently, comparison of the number of neurons and studies of cortical development have improved our understanding of the field. Nowadays, in the era of genomics, new possibilities have arisen for determining changes in gene expression or regulatory activity that underlie the observed differences in phenotypes. This chapter summarizes what is known about the human cerebral cortex. It focuses on the neocortex, which represents about 80% of the human brain mass, and places it into an evolutionary context by considering other hominins, nonhuman primates, and mammals. Finally, it explores the role of genomics in elucidating the shared and unique features of human nervous system development, organization, and function.

## Organization and Development of the Human Cerebral Cortex

The cerebral cortex represents half of the volume of the human nervous system and consists of two main parts: the neocortex and the allocortex (i.e., olfactory system and hippocampus) (Krubitzer and Kaas 2005). The former, the largest section, is involved in higher-order brain functions. Recent studies estimate that the human cerebral cortex contains approximately 16.34 billion neurons that pass signals to each other via synaptic connections (Azevedo et al. 2009;

Sigaard et al. 2016). It has been estimated that there are approximately 149,000 to 176,000 kilometers of myelinated axons connecting these neurons in the adult cerebral white matter (Marner et al. 2003). A total of approximately 164 trillion synapses have been estimated in an adult neocortex (Tang et al. 2001), while the published number of synapses received per neocortical neuron varies from 7,200 to 80,000 (Huttenlocher 1979; DeFelipe et al. 2002; Pakkenberg et al. 2003).

To support the complex circuitry and functions of the cortex, its component neurons are highly specialized and exhibit a variety of axonal projections, dendritic patterns, and electrophysiological properties. As a consequence, there may be thousands of distinct neuronal subtypes in the cerebral cortex, far more than in any other central nervous system structure. However, these subgroups can be broadly classified into two major groups: the glutamatergic excitatory projection neurons and GABAergic inhibitory interneurons (DeFelipe and Fariñas 1992; Petilla Interneuron Nomenclature Group et al. 2008). Accounting for about 80% of cortical neurons, the projection neurons occupy a central position in all cortical circuits. Extending long axonal projections to other structures, both within and beyond the cortex, projection neurons constitute the sole output system of the cortex. They also compose the largest input system of the cortex (i.e., corpus callosum) (Aboitiz and Montiel 2003; Chedotal and Richards 2010) and represent the major target of afferents from other structures of the brain. The great majority of projection neurons are pyramidal neurons, which exhibit a highly polarized morphology with a pyramid-shaped cell body, a long apical dendrite directed to the pia mater, and multiple basal dendrites extending laterally from the cell soma. These dendrites contain numerous spines—bulbous postsynaptic protrusions—along their lengths. The other major class of cortical neurons, the GABAergic inhibitory interneurons, do not have significant numbers of dendritic spines and tend to project locally rather than sending long axons across or beyond the cortex.

A hallmark of the mammalian cerebral neocortex is that its neurons are functionally organized at two levels into six somewhat arbitrary layers and roughly 200 cytoarchitectonically and functionally distinct areas (Krubitzer and Kaas 2005; Glasser et al. 2016). This immensely complex organization of the human cerebral cortex is the result of a prolonged and dynamic development, and will be summarized next (for a comprehensive review, see Geschwind and Rakic 2013; Silbereis et al. 2016). A critical and evolutionary conserved initial step in cortical development involves the establishment of dorsal-ventral, rostral-caudal, and medial-lateral axes within the forebrain anlage. Studies have uncovered a number of transcription factors, morphogens, and receptors involved in defining these axes (Grove and Fukuchi-Shimogori 2003; Sur and Rubenstein 2005; O'Leary et al. 2007). Neuroepithelial cells in the ventricular zone (VZ) serve as the stem or founder cells of the nervous system and divide predominately symmetrically to generate two progenitor daughter cells, thus exponentially expanding the pool of progenitor cells. After the onset of neurogenesis, these progenitors transform into apical radial glial cells (aRGCs), which have the capacity to divide asymmetrically and generate two different cell types. An identical aRGC, a neuron, or a basal progenitor cell (i.e., intermediate progenitor, basal radial glial cell, bRGC, or others) are possible outcomes of mitosis occurring in an apical radial glial cell (Lui et al. 2011; Hevner and Haydar 2012; Taverna et al. 2014). Different subtypes of these basal progenitors lie in an incipient subventricular zone (SVZ) and are also capable of differentiating. In fact, most cortical neurons are generated from basal progenitors, which are abundant in the human cerebral cortex during cortical development. Once neurons are born, they migrate using radial glial cells as a guiding scaffold toward their final laminar destination, forming the subplate and layer 1, and then, consecutively, prospective layers 6 to 2 in the cortical plate (Molnár and Clowry 2012; Geschwind and Rakic 2013; Guo and Anton 2014; Johnson and Walsh 2017). After neurogenesis, gliogenesis starts as RGCs lose their apicobasal polarity and differentiate into different types of glial cells, followed by synaptogenesis, myelination, and the synaptic pruning that continues through adolescence (Silbereis et al. 2016). These neurons, which originate in the VZ/SVZ and radially migrate into the developing cortical plate, are primarily projection neurons (Leone et al. 2008; Greig et al. 2013; Han and Sestan 2013; Lodato and Arlotta 2015; Dwyer et al. 2016; Jabaudon 2017); in contrast, cortical interneurons, which are notoriously diverse in morphology and function, originate largely in the ganglionic eminences and migrate tangentially into the cortical plate (Wonders and Anderson 2006; Fertuzinhos et al. 2009; Bartolini et al. 2013; Hansen et al. 2013; Ma et al. 2013; Hu et al. 2017; Wamsley and Fishell 2017).

## Evolution of the Human Cerebral Cortex

Comparison of brain features between humans and related species is essential to understand which traits are specific to humans, as well as to develop an understanding of the brains of common ancestors. Humans are hominins, a subgroup of primates which, in turn, are a subgroup of mammals. About 340 million years ago (mya) the early amniotes emerged from amphibians and divided into sauropsids, which evolved into extant reptiles, birds, and synapsids, leading to modern mammals. A basic brain structure of an external cortex and a few subcortical nuclear structures with olfactory, sensory, and memory functions are shared between reptiles, birds, and mammals (Krubitzer and Kaas 2005). However, whereas reptiles have a thin dorsal cortex with a single layer of pyramidal neurons and fewer inhibitory neurons, all extant mammals have a more complex six-layer neocortex (Krubitzer and Kaas 2005). In fact, although the time and manner in which the neocortex appeared remains unknown, this trait seems to be specific to mammals, as it has not been found in nonmammals. Over 4,600 present-day species of mammals, which are greatly diverse in terms of absolute and relative brain size, number of neurons and areas, and even gyrencephalization, have evolved from early mammals that inhabited the earth ~200 mya. These animals were probably small in size with a small neocortex organized into few areas, about 15 to 20 mostly sensory areas. Motor and premotor cortex emerged in placental mammals at most 125 mya. Early primates, in turn, were small, nocturnal, and arboreal animals that emerged about 80 mya with an already elaborated visual and motor cortex. Indeed, the visual cortex in all primates is remarkably complex. Finally, the number of specialized areas in primate neocortex is higher than in nonprimates.

Compared to nonhuman primates, the human brain is larger in size and in number of neurons; indeed, it is about three times the size and has about twice as many neurons as the chimpanzee brain (Collins et al. 2016). Compared to Neanderthals, human brain volume is similar, although the parietotemporal lobe is bigger in the modern human brain (Balzeau et al. 2012). Importantly, the human neocortex has dramatically increased in size as compared to more distantly related mammalian species—a 1,000-fold increase in comparison with mouse neocortex (Geschwind and Rakic 2013). That expansion, though, was not accompanied by a similar change at the whole brain level. Instead of augmenting in thickness, cortical expansion is the result of (a) an increase of the number of progenitors in the VZ and (b) indirect neurogenesis from aRGCs via basal progenitors, especially bRGCs, which are highly abundant and proliferative in humans. The latter is related to the expansion of the primate SVZ, particularly the outer SVZ (oSVZ), where bRGCs are located. While brain size and number of neurons in primates is a general predictor of cognitive abilities, these features are not fully sufficient to explain human cognitive capabilities; other animals (e g., elephants, certain whales) surpass humans in each of these characteristics (Sousa et al. 2017a). Similarly, the expansion of the oSVZ and an abundance of bRGCs are common in species with a large neocortex (Reillo et al. 2011). In addition, there is no conclusive evidence of any brain area, cell type, or neural circuit that is completely new in humans (i.e., not present in other primates). However, there is increasing evidence for potential and confirmed changes in the developmental features, morphology, molecular profiles, and quantity of particular neural cell types in the human lineage (Elston et al. 2011; Kwan et al. 2012a; Bianchi et al. 2013; Sousa et al. 2017b). For instance, a subtype of pyramidal neurons, the von Economo cells posited to promote rapid neuronal communication, are found in several primates, elephants, and cetaceans but are particularly large and abundant in humans (Nimchinsky et al. 1999). Small changes like these in the connectome (neural circuits and networks formed by different neural cell types and their synaptic connections) can lead to functional changes. In fact, several studies suggest that neural circuits have undergone structural, molecular, and functional reorganization in the human lineage (Sousa et al. 2017b). Notably, the singular complexity of the human brain takes over two decades to be fully constructed, from neurogenesis to neural circuit assembly and maturation. This translates into a particularly long gestational time, infancy, childhood, and adolescence compared to other mammals (Zhu et al. 2018).

## The Genomic Revolution

A key challenge is to understand the countless molecular and cellular processes as well as the precise system that regulates those processes, which are activated in humans over the long period of time in which the complete formation of the human cerebral cortex occurs. Over the last decades, with the advent of genetic and genomic technologies and the sequencing of large amounts of DNA, some progress has been made (Figure 3.1). The availability of genetic sequences from different species has had a direct impact on the improvement of our knowledge about phylogenetic relationships in mammals and, consequently, on the feasibility of effective comparative studies. The whole genome sequencing of the mouse and human genomes (Lander et al. 2001; Mouse Genome Sequencing Consortium 2002), followed a few years later by that of the chimpanzee (Waterson et al. 2005), enabled the first comparative genomic studies at a genome-wide level. As a result of a substantial reduction in the cost of genomic sequencing, massive sequencing has become feasible, and we quickly transitioned from one or a few individual whole genome sequences to thousands of human individuals sequenced. It is noteworthy that bioinformatics tools as well as increasingly complex methodology are being implemented in parallel to process these massive amounts of data.

Multiple sequence alignments among primates revealed a plethora of human-specific substitutions altering protein-coding sequences. The majority of those human-specific variants have not been functionally characterized. One example, *FOXP2*, has been intensively analyzed due to its relationship with language. It is a highly conserved gene during mammal evolution, but it presents amino acid substitutions in the lineage leading to humans that have been related to deficiencies in speech. These variants were introduced in mice promoting a change in neural-restricted phenotypes, including dendrite morphology and transient differences in ultrasonic vocalization (Enard et al. 2009). In contrast to genetic substitutions, gene duplication provides new genetic material for evolutionary forces to act upon, and thus represents a major source of evolutionary novelty, as Susumu Ohno (1970) established in his seminal book *Evolution by Gene Duplication*. To study gene families that have undergone recent gene duplication presents a further difficulty due to the similitude between paralogous copies. Nonetheless, there are at least three gene families with human-specific expansions that have been related with neurodevelopmental functions:

1.   The *SRGAP2* gene has several duplications in humans that arose after the separation with chimpanzees (Dennis et al. 2012). One duplication

36

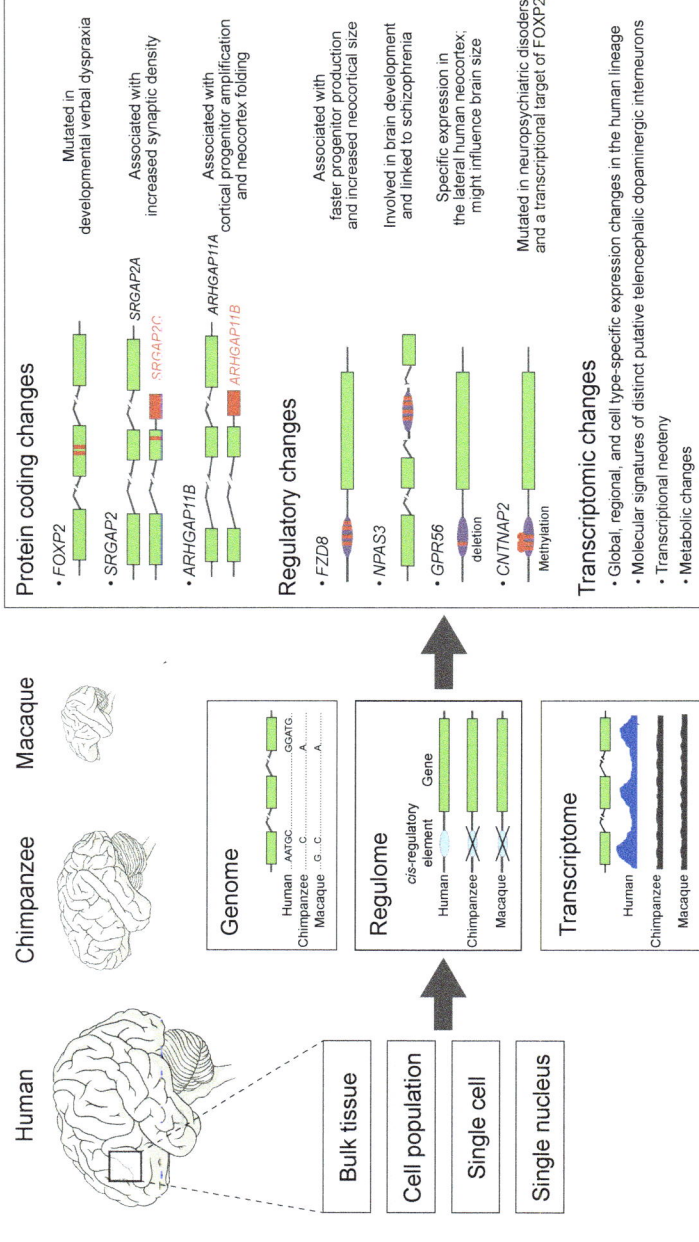

**Figure 3.1** Evolutionary changes in the human cerebral cortex identified with genomics studies. Analysis at different levels can be done to unravel the genetic and molecular causes of brain phenotypic differences between species. General approaches are represented here as are examples of human-specific features (marked in red).

is fixed in all human populations analyzed and interferes with the function of the ancestral copy of this gene, resulting in increased synaptic density in neocortical pyramidal neurons and prolonged spine maturation (Charrier et al. 2012; Fossati et al. 2016).

2. Studies in mice demonstrate a second human-specific gene, *ARHGAP11B*. Highly expressed in radial glial cells, *ARHGAP11B* promotes aRGCs to undergo symmetric divisions with two basal progenitors as outcome, and is capable of producing neocortical folding in otherwise lissencephalic mouse brains (Florio et al. 2015).

3. *DUF1220* is a domain that has expanded in the human genome extraordinarily, mainly in the *NBPF* gene family (Popesco et al. 2006; O'Bleness et al. 2012). Deletions and duplications of this domain have been associated with micro and macrocephaly (Brunetti-Pierri et al. 2008) as well as correlated with brain size (Zimmer and Montgomery 2015), and it is putatively expressed in the VZ early during cortical neurogenesis, promoting proliferation in neural stem cells (Keeney et al. 2015).

As early as 1975, Mary-Claire King and Allan Wilson (1975) hypothesized that protein sequence-altering mutations could not account for the full range of phenotypic differences observed between humans and nonhuman primates. Since then, many advances have been made in the study of the impact of gene regulation on human brain evolution. Discussed in depth below, we first need to understand the relevance of gene expression changes during cortical development. Microarrays, RNA sequencing (RNA-seq), and other technologies allow for the transcriptomic profiling of particular tissues or specific regions of a given tissue. Some early comparison studies of gene expression among tissues reported that genes expressed in the brain have evolved at higher rates than those expressed in other tissues, but other studies have argued the contrary (Bustamante et al. 2005; Nielsen et al. 2005; Wang et al. 2007a). Recent comprehensive analyses of the human brain and particular neocortical areas across time have revealed that an immense number of coding and noncoding RNAs are involved in neuronal functions. In other words, the human brain is dynamically regulated across regions and time, more pronouncedly during early and mid-fetal development (Johnson et al. 2009; Kang et al. 2011; Miller et al. 2014; Li et al. 2018). Other genomic studies have been able to identify human genes displaying species-specific expression patterns that may contribute to evolutionary changes in cortical development, structure, and function (see reviews by Somel et al. 2013; Silbereis et al. 2016; Sousa et al. 2017a). Recent studies of gene expression profiles between adult brain regions of human, chimpanzee, and macaque monkey revealed extensive transcriptional differences between homologous brain regions and cell types in the human lineage (Sousa et al. 2017b; Zhu et al. 2018). Many of the genes that exhibit human-specific global, regional, and cell type-specific expression patterns encoded transcription factors, ion channels, and neurotransmitter enzymes and receptors. Changes in the expression of these proteins can affect function of neural circuits by altering transcription, electrophysiological properties, or neurotransmission. Furthermore, the same study showed that humans may have evolved a distinct dopaminergic neuron system in the cerebral cortex and striatum. Specifically, Andre Sousa, Ying Zhu, and colleagues characterized a rare and unusual population of subpallium-derived putative dopaminergic interneurons enriched in the human striatum and present in the human neocortex but not in the neocortices of African great apes (i.e., common chimpanzee, bonobo, and gorilla), which are our closest living relatives. Moreover, they uncovered evidence that these interneurons likely switch their transmitter system from somatostatin to dopamine, a phenomenon that has never been reported in the cerebral cortex.

Because bulk RNA-seq experiments aggregate data across the diverse range of cell types found in the brain, it is essential to analyze gene expression within individual cell populations. Thus, the ability to determine the gene expression pattern of a small number of cells or single cells is of great importance for resolving a variety of problems in the context of human brain evolution. In fact, the last decade has witnessed a rapid advance in single-cell RNA-seq technologies (Johnson and Walsh 2017; Lein et al. 2017), which represents an unbiased approach to explore evolution and development within the context of specific and defined cell types. An increasing number of cell types and subtypes in the developing and adult human cerebral cortex have been characterized in a systematic manner at an unprecedented pace using single-cell or nucleus RNA-seq (Florio et al. 2015; Pollen et al. 2015; Lake et al. 2016; Onorati et al. 2016; Nowakowski et al. 2017; Li et al. 2018; Zhong et al. 2018). These studies not only confirmed the transcriptomic and developmental signatures of well-established cell types, they discovered previously unknown gene expression patterns in individual human neural cell types. With increasing reproducibility and scalability, single-cell RNA-seq technology has undoubtedly become, and will remain, an important experimental platform for cell type discovery and is poised for cross-species comparison at cellular resolution.

## Gene Regulation and Human Accelerated Regions

The hypothesis of Mary-Claire King and Allan Wilson has been supported by many studies utilizing whole genome sequencing and chromatin immunoprecipitation with sequencing (ChIP-seq). In fact, a large-scale study of the impact of positive selection on coding and noncoding regions in the human genome emphasized not only the importance of noncoding regions in the evolution of human-specific traits, but also showed that noncoding changes have played a particularly large role in the evolution of the human nervous system (Haygood et al. 2010). This idea has led the way for a plethora of research delving into the role of noncoding DNA and gene regulation in the evolution of the human cortex. These studies are very diverse, given that gene regulation is a highly complex process that occurs at several levels. Underlying sequence changes (including substitutions, deletions, and duplications) can influence the regulatory activity of proximal regulatory elements, or promoters, as well as distal elements such as enhancers. However, gene regulation is also dictated by epigenetic factors (e.g., histone modifications and DNA methylation). Here we focus on examples of evolutionary changes in gene regulation at each of these levels, and their potential impact on the development of unique aspects of the human brain.

Sequencing-based comparative genomics studies have allowed for the identification of regulatory regions that have acquired an excess of substitutions in the human lineage, termed human accelerated regions (HARs). Scanning the genome for regions that are highly conserved in other species, while containing an enrichment of human-specific substitutions, has resulted in the identification of over 2,000 HARs. It is hypothesized that HARs are responsible for producing human-specific gene expression patterns and, thus, human-specific phenotypes and traits (Pollard et al. 2006a; Prabhakar et al. 2006; Bird et al. 2007). Further analysis of HARs regions utilized additional genomic data, including chromatin state and transcription factor binding sites, and found that at least 30% of HARs are predicted to act as developmental enhancers (Capra et al. 2013). Several subsequent studies have shown via transgenic mouse enhancer assays that HARs drive distinct reporter gene expression patterns in the developing mouse brain as compared to orthologous sequences from other species; this indicates that HARs may play a specific role in the evolution of the human brain (Capra et al. 2013; Kamm et al. 2013a; Boyd et al. 2015). Finally, the functional relevance of HARs to brain development is further indicated by associations between mutations in HARs and human-specific diseases, such as schizophrenia and autism spectrum disorder (ASD).

Several lines of evidence have indicated that noncoding HARs may be particularly relevant to the development of the human nervous system. For example, HARs are often located near genes that have been implicated in neural development, or genes that show differential expression between developing brain regions (Haygood et al. 2010). The first HAR that was identified, termed HAR1, is located within two overlapping noncoding RNA genes and has been implicated in cortical development (Pollard et al. 2006b). *HAR1F*, one of the noncoding RNAs spanning HAR1, is highly specifically expressed in Cajal-Retzius neurons during neocortical development. Its expression in the adult brain, however, is more diffuse, implying that HAR1 plays a role in regulating a developmentally dynamic gene. In 2015, Boyd et al. (2015) identified that the HARE5 region acts to enhance *FZD8*, a gene that encodes a receptor involved in the Wnt signaling pathway. Interestingly, HARE5 displays species-specific enhancer activity, acting earlier and more robustly in the human neocortex than that of the chimpanzee. Transgenic mouse assays demonstrated that the human

HARE5 sequence directs faster progenitor production and larger cortical size as compared to the chimpanzee HARE5 sequence. Despite this compelling evidence for the role of HARs in human brain development, their function and the consequence of these human-specific sequences changes is still unknown.

HARs are being increasingly implicated in human-specific diseases. For example, the highest density of HARs occurs within the gene *NPAS3*, a transcription factor involved in brain development that has been implicated in schizophrenia (Kamm et al. 2013b). Fourteen HARs occur within the introns of *NPAS3*, 11 of which were shown to drive reporter gene expression in the central nervous system in a zebrafish transgenic assay. Kamm et al. (2013a) further explored one of those regions and showed that orthologous mouse and chimpanzee sequences drove similar LacZ expression patterns, while the human sequence drove a more extensive pattern of expression that included the developing anterior telencephalon, lending evidence for human-specific regulatory activity of this region. Recent studies have been able to pinpoint specific mutations within HARs that might have an influence on human-specific disease phenotypes. For example, a rare homozygous mutation was identified in some unrelated individuals with ASD. This mutation falls within HAR246, a noncoding region upstream of *CUX1*, a gene involved in the morphology of cortical neurons and the formation of synaptic spines (Doan et al. 2016). It was shown that the mutated version of HAR246 displays increased enhancer activity compared to wild type, indicating that this mutation could disrupt cortical neuron morphology and signaling and contribute to the human-specific ASD-related phenotypes observed in individuals with this mutation. Together, these links between HARs and human-specific diseases suggest a role for HARs in the evolution of human-specific cognitive and social traits.

## Duplications and Deletions in Regulatory Regions

In addition to identifying noncoding regions with human-specific substitutions, whole genome sequencing has led to the identification of regulatory regions with human-specific deletions or duplications. Over 500 noncoding regions that are highly conserved in other species, but deleted in humans, have been identified (McLean et al. 2011). As with HARs, transgenic assays have provided evidence for the role of these deleted regions in brain development. For example, there is a human-specific deletion of an enhancer near *GADD45g*, a tumor suppressor gene that is thought to repress proliferation and is expressed in the developing mouse neocortex. Further characterization of the function of this locus is needed, but it is possible that loss of this enhancer could increase proliferation in the developing brain and therefore contribute to the expansion of the human neocortex.

In addition, duplications or insertions in regulatory elements, as seen at the gene *GPR56*, have been implicated in brain development (Bae et al. 2014).

Mouse transgenic assays showed that the human version of a cis-regulatory element of *GPR56*, containing several insertions and duplications, drives specific expression in the lateral neocortex, whereas the mouse element drives a broader expression pattern across the whole neocortex. Given that *GPR56* has been shown to play a role in progenitor proliferation and cortical patterning, and the fact that a deletion within the 5′ promoter of *GPR56* causes cortical polymicrogyria, it is speculated that alterations of the regulation of *GPR56* could also influence human brain size. These examples illustrate the utility of functional characterization of human-specific sequence changes in regulatory regions and provide a framework for similar future studies that will doubtlessly continue to improve our understanding of evolution and cortical development.

## Evolutionary Changes in Regulatory Activity

As demonstrated above, sequencing-based studies have added substantially to our understanding of the role of noncoding regions and gene regulation in the evolution of the human brain. Contemporary techniques, such as ChIP-seq, have allowed us to delve into this field even further by performing comparative studies of the regulatory activity of noncoding DNA between species. Several groups have performed global comparisons of regulatory activity in humans and closely related species, revealing complicated evolutionary dynamics. By profiling active regulatory regions, marked by H3K27ac and H3K4me3, during early cortical neurogenesis in humans, macaques, and mice, Reilly et al. (2015) showed that embryonic enhancers with human-specific gains or losses of activity regulate genes that are enriched in co-expression modules associated with neuronal proliferation and differentiation. A subsequent study compared active regulatory regions in a variety of brain regions in the human, chimpanzee, and macaque (Vermunt et al. 2016). This comparison emphasized the importance of including closely related species in this kind of analysis; while about 1,400 enhancers and 90 promoters were specifically enriched in the adult human brain compared to macaque, only 193 enhancers and 17 promoters continued to show human-specific gain of activity when compared to chimpanzee. Thus, a similar comparison of prenatal regulatory activity would be necessary to identify high-confidence human-specific developmental regulatory programs, but due to the unavailability of prenatal chimpanzee data, this comparison has been heretofore impossible.

Interestingly, these studies of the epigenetic landscape of the human brain all demonstrate that the positions of promoters and enhancers are largely conserved between humans and nonhuman primates, despite underlying sequence divergence (Cotney et al. 2013; Prescott et al. 2015; Vermunt et al. 2016). In addition, a subset of these positionally conserved regulatory regions exhibit tissue- and/or species-specificity, as they show evidence of activity in disparate tissues or across different brain regions in different species (Cotney et al. 2013;

Vermunt et al. 2016). Interestingly, these regulatory regions can also show cell type-specific patterns of conservation. For example, Shulha et al. (2012) showed that in the prefrontal cortex, neuronal epigenomes are more similar between humans, macaques, and chimpanzees than they are across cell types (compared to non-neuronal cell types) within the same species. Overall, these studies have identified sets of putative regulatory regions that likely contain a plethora of information about the evolution of the human brain. The human-specific enhancers identified in these studies did show some overlap with the HARs discussed above, but an overall enrichment was not observed. Thus, these data sets of putatively human-specific regulatory regions displaying specific activity in various brain regions, developmental stages, or cell types provide a novel framework for future studies linking genes to human-specific traits and developmental phenotypes.

## Gene Regulation via DNA Methylation

DNA methylation provides another layer of epigenetic gene regulation. Methylation can occur at any of the ~1 billion cytosines in the human genome but is most often found at the ~28 million CpG sites. Changes in methylation are driven by alteration of the expression of DNA methyltransferases and can direct species-, tissue-, or cell type-specific patterns of expression (Hernando-Herraez et al. 2015). Specifically, promoter methylation is associated with reduced gene expression and is thought to play a large role in the regulation of gene expression. For example, a comparison of liver, heart, and kidney between humans and macaques showed that 12–18% of the differential expression between the species could be explained by differences in promoter methylation (Pai et al. 2011). Studies of methylation levels in the brain have failed to produce large-scale patterns but have identified specific examples of important methylation changes.

In a study of putative regulatory regions of 36 genes, it was shown that CpG sites tend to be more methylated in human brain than in chimpanzee (Enard et al. 2004). A similar study showed, however, that promoter regions of several genes were significantly less methylated in human brain than chimpanzee brain (Zeng et al. 2012). In studies more specific to brain development, methylation differences in genes involved in neuronal function have been reported both at the level of DNA (Farcas et al. 2009; Schneider et al. 2012) and histones (Shulha et al. 2012). An example of a gene with species-specific methylation patterns is *CNTNAP2*, a transcriptional target of *FOXP2*. *CNTNAP2* has been implicated in neurological and psychiatric disorders, including language impairment (Rodenas-Cuadrado et al. 2014). *CNTNAP2* was shown to be differentially expressed in humans compared to chimpanzees, which could be attributed to changes in *CNTNAP2* isoforms and widespread gene methylation differences (Schneider et al. 2014). Thus, it has been hypothesized that differential methylation of *CNTNAP2* could ultimately be responsible for differential expression of the gene in humans, which is thought to be related to the development of human-specific language abilities. Finally, Gokhman et al. (2014) have shown that DNA methylation varies between Neanderthals and Denosivans, via an analysis that uses C→T ratio as an indicator of the amount of methylated cytosines that have decayed in a sample over time. This variance indicates that alterations in regulatory landscapes have occurred recently within human evolution and leaves room for novel studies of the evolution of higher cognitive functions within the recent human lineage.

## Conclusions

Over the last several centuries, humans have devoted extensive time and effort to the study of the cerebral cortex, the source of our extraordinary cognitive abilities. Through these studies, we have developed a rich understanding of the function and development of the cerebral cortex and observed the vital role that genetics and gene regulation play in these processes. The link between genes (both mutations and expression) and phenotypes, however, is still very poorly understood and difficult to study. The examples that we have described here demonstrate how genomics can be used to begin to bridge this gap in understanding, in an effort to ultimately elucidate the evolutionary mechanisms underlying human brain development. To further our understanding of the fascinating human cortex, future genomic studies need to prioritize the functional characterization of genomic elements, especially noncoding regulatory regions.

## Acknowledgments

Work in the authors' laboratory on the topic of this article is supported by grants from the National Institutes of Health, Kavli Foundation, and Simons Foundation. We apologize to all colleagues whose important work was not cited because of space limitations.

# 4

# What Happens When It Goes Wrong?

## Using Human Genetics to Understand Human Brain Development and Evolution

Michael E. Coulter and Christopher A. Walsh

## Abstract

This chapter explores what happens when the development of the cerebral cortex goes awry. It presents results on work with *CHMP1A* mutations, which highlight the importance of specialized cell-to-cell communication via extracellular vesicles in cortical development and function. It reviews genetic causes of microcephaly, with an emphasis on centrosomal proteins, and presents novel insights about cortical evolution shown using a ferret model of microcephaly caused by *ASPM* loss of function. It reviews recent work to identify noncoding mutations that cause brain malformations, which has expanded understanding of cortical development beyond protein-coding genes. These three examples illustrate general principles of cortical growth and function (cellular communication and synaptic plasticity, evolution, and utilization of large data sets), made possible by recent advances in DNA sequencing technology.

## Introduction

The human genetics of cerebral cortical malformations and developmental disorders has provided a powerful tool to identify genes essential for cortical development. Since the Dahlem Workshop some thirty years ago, and rapidly accelerating since 2001 with completion of the Human Genome Project, dozens of cortical development genes with diverse functions have been identified using this approach (Zhang et al. 2014). Over the last decade, it has been appreciated that those genes essential for human brain development represent a rich source of genes modified during the evolution of humans to provide the human brain with its large size and unique network properties. A handful of human genes show evidence of having been added during the human lineage, whereas others show evolutionary selection at the level of the protein-coding sequence, where changes in amino acid sequence can result in changes in biochemistry. Only 2% of the human genome represents protein-coding genes (Gregory 2005). Larger numbers of genes show evidence of evolutionary changes that occurred to their noncoding sequences, which have the potential to alter patterns or timing of gene expression (Reilly et al. 2015). Recently, knowledge of sequence variation and presumptive functions of the noncoding genome has permitted initial insights into how changes in the noncoding portions of the genome might relate to human disease or to evolutionary change.

## Genetics of Developmental Disabilities of the Brain

Identifying the genetic causes of human cortical malformation disorders is a powerful tool for revealing critical mechanisms of cortex development and function. With a total human population of over 7.5 billion, it is likely that every gene has mutated multiple times in humans, and that every gene has mutated at least once across all cells in any individual (Bernards and Gusella 1994; Walsh 1999; Brenner 2003; Walsh and Engle 2010). As a result, individuals with disorders of cortical development represent an unbiased screen for genes that are essential to that process. Catalyzed by publication of the human genome sequence 18 years ago (Lander et al. 2001) and by widespread availability of high-throughput DNA sequencing technology about ten years ago, dozens of genes that cause such developmental disorders when mutated have been identified through the sequencing of single disease-linked genes, whole exome sequencing (which sequences all protein-coding genes), and whole genome sequencing (which sequences all DNA, both coding and noncoding).

The following molecular mechanisms of cortical development have been discovered by identifying genes mutated in cortical malformations:

- The role of centrosomes and mitotic spindles in cortical neurogenesis: *ASPM, CDK5RAP2, WDR62, NDE1*, KATNB1, *CEP63* (Bond et al. 2002, 2005; Shen et al. 2005; Hassan et al. 2007; Nicholas et al. 2009, 2010; Bilgüvar et al. 2010; Yu et al. 2010; Alkuraya et al. 2011; Bakircioglu et al. 2011; Sir et al. 2011; Hu et al. 2014b; Mishra-Gorur et al. 2014)
- The role of extracellular matrix proteins and G-protein coupled receptors in maintaining integrity of the pial surface: *POMT1, FKTN, FKRP, GTDC2, POMK, GPR56* (Beltrán-Valero de Bernabé et al. 2002; Silan et al. 2003; Beltrán-Valero de Bernabe 2004; Piao 2004; Manzini et al. 2012; Bae et al. 2014; Di Costanzo et al. 2014)

- Regulation of microtubule dynamics during neuron migration: *DCX, LIS1, TUBA1A, TUBB3, TUBB5, DYNC1H1, KIF5C, KIF2A* (Reiner et al. 1993; Lo Nigro et al. 1997; des Portes et al. 1998; Gleeson et al. 1998; Keays et al. 2007; Poirier et al. 2010, 2013; Breuss et al. 2012)
- The role of amino acid synthesis and metabolism: *QARS, AMT* (Yu et al. 2013; Zhang et al. 2014)
- The role of DNA damage repair: *NBN, PNKP* (Varon et al. 1998; Shen et al. 2010)
- The role of transcriptional regulation: *MECP2, ZNF335* (Amir et al. 1999; Yang et al. 2012)

These mechanisms span the cell types of cortical development from undifferentiated progenitors to radial glial cells, to committed neural progenitors, to postmitotic neurons. They also span cellular processes from progenitor proliferation, to neuron differentiation, to neuron migration. This diverse and widespread list of affected processes highlights the unbiased and saturating nature of cortical malformation human genetics, illustrating the extent to which human genetics can systematically identify mechanisms underlying key steps of normal development.

## Extracellular Vesicles in Cortex Formation and Function

An example of the surprising novel mechanisms that can be identified by human genetic screens involves the recent analysis of *CHMP1A* mutations that cause microcephaly and cerebellar hypoplasia (Mochida et al. 2012), which unexpectedly implicates small extracellular vesicles (EVs) in cortical and cerebellar development. EVs are small membrane-bound vesicles that are released by many cell types for specialized cell-to-cell communication through transfer of unstable molecules, such as RNA (Tietje et al. 2014), or hydrophobic proteins, such as transmembrane proteins and some growth factors (Korkut et al. 2009; Budnik et al. 2016). EVs are released by neurons and glia and may have many roles in the nervous system (Amir et al. 1999; Lachenal et al. 2011; Frühbeis et al. 2013). EVs, for example, have been implicated in wingless secretion during neuromuscular junction synapse growth (Koles et al. 2012) and Synaptotagmin 4 secretion in retrograde signaling (Korkut et al. 2009) through *in vivo* experiments in *Drosophila* as well as in synaptic strength modulation (Lachenal et al. 2011), and prion-like protein and Tau secretion through experiments in cultured mammalian neurons (Asai et al. 2015). Indeed, newly published work directly links EVs to synaptic plasticity through *Arc*, a master regulator of activity-dependent glutamate receptor trafficking (Pastuzyn et al. 2018). In *Drosophila*, EVs enable Arc1 protein transfer between neurons and muscle cells during neuromuscular junction synapse maturation; in mammalian neurons, EVs enable transfer of *Arc* mRNA between neurons allowing localized Arc translation in the recipient cell that modulates synaptic strength (Ashley et al. 2018; Pastuzyn et al. 2018).

Recently, we have identified a new role for EVs in cortex and cerebellum development by creating a mouse model of a human microcephaly gene, *CHMP1A* (Coulter et al. 2018). Human loss-of-function (LOF) mutations in *CHMP1A* cause recessive microcephaly with severe cerebellar hypoplasia (Mochida et al. 2012), and we found that a *Chmp1a* null mouse model recapitulated this phenotype. Investigating the mechanism of microcephaly in *Chmp1a* null mice, we found that secretion of the hydrophobic growth factor sonic hedgehog (*SHH*) is substantially reduced in the embryonic cerebral spinal fluid (CSF). CHMP1A is a member of the ESCRT protein complex. Since one of its functions involves the formation of multivesicular bodies (MVBs) and EV secretion, we examined MVBs in *Chmp1a* null mice. We found that MVBs were abundant in choroid plexus epithelial cells and in cerebellar Purkinje cells, two sources of SHH during brain development, and that in the absence of *Chmp1a*, each MVB had fewer luminal vesicles. To test the hypothesis that *CHMP1A* regulates SHH secretion via EVs, we turned to *in vitro* experiments and found that in the absence of *CHMP1A*, SHH-positive EV secretion was impaired. We characterized these EVs using protein mass spectrometry and found they are a new subtype of EVs, which we call ART-EVs, that carry SHH protein. Intriguingly, we found that SHH-positive ART-EVs exist in adult human CSF and that MVBs are abundant in adult mouse cortical pyramidal cells; together, this suggests a continued role of EVs in adult cortex function. Serial TEM reconstruction showed that MVBs are often located near synapses in the dendrites of pyramidal neurons (Figure 4.1); together with recently published work showing that *Arc* mRNA and protein is transferred between neurons via EVs (Ashley et al. 2018; Pastuzyn et al. 2018), these findings highlight the likely importance of EVs in mechanisms

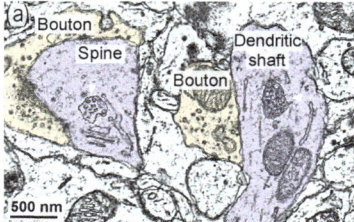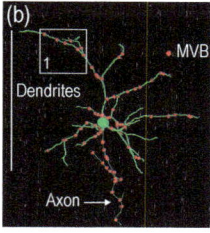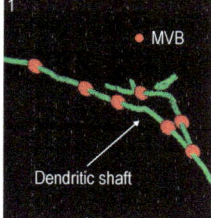

**Figure 4.1** Multivesicular bodies (MVB) near synapses in dendritic tree of mouse cortical pyramidal neuron: (a) High magnification EM image of mouse cortex neuropil showing MVBs (white arrows) near synapses. Two synapses are shown with the presynaptic bouton (yellow) and the postsynaptic dendrite (purple). (b) Three-dimensional reconstruction of a single pyramidal cell from mouse cortex serial EM showing >80 MVBs (red) distributed throughout the dendritic tree and the axon. Inset 1 shows an enlarged section of the dendritic tree from the reconstructed pyramidal cell with MVB locations labeled in red. Data in figure from Lee et al. (2016).

of cortical processing such as LTP or LTD. Further, during late cortical development, SHH protein secretion by layer V corticofugal projection neuron dendrites is required for local synapse formation with callosal projection neurons (Harwell et al. 2012), thus suggesting a potential additional role for EV-mediated SHH secretion in spatially restricted neuronal connectivity. The *Chmp1a* null mouse provides a tool for future experiments to explore more comprehensively the role of EVs and this form of cellular communication in development and function of the cortex.

## Mitotic Spindle and Centrosome Mutations Are a Common Cause of Microcephaly

Two of the earliest genes cloned in recessive microcephaly, *APSM* and *CDK5RAP2*, encode proteins that localize to the centrosome and are required for mitotic spindle organization in cortical progenitors (Bond et al. 2002, 2005; Hassan et al. 2007; Nicholas et al. 2009; Pagnamenta et al. 2012). Since then, several additional microcephaly genes have been shown to localize to the mitotic spindle or to the centrosome and centrioles, showing that these structures have a central role in the development of the cortex (Hu et al. 2014a). The centrosome is a microtubule-organizing protein complex that is present in cells during interphase (reviewed in Fu et al. 2015). It is composed of two smaller structures called centrioles; one centriole, the mother, has distal and subdistal appendages that enable it to form the basal body of the primary cilium during G1. During the DNA replication (S) phase of mitosis, the centrosome is duplicated (creating 4 centrioles); then, during metaphase, the two centrosomes migrate to opposite ends of the cell and form the poles of the mitotic spindle. As spindle poles, they recruit and organize microtubules to become the spindle. Finally, following cytokinesis, each daughter cell receives one of the two spindle poles, which become centrosomes again in the new cells (O'Connell et al. 2001; Balestra et al. 2013). Based on these roles, it is no surprise that disruptions of centrosome number, structure, or function impair cell division and neurogenesis. In fact, microcephaly proteins have been discovered that impair each step in the centrosome cycle.

*KATNB1*, encoding a microtubule-severing protein in which LOF mutations cause severe microcephaly, controls centrosome duplication and when *KATNB1* is absent, cell division is impaired on account of supernumerary centrosomes and disordered mitotic spindles (Hu et al. 2014b; Mishra-Gorur et al. 2014). *WDR62*, in which LOF mutations cause microcephaly, encodes a protein that localizes specifically to the mother centriole and is required for centriole and centrosome duplication. In the absence of *WDR62*, centrosomes fail to duplicate during S-phase and mitosis is subsequently impaired (Bhat et al. 2011; Jayaraman et al. 2016). Interestingly, milder, partial LOF mutations in *WDR62* cause other cortical malformations, without microcephaly, suggesting centrioles also play an important role in cortex development outside of progenitor proliferation (Murdock et al. 2011). *ASPM* human mutations cause severe primary microcephaly (where brain size is decreased but body size is normal), and in the absence of *ASPM* there is partial loss of centriole duplication (Bond et al. 2002; Jayaraman et al. 2016). *CDK5RAP2* also encodes a centrosomal protein and is found mutated in patients with microcephaly. During mitosis of cortical progenitors, the loss of *CDK5RAP2* creates extra spindle poles that disrupt progenitor cell division and lead to abundant cell death (Lizarraga et al. 2010; Pagnamenta et al. 2012).

Recent work reveals physical and genetic interactions between many microcephaly-related centriole proteins. For example, *WDR62/ASPM* double knock-out (KO) mice show more severe centriole duplication defects than single KO of either gene alone (Jayaraman et al. 2016). In addition, losing a single *WDR62* allele on an *ASPM* KO background produced an intermediate phenotype. ASPM and WDR62 proteins interact physically, and WDR62 is required to recruit ASPM to the centrosome; these two proteins form part of a larger complex that includes CDK5RAP2, CENPJ, and CEP63. Together, these findings suggest a model of centrosome protein recruitment that occurs in a specific order, with WRD62 recruited before ASPM (Kodani et al. 2015; Jayaraman et al. 2016).

## Evolutionary Mechanisms from Cortical Development Disorders

One of the most striking features of the cerebral cortex is the enormous expansion in size as well as regional and cellular complexity throughout the course of mammalian evolution. The cortex has increased relative to body size from mice to humans with a particular increase in the frontal cortex (Rakic 2009). Interestingly, evolution of microcephaly and developmental disorder genes contribute to the genetic mechanisms driving these changes. For example, *FOXP2*, a highly expressed transcription factor in human cortex, was mutated in a British family with severe language impairment (Lai et al. 2001). *FOXP2* shows evidence of human-specific evolution because the few amino acid changes between mouse, primate, and human dramatically alter the array of *FOXP2* transcriptional targets (Konopka et al. 2012), and because mice expressing humanized *FOXP2* exhibit accelerated learning and increased vocalizations (Enard et al. 2002; Fujita et al. 2008). This suggests that the evolutionary changes in *FOXP2* contribute critically to language development, a function unique to human cortex.

There is evidence of positive selection across mammals and an association with increased brain mass in additional cortical development disease genes, including *CDK5RAP2* and *ASPM* (Zhang 2003; Kouprina et al. 2004; Montgomery and Mundy 2014). *ASPM* is an interesting example because both its coding sequence and protein length have increased consistently over evolution. ASPM protein is composed of two N-terminal CH domains and a variable number of C-terminal IQ domains. *Caenorhabditis elegans ASPM* has 2 IQ domains, *Drosophila* has 24, mouse has 55, and human has 63 (Bond et al. 2002; Johnson et al. 2018). Although the shorter sequence of ASPM in mice compared to humans was originally thought to represent a length increase from mice to humans, direct analysis has shown that in fact rodents are outlier mammals, with an unusually short ASPM protein (Zhang 2003): this suggests a potential contribution of this ASPM shortening to the unusually small cortex which characterizes rodents. Interestingly, *ASPM* null mice have very mild microcephaly and thus model the human phenotype only very poorly (Pulvers et al. 2010; Fujimori et al. 2014; Capecchi and Pozner 2015; Williams et al. 2015; Jayaraman et al. 2016). This hypothesis raises the question of whether other mammalian *ASPM* models, whose ASPM sequence more closely resembles that of humans, would exhibit a greater degree of microcephaly in the absence of *ASPM*, and hence provide a better model system.

### *Aspm* Knockout in Ferrets Recapitulates Human Primary Microcephaly

A recent test of the hypothesis that animals with a larger cerebral cortex might better model human microcephaly came by generating *Aspm* KO ferrets through gene editing technology. The ferret cortex is larger than the mouse and, unlike the mouse, is gyrified, like the human cortex. In addition, ferret *Aspm* has 64 IQ domains, similar to 63 in human, and more than the 55 in mouse (Johnson et al. 2018). Moreover, the fetal ferret brain shows a broader diversity of progenitor types than mice, with abundant outer SVZ basal radial glial cells (radial glia lacking an apical process) unlike mice, which virtually lack this progenitor type (Hansen et al. 2010; Reillo et al. 2011). *Aspm* KO ferrets show robust microcephaly with an up to 40% reduction in brain weight and no change in body weight as well as decreased cortical surface area and volume (Johnson et al. 2018; Figure 4.2). *Aspm* KO in ferret provides a much more accurate model of human *ASPM* LOF than mouse and likely reflects the active evolution of *ASPM* in the mammalian lineage that increases the gene length from mice to ferrets to humans.

Ferret and human cortical neurogenesis is driven both by progenitors at the ventricular surface (apical progenitors) and by progenitors above the SVZ (basal progenitors or outer radial glia), whereas outer radial glia are rare or absent in mice (Fietz et al. 2010; Reillo et al. 2011; Johnson et al. 2015). Interestingly, *Aspm* KO ferrets showed an increased number of proliferating cells (Ki67+) in the basal SVZ and intermediate zone compared to wild type, the location of outer radial glia (Johnson et al. 2018). In *Aspm* KO cortex, excess basal proliferative cells formed discontinuous clusters accompanied by reduced thickness of the corresponding ventricular zone, suggesting that premature withdrawal of progenitors from VZ into the oSVZ is the cellular mechanism driving microcephaly (Johnson et al. 2018). oSVZ progenitors are

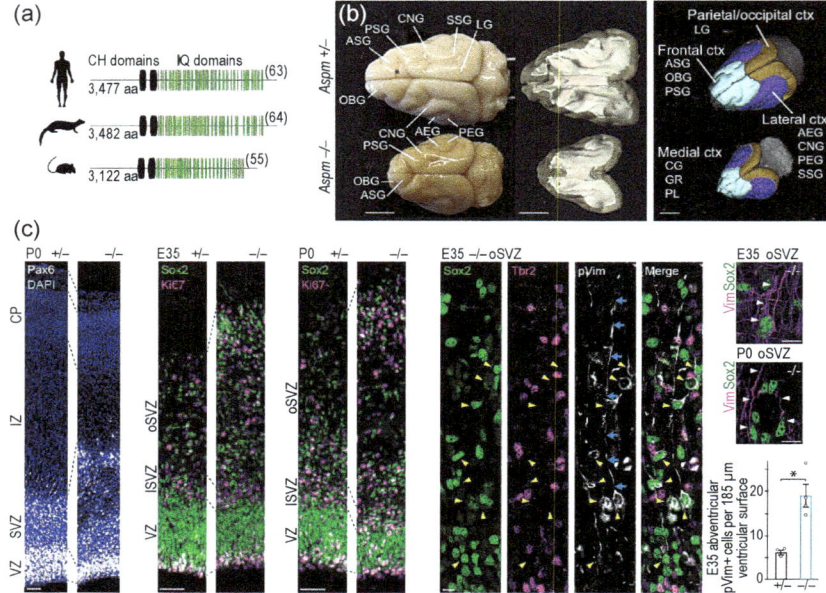

**Figure 4.2** *Aspm* KO causes microcephaly in ferret: (a) Protein structure of ASPM in mouse, ferret, and human shows that the number of IQ repeats increases farther in the evolutionary tree. (b) Photograph (left) and MRI reconstruction (right) reveals decreased ferret brain size in *Aspm* –/– ferret compared to *Aspm* +/– control. Gyrification pattern is preserved. Labeling of cortical regions (far right): frontal cortex shows greatest reduction in size of *Aspm* –/– ferret. (c) The outer subventricular zone (oSVZ) is expanded in *Aspm* –/– ferret at birth and in late gestation. Pax6 and Sox2/KI67 stained layer above ventricular zone (VZ) is larger in *Aspm* –/–. In *Aspm* –/– ferret, oSVZ has outer radial glial cells defined as Sox2/Vimentin+ and Tbr2– cells (yellow arrowheads), one of which has a clear process extending toward pial surface (blue arrows). Quantification of increase in Vimentin+ cells in oSVZ in *Aspm* –/– ferret. Adapted from Johnson et al. (2018).

predicted to give rise to fewer postmitotic neurons than apical progenitors; thus, a premature switch to outer radial glia will result in decreased overall neurogenesis.

The essential role of *Aspm* in regulating the switch from apical RG to outer RG provides an evolutionary mechanism that might dynamically control cerebral cortical size, in which subtle variation in *Aspm* sequence or structure might alter the number of apical RG, which in turn would dictate cerebral cortical surface area. This key cellular feature of human cortex development (i.e., neurogenesis via outer RG) is not modeled in mice where this progenitor type is so rare. Hence, ferrets may become a powerful model in the future to understand mechanisms of many human cortical disorders that are poorly modeled in mice.

## Coding and Noncoding Mutations in Human Disease

Although 98% of the human genome is noncoding DNA (i.e., DNA that does not encode a protein), recent work has started to identify human cortical disease mutations in noncoding DNA. Noncoding DNA includes introns, unique regulatory elements, transposable elements, and repetitive DNA not related to transposable elements (Gregory 2005). Noncoding cis-regulatory sequences such as promoters and enhancers have been known for several years, and they have been shown to modulate gene expression through transcription factor binding sites as well as steric interactions and DNA folding (Nobrega 2003; Pennacchio et al. 2006). Although noncoding DNA does not directly create protein products, it regulates gene expression and encodes active RNA molecules (Reilly et al. 2015). Indeed, it has been hypothesized that tissue-specific expression and different expression levels dictated by noncoding DNA greatly increases the complexity of the human transcriptome and proteome, even with a relatively small number of coding genes (~19,000), and that this complexity is a key feature of human evolution (Geschwind and Rakic 2013; Kellis et al. 2014; Reilly et al. 2015). Mutations in noncoding DNA have now been identified in diseases of cortical development.

## Noncoding Mutations Mimic Phenotype of Loss of Function Mutations

Heterozygous LOF mutations in the growth factor *SHH* cause holoprosencephaly in humans, a syndrome of incomplete separation of the two cerebral cortical hemispheres that results in a single lateral ventricle and craniofacial anomalies (Roessler et al. 1996). Recent work identified a heterozygous mutation in a conserved noncoding element 460 kilobases upstream of *SHH* called Shh brain enhancer-2 (SBE2) (Jeong et al. 2008). Jeong et al. found a single base substitution in a 10 basepair (bp) sequence of SBE2 highly conserved across species, and then showed that this sequence binds the transcription factors (TF) *Six3* and *Six6* and that TF binding was largely abolished by the mutation. In addition, expression of lacZ in developing mouse embryo, driven by wild-type or mutant SBE2, showed that expression in the developing brain was reduced with the patient mutation (Jeong et al. 2008). These findings illustrate how a noncoding DNA mutation can cause neurodevelopmental disease by reducing expression of an essential, dosage-sensitive growth factor in the brain.

## Some Noncoding Mutations also Highlight Evolutionary Mechanisms

LOF mutations in the G-protein coupled receptor, *GPR56*, cause a brain malformation syndrome called polymicrogyria (PMG), in which the cortical surface is covered in numerous small gyri (Piao 2004). Recently, a family was identified with a recessively inherited variant form of PMG in which the cortex surrounding the Sylvian fissure was strongly affected with PMG, but the rest of the cortex had normal gyrification (Bae et al. 2014; Figure 4.3). The syndrome showed strong linkage to *GPR56* locus; however, no mutations were identified in the *GPR56* coding sequence. Instead, sequencing of 38 conserved noncoding elements upstream of the first exon revealed a homozygous 15-bp deletion, which segregated with disease. This deletion is located about 150 bps upstream of a noncoding alternative start exon for *GPR56* (e1m). Remarkably, when a large (23 kb) region containing these upstream elements was expressed

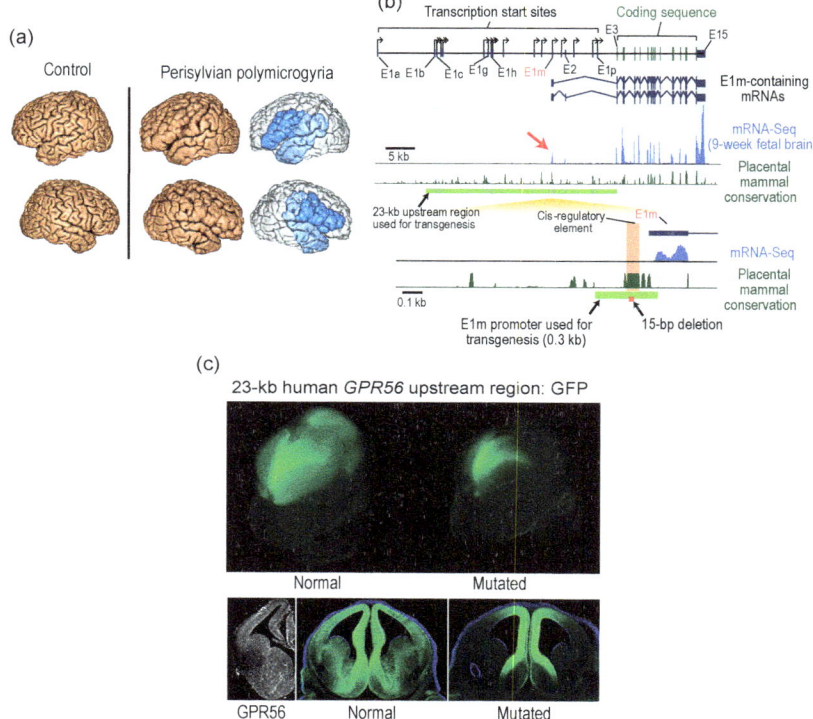

**Figure 4.3** Human *GPR56* noncoding deletion causes perisylvian polymicrogyria (PMG): (a) Three-dimensional reconstruction of MRI from patient with GPR56 noncoding deletion shows perisylvian PMG (highlighted in blue). Left image shows reconstruction of a normal brain MRI. (b) 15-bp patient deletion is found near a 5′ UTR noncoding exon in human *GPR56*. This exon, labeled E1m, is highly expressed in human fetal brain (blue trace) and has a highly conserved cis-regulatory DNA element just upstream (lower panel, green trace and highlighted in orange) which contains the 15-bp deletion. (c) Transgenic green fluorescent protein (GFP) expression of 23-kb human *GPR56* upstream region containing E1m promoter [light green bar in (b)] in developing mouse brain, eliminated lateral cortical expression of *GPR56* while preserving medial expression. Adapted from Bae et al. (2014).

transgenically in developing mouse brain, the 15-bp deletion eliminated lateral cortical expression of *GPR56* while medial expression was preserved. This finding suggests that this noncoding conserved element is required for expression of *GPR56* in the perisylvian cortex via cis-acting activation of e1m transcription. In this case, spatially localized expression of a cerebral cortex gene is regulated by a noncoding DNA element. This illustrates the importance of gene expression refinement by noncoding DNA and shows how mutation of such DNA can produce a spatially restricted cortical malformation syndrome.

Interestingly, this *GPR56* mutation also illustrates an important concept: in addition to protein-level evolution, as discussed above in *ASPM*, evolution also changes noncoding DNA. Noncoding DNA in the *GPR56* locus is actively evolving and has greatly expanded between mouse and human. In particular, a number of new untranslated exons, alternative promoters, and other noncoding regulatory elements are found in *GPR56* only in the primate linage, including humans (Bae et al. 2015). These additional noncoding elements enable *GPR56* to be expressed with more regional and temporal precision in humans and this increased repertoire of expression may drive the greater complexity and capabilities of the human cortex.

## Discovery and Analysis of Human Accelerated Regions

There is a class of noncoding DNA elements defined by human-specific evolution, called human accelerated regions (HARs). HARs are regions of DNA that are highly conserved in most mammals but which show strong, specific sequence divergence in humans (Pollard et al. 2006a). The genetic differences in HARs in humans suggest that HARs are under recent evolutionary selection (thus, the name "accelerated") and that they represent essential functional sequences whose precise function may have changed between nonhumans and humans. Evidence suggests HARs have varied functions, including expression of RNA (Pollard et al. 2006b) and as transcriptional enhancers through physical interaction with promoter DNA (Capra et al. 2013). Epigenetic signatures suggest ~30% of HARs are active during embryonic development in the limbs, heart, and brain (Capra et al. 2013).

Recent work examining the genetic causes of developmental disorders has found that HARs are essential for normal brain development. Enrichment analysis suggests HARs may have roles in neurologic disease. HARs are enriched near haploinsufficient genes, raising the possibility that gene expression changes resulting from mutations within HARs may cause disease. Contributing to this hypothesis is a recent study, which reported that several HARs are within linkage regions for schizophrenia identified by genome-wide association studies (Xu et al. 2015). Of particular interest in neurodevelopment, HARs are enriched near genes associated with autism spectrum disorder (ASD), suggesting that they may play a role in ASD pathogenesis (Doan et al.

2016). Indeed, single nucleotide variants found in HARs in ASD patients high-light the importance of noncoding mutations in neurodevelopmental disease. Doan et al. studied a cohort of ASD patients with no coding mutation or copy number variant identified through whole genome sequencing and examined the HAR sequences in each patient. Compared to neurologically normal controls, they showed that in ASD patients there was a significant enrichment of biallelic point mutations in HARs. Several mutated HARs were shown to interact with brain-expressed genes, in particular *MEF2C*, *CUX1*, *TMEM161B*, *PTBP2*, *GPC4*, *CDKL5*, *USP32*, and *DAB2* (Doan et al. 2016). These HARs have pre-dicted enhancer activity, and transgenic expression of the mutated HAR in developing mouse embryos showed changes in expression of the target gene.

In one example, a homozygous point mutation was identified in HAR426, located 200 kb upstream of the *CUX1* promoter (Figure 4.4). This mutation was

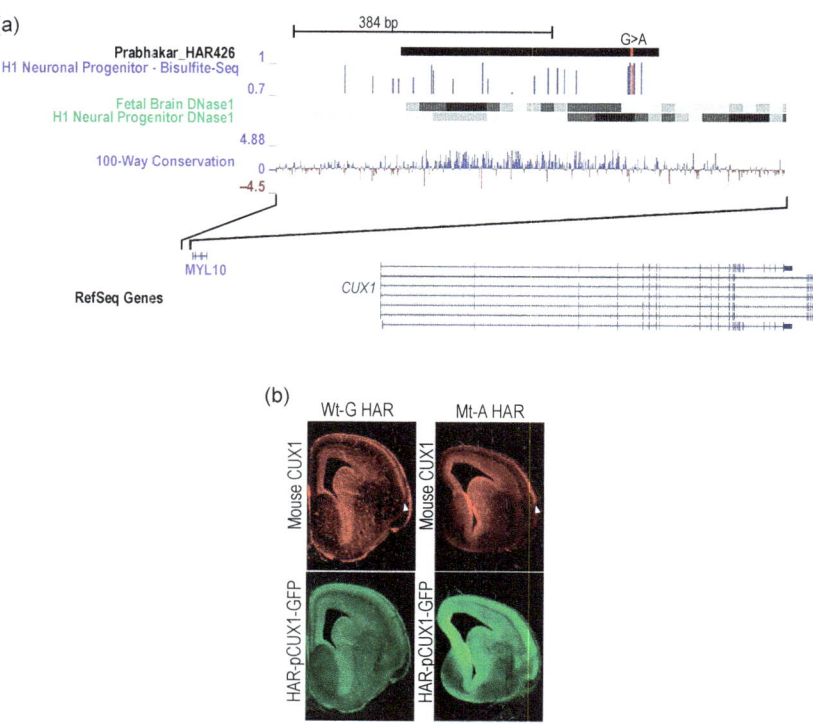

**Figure 4.4**   Human mutation in autism spectrum disorder (ASD) in HAR426 alters *CUX1* expression: (a) Annotated HAR426 (black bar) with G>A point mutation found in a patient with ASD. This HAR has multiple bisulfite peaks in DNA from neuronal progenitors and shows DNase1 activity in fetal brain and progenitors. This HAR se-quence is highly conserved among 100 species. Bottom panel shows that this HAR is upstream of *CUX1*. (b) Wild-type (Wt) and mutant HAR426 driving green fluorescent protein (GFP) expression in developing mouse cortex shows increased transcriptional activity of mutant HAR426. Adapted from Doan et al. (2016).

found in three individuals from two families with ASD and intellectual disability. The mutation was predicted to create a new transcription factor binding site, and expression of the mutated HAR in mouse embryos showed increased expression compared to expression of the wild-type HAR. Overexpression of *Cux1* in cultured cortical neurons showed increased spine density (Cubelos et al. 2010), suggesting that HAR426 mutation may interfere with normal spine refinement (Doan et al. 2016).

In a second example, a 5-bp homozygous insertion/deletion in HAR169 near *PTBP2* was identified in two brothers with ASD and intellectual disabilities. *PTBP2* encodes a brain-specific splicing protein that regulates neuron differentiation (Licatalosi et al. 2012; Li et al. 2014). Chromatin interaction measurement showed that HAR169 binds the *PTBP2* promoter. The patient mutation is predicted to disrupt TF binding: a luciferase expression assay showed that the mutation decreased the enhancer activity of HAR169 by 50% in cells in a neuron-like state, and an expression analysis showed it decreased enhancer activity by 40% in primary mouse neurospheres (Doan et al. 2016). Together, HAR mutations in ASD patients illustrate how noncoding mutations can disrupt normal brain development by altering expression of neuronal genes, and, more importantly, how the essential functions of these evolutionarily important sequences can be analyzed through larger-scale application of human genetics of neurological disorders.

## Decreased Sequencing Cost Will Accelerate Identification of Noncoding Mutations

The previous three examples of noncoding mutations causing neurodevelopmental disease were reported over the last ten years (in 2008, 2014, and 2016, respectively) and illustrate the power of recent advances in DNA sequencing technology. The amount of DNA sequenced increased across these studies, from targeted sequencing of a 1,000 bp enhancer in 2008, to a collection of 38 noncoding elements totaling 5,000 bp in 2014, to extraction of HAR sequences from 3,000,000,000 bp of whole genome sequencing in 2016. In parallel, across this same time span, the cost of sequencing a single human genome decreased 1,000-fold, according to the National Human Genome Research Institute: from $1,300,000 ($15/Mb) in 2008 to about $1,000 ($0.01/Mb) in 2017. These studies illustrate how the decreasing cost of DNA sequencing, through introduction of new technology for high-throughput sequencing, has enabled ever more complete examination of noncoding DNA, which comprises 98% of the human genome. Over the next few years, sequencing costs are predicted to continue to fall, which will make it feasible to perform whole genome sequencing on an increasing number of patients. This, in turn, will further expand our ability to identify patients with mutations in noncoding DNA and add to the rich diversity of genetic causes of neurodevelopmental disease. In addition, greater understanding of noncoding variants in disease will increase our understanding of noncoding DNA in normal brain development. Noncoding DNA includes gene enhancers and repressors, noncoding RNAs, micro RNAs, inserted retrotransposons, and additional regulatory sequences; each of these categories has increasingly appreciated functions in brain development.

## Conclusion

In this chapter we have presented three lessons that can be learned by studying the genetic causes of cortical development that has gone awry. *CHMP1A* mutation highlighted the function of EVs and specialized cellular communication in cortical development and adult function. Microcephaly in *Aspm* KO ferrets illustrated the active role that evolution plays in cortical development and the advantages of higher-order model organisms. Finally, noncoding mutations in *GPR56* and HARs demonstrated new appreciation for genetic regulation of cortical development beyond protein-coding genes. These three discoveries were made possible by recent advances in DNA sequencing technology, and we hope they will raise new questions that drive us to continue advancing our understanding of the cerebral cortex in the years to come.

## Acknowledgments

M. E. C. was supported by F30 MH102909, a Howard Hughes Medical Institute Medical Student Fellowship, and a Nancy Lurie Marks Family Foundation Medical Student Fellowship. C. A. W. was supported by R01-NS35129 and R01-NS032457 from the NINDS, U01 MH106883 from the NIMH, and the Allen Discovery Center Program through the Paul G. Allen Frontiers Group. C. A. W. is an Investigator of the Howard Hughes Medical Institute.

# 5

# Evolution and Ontogenetic Development of Cortical Structures

Debra L. Silver, Pasko Rakic, Elizabeth A. Grove,
Tarik F. Haydar, Takao K. Hensch, Wieland B. Huttner,
Zoltán Molnár, John L. Rubenstein, Nenad Sestan,
Michael P. Stryker, Mriganka Sur,
Maria Antonietta Tosches, and Christopher A. Walsh

## Abstract

The cerebral cortex controls our unique higher cognitive abilities. Modifications to gene expression, progenitor behavior, cell lineage, and neural circuitry have accompanied evolution of the cerebral cortex. This chapter considers the progress made over the past thirty years in defining potential mechanisms that contribute to cortical development and evolution. It discusses the value of model systems for understanding elaboration of cortical organization in humans, with an emphasis on recent technical and conceptual advances. It then examines our current understanding of the molecular and cellular basis for cortical development and evolution; discusses how neuronal fates are specified and organized in lamina, columns, and areas; and revisits the radial unit and protomap hypotheses. Finally, it considers our current understanding of the development, stability, and plasticity of cortical circuitry. Throughout, it highlights the profound impact that new technological advances have made at the molecular and cellular level, and how this has changed our understanding of cortical development and evolution. The authors conclude by identifying critical and tractable research directions to address gaps in our understanding of cortical development and evolution.

---

**Group photos (top left to bottom right)** Debra Silver, Pasko Rakic, Christopher Walsh, Takao Hensch, Elizabeth Grove, Michael Stryker, Tarik Haydar, John Rubenstein, Zoltán Molnár, Mriganka Sur, Nenad Sestan, Maria Antonietta Tosches, Wieland Huttner, Debra Silver, Mriganka Sur, Maria Antonietta Tosches, John Rubenstein, Pasko Rakic, Zoltán Molnár, Nenad Sestan, Michael Stryker

# Introduction

The cerebral cortex is generally considered the biological substrate of our unique cognitive abilities, including memory, complex reasoning, and advanced language. Over the course of evolution, the neocortex has undergone a disproportionate number of changes relative to other brain regions, suggesting that anatomical, cellular, and molecular modifications of the cerebral cortex may have gone hand in hand with human cognition. This great advance in computing power, however, has come at a price, as complex cognitive and psychiatric disorders also appear to be largely unique to humans. The study of cortical evolution is thus crucial as it can inform fundamental principles governing how the brain works as well as elucidate mechanisms relevant to human health.

The cerebral cortex is derived from the dorsal telencephalon or pallium, which has been traditionally divided into medial, dorsal, and lateroventral areas. A distinct feature of mammals is their six-layered cortex, termed "neocortex," considered to be a substrate for our highest cognitive functions, including abstract thinking and language. The medial pallium, archicortex, or hippocampus consists of three layers and is involved in short-term memory and cognitive spatial mapping functions. Some lateroventral pallium areas also contain three layers that receive inputs directly from the olfactory system. The neocortex is organized in the radial dimension into neuronal layers that are further divided into sublayers. Historically, it is well established that the neocortex is tangentially composed of functional areas that control sensory, motor, and cognitive capacities (Brodmann 1909).

Neocortical anatomical features correlate with complex behavior, such as language and an ability to develop and use tools and technology, which distinguishes humans from other species (Geschwind and Rakic 2013; Molnár and Pollen 2013). Relative to nonhuman primates, humans also possess a higher brain to body ratio, more neurons, greater degree of brain lateralization (Lewitus et al. 2014; Sousa et al. 2017a), and a complex pattern of gyri and sulci (Borrell and Götz 2014). In addition, these cortical features are derived in humans during a longer gestational period and an extended adolescence up to the third decade of life (Petanjek et al. 2011).

As we reflected on the question of what is uniquely human, it became apparent that our understanding of human-specific cortical features still remains inadequate. Yet, when compared to the issues discussed at the Dahlem Workshop on neocortex (Rakic and Singer 1988), research over the past thirty years has given rise to enormous progress in our understanding of cortical development and evolution.

## Technological Breakthroughs Advancing Our Understanding of Cortical Development and Evolution

Conceptual advances in cortical development and evolution have coincided with major technological breakthroughs. The first advance, in molecular neuroscience, now enables us to pinpoint genomic, epigenomic, transcriptomic, and proteomic features of development and evolution. Recent implementation of single-cell transcriptomics (scSEQ) has led to comprehensive classification of cortical cell types, progenitor states, and developmental trajectories across species (Camp et al. 2015; Macosko et al. 2015; Bakken et al. 2016; Tasic et al. 2016; Nowakowski et al. 2017; Mayer et al. 2018; Mi et al. 2018; Tosches et al. 2018). A second advance is our ability to manipulate genes, cells, and circuits using various approaches: genomic engineering, viral transduction, electroporation, optogenetics, RNAi, and cell transplantation. More recently, CRISPR/Cas9 technology allows for constitutive and conditional mutagenesis, as well as manipulation of promoters and enhancers to control gene expression precisely (Cong et al. 2013; Wang et al. 2013; Kalebic et al. 2016; Tsunekawa et al. 2016; Yang and West 2016). Major advances in microscopy have further enabled real-time visualization of genetic manipulations. Further, progress in cortical development and evolution has been propelled by new and accessible advances in model systems, including macaques, marmosets, and ferrets, which have a more complex cortical organization than commonly studied rodents (Homman-Ludiye and Bourne 2017; Johnson et al. 2018). Moreover, we now possess the ability to generate brain organoids *in vitro* for diverse species, including the great apes and notably humans (Lancaster et al. 2013; Camp et al. 2015; Mariani et al. 2015; Mora-Bermudez et al. 2016; Otani et al. 2016; Giandomenico and Lancaster 2017). These technological advances have laid the groundwork for thirty more years of deciphering even deeper mechanisms of cortex development, evolution, and human cortical disorders.

## Theories Underlying Human Cortical Evolution

A number of theories have been put forth to explain human cortical evolution (Geschwind and Rakic 2013; Molnár and Pollen 2013). One theory posits that the duration of gestation and infancy can explain cortical differences, due to prolonged neurogenesis and a differential impact of experience. Humans have by far the longest neurogenic period among primates (Petanjek et al. 2011). Mathematical modeling of cortical progenitor lineages suggests that a longer neurogenic period in humans is sufficient to explain the increased cortical neuron number compared to other great apes (Lewitus et al. 2014; Picco et al. 2018). Consistent with this, human babies born prematurely, with a reduced gestation period, are at elevated risk for neurological deficits, including learning and communication disabilities. However, in comparison, synaptogenesis proceeds intrinsically according to the day of conception rather than the birth date (Bourgeois et al. 1989).

A second theory posits that human-specific traits, such as higher cognition and abstract thinking, are associated with a disproportionately large cerebral cortex. It is notable that some cetaceans, such as dolphins, also have extra large and complex neocortices and are considered highly intelligent (Sousa et al.

2017a). Indeed, the importance of brain size for human cognition is not easily reconciled with disorders of microcephaly in which patients have disproportionately smaller brains, yet retain human-like social behavior and in some cases language (Sousa et al. 2017a). Thus, it may be more relevant to consider the extent to which the prefrontal cortex (PFC) is enlarged. Indeed, the PFC, together with some association areas of the parietotemporal lobes, is the most expanded brain structure in primate evolution (Goldman-Rakic 1987), but its relative size in humans is still a matter of debate (Wise 2008; Elston et al. 2011; Gabi et al. 2016). Thus, beyond cortical size, several additional factors, including the pattern of connectivity and large subcortical white matter, have been posited to shape human brain evolution (Rash et al. 2019):

1.  Quantification of neuronal nuclei has shown that relative to chimpanzees and rodents, human brains have more cortical neurons (Herculano-Houzel et al. 2007; Gabi et al. 2016). The embryonic telencephalic vesicles of human and nonhuman primates, however, are disproportionately enlarged relative to rodents, even before the first neurons have been generated (Bystron et al. 2008). This indicates that cortical expansion may initiate in the neuroepithelium (Rakic 2009).

2.  Beyond neuronal number, neuronal diversity and morphological differences distinguish humans and primates. For example, a subtype of enlarged pyramidal neurons, von Economo cells, are enriched in humans and other great apes, and are hypothesized to promote rapid communication (Nimchinsky et al. 1999). In humans, some pyramidal neurons have been described with extensive branching, which could augment neuronal activity (DeFelipe 2011). Rare subpallial-derived interneurons expressing dopamine biosynthesis genes and capable of producing dopamine *in vitro* are also enriched in the human striatum, yet absent in the nonhuman African ape neocortex (Sousa et al. 2017a). In addition to structural differences, homologous human neuronal cell types have undergone molecular changes that may have changed their physiological properties (Sousa et al. 2017a). Also, the number and size of astrocytes and oligodendrocytes are greater in primates compared to rodents (Oberheim et al. 2009).

3.  Primates have thicker and more complex supragranular layers, thought to promote increased connections between cortical regions (Marin-Padilla 2014). Further, humans have robust white matter connecting language regions of the perisylvian cortex, which is smaller or absent in nonhuman primates (Rilling et al. 2008). Likewise, differences in the number and composition of functional areas and their asymmetry may also influence cortical capacity, particularly with regard to language skills (Chance 2014).

4.  Finally, noncortical structures, such as the cerebellum, are greatly expanded in humans, with a disproportionate increase in granule cell

number (Weaver 2005; Ito 2008; Barton 2012; Barton and Venditti 2014; Sokolov et al. 2017). Further, higher-order nuclei within the thalamus are massively enlarged in primates and may mediate cortico-cortical interactions (Sherman and Guillery 2011).

## How Model Systems Have Contributed to Understanding Cortical Development and Species Differences

### Mouse Models

Different animal model systems afford distinct advantages to study cortical development and evolution. Mice have been a historical model of choice, in large part due to their genetic tractability and the fact that they share key features with humans (Clowry et al. 2010). During development, both species undergo similar cellular processes with comparable temporal progression. They also have homologous cell types and, in many cases, utilize identical molecular programs.

The mouse cortex, however, is the product of its own unique evolutionary forces that resulted in a small body and lissencephalic brain, a nonlaminated lateral geniculate nucleus, and lateral-set eyes with minimal binocular vision. The PFC of mice is limited in size, containing medial, orbitofrontal, and cingulate areas but probably no equivalent of the primate dorsolateral PFC (Preuss 1995). Beyond cortical size and a limited number and diversity of higher-order cortical areas, mouse brains are lissencephalic. In addition, compared to primates which contain >40% white matter, mice only have about 5% (Herculano-Houzel et al. 2010), perhaps due to their smaller brain size and the relatively shorter distances that neuronal signals traverse.

Beyond these structural differences, common inbred lab mice are overfed and understimulated, factors which could affect the precision of cortical responses. Thus, generation of knockout strains may select for the fittest animals and mask biological insights, failing to model disease phenotypes. For example, *ASPM* mutations are associated with severe human microcephaly (Jamuar and Walsh 2015), yet *Aspm* knockout in mice results in mild microcephaly (Pulvers et al. 2010). By contrast, the same genetic perturbation in gyrencephalic ferrets causes profound microcephaly and preferential loss of frontal areas, as seen in humans (Johnson et al. 2018; see also Coulter and Walsh, this volume). Phenotypic discrepancies may be amplified when modeling complex human psychiatric disease. Yet, compared to other model species, such as great apes, which are subject to ethical and legal hurdles, mice are superior for studying behavior. Thus, we emphasize that for elucidating basic principles of development and circuitry, the mouse remains invaluable (Goffinet and Rakic 2000).

**New Primate Models: *In Vivo*, *Ex Vivo*, and *In Vitro* Toolsets**

Much of our classical understanding of cortical function and development has employed nonhuman primates, particularly the Old World primate, the rhesus macaque (Geschwind and Rakic 2013). Relative to thirty years ago, our knowledge base today has been driven by a deeper examination of a range of primate models. Remarkable progress has also been made through direct studies of human fetal and adult samples obtained via surgery, postmortem, or tissue banks. While *ex vivo* human brain slices are invaluable for investigating cellular and molecular aspects of development, they are less amenable to longer-term studies and lack important extrinsic cues. Moreover, access to human tissue remains a major hurdle. Thus, recent efforts toward developing additional primate models, such as marmosets (Homman-Ludiye and Bourne 2017), offer complementary approaches for future studies.

The ability to generate induced pluripotent stem cells (iPSCs) has transformed traditional primate model approaches, enabling investigation of evolutionary differences within species-specific contexts. Several groups have established iPSC lines from human and nonhuman primate somatic cells, which can be readily directed toward a neural fate to model early developmental stages (Eiraku et al. 2008; Lancaster et al. 2013; Marchetto et al. 2013; Gallego Romero et al. 2015; Heide et al. 2018). Three-dimensional organoids have revealed new evolutionary differences between humans and nonhuman primates (Camp et al. 2015; Mora-Bermudez et al. 2016; Otani et al. 2016). For example, such comparisons led to the identification of the first differences concerning cortical progenitor cell behavior between human and other great apes—the specific lengthening of metaphase during human apical progenitors mitosis (Mora-Bermudez et al. 2016). However, there remain limitations with current protocols for generating organoids, including lack of vasculature for long-term culture, lack of basement membrane and cerebral spinal fluid, as well as lack of standardization across labs. Given the current pace of research, continued optimization of organoid protocols will likely overcome many of these technical hurdles.

## Which Molecular and Cellular Processes Shape Development and Are Evolutionarily Divergent?

**Genetic Basis for Cortical Development and Evolution**

With the complete sequencing of the genomes of humans and most major mammalian species comes the promise of discovering specific molecular changes that make each species unique. Delivering on this promise, however, remains a complex, multifaceted challenge. Genome-wide approaches have collectively uncovered human-specific features including structural variations (e.g., chromosomal deletions and duplications) and point mutations in coding and noncoding regulatory regions (Lui et al. 2011; Borrell and Reillo 2012; Geschwind and Rakic 2013; Dennis et al. 2017; Florio et al. 2017; Sousa et al. 2017a). Many of these have been empirically demonstrated to affect protein structure, gene function, and/or expression as well as to influence diverse aspects of the neocortex (see Lorente-Galdos et al., this volume).

Changes in gene regulatory regions are strongly linked to brain evolution (King and Wilson 1975). For example, the vast majority of 510 annotated human-specific deletions reside within noncoding regions (McLean et al. 2011). Evolutionary changes to noncoding regulatory elements are frequently located near genes implicated in neural development, whereas coding changes do not show the same bias (Haygood et al. 2010). Over 3,000 human accelerated regions (HARs), sequences that have undergone rapid positive selection in humans, reside mostly in regulatory elements (Pollard et al. 2006a, b; Lindblad-Toh et al. 2011; Capra et al. 2013). To date, HARs have been implicated in diverse aspects of cortical function, ranging from progenitor proliferation to control of neuronal spine density (Capra et al. 2013; Boyd et al. 2015; Reilly et al. 2015; Doan et al. 2016). While HARs are poised to fine-tune human cortical development, annotated functions for the vast majority of HARs and other human-specific coding and noncoding elements are lacking.

Approximately one-third of mouse and human enhancers are predicted to diverge between species (Nord et al. 2013; Reilly et al. 2015; de la Torre-Ubieta et al. 2018). Indeed, epigenetic profiling for histone acetylation and methylation marks in humans, macaques, and mouse neocortices reveal promoters and enhancers that have gained human-specific activity (Silbereis et al. 2016; Mitchell and Silver 2018). Importantly, more than 4,600 human telencephalic enhancers have been identified, but only a subset have demonstrated activity (Visel et al. 2013).

In comparison, transcriptional circuits have been well defined in the developing mouse neocortex. These circuits highlight both individual and redundant transcriptional regulation. For example, the transcription factors FEZF2, SOX5, SATB2, and TBR1 control specification of different subtypes of excitatory projection neurons, whereas multiple Dlx factors are present during development of perhaps all forebrain GABAergic neurons, projection and local circuits (Leone et al. 2008; Kwan et al. 2012b; Greig et al. 2013; Hu et al. 2017). However, while the human genome is estimated to harbor 400,000 enhancers and 70,000 promoters (ENCODE Project Consortium 2012), it remains largely unclear how transcriptional networks control 20,000 protein-coding genes, including human-specific cortical development (Nord et al. 2015; Emera et al. 2016).

Transcriptome comparisons of developing and early postnatal human, chimpanzee, and macaque brains indicate prevalent, global differences in gene expression and splicing among primates (Johnson et al. 2009; Kang et al. 2011; Fietz et al. 2012; Konopka et al. 2012; Lui et al. 2014; Miller et al. 2014; Pletikos et al. 2014; Sousa et al. 2017a). Sequencing of isolated cells has further reinforced these differences (Pollen et al. 2014, 2015; Florio et al. 2015; Johnson et al. 2015). For example, a recent *in silico* screen of five transcriptome data sets led to the identification of 15 human-specific genes with preferential expression in progenitors (Florio et al. 2018). One of these human-specific genes, *ARHGAP11B*, has been shown to amplify basal progenitors in the mouse neocortex and is implicated in cortical folding and expansion (Florio et al. 2015, 2016). Thirty-five human genes with progenitor-enriched expression have primate-specific orthologs, constituting a resource of candidates which may exert key roles in neocortical development during human evolution (Florio et al. 2018). Several prominent signaling pathways (STAT, mTOR, Notch, WNT, FGF, SHH) have also been implicated in human-enriched progenitors (Lui et al. 2014; Pollen et al. 2014, 2015; Nowakowski et al. 2017). Taken together, these discoveries indicate molecular support for species-specific patterns of gene expression and give us the ability to interrogate functionally a finite number of genes. Future progress will rely on our ability to exploit this knowledge base to understand cell fate specification, heterogeneity, and circuitry across development and evolution.

## Neural Progenitors: Building Blocks for the Cortex and Underlying Cortical Evolution

Initial stages of corticogenesis involve early patterning of the neural plate and early neurula. As of 25 days postcoital, species differences are already visible. At this stage, relative to the posterior neural plate, the anterior neural plate is larger in humans than in mice (Bystron et al. 2008). This suggests that the human neural plate, the anterior neural ridge, the rostral patterning center, or the prechordal plate may secrete factors to control specification of anterior structures.

The founder population for the neocortex is neuroepithelial progenitor cells (NPCs/NECs), which are arguably the most impactful progenitor for brain expansion (see Figures 5.1a and 5.2). They are critical for amplifying the precursor pool via symmetric proliferative divisions near the ventricular cavity, termed the ventricular zone (VZ) (Rakic 1972). The proliferative pool is further augmented by the emergence of the subventricular zone (SVZ), which is particularly enlarged in primates, including humans, where it was initially identified and named (Bystron et al. 2008). Empirical support for NPC function was shown by increasing proliferation or decreasing apoptosis of founder cells in mice, which causes enlargement of the cortical surface and convolutions (Kuida et al. 1998; Chenn and Walsh 2002). Genetic manipulation that increases founder cells in mice also can enlarge frontal brain regions (Assimacopoulos et al. 2012).

The radial unit model, put forth more than thirty years ago (Rakic 1988b), proposes that increasing the size and proliferative capacity of neural precursors close to the ventricular cavity helps explain the beginning of the

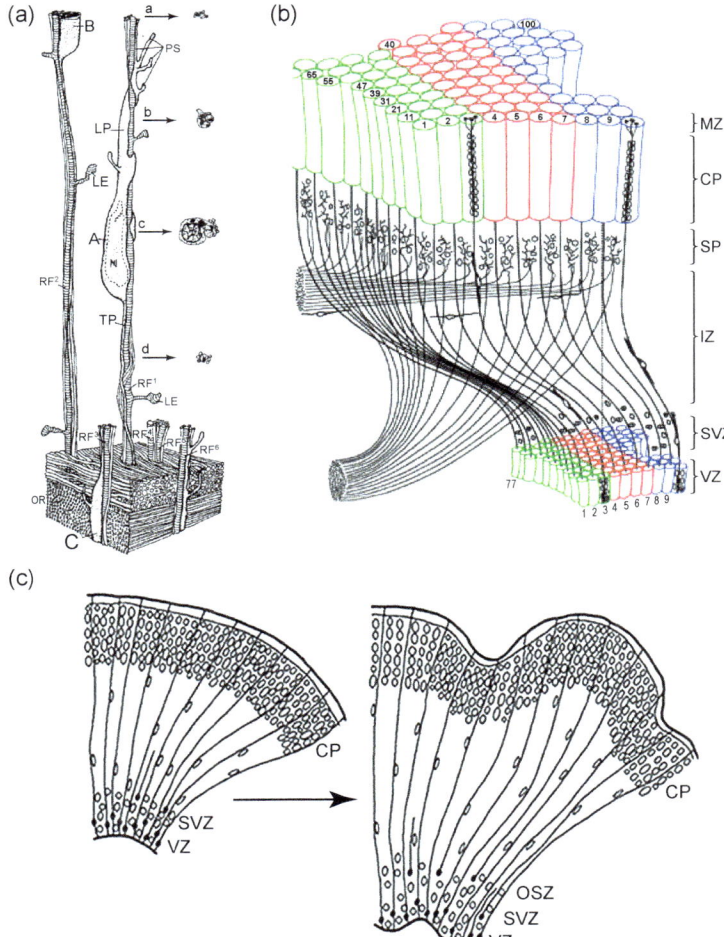

**Figure 5.1**  Interpretation of the radial unit hypothesis. (a) Three-dimensional reconstruction of migrating neurons, based on electron micrographs of semi-serial sections of the occipital lobe of the monkey fetus; reprinted with permission from Rakic (2003). (b) Representation of the radial unit hypothesis; reprinted with permission from Rakic (1988b). (c) Graphic explanation for cortical expansion and elaboration during evolution. An expanded cellular sheet due to increased proliferation or decreased cell death of radial units is associated with transformation from a lissencephalic (left) to gyrencephalic brain; reprinted with permission from Geschwind and Rakic (2013).

distinct cytoarchitecture and enlarged neocortical size of higher mammals (Figure 5.1b). A larger progenitor pool would ultimately generate more cortical neurons and a bigger brain (Bystron et al. 2008). The radial unit model is supported experimentally (Kuida et al. 1998; Chenn and Walsh 2002; Pattabiraman et al. 2014) and provides a basic cellular explanation for how

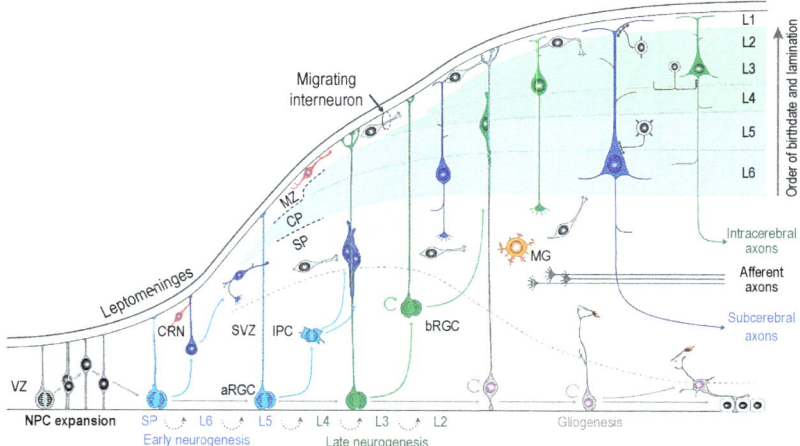

**Figure 5.2**  Illustration of the development of cerebral neocortex with diverse progenitors. Prior to the onset of neurogenesis, neuroepithelial progenitor cells (NPCs) in the ventricular zone (VZ) of the developing neocortex divide symmetrically to expand the progenitor pool. Later, NPCs transform into apical radial glia cells (aRGCs), which line the VZ; from there they extend a long radial process stretching to the basal surface. aRGCs asymmetrically divide to generate another aRGC and either a nascent projection neuron, basal intermediate progenitor cell (IPC), or basal radial glial cell (bRGC). Basal progenitors in the subventricular zone (SVZ) also generate neurons. The nascent projection neurons migrate radially from the VZ along the RGC basal process into the cortical plate (CP). The earliest born neurons migrate to form the preplate. Later-migrating neurons split the preplate into the marginal zone (MZ) and subplate (SP). The MZ also consists of Cajal-Retzius neurons (CRNs), which originate from multiple sites in the forebrain. As neurogenesis proceeds, diverse subtypes of neurons are generated through successive asymmetric divisions of RGCs. Early-born projection neurons settle in the deep layers (layers 5 and 6) and later-born projection neurons migrate past older neurons to form more superficial layers. Thus, radial neuronal migration in mammals occurs in an inside-first, outside-last manner. Mature subcerebral projection neurons extend axons to the striatum, thalamus, brainstem, and spinal cord, and mature upper layer projection neurons project axons within the cerebrum. In contrast, cortical interneurons originate in the subcortical forebrain and tangentially migrate in the MZ, intermediate zone, and SVZ. At the end of neurogenesis, the radial scaffold is dismantled and most of the RGCs become gliogenic, generating cortical and subependymal zone astrocytes and giving rise to a layer of ependymal cells.

the cerebral cortex expands in surface area as a sheet during development and evolution (Figure 5.1c). In light of this model, we consider the current knowledge base for cortical specification; for further details, see Tyler and Haydar (this volume).

After the neural plate closes, cortical neurons begin their genesis from populations of neural progenitors (Figure 5.2) (Lui et al. 2011; Taverna et al. 2014; Matsuzaki and Shitamukai 2015). NECs give rise to apical (ventricular) radial glial cells (aRGCs), which produce cortical neurons, largely via indirect divisions (Malatesta et al. 2000; Noctor et al. 2001; Tamamaki et al. 2001). After the last division, newborn neurons migrate through the expanding intermediate zone (IZ) to enter the cortical plate and settle in an inside-to-outside pattern (Angevine and Sidman 1961; Rakic 1974 reviewed in Bystron et al. 2008).

A few molecular markers, such as GFAP, faithfully distinguish the highly related aRGCs from NECs (Choi and Lapham 1978; Levitt and Rakic 1980). Both NECs and aRGCs exhibit epithelial features, apical-basal cell polarity, and contact the ventricular surface and the basal lamina. However, interkinetic nuclear migration is distinct between NECs and aRGCs (Taverna and Huttner 2010). At the neuroepithelial stage, there is essentially one zone of cells, and NEC nuclei migrate between the ventricular surface and the basal lamina, in concert with cell-cycle progression (Sidman and Rakic 1973). In contrast, at the aRGC stage, aRGCs span several zones with a basal process that emerges from the cell body in the VZ. Interkinetic nuclear migration of aRGCs remains confined to the VZ (Lui et al. 2011; Geschwind and Rakic 2013). Thus, the absence or presence of a basal process distinguishes NECs and aRGCs, respectively (Delaunay et al. 2016). Notably, the plasma membrane composition of the aRGC basal process is distinct from the aRGC apical process (Taverna et al. 2016).

An important concept in the cortical development and evolution field is the granular classification and diversification of progenitor cell types and, in particular, the enlarged pool of basal progenitors (Fietz et al. 2010; Hansen et al. 2010; Shitamukai et al. 2011; Betizeau et al. 2013; Pfeiffer et al. 2016). aRGCs generate basal progenitor populations composed of basal/outer radial glia cells (bRGCs/oRGs), basal intermediate progenitors (bIPCs), and aIPCs (Stancik et al. 2010). As suggested by the name, aRGCs divide in the VZ whereas basal progenitors divide in the SVZ (Bystron et al. 2008). aRGCs can also become detached from basal and apical surfaces as human cortical development progresses (Sidman and Rakic 1973; Smart et al. 2002; Lukaszewicz et al. 2005; Nowakowski et al. 2016). It has been suggested that human and rodent aRGCs may be structurally and genetically different (Rakic 2003).

Evolutionary differences suggest that bRGCs have a significant role in neuronal production in primates. Initially observed by Golgi staining of human and monkey embryos (Schmechel and Rakic 1979), bRGCs are characterized morphologically by a basal process and, in some cases, by a short apically directed process which does not reach the ventricular surface (Betizeau et al. 2013; Florio and Huttner 2014; Rash et al. 2019). Primates possess an expanded and elaborate SVZ composed of outer and inner SVZs (oSVZ and iSVZ, respectively) and containing about 50% bRGCs and 50% bIPCs (Smart et al. 2002). Gyrencephalic nonprimate mammals, such as sheep, ferrets, and cats, also tend to have an expanded SVZ with significantly more bRGCs than lissencephalic mammals. Nevertheless, both the lissencephalic primate marmoset and the gyrencephalic rodent agouti also possess abundant bRGCs (Garcia-Moreno et al. 2012). In contrast, mice contain few bRGCs, which show markedly reduced proliferative capacity compared to human bRGCs (Wang et al. 2011a;

Wilsch-Bräuninger et al. 2016). Instead, in mice, neurons are produced primarily from aRGCs and bIPCs (Vasistha et al. 2015). Indeed, analysis of basal progenitors is an area where there was consensus among us that mouse is a poor model (Liu et al. 2014).

Importantly, although basal progenitors are considered the predominant neurogenic cell, the proportion of neurons born from bRGCs or bIPCs in humans is unknown. *In vitro* studies demonstrate that human bRGCs undergo expansive symmetric proliferative divisions before producing neurons destined for layers II/III as well as the majority of astrocytes and oligodendrocytes (Pollen et al. 2015). In the nonhuman primate macaque, bRGCs initially produce neurons which increase the thickness and complexity of superficial layers. However, after cortical neurogenesis stops, around E100, bRGCs produce astrocytes and oligodendrocytes (Rash et al. 2019).

Above the radial glia, at the pia mater, a milieu of basement membrane and meninges reside. ECM-mediated activation of integrins promotes basal progenitor proliferation (Fietz et al. 2010; Stenzel et al. 2014). In fact, some differences in proliferative potential between mouse and human cortical progenitors may derive from differential expression of genes encoding components of the extracellular matrix (ECM) (Fietz et al. 2012). Both mouse and human aRGCs, which are endowed with proliferative potential, show substantial endogenous expression of ECM genes. In contrast, highly proliferative human basal progenitors, but not mouse bIPCs, sustain expression of ECM genes.

## Progenitors and Evolution: Where Do We Go Next?

Single-cell profiling of cortical glutamatergic neurons has uncovered a remarkable diversity both within cortical layers and across cortical areas, including at least 13 transcriptomically defined glutamatergic types in layer 5 of the mouse visual cortex (Tasic et al. 2016). It remains unclear, however, whether these glutamatergic types are produced by distinct sets of progenitors which change over the course of neurogenesis, and/or if they are predisposed to form gyri. Another question is whether cell cycle plays an instructive role in progenitor fates in primates. Indeed, progenitor cell cycle diverges across cortical areas in monkeys (Lukaszewicz et al. 2005) and varies between humans, nonhuman primates, and rodents (Kornack and Rakic 1998; Dehay and Kennedy 2007; Geschwind and Rakic 2013). This is relevant since in mice, cell cycle can modulate progenitor symmetric versus asymmetric divisions (Lange et al. 2009; Pilaz et al. 2009; Arai et al. 2011; Okamoto et al. 2016; Pilaz et al. 2016).

To what extent do differences in progenitor heterogeneity influence neuronal diversity and cortical traits? The historical view of multipotent progenitors has also been challenged with recent evidence of fate-specified progenitors driving expression of specific neuronal subtypes (Franco et al.

2012; Garcia-Moreno and Molnár 2015; Gil-Sanz et al. 2015). Further, fate-mapping technologies and abilities to manipulate gene expression in mice have demonstrated that different precursor types can generate neurons of similar cortical layers (Tyler et al. 2015). Recent scSEQ studies have also uncovered molecules unique to aRGCs, bRGCs, and IPCs in humans (Pollen et al. 2014; Florio et al. 2015; Johnson et al. 2015; Pollen et al. 2015; Johnson and Walsh 2017; Nowakowski et al. 2017). Using new technologies to interrogate cellular morphology, profile gene expression, and perform lineage analyses in nonhuman primates as well as humans, we are optimistic that key questions can be addressed.

Given their clonal capacity, bRGCs are predicted to play a significant role in increasing the size and complexity of superficial layers II/III and may contribute to the expansion of cortical surface and formation of convolutions in humans (Wilsch-Bräuninger et al. 2016). However, most studies of bRGC clonal output have been done *in vitro*; thus in future studies, it will be critical to assess neurogenic potential of bRGCs *in vivo* (Mariani et al. 2012; Pollen et al. 2015). Likewise, it will be valuable to understand the nature of the few bRGCs found in mice. In this light, recent findings suggest the mouse dorsomedial telencephalon contains an oSVZ with abundant bRGCs (Huttner and colleagues, unpublished).

Thus far there is no correlation between the presence of convolutions and an oSVZ, as convolutions develop in species lacking an oSVZ and, likewise, some species with gyri and sulci lack an oSVZ (Garcia-Moreno et al. 2012; Hevner and Haydar 2012). Further, the thousandfold increase in the surface area of the human cerebral cortex occurs across all layers without a comparable expansion of thickness (Bystron et al. 2008; Geschwind and Rakic 2013). Notably, there is evidence that neurogenesis is complete in the Macaque even prior to gyrification (Rash 2019). Additionally, secondary and tertiary gyri form postnatally, long after neuronal generation and migration is complete. Therefore, the subsequent enlargement of cortical surface is likely due to neuronal growth in volume and their dendrites, formation of neuropil, and addition of protoplasmic astrocyte and oligodendrocytes (Rakic 2009; Rash et al. 2019). Thus, the extent to which bRGCs contribute separately to expanded cortical surface has been debated by the examination of its evolutionary history (Hevner and Haydar 2012).

How can we explain cortical surface expansion? It has been suggested that in organisms with larger and gyrencephalic brains, the radial unit is more conical or wedge shaped (Fietz and Huttner 2011). However, this model may not fully consider the fact that for each conical summit with larger superficial layers and larger surface area, there is a valley where the situation is reversed. Thus, enlargement of the oSVZ, which generates mostly layers II and III, is important but not sufficient to explain expansion of cortical surface during human evolution.

## Radial Neuronal Migration and Clonal Dispersion

The basal processes of aRGCs and bRGCs provide scaffolds for newborn glutamatergic neurons to migrate into the cortical plate. The increasing distance and emergence of primary convolutions add to the importance of RGC scaffolding to guide neurons to their proper columnar and areal positions (Rakic 1988b) (Figure 5.1). In humans, neurons migrate centimeters, whereas in mice this distance is much shorter. During their radial migration, neurons undergo polarity changes, transitioning from a multipolar to a bipolar morphology (Gal et al. 2006). Neurons migrate in an inside-out fashion such that the earliest born neurons ultimately reside within the deepest cortical layers while later born neurons form superficial layers (Angevine and Sidman 1961; McConnell and Kaznowski 1991). Histological studies and thymidine labeling suggest that inside-out corticogenesis is conserved among mammals and that the relationship between time of origin and cell position is sharply defined (Rakic 1974). Nonetheless, we still do not fundamentally understand why neurons migrate in an inside-out fashion and segregate into different laminar structures in the cortical plate.

The past thirty years have, however, yielded important new insights into neuronal dispersion. Lineage-tracing studies performed in mice, including using a technique called mosaic analysis with double markers (MADM), show that labeled clones of excitatory neurons are relatively constrained (Gao et al. 2014; Hansen et al. 2017) (Figure 5.3a). Columnar or cylindrical territories contain clones of up to 100 cells measuring 500 µm or less. In the future it will be valuable to measure more cells by inducing recombination at earlier developmental stages (i.e., prior to E10). Importantly, MADM studies are consistent with investigation of Tis21+ biPCs and their progeny, which distribute across all cortical layers (Kowalczyk et al. 2009). Likewise, they corroborate experimental manipulation of ephrin molecules in mice, which show that clonal intermixing occurs within functional columns but that large dispersion may be relatively rare in monkeys (Torii et al. 2009) (Figure 5.3b).

In contrast, recent preliminary studies in the human cortex reveal evidence of a small, but significant amount of dispersion of newborn cells (Lodato et al. 2015; Woodworth et al. 2017) (Figure 5.3c). These studies used long interspersed nuclear elements (LINEs) which spontaneously transposed in about 5% of cells, followed by whole genome sequencing of areas 17 and 18. The researchers detected LINE-induced mutations in adjacent cells and found that a very small number of cells (<1%) were labeled but primarily localized in columns, with some clonal intermingling. Once labeling was achieved in 1–3% of cells, however, there was more clonal dispersion. Going forward, somatic clonal studies may allow a quantitative grasp on neuronal distribution in the brain and on patterns of cell division. In some mammals, such as ferrets, clonal dispersion may be more profound (Reid et al. 1997; Ware et al. 1999; Reillo et al. 2011). Since these species diverged from the human phylogenetic tree millions of years before rodents, their evolution may have proceeded differently. Regardless, direct comparisons of clonal dispersion, using assays such as piggyback or CRISPR to label cells, could be performed in parallel in mice and nonhuman primates.

## Interneuron Generation and Migration

An additional concept in our understanding of cortical development is the recognition that inhibitory GABAergic interneurons have a different pattern of neurogenesis and migration than excitatory neurons. They are generated primarily in the medial (MGE) and caudal (CGE) ganglionic eminence, as well as in the preoptic area and lateral (LGE) regions, which produce predominately olfactory bulb interneurons (Anderson et al. 1997; Wonders and Anderson 2006; Batista-Brito and Fishell 2009; Gelman and Marin 2010; Welagen and Anderson 2011). As in the dorsal telencephalon, GABAergic interneurons are generated via radial glia and transit amplifying progenitors (Turrero Garcia and Harwell 2017). Initially, newborn interneurons undergo tangential migration, moving along distinct routes from the LGE/CGE to the dorsal telencephalon via the SVZ, IZ, and marginal zone (MZ), before migrating radially to their final positions in the cortical plate (Glickstein et al. 2007; Brown et al. 2011; Sultan et al. 2014, 2016; Harwell et al. 2015; Mayer et al. 2015; Tischfield et al. 2017). Interneuron migration is governed by molecular cues including Semaphorin3a/f, CXCL12, Neuregulin1, Robo, and Ephrin (Marin et al. 2001; Flames et al. 2004; Andrews et al. 2007; Sanchez-Alcaniz et al. 2011; Wang et al. 2011b; Steinecke et al. 2014).

As MGE-derived cortical interneurons mature, they gain morphological and molecular diversity to become two main subclasses that express either parvalbumin (PV) or somatostatin (SST) (Wamsley and Fishell 2017). PV interneurons primarily target the cell body and axonal initial segment, whereas SST interneurons selectively target dendrites. CGE-derived interneurons, expressing VIP and Reelin, typically innervate other interneurons. The fate of the target cell may determine how inhibitory circuits are ultimately wired (Lodato et al. 2011; Ye et al. 2015).

Over the past thirty years, scSEQ and genetic studies in mice have significantly informed our understanding of interneuron fate specification. For example, progenitors and newborn neurons express markers characteristic of distinct interneuron subtypes (Mayer et al. 2018; Mi et al. 2018). Indeed, interneuron fates are specified by combinations of transcription factors (Hu et al. 2017). Additionally, and in contrast to excitatory neurons, sequencing experiments suggest the molecular identity of cortical GABAergic neurons does not depend on the cortical area in which they reside (Tasic et al. 2016).

To what extent are there species-specific interneurons or progenitors? Recent reports indicate that interneuron migration occurs in human brains for several months after birth, unmasking a novel population of late-born

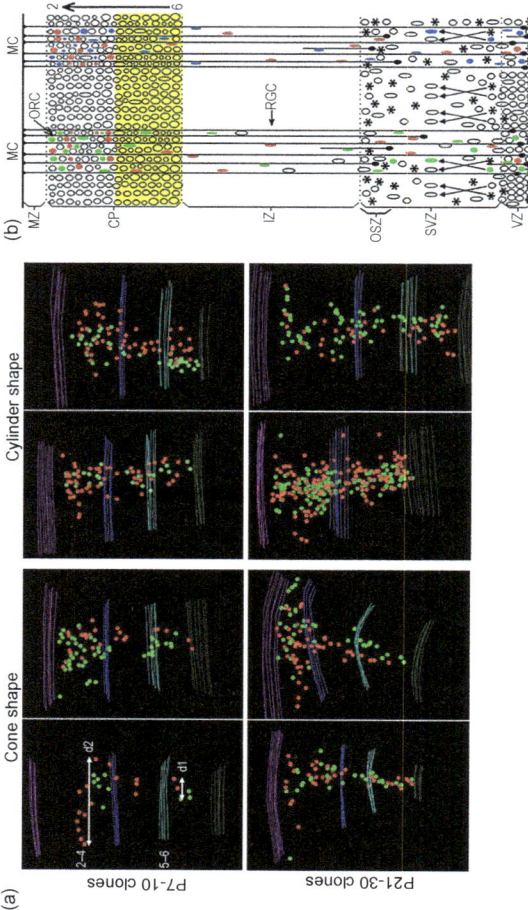

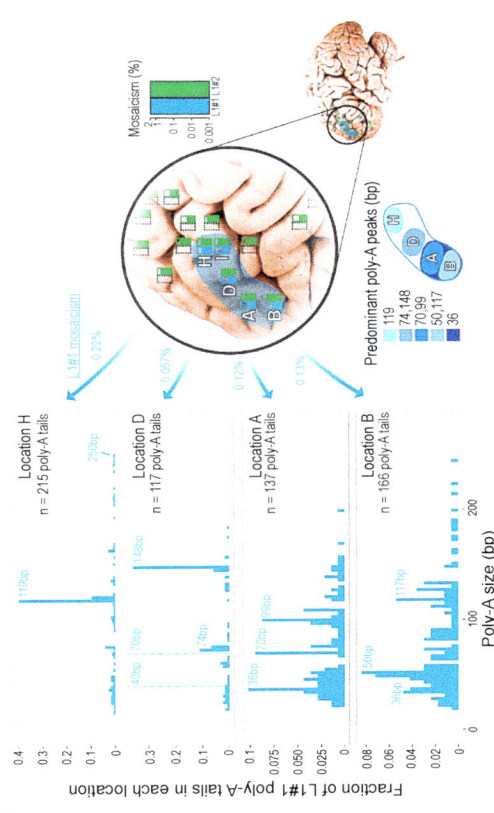

**Figure 5.3** Clonal dispersion in mouse and human cortical development. (a) Clones of excitatory neurons in the mouse cortex, labeled using the MADM technique; clones (labeled in red and green) adopt conical or cylindrical shapes, ranging from 100–400 μm across at their widest point; reprinted with permission from Gao et al. (2014). (b) Clonal heterogeneity (red and green cells) within minicolumns (MC) due to intermixing within the subventricular zone (SVZ, arrows) before neuronal migration into the cortical plate (CP); reprinted with permission from Geschwind and Rakic (2013). (c) Two clones in the human cerebral cortex, labeled by spontaneous mobilization and insertion of two LINE elements. One LINE insertion, illustrated in blue, labels a clone limited to a gyrus, and about 1 cm across, labels from .05%–0.4 % of the neurons within this region. A second LINE insertion, labeled in green, is dispersed across the entire cortex among about 1–3% of neurons. The blue clone shows multiple somatic mutations of the polyA sequence of the LINE element, suggesting further subdivisions of the major clone with even more limited distribution in cortex, though each "subclone" again would include a small fraction (<0.4%) of neurons in the regions in which it is present. What remains to be understood is whether these lineage patterns represent fundamentally conserved patterns of clonal arrangement between species or fundamentally different patterns of dispersion; reprinted with permission from Evrony et al. (2015).

inhibitory neurons (Paredes et al. 2016). Distinct GABAergic neuronal populations have also been described in humans (Raju et al. 2018). There are also reports of isolated GABAergic neurons in human and nonhuman primates that originate in proliferative zones of the dorsal telencephalon (Howard et al. 2006; Fertuzinhos et al. 2009; Radonjic et al. 2014). Consistent with this notion, in postnatal mice, a subpopulation of olfactory bulb interneurons is produced from aRGCs (Kohwi et al. 2007). This raises the interesting question of whether neocortical GABAergic interneurons could be produced by pallial progenitors. To date, however, human clonal studies have been biased toward investigating excitatory neurons. Thus we know very little about interneuron origin and dispersion and what defines their stop signals. The advent of scSEQ, human organoids, and human fetal tissue (e.g., Marinai et al., 2014) affords new methodologies to investigate human-specific features of inhibitory neuron development (Laclef and Metin 2018).

## Transient Cells: Subplate Neurons and Their Contribution to Brain Development and Evolution

In mammals, the first postmitotic cell layer is the preplate (primordial plexiform, PP, zone), described first from Golgi analysis in cats and hypothesized to relate to the amphibian and reptile cortex (Marin-Padilla 1971). It also contains the first (pioneer) cortical neurons in humans (Kostovic and Rakic 1990; Meyer et al. 2000; Bystron et al. 2006) (Figure 5.4). Based on comparative anatomical studies, Marin-Padilla (1971) suggested that the PP layer later split, by the growing cortical plate, into layer 1 and subplate. The subplate, first discovered by Kostovic in humans (Kostovic and Molliver 1974), is a transient zone situated below the cortical plate, above the IZ (Rakic 1977; Bystron et al. 2008). While almost undetectable in marsupials, the subplate is a thin, distinct layer in mouse and rat (Rickmann et al. 1977), and a larger layer in carnivores (Luskin and Shatz 1985). The subplate zone is most expansive in human (Molliver et al. 1973) and nonhuman primates (Rakic 1977), particularly subjacent to the prospective association areas (Duque et al. 2016) (Figure 5.4). It has been proposed that later-born neurons split the MZ, subplate, and cortical plate (Marin-Padilla 1971). However, recent studies in primates, using H3-thymidine and BrdU to label cells at their birth and monitor their eventual positions, show a more complex picture in which subplate neurons are displaced by the arrival of new neurons (Duque et al. 2016). This enables the subplate to provide a constant platform upon which cortical afferents line up while the cortex is constructed.

During development of mammalian brains, subplate neurons form transient connections with the thalamus to establish cortical circuits. Over the last several decades, knowledge about the subplate has extended to include functional and molecular properties pointing to a structure with heterogeneous cell populations and a highly dynamic ontogeny (Antonini and Shatz 1990;

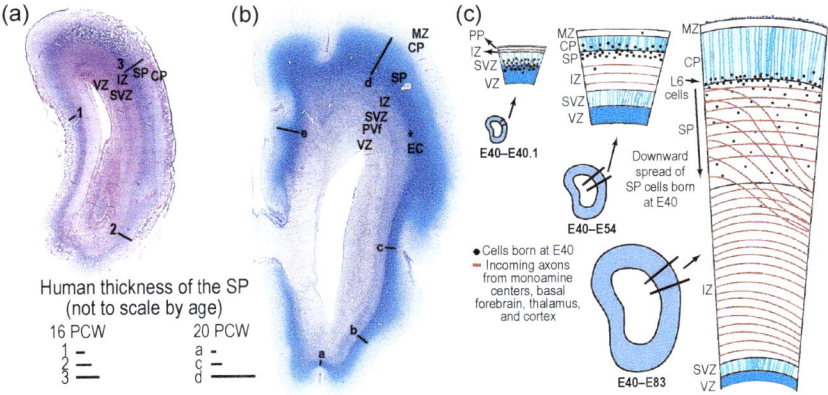

**Figure 5.4** Comparative subplate (SP) development in mice and nonhuman primates: (a) and (b) show images of the human SP at 16 and 20 postconceptional week (PCW), depicting nonhomogeneous thickness of this structure. (c) Model of secondary expansion in the transient SP zone over the course of development in the Rhesus macaque. Reprinted with permission from Duque et al. (2016).

Hoerder-Suabedissen et al. 2009; Oeschger et al. 2012). H3-thymidine labeling in primates shows subplate neurons originate in the VZ (Duque et al. 2016); some glutamatergic subplate neurons, however, can also derive from the rostral medial telencephalic wall in mice (Pedraza et al. 2014). Further, gene expression patterns show homologies of cellular morphology, birthdating, and homology in the dorsal cortex/dorsal pallium of several amniote species. Thus, the subplate is hypothesized to contain both ancestral and newly derived cell populations (Montiel et al. 2011).

Postnatally, a large proportion of subplate neurons die; a small fraction, however, survive and become scattered below the cortex in the fiber layer in humans, or form a thin band of cells (layer 6b) in mouse (Hoerder-Suabedissen and Molnár 2015). In humans these cells are implicated in cognitive developmental disorders, such as autism and schizophrenia. Distinct subgroups of 6b neurons connect to select thalamic targets with known functions, providing a means to investigate subplate function (Hoerder-Suabedissen et al. 2009). Indeed, deep recording through all cortical layers suggests that in addition to neurons in cortical layers 1–6, subplate neurons also exhibit stimulus-driven responses (Pho et al. 2018). These neurons receive inputs from diverse brain regions and may project to nonprimary thalamic targets, thus suggesting that they represent a system for modulation of visual processing and brain states which persists into adulthood.

Beyond subplate neurons, additional transient glutamatergic neuronal populations include the Cajal-Retzius cells, pioneer neurons (Bystron et al.

2008), and cortical-plate transient cells (Barber and Pierani 2016) (Figure 5.2). Lineage-tracing experiments have shown that Cajal-Retzius cells derive from the cortical hem, pallial septum, and pallial-subpallial boundary. They tangentially migrate into the cortex within the preplate and MZ, where they are implicated in radial neuronal migration, cortical lamination, and radial glia morphology via secretion of the extracellular matrix protein Reelin (Tissir and Goffinet 2003). In humans and nonhuman primates, Cajal-Retzius cells show great complexity and increased Reelin expression.

## Organizational Features of the Neocortex: Areas and Layers

### Debate of Protomap versus Protocortex Models of Arealization Enhanced Understanding of Cortical Development and Evolution

Three decades ago neuroscientists were debating the protocortex and protomap hypotheses. The protocortex (also called "tabula rasa") hypothesis proposes that the cortical plate initially has the same potential and that regionalization is controlled by external influences, such as axonal inputs from the thalamus (Creutzfeldt 1977). Today, the tabula rasa hypothesis has been largely disproven (e.g., O'Leary et al. 2013). Indeed, modern evidence indicates that cells generated within the VZ contain intrinsic information about their prospective laminar and areal fates. There are many levels of evidence, but one simple experiment is that X-ray irradiation, which ablates cells of the VZ/SVZ on a given day, produces a cortex in which neurons are missing from specific layers, with a sharp border between areas 17 and 18 (Algan and Rakic 1997). This shows that neurons are dedicated to specific layers and areas at the time of their genesis. When the protomap hypothesis, now accepted as the protomap model, was proposed, we did not know how initial positional information was imposed at the molecular level. This has changed significantly over the past thirty years.

To reconcile older descriptive and new experimental data, the protomap hypothesis suggests that the basic pattern of species-specific cytoarchitectonic areas emerges through synergistic, interdependent interactions between developmental programs intrinsic to cortical neurons that can be modified by extrinsic signals. Such signals can arise at later stages, supplied by specific inputs from subcortical structures (Rakic 1988b). Thus, neurons in the embryonic cortical plate—indeed in the proliferative VZ where they originate—set up a primordial map that preferentially attracts appropriate afferents and has the capacity to respond specifically to these inputs. Importantly, the protomap hypothesis is named "proto" (meaning modifiable) because it is not a fate map, since specific thalamic inputs, which arrive at given areas, are essential for proper cortical differentiation (Rakic 1988b). It is thought that species-specific functionally specialized areas are arranged in protomaps (Clowry et al. 2018).

According to the protomap model, newborn radially migrating neurons carry positional information inherited from progenitors into the cortical plate, where subsequently thalamocortical axons and other afferents arrive in the cortex. Consistent with this, a mouse mutant with little or no thalamocortical innervation shows regional gene expression that marks prospective area borders (Miyashita-Lin et al. 1999). As further evidence of positional information, when *Emx2* expression is deficient in cortical progenitors, rostral areas expand and caudal areas contract, reflecting the low to high rostral-caudal gradient of *Emx2* expression. Loss of *Pax6*, expressed in the opposing gradient, has the opposite effect on area size (Bishop et al. 2000). Formation of specialized areas may also depend upon progenitor cell-cycle differences present between areas 17 and 18 (Lukaszewicz et al. 2005).

Another key finding was the recognition that the cerebral cortex is initially patterned by signaling centers which release diffusible proteins, including members of the fibroblast growth factor (FGF), Wnt, bone morphogenetic protein (BMP), and SHH families. FGFs promote telencephalic and neural identity in the neural plate in part by driving expression of the Foxg1 (BF1) transcription factor (Shimamura and Rubenstein 1997). Subsequently, a rostral telencephalic source of FGF8 patterns the area map along its rostral-caudal axis (Fukuchi-Shimogori and Grove 2001; Garel et al. 2003). FGF17 further promotes prefrontal and frontal area identity (Cholfin and Rubenstein 2007, 2008). As further evidence, an exogenous caudal FGF8 source induces duplicate sensory areas, which are oriented in mirror image to the endogenous areas (Assimacopoulos et al. 2012). Finally, a dorsal telencephalic signaling source, termed the cortical hem, rich in Wnts and BMPs, influences the dorsal to ventral axis (Caronia-Brown et al. 2014).

FGFs, Wnts, and BMPs operate in part by altering expression gradients of transcription factors such as CoupTF1, Emx2, Lhx2, Pax6, Pbx1, and Sp8 (Grove and Monuki 2013; O'Leary et al. 2013). Modifying the dosage of these transcription factors modulates the relative sizes of cortical areas (Garel et al. 2003; Hamasaki et al. 2004). Studies using advanced genomic methods have demonstrated the embryonic cerebral wall can be labeled by the activity of small enhancer elements in specific cortical progenitor domains. Relevant to this expression gradient, the activity of enhancer-like regulatory elements can be localized in small domains with sharp borders (Pattabiraman et al. 2014). Fate-mapping from these small domains provides evidence for a protomap of the cortex that is encoded by the integration of transcriptional information processed by gene regulatory elements (Pattabiraman et al. 2014). Cells are arranged largely in radial patterns from the ventricular to the pial surface (Figure 5.5).

There is broad agreement that thalamus inputs are essential for arealization and secondary area formation of cortical functional columns (Geschwind and Rakic 2013). Thus, the arrival of thalamic axons induces functionality, including, for example, anatomical barrels in somatosensory cortex, the morphology of layer 4 neurons, and some differential gene expression between areas.

82

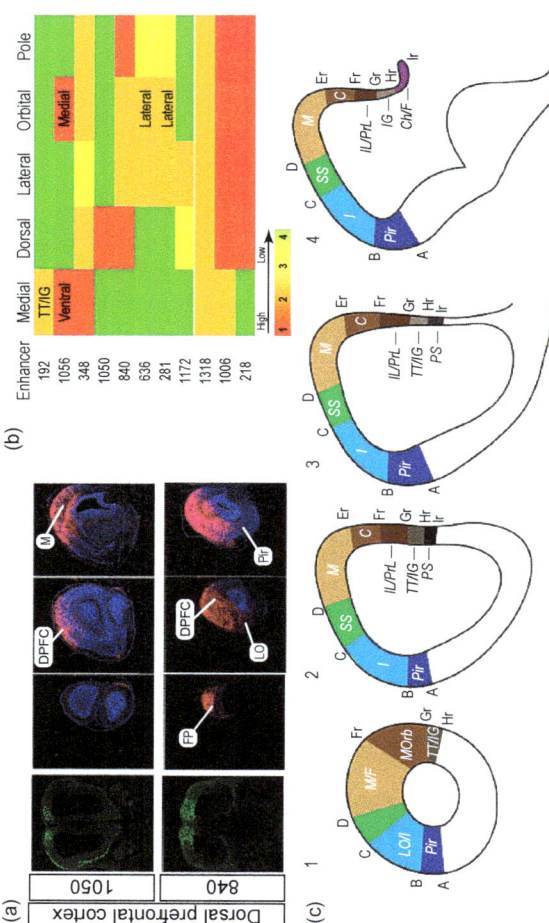

**Figure 5.5** Fate-mapping analysis of enhancers with regional activity in the primordium of the mouse embryonic frontal cortex. (a) Coronal sections depicting two enhancers with activity in pallial progenitors that fate map to prefrontal cortex subdivisions. Shown are green fluorescent protein expression at E11.5 (left) and fate-mapping with tdTomato expression in an E17.5 rostrocaudal series (right). (b) Annotation of fate-mapping results from selected enhancers (y axis) in five regions of the frontal cortex (x axis). Different levels of density of tdTomato expression are estimated and described as high density (red), medium density (orange), low density (yellow), and negligible density (green). In some cases, subdomain expression is noted. (c) Deduced progenitor domain organization of the rostral E11.5 pallium. Abbreviations: C, cingulate gyrus; CR, Cajal-Retzius cells; DPFC, dorsal prefrontal cells; Er, entorhinal; FP, frontal pole; F, frontal; IG, indusium griseum; IL, infralimbic; LO, lateral orbital; M, motor; MOrb, medial orbital; Pir, piriform; PS, pallial septum; SP, subpallium; SS, somatosensory; TT, tenia tecta. Reprinted with permission from Pattabiraman et al. (2014).

FGF8 can also induce thalamocortical projections, and when the cortex contains extra Emx2, which regulates FGF8, this alters the areal map (Fukuchi-Shimogori and Grove 2003). Integration of thalamic afferent inputs influences primary and secondary visual areas (Chou et al. 2013). Further, thalamocortical projections are attracted to area 17 and control barrel field formation but do not induce area 17 (Rakic et al. 1991). Likewise, an experimental decrease of geniculocortical afferents does not diminish area 17 but instead induces formation of a novel area with abnormal architecture and a sharp border to area 18 (Rakic 1988b; Rakic et al. 1991). This supports the notion that progenitors within areas 17 and 18 may be genetically predisposed to induce secondary areas. Indeed, progenitors under the influence of thalamic inputs show altered proliferation, which can influence secondary zone formation (Dehay et al. 2001).

A critical next step is to determine whether the area-patterning model worked out thus far in the mouse generalizes to other mammalian species. At similar developmental stages just after neural tube closure, the cortical primordium in ferret is equivalent to that of the mouse, and gene expression indicates similarly positioned sources of FGFs and Wnts (Grove and Jones, unpublished). Likewise, humans with mutations in FGF receptor 3 show abnormal cortical patterning, suggesting that areal patterning mechanisms by FGF may be conserved (Hevner 2005). It is clear that existing areas can be postnatally refined to accommodate new functional needs or opportunities. Qualitative differences in maps could be related to differential experience. For example, studies of the prehensile grip of monkeys have shown that manipulating biomechanics can modulate behavior and induce area 5 in new locations (Krubitzer and Stolzenberg 2014). Interestingly, functional area size can vary across humans, as exemplified by the visual cortex which shows threefold variation (Andrews et al. 1997). Novel maps may also be added during evolution by introducing new cells as the cortex expands. In more complex organisms, cortical expansion would thus coincide with new modules, which can serve as receptive units to process additional or slightly refined tasks.

## Lamination

Cortical neurons are organized into six distinct layers which include layers I–III (supragranular layers), layer IV (internal granular layer), and layers V/VI (infragranular layers). Over the past three decades we have learned a great deal about molecular features of mammalian neuronal layers (Greig et al. 2013), including key transcription factors and gene expression networks which control both specification and maturation of glutamatergic type neurons. Remarkably, there is also a window of postmitotic development during which neuronal fates can be reprogrammed *in vivo*, as shown by manipulations of transcription factors in the mouse (Rouaux and Arlotta 2013; Lodato et al. 2015). Studies in mice have shown that sister neurons can be connected by gap junctions, but whether this is borne out in other species is unclear (Yu et al. 2012). In addition, it is enigmatic what drives distinct laminar organization across areas where neurons exhibit inherently different axonal projections. For example, in primates, layer IV is significantly thicker in the visual cortex relative to adjacent areas (Rakic 2009). This question is also relevant for understanding evolution, as relative to other primates, layers II/III are thicker and cytologically more diverse in humans, which could enable increased cortical-cortical connections (Hill and Walsh 2005; Rakic 2009).

## Gyrification

Higher-order mammalian brains are gyrencephalic, having acquired cortical folds called gyri and sulci (Welker 1990), which allow expanded brain surface area. Gyrification is thought to require further expansion of initially larger cortical plate formed by proper areal distribution of neurons via radial migration (Rakic 2009). It also involves development of subcortical white matter that consists of various axonal bundles and numerous glial cells, which are particularly large in primates, including humans (Rash et al. 2019). The mechanisms of gyrification are evidenced by studies of the adhesion molecule FLRT, which is differentially expressed in humans (Del Toro et al. 2017). In developing ferrets, FLRT1/3 show differential expression in prospective gyri and sulci, and *FLRT1/3* double KO mouse have aberrant neuronal migration, associated with increased formation of gyri and sulci. Thus, reduced expression of specific genes which modulate neuronal migration may promote gyrification. Likewise, ECM components influence initial folding of the fetal human neocortex. Specifically, the ECM components HAPLN1, lumican, and collagen I cause a hyaluronic acid-dependent folding of fetal human neocortex tissue in an *in vitro* system (Long et al. 2018). Beyond gene expression, some modeling studies suggest that physical forces could also promote gyrencephaly (Tallinen et al. 2014).

Secondary and tertiary gyri in humans develop after neurons have been generated and have attained their final areal and laminar positions (Welker 1990; Kroenke and Bayly 2018). This suggests that human secondary and tertiary convolutions could be independent of increasing neuron number but rather an effect of neuronal enlargement and expanded neuropil and glial cells. In ferrets, which separated from the human phylogenetic tree before rodents, gyri develop postnatally (Kroenke and Bayly 2018). Thus, convolutions in carnivores and primates may be an example of analogy rather than homology. The most recent ancestor to all mammals is assumed to have already been gyrencephalic (gyrencephalic index of 1.3–1.4) (Lewitus et al. 2014). Hence, lissencephaly happens secondarily and is often associated with evolutionary dwarfism (e.g., the mouse is lissencephalic but originated from a larger and gyrencephalic ancestor).

## Cortical Evolution and Lessons from Nonmammalian Vertebrates

An invaluable path to understand unique structural and functional features of the cerebral cortex is through comparative investigation with birds and reptiles. The comparison of amphibian, reptilian, bird, and mammalian embryos supports the hypothesis that the embryonic pallium is subdivided in homologous (medial, dorsal, lateral, and ventral) sectors that are demarcated by the co-expression of developmental transcription factors (Puelles et al. 2000; Brox et al. 2004). Although this early developmental body plan is conserved, differences of the reptilian, bird, and mammalian adult telencephalon suggest that developmental programs diverge at later stages (Figure 5.6).

Unlike fish and amphibians, a large portion of the reptilian pallium has a three-layered organization which emerged about 320 million years ago in

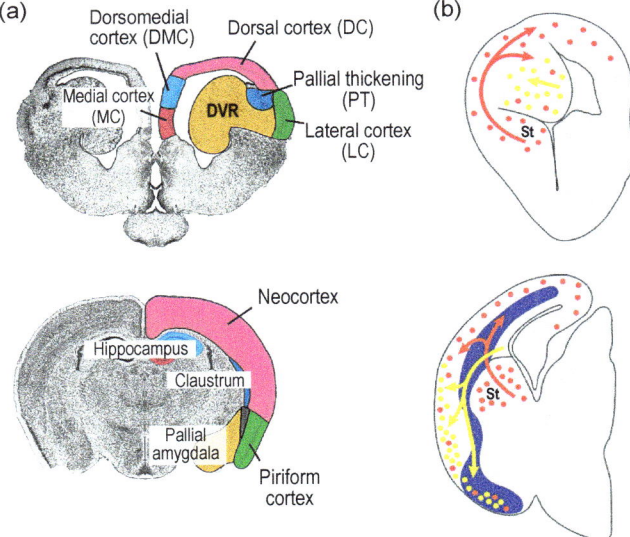

**Figure 5.6** Reptile/mammalian homologies and differences in cortical organization and neuronal migration. Pallial regions in turtle (top) and mouse (bottom). (a) These regions are defined by neuroanatomy and transcriptomics. Colors represent proposed homologies, on the basis of current anatomical, developmental and transcriptomic data. (b) In mammals, the Pax6 territory is indicated in dark blue. Inhibitory, GABAergic neuronal precursors (red dots) originate from subpallial sources and migrate tangentially into the pallium in both mammals and sauropsids. Excitatory, pyramidal-type neuronal precursors (yellow dots) of the lateral migratory stream traverse the Pax6 territory to reach lateral pallial regions in mammals but remain *in situ* within the dorsal ventricular ridge (DVR) in sauropsids. Despite its extensive target area, the lateral migratory stream is considered to be a subset of the radially migrating pallial neurons. Interestingly, the tangentially migrating GABAergic cells have similar origin from Dlx gene expression territories from the medial ganglionic eminence (origin of red arrows) and they migrate dorsal to the cortex in mammal and DVR and dorsal cortex in reptiles.

the amniote ancestor of mammals and reptiles. Genetic fate-mapping of the mammalian neocortex (Pattabiraman et al. 2014) supports the notion that this structure and the reptilian dorsal cortex develop from homologous embryonic regions (the dorsal pallium). The reptilian pallium harbors a nonlaminated region, called the dorsal ventricular ridge (DVR), which is derived from the ventral pallium (Karten 1969; Puelles et al. 2000; Butler and Molnár 2002; Tosches et al. 2018). This suggests that the DVR is unrelated to the neocortex. The DVR, however, harbors neocortical-like circuits, including the existence of thalamo-recipient neurons (Calabrese and Woolley 2015). This led to the "equivalent circuits" hypothesis, stating the homology of anterior DVR and neocortical layer 4 neurons (Karten 1969). Molecular analysis supports this idea (Dugas-Ford et al. 2012), indicating that different pallial regions expanded independently in the reptilian and mammalian lineages—ventral pallium (anterior DVR) versus dorsal pallium (neocortex)—resulting in the convergent evolution of gene expression and circuit architecture.

Notably, the reptilian cortex, which contains only a VZ, develops in an outside-in fashion (Blanton and Kriegstein 1991), in sharp contrast to the inside-out development of the mammalian neocortex (Angevine and Sidman 1961; Rakic 1974; McConnell and Kaznowski 1991). Further, there is an inversion of the corticogenesis gradient. In addition, recent evidence suggests that the tangentially migrating glutamatergic neuronal populations, such as Cajal-Retzius cells, subplate cells, and Dbx1 positive cortical neurons found in mammalian brains, do not exist in developing avian brains. This suggests that the emergence of these early neuronal populations in mammalian ancestors might have played a role in shaping the early development of dorsal pallium and could have triggered the evolution of the mammalian neocortex (Garcia-Moreno et al. 2018). Interestingly, inhibitory GABAergic neurons show similar tangential migratory behaviors in the reptiles and mammals (Cobos et al. 2001; Metin et al. 2007). However, excitatory pyramidal neuronal precursors of the lateral migratory stream traverse the Pax6 territory to reach lateral pallial regions in mammals but remain *in situ* within the DVR in sauropsids (Figure 5.6b). A few transcription factors are differentially expressed in the mammalian and bird ventral pallium, which might be responsible for the different migratory behaviors of ventral pallial derivatives (Garcia-Moreno et al. 2018; Yamashita et al. 2018).

Changes at the pallial-subpallial boundary might have participated in rerouting thalamocortical projections in our mammalian ancestors. In sauropsids, thalamic fibers reach the dorsal pallium through an "external" path, which traverses the ventral pallium. Conversely, mammalian thalamocortical projections arrive at the neocortex via the internal capsule (Bielle et al. 2011). Neuronal migration from the ventral pallium to the mammalian lateral amygdala, endopyriform nucleus, and claustrum present a rather difficult territory for thalamocortical projections. Early corticofugal projections which cross the pallial-subpallial boundary are thought to be important, as postulated in the handshake hypothesis (Molnár and Blakemore 1995). Further, the subplate/layer 6 handshake and basal ganglia play roles in directing corticothalamic topography and connectivity (Garel and Rubenstein 2004).

Genomic studies have enabled a comparison of glutamatergic cell types developing from the dorsal pallium (neocortex in mammals, dorsal cortex in reptiles) to clarify the evolution of neocortical layers (Tosches et al. 2018). ScSEQ data do not indicate simple one-to-one homologies between turtle cell types and individual cell types (e.g., layers) of the mammalian neocortex (Nomura et al. 2018). However, the turtle dorsal cortex contains cell types broadly similar, at the molecular level, to mammalian upper and deep layers. Likewise, turtle upper and deep layer neurons stratify according to birth order, with the former being deeper and the latter superficial (Ulinski 1986; Blanton and Kriegstein 1991).

These data suggest that neocortical glutamatergic neurons are new cell types that arose through the diversification of preexisting ones. Callosal projection neurons are an example of these new mammalian glutamatergic types (Garcia-Moreno and Molnár 2015). In mammals, these neurons originate from Emx2+ progenitors. Although less well defined, there is evidence that Emx2+ progenitors may exist in chick (Crossley et al. 2001). It thus remains possible that dorsal pallium progenitors vary across species, at least in part, at the level of gene regulatory networks, progenitor behavior, and heterogeneity (Garcia-Moreno and Molnár 2015). Notably, comparison of turtle and mouse data shows that the same classes of GABAergic interneurons exist in both species: MGE- and CGE-derived interneurons, including SST, PV-like, and VIP-like types (Tosches et al. 2018). This stands in stark contrast with the diversification of glutamatergic types and might reflect the existence of developmental constraints in the subpallium, where interneurons are born.

## Synaptic Connectivity and Plasticity in Cortical Development and Evolution

Achieving the adult pattern of connections in each individual and species is activity dependent, wherein inhibition plays a key role. In the retina, for instance, cell connections are influenced by gradients of ephrins (Triplett and Feldheim 2012). However, ephrins play a crucial role in the formation of functional columns even before birth (Torii et al. 2009). Thus, before birth, innate patterns of neural activity, independent of sensory stimulation, set the stage for circuitry organization, similar to the influence of patterning centers.

The number of synapses in the developing human cerebral cortex is much higher than in adults (Huttenlocher and de Courten 1987), and similarly large overproduction occurs in developing nonhuman primates (Rakic et al. 1986). In the developing macaque, for example, the PFC contains as many as 60% more synapses than in adults (Bourgeois et al. 1994). This stage of exuberant synaptogenesis is followed by pruning which is prominent during puberty; in humans, however, this proceeds until the third decade of life (Petanjek et al. 2011). This might seem like an inefficient way to build a cortex, but recent theoretical analysis of network construction has shown that, paradoxically, this strategy leads to a more efficient network design compared to algorithms that do not depend on pruning (Navlakha et al. 2015). Once synapses form and circuits become functional, experience in the form of neural activity reshapes connectivity dramatically. This process of synaptic pruning followed by maturation is particularly potent during critical periods of brain development and is sensitive to environmental context. Great progress in cellular and molecular understanding has been made mainly in the mouse sensory cortex, where genetic manipulation has become a powerful dissection tool (Figure 5.7).

Proliferation and pruning of synapses is a hallmark of these late developmental stages from mouse to human (Rakic et al. 1986). In response to sensory deprivation, a gradual loss of dendritic spines is followed by their regrowth. Interestingly, spine motility is elevated by deprivation in a lamina-specific manner initially outside layer IV. Such events are likely enabled by extracellular proteases, such as tPA, and activated microglia. The subsequent regrowth and homeostatic strengthening of synaptic input may instead involve brain-derived neurotrophic factor (BDNF) and tumor necrosis factor secreted from neighboring neurons and astrocytes, respectively. Ultimately, synapses are largely converted from silent (NMDA only) to functional through PSD95-mediated AMPA receptor insertion and stabilization (Takesian and Hensch 2013).

Timing of the critical period is instead determined by the maturational state of fast-spiking, PV-positive inhibitory basket cells. Strengthening GABAergic synapses (by GAD65 expression, benzodiazepine exposure, BDNF overexpression, loss of PSA-NCAM, or Mecp2) can trigger premature plasticity. Slowing PV cell maturation by genetic deletion (Clock or a variety of

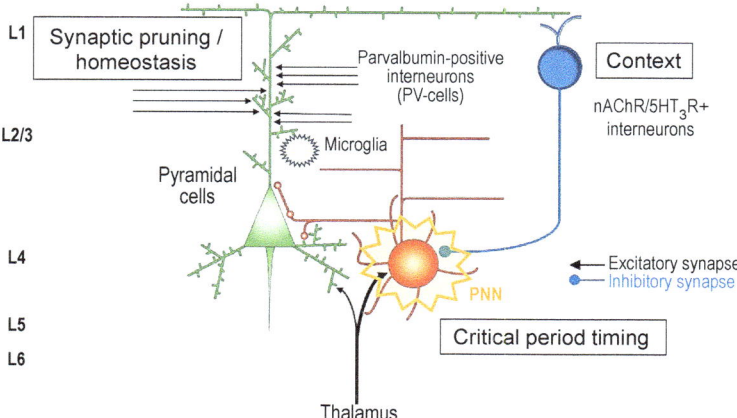

**Figure 5.7** Overview of understanding of cortical circuitry. Findings over the last thirty years have given rise to details regarding synaptic pruning and maturation, critical period timing, and context in shaping cortical circuitry.

autism-related proteins) delays plasticity. These findings reveal that critical period timing is itself plastic and not strictly determined by the age of the animal (Espinosa and Stryker 2012; Takesian and Hensch 2013; Sahin and Sur 2015). Transplantation of MGE-derived inhibitory precursors can reintroduce a second critical period later in life. As PV cells mature through the natural critical period, they enwrap themselves in a specialized extracellular matrix, the perineuronal net (PNN). This traps impinging synaptic boutons, such as thalamic and reciprocal inhibitory inputs, as well as a variety of noncell autonomous maturation and maintenance factors. Removal of the PNN can reopen critical period plasticity in adulthood (Hensch and Quinlan 2018). There are important homeostasis controls of excitation, as the firing of excitatory neurons can occur at the expense of inhibitory neurons.

Enriched environments can extend the critical period duration, while early life stress may accelerate closure. If mice are raised in an enriched environment, for example, the critical period of plasticity can be lengthened (Kaneko and Stryker 2017; Hensch and Quinlan 2018). Such ambient conditions may delay or accelerate the emergence of brake-like factors, such as PNNs, or increase neuromodulatory tone. Upper layer interneurons are enriched in ionotropic receptors for serotonin or acetylcholine (nicotinic) and send narrowly columnar input preferentially onto PV cells in layer 4 below. They also gradually express Lynx1 which dampens the action of nicotinic receptors after the critical period (Morishita et al. 2010; Takesian et al. 2018). This molecular brake can be overridden by gene deletion, acetylcholinesterase treatment, or exercise (running) to boost acetylcholine levels that enables plasticity in adulthood (Takesian and Hensch 2013).

Emergence of PNNs may serve as predictive biomarkers for novel critical period closure across brain regions (Takesian and Hensch 2013; Hensch and Quinlan 2018). Interestingly, localization of PNN components to astrocytes or non-PV cells in higher-order human brain regions in psychosis may not be captured in the mouse. Their molecular absence from more plastic, higher-order associational areas that are most vulnerable to Alzheimer's degeneration suggests a neuroprotective role for critical period closure. Similarly, by one year of age, Lynx1-deficient mice suffer neurodegeneration, a condition not typically seen in mice (Miwa et al. 2006). In the short term, however, pharmacological approaches to reopen critical periods may serve as a therapeutic strategy when increased plasticity may be desirable (e.g., stroke, recovery from brain injury).

Multiple lines of evidence suggest that neurons in the superficial layers of visual cortex in mice have prolonged plasticity after monocular deprivation, extending into adulthood. While neurons in the middle and deep layers of the cortex decrease responses to the closed eye and increase responses to the open eye, mainly during a well-defined "critical period" for ocular dominance plasticity, neurons in the superficial layers continue to exhibit plasticity until later (Frenkel et al. 2006; Espinosa and Stryker 2012). This increased propensity for plasticity may reflect layer-specific mechanisms of plasticity in cortex (McCurry et al. 2010).

How do molecular differences in species influence circuitry? Recent studies indicate that genetic modifications may modify circuitry over the course of evolution. For example, human-specific *SRGAP2* is implicated in control of the development of excitatory and inhibitory synapses. Notably, expression of the human ortholog in mice results in denser dendritic spines and delayed spine maturation (Charrier et al. 2012; Fossati et al. 2016). This may result from functional inhibition of ancestral *SRGAP2*. These modifications are thus hypothesized to impact circuit formation, cognition, and memory relevant for some developmental disorders.

## Conclusion

In summary, over the last thirty years the following concepts emerged:

- Novel genomics, cell biological, and imaging approaches have led to recognition of additional distinctions between apical and basal progenitors during cortical evolution.
- During cortical development, genes are expressed at low resolution whereas enhancers define sharp expression boundaries.
- New technologies have enabled an unprecedented investigation of clonal dispersion in mice and humans.
- The protomap model has been realized with the discovery of thalamic and gradients of patterning factors, giving a mechanistic understanding of how patterning is established.
- Modules in evolution have enabled growth and duplication of cortical areas via transcription factor codes.
- Innervation and cortical circuitry are established via inhibitory–excitatory balance.
- Transcriptomic analyses of cerebral cortex have revealed convergent evolution of gene expression and circuit architecture.

To chart the way forward, we suggest the following approach:

### Connect Genotype to Phenotype in an Evolutionary Context

Molecular neuroscience has enabled an understanding of genomic, epigenomic, transcriptomic, and proteomic features of cortical development and evolution. Yet linking these molecular changes to cortical phenotypes remains a significant challenge. These human-specific and evolutionary divergent cortical features include cortical areas, neuronal circuits, cortical neurons, neuronal processes (axon, dendrites), features of synapses, subcellular features within cortical neurons, cellular processes, reactions within cortical neurons, human-specific genes, RNAs, proteins, lipids, and carbohydrate structures. We propose that future research should focus on the following:

- Exploit and interrogate evolutionary differences between related species, such as the mouse and rat. Clear gene expression and cellular differences between mouse and rat make this a tractable approach in which to define the genetic underpinnings of evolution.
- Investigate different strains of the same species to correlate quantitative trait loci with transcriptomes, enhancer activity, and cellular and behavioral states. For example, the Peromyscus strains of deer and beach mice evolved differently, with the latter strain possessing a bigger PFC (Hu and Hoekstra 2017).
- Study genetics of human disorders, neurological features, and behaviors. By identifying and prioritizing variants associated with specific neurological features, this may inform evolution and be valuable for human health.
- Exploit organoid models to interrogate evolution "in a dish." This would enable species comparison of cortical development between human and other great apes, and experimental investigation of disorders of human neocortical development.
- Interpret findings in the context of sample origin. When studying human tissue, it is critical to ensure that human-specific differences are not due to technical reasons, such as using brains sourced from diseased or abnormal embryos or adults.

**Interpret Gene Expression Differences to Understand Progenitor Cell Types, Lamination, and Arealization**

Human-specific gene expression is driven by chromosomal deletions and duplications, alterations to coding and noncoding regions, and modifications to enhancer activity, but functional relevance of these human-specific changes remain to be elucidated. Great headway has been made in understanding progenitors, yet many questions remain unanswered, as noted in this chapter. Thus, we propose the following goals for future research:

- Identify additional features underlying the evolutionary increase in basal progenitor proliferative capacity, and clarify if bRGCs introduce an additional layer of clonal dispersion.
- Define progenitor heterogeneity both temporally and spatially. Use of single-cell omics technologies (transcriptomics, splicing, epigenetics, and proteomics) will be valuable toward understanding progenitor and cell identity and understanding human-specific aspects.
- Define how excitatory and inhibitory progenitors specify cell fate via both proliferative and neurogenic divisions. Investigate how neuronal

migration is coordinated with progenitor proliferation. Clarify if clonal dispersion influences disease etiology and is expanded across primates.

- Link positional information at the level of progenitors to areal demarcation and cortical function: Are conserved or divergent signaling and patterning factors involved in control of primary and secondary maps?
- To fill in molecular details in support of the protomap, one needs to define factors downstream of FGF8, determine what axon guidance cues influence thalamic afferents, and whether these factors are at play in other species, including those with gyrencephalic brains. Do signals in the cortical primordium generate boundaries between primary and secondary areas?

### Evolution of Ontogenetic Columns, Layers, Areas, and Gyrification

What are quantitative or qualitative differences in cortical area specification and function in humans versus other primates? To move beyond using cytoarchitecture to define areas, goals for future research include the following:

- Investigate the role of noncortical inputs in cortical development and evolution (e.g., vasculature, ECM, meninges, cerebellum, gut–brain axis).
- Despite access to new sequencing and imaging technologies, we lack a clear understanding of species anatomical differences. To compare cortical areas between humans and macaques, we need to explore using comparative MRI and electrophysiology, and take advantage of brain banks.
- To understand species differences in synapse complexity, one could use high throughput electron microscopy or determine synapse density per neuron using biochemical methodologies not reliant solely on immunohistochemistry or EM. This could employ quantification of NeuN+ neuronal nuclei and synaptophysin+ synaptosomes in cerebral cortex homogenates of humans and other species.
- Another important issue concerns the decreasing rate of adult neurogenesis during evolution (e.g., Arellano et al. 2018). We need to understand these differences and clarify why neurogenesis diminishes or does not occur in humans in order to understand the human capacity for retention of memory over many decades of life (Rakic et al. 1986).
- Gyrification allows the enormous increase of cortical surface and hence is a fundamental feature of cortical development and evolution, yet we know little about the mechanisms and specific-specific differences and similarities.

### Understanding How Cortical Circuits Develop

For both cell- and position-specific circuitry, how do these develop, and what is the role of genes and activity? In addition, not only the quantity but quality of connections is important, as is the relationship between neuron number and types of connections. We propose that future research needs to pursue the following lines of enquiry:

- Determine if there is experience-dependent wiring among different species and if genes related to plasticity are upregulated or brake-like factors are absent in humans.
- Perturb circuits in mice to investigate function and likewise interrogate human neural function by longitudinal recordings. Overlapping genetics may enable clarification of common circuits.
- Define a comprehensive whole brain connectivity map in 200 µm patches. Use high throughput means (such as mapseq) to map the human brain by transfections of bar-coded factors (Kebschull et al. 2016).

# The Cortical Connectome

# 6

# Brain Networks

## How Many Types Are There?

Marcus E. Raichle, Ryan V. Raut, and Anish Mitra

### Abstract

Unraveling the organizational structure of the brain has, in large measure, been reductionist in nature. While this has revealed, in ever-increasing detail, the fine structure of the brain, it does leave less directly addressed the beautifully integrated nature of brain function. Views of the functional organization of the brain should include a unitary perspective, despite the diversity of its constituent parts. This chapter focuses on recent observations from the authors' laboratory, which point to the value of an integrated approach as well as to answer the assigned title question: arguably, the brain consists of a single network with functional diversity.

### Introduction

Categorizing network types in the brain requires knowledge of the context in which the term *networks* is applied. Generally, the term emerges from research that associates component operations and behavior with identifiable brain parts and their interactions, ranging from the cellular and molecular to a whole brain level of analysis (Bassett and Sporns 2017). The breadth of extant work on this subject is exemplified by the observation that a current PubMed search of the terms "brain" and "networks" presently yields over 33,000 citations. Taking a whole-brain perspective, restricting the search to "brain networks" and "fMRI" yields approximately 1,500 citations per year, over the past three years. Remarkably, this represents ~40% of the total citations for "brain" and "networks" over the same period of time. Here, we focus on this whole-brain or large-scale perspective that now characterizes a significant fraction of human brain imaging research and, more generally, research on functional brain organization.

From this large-scale perspective, a generally agreed upon list of cortical networks has emerged with names such as somatomotor, visual, dorsal, and ventral attention, cingulo-opercular and frontoparietal control, salience, and default mode (e.g., Power et al. 2011; Yeo et al. 2011; Hacker et al. 2013). Research seeking an understanding of the large-scale organization of these networks in the mammalian brain has included anatomically and theoretically based network analyses (e.g., van den Heuvel et al. 2016) and genetics (Richiardi et al. 2015; Ge et al. 2017) as well as resting-state functional magnetic resonance imaging (fMRI) in humans (Raichle 2011), nonhuman primates (Vincent et al. 2007), and rodents (Lu et al. 2012; Stafford et al. 2014). While the nature of the relationships within and among these networks has not been ignored, the primary emphasis has been on their parcellation; that is, how crisp the boundaries delineating the networks are and how many cortical parcels exist (e.g., Power et al. 2011; Glasser et al. 2016).

Our objective is to outline a unifying, functional framework within which individually identifiable components (*networks/systems*) arise and communicate with one another. There are two major elements to our approach.

The first element is to specify the biological underpinnings of the fMRI blood oxygen level dependent (BOLD) signal. The logic behind doing so is that this signal has become an extremely attractive window on the brain's large-scale, functional organization. Nonetheless, an agreed upon understanding of its underlying neurophysiology has been lacking. Many have asserted that infra-slow ($< 0.1$ Hz) fMRI signals are simply a low-pass filter of the brain's overall neurophysiology (e.g., de Zwart et al. 2005; Logothetis 2008). In contrast, we have recently shown (Mitra et al. 2018) that the signal is a very specific representation of the brain's infra-slow activity, and thus it offers a unique window on an element of brain neurophysiology that is critical for maintaining and orchestrating the brain's large-scale, functional organization.

The second element of our approach is to utilize functional information available from the spontaneous, ongoing activity of the brain. The logic behind this choice is that most of brain energy resources are devoted to this activity, well over 90% (Raichle and Mintun 2006). Furthermore, it has become a major source of information related to the functional organization of the mammalian brain in health and disease.

## The fMRI BOLD Signal

The fMRI BOLD signal has a long and storied history. It is based on the properties of oxygenated hemoglobin in a magnetic field, a property first hinted at by Michael Faraday in 1846 (Faraday 1933), formally discovered by Linus Pauling and Charles Coryell in 1936 (Pauling and Coryell 1936), and reintroduced by Keith Thulborn and colleagues in 1982 (Thulborn et al. 1982). Deoxyhemoglobin, being paramagnetic, disrupts a magnetic field and causes a loss of signal in an MRI scanner. Therefore, veins will prominently appear in MRI images as areas of signal loss. Task-induced increases (decreases) in regional brain blood flow are not accompanied by proportional changes in oxygen consumption (Fox and Raichle 1986), thus producing localized decreases (increases) in deoxyhemoglobin. Combining this knowledge with the effect of deoxyhemoglobin on the MRI signal, Seiji Ogawa and colleagues proposed at the Bell Laboratories an *in vivo* MRI strategy for brain mapping based on what they dubbed the BOLD signal (Ogawa et al. 1990). Their proposal launched fMRI as the primary tool in cognitive neuroscience.

The physics of the fMRI BOLD signal is well understood. The fMRI signal is, quite simply, based on the ratio of oxy- to deoxyhemoglobin in the brain vasculature, which varies both spontaneously, reflecting the brain's ongoing intrinsic activity, and predictably in response to task-induced changes in brain activity (see figure 6 in Raichle and Mintun 2006). The relationship of the BOLD signal to the underlying neurophysiology of the brain, however, has been a matter of considerable debate. At the heart of this discourse has been the variably articulated idea that the fMRI BOLD signal is a vascular, low-pass filter of brain neurophysiology writ large (e.g., de Zwart et al. 2005; Logothetis 2008). The slow temporal dynamics of the BOLD signal (stimulus to onset ~2 sec and spontaneous frequency of < 0.1 Hz) has been attributed to the response time of the vasculature, a phenomenon often referred to as *neurovascular coupling*.

The debate over neurovascular coupling frequently ignores the possibility that there is an element of the brain's neurophysiology that does correspond to the temporal scale of the fMRI BOLD signal, which is predominantly < 0.1 Hz. This has been dubbed infra-slow activity (ISA) (for a superb scientific and historical review, see Palva and Palva 2012). In support of this view, recent studies in humans (Mitra et al. 2014; Mitra et al. 2015) and mice (Matsui et al. 2016; Vanni et al. 2017) report that ISA, as measured by BOLD or calcium imaging, travels slowly through the cerebral cortex along stereotypical spatiotemporal trajectories. Spontaneous BOLD signals have also been linked to ISA in local field potentials (Leopold et al. 2003; He et al. 2008; Pan et al. 2013). Together, these findings suggest the possibility of a distinct ISA process that moves dynamically through the brain to establish a systems-level organization that is captured in the resting-state BOLD signal.

Key questions, however, remained unanswered. Is ISA, especially its spatiotemporal trajectory through the cortex, distinct from other frequencies, such as delta activity (1–4 Hz)? Do the spatiotemporal trajectories of BOLD signals correspond specifically to ISA or do they represent higher frequencies as well? Finally, does ISA travel through specific cortical layers as do other distinct spectral bands, such as gamma (> 40 Hz), alpha (8–12 Hz), and delta?

Recently we addressed these questions in mice (Mitra et al. 2018) using whole cortex, calcium/hemoglobin imaging, and laminar electrophysiology (Figure 6.1). With calcium/hemoglobin imaging we showed that ISA in each of these modalities travels through the cortex along stereotypical spatiotemporal trajectories that are state dependent (wake versus anesthesia) and distinct

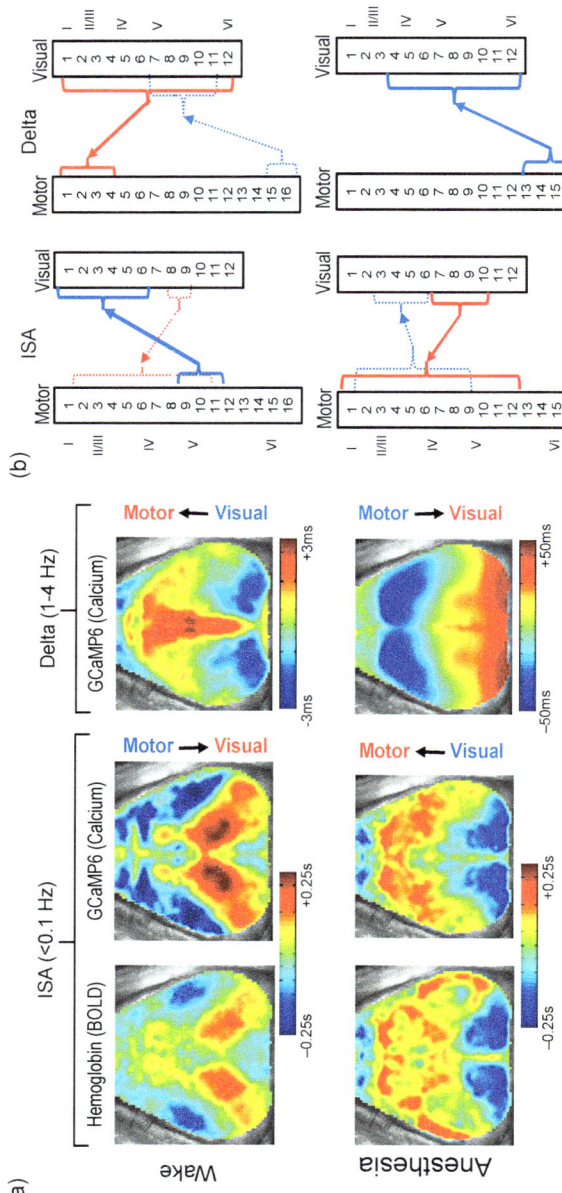

**Figure 6.1** Utilizing whole cortex calcium fluorescence and hemoglobin absorbance optical imaging (a) and cortical, laminar electrophysiology (b) in awake (top) and anesthetized (bottom) mice, we explored the temporal and spatial dynamics of the brain's infra-slow activity (ISA) (frequencies < 0.1 Hz) and compared ISA to activity in the delta frequency range (1–4 Hz). Hemoglobin imaging employed here is sensitive to the ratio of oxy- to deoxyhemoglobin and thus equivalent to the fMRI BOLD signal. (a) In the wake state, ISA travels from the motor cortex (anterior) to the visual cortex (posterior) in a pattern revealed almost identically by calcium fluorescence and hemoglobin absorbance. Under general anesthesia this pattern reverses direction. This is to be contrasted with calcium fluorescence in the delta frequency range, which reveals movement in a direction opposite to that in the ISA range in the wake and anesthetized state. (b) Laminar physiology further reveals the unique distinctions between ISA and delta activity in the mouse cortex. Complete experimental details behind the material depicted in this figure are available in Mitra et al. (2018), from which this figure was adapted.

from trajectories in delta (Figure 6.1a). This confirmed our earlier work in humans which compared wake and sleep states (Mitra et al. 2016). Moreover, our mouse laminar electrophysiology reveals that ISA travels through specific cortical layers and exhibits cross-laminar temporal dynamics distinct from higher-frequency local field potential activity (Figure 6.1b). A corollary to this latter observation is the possibility that resting-state fMRI reflects heretofore unsuspected frequency and laminar specificity.

From the perspective presented above, we now turn to a discussion of research that has utilized resting-state fMRI BOLD imaging of spontaneous brain activity to delineate the functional organization of the human brain. We posit that what is being revealed is the role of a unique component of brain neurophysiology, namely ISA.

## Resting-State Functional Connectivity

In 1995, Bharat Biswal et al. (1995) reported that spontaneous fluctuations in the fMRI BOLD signal in the motor hand area of one cerebral hemisphere correlated with spontaneous activity in the motor hand area of the other hemisphere. Despite earlier work that made the findings of Biswal and colleagues plausible (e.g., Vern et al. 1997), there were initial doubts about the importance of their findings. Gradually, the skepticism abated in the face of a flurry of observations that this strategy, when applied to other areas of the brain, revealed a large-scale functional organization that mirrored that known from task-based fMRI and its predecessor, positron emission tomography or PET (Figure 6.2). This represented a paradigm shift in the imaging of the human brain in health and disease across the life span (for reviews, see Fox and Raichle 2007; Raichle 2009). The work has now been extended to nonhuman primates (Vincent et al. 2007) as well as other species, including rodents (Lu et al. 2012; Stafford et al. 2014).

The stunning appeal of the maps of resting-state functional connectivity (Figure 6.2c) led to questions about their stationarity and whether it might be possible to understand how the various systems communicated with each other in various states (e.g., sleep versus wake as well as during task performance). As depicted in Figure 6.2d, not only do individual systems exhibit strong internal correlation structures but in the off-diagonal elements of this correlation matrix there are, not surprisingly, definite hints of relationships among the various systems. For example, the well-known anticorrelations (Fox et al. 2005) between the dorsal attention network (DA in Figure 6.2) and the default mode network (DM in Figure 6.2) can be faintly seen in the upper right-hand corner of the matrix. Animating this matrix produces a very seductive picture of changing relationships within and among the constituent systems over time. Not surprisingly, this spawned an active area of research, known as *dynamic functional connectivity* (Hutchison et al. 2013), which has been criticized for

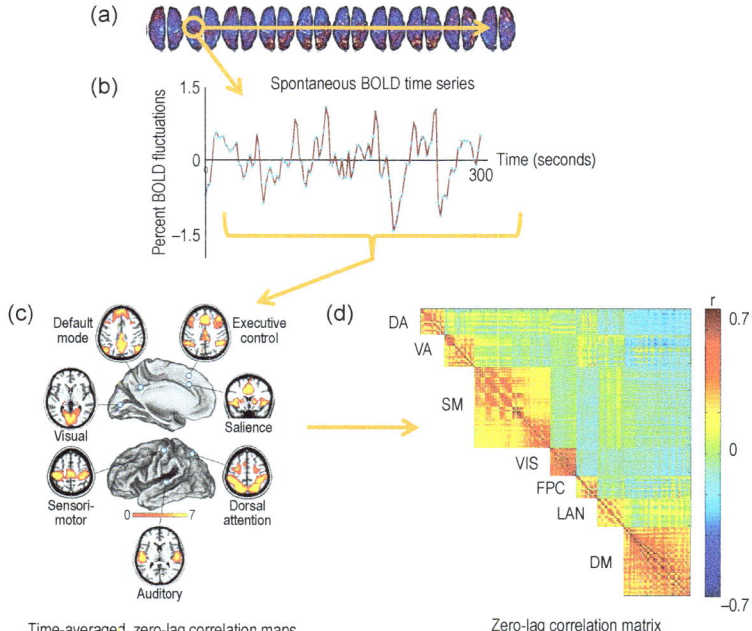

Time-averaged, zero-lag correlation maps          Zero-lag correlation matrix

**Figure 6.2** From a series of fMRI BOLD images obtained every 2.3 sec from individuals in a relaxed but awake state (a), one can obtain a time-activity curve (b) from selected brain regions. When the region of interest lies within a known brain network (c), correlations with this time-activity curve outside of this region of interest delineate the spatial topography of the network. This is known as resting-state functional connectivity. A symmetric correlation matrix (d) of these relationships can be constructed exhibiting correlations within networks along the diagonal and correlations among networks in the off-diagonal blocks. The network abbreviations utilized in this and subsequent figures include the dorsal attention (DA), ventral attention (VA), somatomotor (SM), visual (VIS), frontoparietal control (FPC), language (LAN), and default mode (DM). Elements of this figure were adapted from Raichle (2011).

being very artifact prone due to such things as subject movement, to which resting-state functional connectivity is exquisitely sensitive (Power et al. 2012; Laumann et al. 2016; Liegeois et al. 2017). Because of these concerns, we elected to approach the question of the spatial and temporal aspects of relationships within and among resting-state networks in a different manner.

## Lag Structure in Resting-State fMRI

To explore how spatially segregated networks such as those illustrated in Figure 6.2c communicate, we elected to examine the latency structure of the spontaneous, correlated fluctuations in the fMRI BOLD signal among nodes within and between networks (Mitra et al. 2014). This deviates from the standard practice of assuming no latency, or *zero lag functional connectivity*. Despite the stunning results obtained with the zero lag functional connectivity approach, it tacitly ignores the existence of a temporal component within the spatial structure of these correlations. Our approach was to seek evidence of this temporal component.

Studying the interregional lags of a poorly sampled signal, such as spontaneous fluctuations in fMRI BOLD, might seem like a dubious undertaking. However, as illustrated in Figure 6.3a, as well as in articles that delineated and defended the details of our approach (Mitra et al. 2014; Mitra et al. 2015), it worked out quite well. One of the keys to our success was the availability of a very large, high-quality data set, The Brain Genomics Superstruct Project (Buckner et al. 2014). Our work revealed that intrinsic activity propagates both through and across networks on a timescale of ~ 1 sec (Figure 6.3b), such that no network is entirely early or late compared to the others (Figure 6.3c). Instead, each network has components that send signals to the rest of the brain as well as components that receive signals from the rest of the brain.

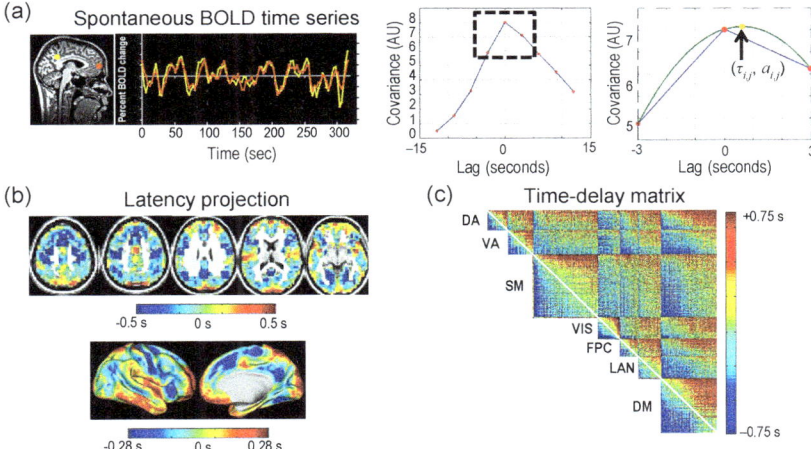

**Figure 6.3** Calculation of a pairwise time series lag or latency uses cross-covariance and parabolic interpolation (green line) as illustrated in (a) where two fMRI BOLD time series extracted from 2 brain loci are exhibited. The lag between the time series is the value at which the absolute value of the cross-covariance function is maximal (yellow dot, far right). This extremum can be determined at a resolution finer than the temporal sampling density. (b) Three-dimensional (top) and surface-based (bottom) latency projection maps from 692 subjects. Projection maps are computed by taking a column-wise average of the full time-delay matrix, shown in (c); thus, the value at each region indicates the region's mean temporal relationship (blue, early; red, late) with the rest of the brain. (c) The relationship of latency to resting-state networks (abbreviations as in Figure 6.2) shown in a matrix format ordered by resting-state membership. Note the wide range of latencies within resting-state networks as well as among networks as depicted in the off-diagonal blocks. Adapted from Mitra et al. (2014) and Mitra and Raichle (2016).

As stressed above, zero lag resting functional connectivity reveals, in a remarkably consistent manner, the basic large-scale network structure of the human brain (Figure 6.2c). Examining the latency structure of the very same signal that gave us this result (Figure 6.3b), we observe propagation of this signal both within and among networks that appears to cross network boundaries (Figure 6.2c). How can this be?

Demystifying the propagation pattern of the spontaneous fMRI BOLD signal took a major step forward through the discovery of *lag threads* (Mitra et al. 2015). The spontaneous fMRI BOLD signal that is depicted in Figure 6.3b and 6.3c, in terms of its spatial latency, consists of an estimated eight orthogonal components, which we dubbed lag threads. Using a data set of 1,376 subjects (Buckner et al. 2014) which were randomly assigned to two groups of 688 subjects, and employing preprocessing and computational methods detailed elsewhere (Mitra et al. 2015), we were able to show, reproducibly, that there are at least eight lag threads characterized by distinct "sources" and "sinks."

Still, the fact that the lag structure of spontaneous fMRI BOLD signal was multidimensional did not fully explain how this signal could functionally delineate several spatially non-overlapping networks of correlated activity, such as those shown in Figure 6.2c. What was the missing feature of the lag threads? The answer was what we termed *motifs*: sets of regions whose temporal ordering is consistent across lag threads. Defined in this way, such motifs were found to correspond to conventional resting-state networks. This implies that large-scale networks are characterized by unidirectional propagation. Metaphorically speaking, motifs and the networks they delineate represent *one-way streets* for ISA. From this perspective it can be convincingly shown that the zero lag temporal correlation network structure of resting-state fMRI (Figure 6.2d) can arise from the structured, unidirectional propagation of ISA through specific sets of regions (networks); sets of regions that do not follow such structured ordering across lag threads do not manifest strongly correlated signals (Mitra et al. 2015). The sources and sinks of fMRI BOLD and ISA assume added significance in considering the relationships among functionally defined networks.

## Communication among Networks

If motifs describe the movement of ISA within functionally identified brain networks, how can we characterize the relationships among networks? Although the precise boundaries of networks are arbitrary, functional networks by definition comprise positively correlated regions. Thus, we may decompose the full time-delay matrix (Figure 6.3c) into region pairs that are positively or negatively correlated (Figure 6.4a, b); in doing so, we are left with, respectively, predominantly within-network relationships and exclusively between-network relationships. At one extreme is the purely anticorrelated relationship between

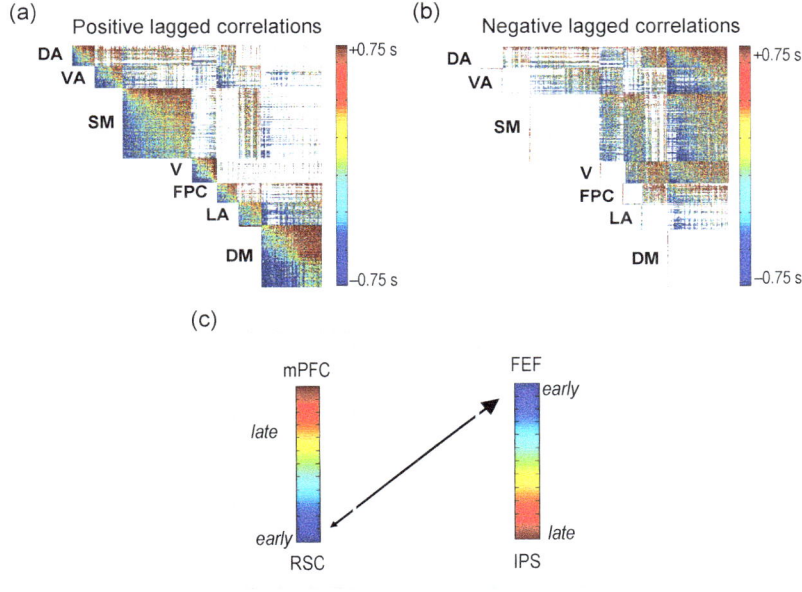

**Figure 6.4** Decomposing the time-delay matrix depicted in Figure 6.3c into positively (a) and negatively (b) correlated voxel pairs provides additional details about the spatiotemporal relationships of the spontaneous fMRI BOLD signal within (diagonal) and among (off-diagonal) large-scale brain networks. Notice that resting-state networks contain only positive correlations, by definition. Positive correlations also exist outside of the resting-state networks as inter-network relations comprise both positive and negative correlations; however, focusing exclusively on lags among negatively correlated regions (b) permits analysis of strictly inter-network signaling (c). Illustrated here is the general relationship of lag trajectories within and between the default mode network (DM) and the dorsal attention network (DA). Within each network there is an early and late temporal gradient. In the DM this goes from the retrosplenial cortex (RSC; posterior) to the medial prefrontal cortex (mPFC; anterior). In the DA it extends from the frontal eye fields (FEF; anterior) to the intraparietal sulcus (IPS; posterior). Finally, the communication between the two systems occurs via their earliest components (i.e., RSC to FEF). More generally, the similarity of on- and off-diagonal blocks within each column of the time-delay matrix in Figure 6.3c reveals shared trajectories of within- and between-network signals. Adapted from Mitra and Raichle (2016).

the default mode network and the dorsal attention network, consistent with previous work on their relationship (Fox et al. 2005). More generally, however, the relationships among conventional networks are a combination of positive and negative correlations.

An intriguing feature of activity shared between networks is that it follows a similar spatiotemporal trajectory to within-network propagation. Thus, cross-network signals begin at the earliest nodes of each network (i.e., within network "sources" are nodes of communication with other networks) and subsequently propagate through each network involved. This is graphically depicted in Figure 6.4c, where the retrosplenial cortex of the default mode network communicates with the frontal eye field component of the dorsal attention system. The full significance of the correspondence between within- and between-network spatiotemporal trajectories and how cross-network signaling is implemented remains to be understood. When this understanding is accomplished, however, it will begin to reveal how information is *integrated* among the large-scale functional networks, which resting-state fMRI has been so instrumental in defining. Characterizing such integration is crucial for revealing how these components together produce a unitary brain network, whose function and dysfunction may be understood best at this emergent level.

## Summary

The above material gives a broad overview of work that has occupied us over the past several years, nourished, of course, by the work and advice of many others. Viewed in the context of the question posed to us—How many types of brain networks are there?—we must conclude that seen "from the top," the brain exhibits a remarkable degree of integration, so much so that the answer could arguably be "one." At the very least, the brain operates on a background of highly integrated and energetically costly activity represented, in part, by ISA. This activity provides a tapestry upon which the contributions of the more spatially and temporally granular elements of the brain are coordinated in the execution of their unique contributions to brain function. Many issues remain to be explored, which we highlight below.

### Cross-Frequency Coupling

If the fMRI BOLD signal specifically represents ISA as we have demonstrated (Mitra et al. 2018), how do other frequencies fit in to this perspective? The idea of cross-frequency coupling is certainly not new (e.g., Monto et al. 2008). Conceptually, the phase of lower frequencies (e.g., delta or ISA) modulates the power of higher frequencies. This has been our experience, both in our work in humans (Mitra et al. 2016) as well as in rodents (Mitra et al. 2018). It is therefore important, when studying conventional electrophysiological activity (e.g., spiking and high-frequency local field potentials), to keep in mind the context—in a neurophysiological sense, provided by lower frequency activity—in which such phenomena occur. As Rodolfo Llinás so aptly said: "the significance of incoming sensory information depends on the preexisting functional disposition of the brain, [and] is a far deeper issue than one gathers at first glance" (Llinás 2001:8).

**Task-Evoked Activity**

It is a strongly embedded tradition in cognitive neuroscience to refer to local, task-evoked changes in the fMRI BOLD signal as representing "activations." Broadly translated, this means to most that one is observing a low-pass filter of the brain's neurophysiology. That is clearly not the case. Considering the specificity of the fMRI BOLD signal, we need to reconsider how we interpret "activations" and, for that matter, "deactivations." Could it be, as the work of Schroeder and colleagues suggest for delta activity (Schroeder and Lakatos 2008), that changes in the fMRI BOLD signal similarly represent phase resetting of ongoing ISA? This is an important idea that deserves our attention as we attempt to understand better the brain's remarkable capacity to predict and prepare for future events. Hints of what to expect are already present in extant data (Ress et al. 2000; Sirotin and Das 2009; Cardoso et al. 2012).

**Cellular Origins of ISA**

Many discussions of brain function refer generically to "neurons" without being clear about which type. In considering something as potentially complex as ISA, it is likely that interneurons play a role that has yet to be defined. But our vision should broaden even further to include the glia, particularly the astrocytes. As Poskanzer and Yuste (2011) pointed out, astrocytes have a direct role in inducing "up states" in neurons. If, as we suspect, ISA represents large-scale changes in neuronal excitability, then astrocytes need to be factored into the equation.

**Metabolism**

We suspect that many in neurobiology would be surprised to know that cellular metabolism, particularly glycolysis, is rhythmic in every cell system in which it has been studied (Goldbeter 1996). These rhythms have a frequency remarkably similar to ISA and are intimately related to cellular excitability and action potentials (Bertram et al. 2007). Furthermore, cells form communities on the basis of these rhythms (Campbell et al. 2015)! The close relations between cerebral blood flow, metabolism, and functional brain imaging signals command attention to the relationship between metabolic activity and ISA. More generally, uncovering potential consequences of metabolic rhythms on electrical excitability in the brain will be valuable as we strive to achieve a more broadly based understanding of brain function.

**Neuromodulation**

Finally, as we consider the mechanisms behind state changes in ISA, be it sleep versus wake in humans (Mitra et al. 2016) or anesthesia versus wake in laboratory animals (Mitra et al. 2018), it is important to consider the role of neuromodulators in rebalancing relationships within and among brain systems (Bargmann and Marder 2013) that are being mediated through ISA.

## Acknowledgments

This work was funded by the NIH via NS080675 to M. E. R., MH106253 to A. M. and NSF via DGE-1745038 to R. V. R. We thank our many colleagues, collaborators, and friends for productive discussions over many years and apologize to those whose influential papers were not cited because of space limitations.

# 7

# Network Models in Neuroscience

Danielle S. Bassett, Perry Zurn, and Joshua I. Gold

## Abstract

From interacting cellular components to networks of neurons and neural systems, interconnected units comprise a fundamental organizing principle of the nervous system. Understanding how their patterns of connections and interactions give rise to the many functions of the nervous system is a primary goal of neuroscience. Recently, this pursuit has begun to benefit from the development of new mathematical tools that can relate a system's architecture to its dynamics and function. These tools, stemming from the broader field of network science, have been used with increasing success to build models of neural systems across spatial scales and species. This chapter discusses the nature of network models in neuroscience. It begins with a review of model theory from a philosophical perspective to inform our view of networks as models of complex systems, in general, and of the brain, in particular. It summarizes the types of models that are frequently studied in network neuroscience along three primary dimensions: from data representations to first-principles theory, from biophysical realism to functional phenomenology, and from elementary descriptions to coarse-grained approximations. Ways to validate these models are then considered, with a focus on approaches that perturb a system to probe its function. In closing, a description is provided of important frontiers in the construction of network models and their relevance for understanding increasingly complex functions of neural systems.

## Introduction

The brain is composed of intricate networks that operate at many different levels of organization. At small spatial scales, gene regulatory networks direct neuronal cell fate, and both chemical and electrical synapses define the accessible routes of information transmission between neurons (Francis et al. 2003). At intermediate spatial scales, laminar architecture in cortex is accompanied by stereotyped interlaminar connectivity thought to support ensemble dynamics and resultant computations (Sherman et al. 2016). At even larger spatial scales, the anatomical locations of inter-areal projections display a precise spatial arrangement associated with a diverse repertoire of functional processes (Betzel and Bassett 2018; Betzel et al. 2018).

Although networks are fundamental to brain structure, the complexity of these networks poses challenges to understanding their function. Unlike a sphere, which we can quickly guess to have the capacity to be rolled or thrown, a network—with its tangle of wires—defies any simple conspectus. Thus, even though more than a century has passed since Camillo Golgi, Santiago Ramón y Cajal, and other neuroanatomists introduced to the world the intricate beauty of the networks of neurons that comprise our nervous system, our understanding of how those networks give rise to perception, learning, memory, cognition, action, and other aspects of brain function remains incomplete.

To understand relationships between the brain's networked architecture and its many functions, one fruitful set of approaches stems from the emerging field of network science. This discipline addresses the study of systems whose structure, function, or dynamics depend upon the pattern of interconnections between units (Albert and Barabasi 2002). Network science is inherently interdisciplinary, drawing on and integrating among recent advances in mathematics, physics, computer science, and engineering (Newman 2010, 2011). Although early work in the field was largely devoted to the study of social systems, efforts over the last decade have focused increasingly on the study of neural systems across spatial scales, temporal scales, and species (Bullmore and Sporns 2009; Fornito et al. 2016; van den Heuvel et al. 2016). These newer efforts, collectively referred to as *network neuroscience*, model neural systems as networks to distill the dependence of brain function and dysfunction on interconnection architecture (Bassett and Sporns 2017).

Here, we review recent work in network neuroscience that has been applied to our collective quest to understand the brain, and emphasize the diversity of approaches that now fall under this general framework (Bassett et al. 2018). Because network neuroscience is fundamentally a modeling endeavor, we begin with a broad philosophical perspective on model theory. We then consider how networks are models, before turning to a discussion of the types of network models that are commonly used in neuroscience. Motivated by approaches to validate network models via prediction, we discuss the importance of perturbation-based techniques for understanding network function. We close by outlining important directions for future work to build, use, and validate network models in neuroscience.

## Model Theory: A Philosophical Perspective

The term "model" can evoke quite different pictures in the mind's eye:

- a small, inexact replica of a 1910 Schacht Roadster
- a miniature cityscape coarsely true to the form of Cambridge, England

- Rodin's "The Thinker"
- a contemporary reworking of a piece from Greek mythology (Eugenides 2002)
- a person one wishes to emulate or an ideal one hopes to become
- IBM's Watson
- a cerebral organoid—miniature organ *in vitro* (Huang et al. 2017)—or a human blinking eye-on-a-chip (Chan et al. 2015)
- a set of interdependent partial differential equations producing dynamics reminiscent of a real-world system

In mentally traversing these diverse examples, one immediately realizes that the space of model types is exceedingly large, and one wonders whether it is even possible to define what a model actually is, or what it is not.

The question of how to define the term "model" is the focus of a branch of philosophy known as model theory, which aims to identify the essential elements that make models what they are and to disambiguate the characteristics that distinguish different types of models from one another (Gelfert 2016). At its most basic level, a model is a representation of one or more aspects of the world. It aims to increase understanding of what something is by measuring and imaging what something does. As such, models inherit a basic philosophical conundrum (Papineau 1987): What precisely is their relationship to the target systems they model, and from whence do they derive their truth value? Must they simply evidence functional coherence and have pragmatic purchase, or must they also meaningfully correspond to what they represent? If so, how is that meaningfulness determined?

In science, at least four types of models have been recognized, each of which provide different answers to these questions: scalar, idealized, analogical, and phenomenological models (Frigg and Hartmann 2012; Hartmann and Frigg 2012). Scalar models, much like the Roadster replica, either magnify or reduce their target systems. Idealized models abstract and isolate a limited set of features from their target systems. Analogical models highlight relevant similarities between two target systems, whether those similarities are shared properties or comparable structures. Finally, phenomenological models represent only the observable elements of their target systems, without postulating any theoretical explanation as to why those elements are what they are.

Recent work in model theory explores the intriguing possibility that all of these forms of scientific models are heuristic devices not unlike literary fictions (Suarez 2009; Frigg and Hunter 2010; Toon 2012). The models-as-fictions theory reconceptualizes models as fictional entities that aim to narrativize certain features of a target system (Barbrousse and Ludwig 2009; Godfrey-Smith 2009; Frigg 2010a, b, c; Garcia-Carpintero 2010; Toon 2012; Frigg and Nguyen 2016, 2017). Accordingly, models—much like fictions—may imaginatively isolate and abstract or distort and exaggerate certain features of the world in such a way as to facilitate epistemic access (Elgin 2010). They may creatively instantiate either analogical or phenomenological substructures of their target systems in order to crystallize insight. Scientific models are therefore subject to evaluation at the level of both artistry (clarity, elegance, originality) and function. Moreover, just as the literary tradition provides new fiction with meaningful constraints in advance, the scientific community provides the parameters within which new models are developed and applied. Scientific models are, in this sense, accountable to the scientific communities that use them in the exploration of target systems that are already of particular value and interest (Almeder 2007). Finally, different sorts of models can be used together to build a multiscalar narrative architecture, modeling complementary features of a target system beside one another.

Whether used in isolation or in conjunction, scientific models illuminate the overarching structure of a target system precisely through the practice and provocation of creative imagination.

## Networks as Models

The incipient challenge in modeling biological systems is to identify the most meaningful characteristics of the system that are distilled into a sensible representation (Bellomo et al. 2015). That is, biological models are inherently idealized models of complex systems, and their construction requires identifying first the form and degree of abstraction to use. This process requires a set of value judgments (Which characteristics are most meaningful?) and a commitment to epistemic cleanliness (What details of biology can we defensibly ignore?). These principles of valuation and purposeful ignorance are manifest even when exercising simple visual depictions, which arguably comprise the most impoverished of modeling approaches (Tufte 2001). One similarly faces choices of what to depict and what not to depict when building any simple mathematical representation of the system. For example, when building a differential equation to represent a system, one must choose which processes to encapsulate or not in a variable.

The fundamental assumption of network neuroscience is that idealized models of the brain should be constructed using analogical principles that focus on the networked architecture of the nervous system. As Cajal saw under his microscope, the nervous system is composed of individual neurons that are interconnected in complex ways. Accordingly, the earliest network models were idealized versions of this network structure, with nodes representing neurons and edges representing the connections between them. More recently, network models have been developed within and across multiple spatial and temporal scales, at the level of interconnected neurons as well as involving networks of subcellular components, multicellular systems, or both. As detailed below, these models can also be phenomenological, based on measured elements of the nervous system, or more theory driven. Despite this diversity, these models retain key features that can be understood in terms of their basic idealized and analogical structure: an architecture built using interconnected units.

Such an architecture is typically encoded in a graph: an object composed of nodes, representing units of the system, and edges, representing interactions or links between those units (Bollobás 1979, 1985). Studies of graphs can be neatly separated into two categories: those who consider artificial graphs with arbitrary wiring principles (Harary 1969) and those who consider them to reflect the architecture of a real system (Cohen and Havlin 2010). In both cases, one seeks to describe the mathematical properties of the graph with the goal of understanding the function of the system. The patterns of which units can and cannot (or do and do not) interact with one another can allow one to deduce where information might be relatively more densely or relatively more sparsely located, where vulnerability might exist to injury or perturbation, and where circumscribed instances of collective dynamics might emerge (Albert et al. 2000; Cisneros et al. 2002; Gomez-Gardenes et al. 2007; Simonsen et al. 2008).

In simple graphs, all units are represented by identical nodes, and all edges are represented as either existing or not existing (Figure 7.1). These representations can be encoded using a binary weighting scheme. Furthermore, interactions are assumed to be bidirectional: if an edge exists between node $i$ and node $j$, then an edge also exists between node $j$ and node $i$. The very first formal network models of neural systems employed such binary, undirected graphs (Felleman and Van Essen 1991; Young et al. 1994; Scannell et al. 1995; Sporns et al. 2005; Achard et al. 2006; Kaiser and Hilgetag 2006). Nonetheless, it is relatively straightforward to adapt this encoding to a continuous weighting scheme, as well as to specify distinct weights for the edge from node $i$ to node $j$, and for the edge from node $j$ to node $i$. With the continued refinement of empirical measurement techniques, the inclusion of edge weights in network models has become increasingly prevalent, providing richer insights into system function and dynamics (Rubinov and Sporns 2011; Markov et al. 2013; Oh et al. 2014; Bassett and Bullmore 2017; Betzel and Bassett 2018).

Network models are simple constructs. They can be used effectively to study social, biological, technological, and physical systems (Newman 2010). Yet this flexibility is a marked reminder that the intuitions one gains from a network model depend strongly on what the nodes and edges are chosen to represent. A structural motif in a network of humans interlinked by friendships can mean something quite different than the same structural motif in a network of neurons interlinked by synapses. Thus, in any endeavor that translates a complex system into a network model, it is critical to specify exactly what the nodes and edges (or more complicated model components) represent, and to ensure that interpretations are drawn in accordance with those choices (Butts 2009).

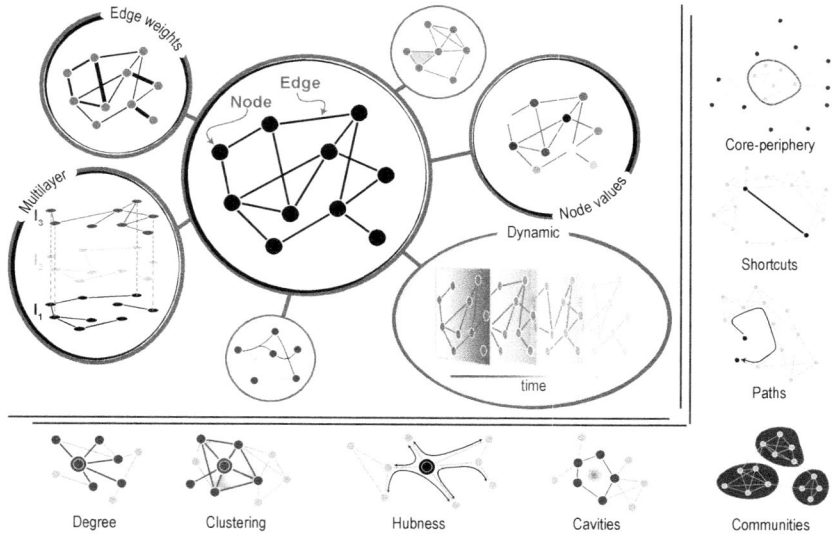

**Figure 7.1** Schematic of network models. Upper left: The simplest network model for neural systems is one that represents the pattern of connections (edges) between neural units (nodes). More sophisticated network models can be constructed by adding edge weights and node values, or explicit functional forms for their dynamics. Multilayer networks can be used to represent a set of interconnected networks; dynamic networks can be used to understand the reconfiguration of network systems over time. Bottom (left to right): Common measures of interest include degree (the number of edges emanating from a node), clustering (related to the prevalence of triangles), hubness (related to a node's influence), cavities (the absence of edges), communities (local groups of densely interconnected nodes), paths (which determine the potential for information transmission), shortcuts (one possible marker of global efficiency of information transmission), and core-periphery structure, which facilitates local integration of information gathered from or sent to more sparsely connected areas.

## Types of Network Models in Neuroscience

As a field, network neuroscience aims to build, exercise, and validate network models of neural systems with the explicit goal of better understanding brain structure and function, as well as cognition, behavior, and disease (Sporns 2014; Stam 2014; Medaglia et al. 2015; Fornito et al. 2017; Braun et al. 2018). The types of network models that are built share a similar analogical basis that emphasizes the importance of network-based architectures across spatial and temporal scales. These models, however, differ from one another in many important ways, which directly impact the sorts of inferences that can be justifiably drawn from them. Here we briefly describe recent efforts to systematize the study of network models in neuroscience by organizing these similarities and differences according to three dimensions (see Figure 7.2) that reflect the model categories described above (Bassett et al. 2018):

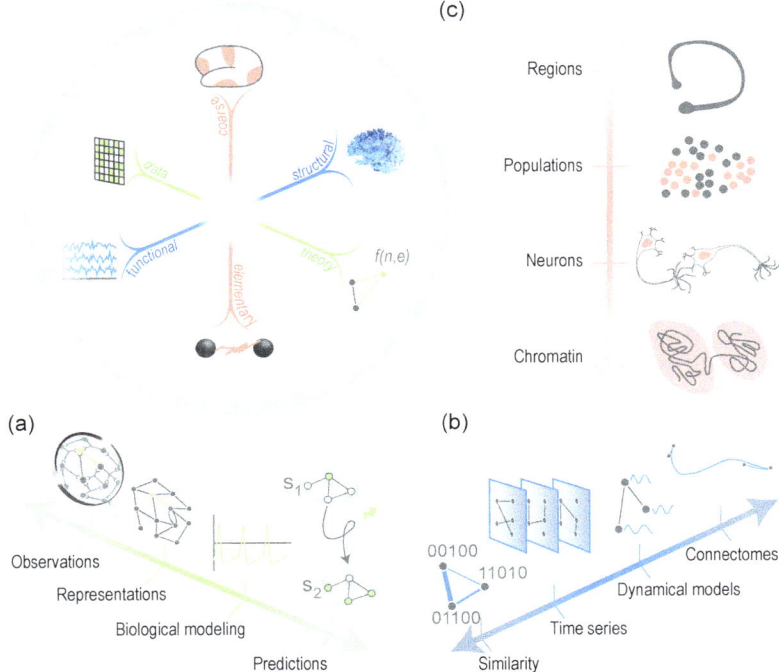

**Figure 7.2** Efforts to understand mechanisms of brain structure, function, development, and evolution in network neuroscience can be organized along three key dimensions of model types. The first dimension (a) extends from elementary descriptions to coarse-grained approximations. The second (b) extends from biophysical realism to functional phenomenology. The third (c) extends from data representation to first-principles theory.

- their phenomenological basis, ranging from representations of measured phenomena to first-principles theory;
- their target of idealization, from biophysical to functional features; and
- their scalar focus, ranging from elementary descriptions to coarse-grained approximations.

The dimension from data representation to first-principles theory is arguably the most fundamental to network modeling efforts in neuroscience (Abbott 2008). Modeling efforts of the former type begin with empirically acquired data. They then seek to build a representation of those data by stipulating which part of the data to represent as a network node, and which part of the data to represent as a network edge. Intuitively, the data representation provides an abstract, nonvisual depiction or description of the system (for examples, see Young et al. 1994; Scannell et al. 1995; Watts and Strogatz 1998; Hilgetag et al. 2000; Stam 2004; Sporns et al. 2005; Achard et al. 2006; De Vico Fallani et al. 2006; Kaiser and Hilgetag 2006; Micheloyannis et al. 2006; Bettencourt et al. 2007). In contrast, to make a prediction about system behavior either now or in the future, one must turn to models that instantiate first-principles theories. These models combine a network with a mathematical expression specifying the dynamics of network nodes, network edges, or collections of nodes and/or edges (see, e.g., Ritter et al. 2013; Roy et al. 2014; Gu et al. 2015; Falcon et al. 2016; Bezgin et al. 2017; Breakspear 2017; Melozzi et al. 2017; Yan et al. 2017; Kim et al. 2018). Data-driven network models enjoy the benefits of biological realism, whereas theory-based models have the capacity to make predictions and unearth function.

The dimension from biophysically to functionally defined features differentiates models that are physical in nature from those that are statistical in nature. Network models with biophysical realism are composed of nodes that represent physical units, including neurons, cortical columns, or Brodmann areas; and of edges that represent physical links, including synapses, projections, or white-matter tracts (see, e.g., Sporns et al. 2005; Kaiser et al. 2009; Bassett et al. 2010; Varshney et al. 2011; Nicosia et al. 2013; Oh et al. 2014). Network models addressing functional phenomenology are comprised of nodes and edges that are not necessarily physically instantiated but may instead be defined as statistical abstractions (Achard et al. 2006; Bettencourt et al. 2007; Chu et al. 2012; Burns et al. 2014; van Diessen et al. 2015; Khambhati et al. 2016). Common examples of such abstract edges are those that offer estimates of effective connectivity or functional connectivity, the latter of which are also referred to as noise correlations (Brody 1999a, b; Friston 2011a). It is often important to distinguish between these two types of models because they have distinct utility in assessing a network's physical constitution versus inferring its functional capacities.

The dimension from elementary descriptions to coarse-grained approximations is critical to support a multiscale understanding of brain structure, function, and dynamics. In general, network models can encode the organization of interconnections among cells, ensembles, cortical columns or subcortical nuclei, and large-scale brain areas. As evidenced by the diversity of scales represented in current empirical and theoretical investigations, no single level of description can provide a complete explanation for cognitive function and behavior. In many cases, however, it is worthwhile or at least practical to consider a single scale for a given study, and then to use insights gained at that scale to inform larger theories of multiscale function. The challenge in developing an appropriate network model at a particular scale is to ensure that the network nodes represent well-defined, discrete, non-overlapping units, and that network edges represent organic, irreducible relations (Butts 2009). Whereas models at the final spatial scale consider elementary building blocks (Brody 1999b; Sautois et al. 2007; Tang et al. 2008; Feldt et al. 2011; Teller et al. 2014; Kim and Lim 2015; Kaiser 2017; Mahadevan et al. 2017; Betzel et al. 2018), models at the coarse spatial scale consider emergent functions (Breakspear 2017).

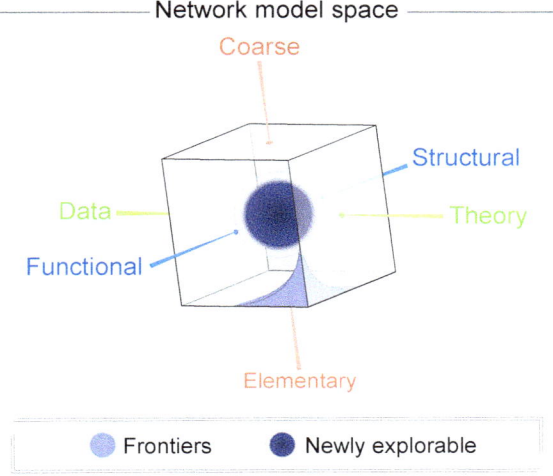

**Figure 7.3** Density of study in the three-dimensional space of network models. Density plot showing varying levels of study across the three-dimensional space of model types in network neuroscience. The center of the three-dimensional space (aubergine) is becoming increasingly accessible due to empirical and computational advances over the past few decades. The least well-studied space (periwinkle) represents first-principles theories of functional phenomenology at the elementary level of description. Clear volumes indicate spaces that are most commonly studied.

Together, these complementary dimensions define a three-dimensional space of network models that can be used to enhance our understanding of brain structure and function (Figure 7.3). Notably, prior modeling efforts have not been pursued with equal vigor in all volumes of this space, partly reflecting historical factors and the changing state of neurotechnologies. Early work focused on large-scale network models of anatomy, which represent coarse-grained data representations with biophysical realism (Bassett and Bullmore 2017; Liao et al. 2017). Less well-studied are first-principles theories of functional phenomenology, particularly among elementary units. With a marked increase in the pace of data acquisition and the capacity for data analysis, we anticipate increasing success developing network models at the center of the volume. In other words, we anticipate a wider array of studies that inform first-principles theories with data, complement physical models with statistical inferences on informational capacity, and build explicit multiscale accounts of network function. These multifaceted network models will enhance our ability to explain different parts, processes, or principles of the nervous system.

## Modeling Perturbations to Networks

When building network models, we are often concerned with demonstrating their validity (Bassett et al. 2018). In prior work, we followed canonical principles for validating animal models of disease to suggest that network models can display three distinct types of validity (McKinney and Bunney 1969; Willner 1984; Shmueli 2010; Belzung and Lemoine 2011): descriptive, explanatory, and predictive. Intuitively, descriptive validity requires that the model resembles the system under study, a concept akin to face validity in animal models (McKinney and Bunney 1969; Willner 1984; Shmueli 2010; Belzung and Lemoine 2011; Willner 2017). For instance, a model with descriptive validity might accurately reflect the specific pattern of nodes and edges observed in anatomical or functional data. By contrast, explanatory validity requires that the model can be used to define statistical tests, for example by assessing causal relations based on the network's architecture. Finally, predictive validity is attained when there is a correlation between a network model's response to perturbation and an organism's response to that same perturbation (Belzung and Lemoine 2011). Such perturbative studies can be operationalized using stimulation, lesion, ablation, or drugs.

Predictive validity is often the final goal of any scientific domain of inquiry (Shneider 2009). Because predictive validity depends on understanding its response to a perturbation, it is of interest to consider the different ways in which a network model can be perturbed. Recent work in the physics and engineering communities has begun to focus on the means by which the architecture of a network determines how perturbations affect its function. A simple way in which to parse these studies is to consider separately perturbations applied: (a) to a single node or to a single edge ("point perturbations"), (b) to a set of nodes or to a set of edges, and (c) across a fixed area or volume of the network's topology. In the context of network neuroscience, these different types of perturbations may be accessible to distinct empirical techniques and collectively could be used to better understand both endogenous and exogenous control, thereby informing clinical intervention (Tang and Bassett 2018).

From a modeling perspective, point perturbations are perhaps the simplest type to study. Initial modeling approaches focused on point perturbations in the form of node or edge removal. Referred to as virtual lesioning at times, this approach was developed to quantify the robustness of a network by estimating the difference between the value of a graph statistic estimated before node or edge removal, and the value of that same graph statistic estimated after node or edge removal (see figure 1b in Dong et al. 2013). When nodes are removed at random, the approach is referred to as a random attack. When nodes are removed based on their topological role in the network, estimated using the values of various graph statistics such as degree and betweenness centrality, the approach is referred to as a targeted attack (Achard et al. 2006). These approaches have recently been used to better understand the impact of regional dysfunction in schizophrenia and stroke (Alstott et al. 2009; Lynall et al. 2010; Lo et al. 2015).

Other approaches address how a perturbation of the activity of a node or edge can change the activity of other parts of the network. This approach is central to network control theory and built on the foundations of linear systems theory (Kailath 1980; Motter 2015). Here one considers the pattern of interconnections between units as well as a model of dynamics that specifies how the activity at one node can travel along edges to other nodes in the graph (Tang and Bassett 2018). By modeling both the connectivity and the dynamics, one can identify "driver" nodes with time-dependent control that can guide the system's activity (Liu et al. 2011). A recent application of these techniques to the connectome of *Caenorhabditis elegans* demonstrated that a particular network model had striking validity in predicting the effects of single-cell ablasions on the organism (Yan et al. 2017). The approach can also be extended beyond the identification of drivers controlling all dynamics to the identification of drivers controlling specific dynamics (Pasqualetti et al. 2014). The specificity of this extension allows for the study of unique control strategies within neural systems. A recent application of this technique provided an explanation for the anatomical location of areas of the brain involved in executive function, as those most capable of enacting modal controllability (Gu et al. 2015).

Despite their analytical tractability, point perturbations can be the most difficult to enact and interpret in the context of real neural systems. On a conceptual level, a single, functional node or edge used in a model might not have an obvious, well-defined anatomical substrate in the brain to target. On a practical level, even given a well-defined target, it may not be possible to perturb cleanly just that target, given the lack of complete specificity associated with current microstimulation, optogenetic, and pharmacological methods. Nonetheless, point perturbations represent a useful starting point in considering the validation of network models.

Moving beyond point perturbations, it is also of interest to consider perturbation to multiple points in the network, or to entire areas or volumes of neural systems. Intuitively, multipoint control is a natural reflection of circuit activity, where several areas may be activated simultaneously to orchestrate a change in communication or dynamics (Palmigiano et al. 2017). Multipoint control could also be fruitfully applied to the development of stimulation therapies to quiet seizure dynamics using implantable devices (Ehrens et al. 2015; Taylor et al. 2015; De Ridder et al. 2017; Jobst et al. 2017). Fortunately, the general network control framework is readily extended to account for the activity of multiple control points simultaneously, and can be used to model the propagation of stimulation directly along white matter tracts to predict distant effects on regional activity (Muldoon et al. 2016; Stiso et al. 2018). Extending these tools to affect control over continuous areas or volumes of a network is more difficult and remains an important area for future work. Progress in this area is critical for extending network models to account for other chemical mechanisms of transcellular communication and the effects of glia, neuromodulatory systems, and other mechanisms on brain function and behavior (Borroto-Escuela et al. 2015; Safaai et al. 2015; Bruinsma et al. 2018; Savtchouk and Volterra 2018).

## Modeling Network Growth and Evolution

The study of network perturbations, while useful for understanding endogenous and exogenous mechanisms of control, is also pertinent to an understanding of how neural systems came to be, how they develop, and how they age. A change in gene expression can alter the natural progression of cell fate from pluripotent stem cell through neuroprogenitor cell and eventually neuron (Mahadevan et al. 2017).

A fluctuation in chemical gradients can comprise a perturbation that alters the course of neuronal migration and, by extension, the location and density of synapses (Wrobel and Sundararaghavan 2014). In fully developed adult neurons, Hebb's rule essentially postulates that perturbations to neuronal firing can alter cellular-level network architecture (Bi and Poo 2001). Even at the large scale in humans, long-term training can induce changes in white matter architecture evident in noninvasive neuroimaging (Scholz et al. 2009). Understanding how perturbations of the organism or part of the organism over both short and long timescales affect network growth and evolution is an important open area of research.

Some progress has been made over the last few years in constructing so-called generative network models. Such models stipulate a wiring rule in the hope of producing a network architecture that displays topological or functional properties that are similar to those observed in the networks representing real systems (Betzel and Bassett 2017). The basic idea is that a wiring rule which produces a network topology similar to that observed in the real system is a candidate mechanism for network generation. The inference is made stronger if the wiring rule also displays characteristics thought to be consistent with biology, such as parsimony and efficiency. A common way of testing the pragmatic utility of the generative network model is to determine if it can be used to make out-of-sample predictions about held-out network data.

Generative network models tend to be built in one of three types: single-shot models, growth models, or developmental models (Figure 7.4). A single-shot model specifies a form for the connection probabilities, from which all edges and their weights are then drawn (Vertes et al. 2012, 2014; Beul et al. 2015, 2017; Betzel et al. 2016; Hilgetag et al. 2016). A growth model specifies a time-dependent wiring rule that indicates how nodes and possibly even edges are added over time (Klimm et al. 2014). Developmental models extend the biological realism of the effort even farther by specifying

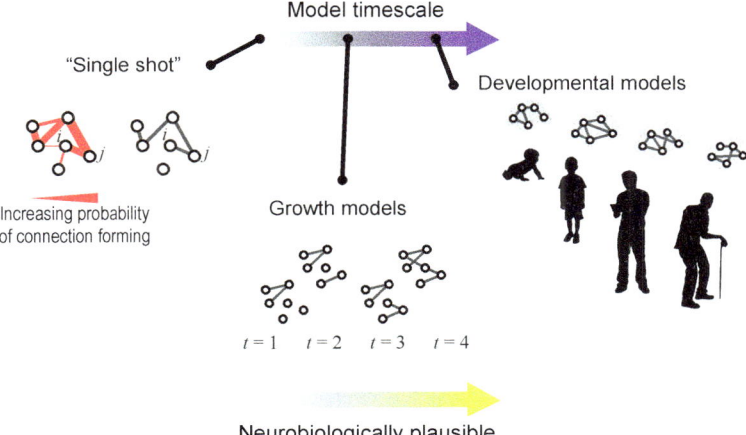

**Figure 7.4** Distinct classes of generative network models. Generative network models exist in three main classes that differ in the timescales over which they operate. Single-shot models specify a functional form for the probability with which any two nodes are linked with one another. Growth models specify rules by which nodes and/or edges are added to the network over time, which is commonly discretized in arbitrary units. Developmental models specify wiring rules with fixed timescales in actual units of seconds, minutes, days, months, years, etc., in an effort to match the true growth mechanisms of an organism better. From Betzel and Bassett (2017).

wiring rules in which the timescales of the model match the timescales of development in the organism under study (Nicosia et al. 2013). Together, these three types of generative network models vary in the timescale over which they operate and in their neurobiological plausibility.

Recently, generative network models have been developed and applied to explain neurophysiological and neuroanatomical data across both elementary descriptions and coarse-grained approximations (Beul et al. 2015, 2017). For example, a particularly striking single-shot model of the neural connectome of the nematode *C. elegans* demonstrated that a wiring rule based on the random outgrowth of axons, in combination with a competition for available space at the target neuron, was able to recapitulate the empirical network's edge length distribution (Kaiser et al. 2009). A developmental model of the same organism combined information regarding the pattern of interconnectivity between neurons with information regarding the birth times of neurons and their spatial locations (Nicosia et al. 2013). This study provided compelling evidence for a trade-off between the network's topology and cost that appears to be differentially negotiated over different developmental time periods. With a few exceptions (Vertes et al. 2012), most generative network models have focused on data representations more so than first-principles theories, and biophysical realism more so than functional phenomenology. Expanding efforts to fill the full space of model types will be an important area for future work in generative modeling.

## Future Directions

In considering the future utility of network models for advancing our understanding of neural systems, it is worth pointing out that the models used to date are relatively simple from a mathematical perspective. It remains an open question whether more complex network models might prove useful or merely obfuscate inference. To address this question, one could rationally assess whether some aspect of a known neurophysiological process remains unaccounted for by existing models. For example, to increase descriptive validity, one might wish to build an annotated network, where nodes can be assigned values or properties, reflecting for example cerebral glucose metabolism estimates from fluorodeoxyglucose (FDG)-positron emission tomography (PET), blood oxygen level dependent (BOLD) contrast imaging, magnetoencephalographic (MEG) or electroencephalographic (EEG) power, gray matter volume or cortical thickness, or cytoarchitectonic properties (Murphy et al. 2016; Newman and Clauset 2016). Furthermore, to increase explanatory validity, one might wish to build multilayer networks where the nodes and edges in each layer are obtained from different types of measurements (Kivel et al. 2014; Tewarie et al. 2014; Yu et al. 2016), and where the architecture of the network is allowed to vary over time in concert with system function (Holme and Saramaki 2012; Kopell et al. 2014; Breakspear 2017; Khambhati et al. 2017; Sizemore and Bassett 2018). Such richer models could allow one to test how network dynamics at one time point or in one modality might cause a change in network dynamics at another time point or in another modality.

A second way in which to address the question of whether more complex network models might prove useful is to consider whether the testing of a particular hypothesis requires a novel network model. For instance, recent efforts have provided initial evidence that some higher-order, non-pairwise interactions occur between neurons and between large-scale brain areas (Ganmor et al. 2011; Lord et al. 2016). Critically, all of the network models we have discussed here are based on pairwise interactions and cannot directly account for non-pairwise interactions (Petri et al. 2014; Sizemore et al. 2018). Tools that have been developed in the applied mathematics community that can encode and characterize higher-order relations include hypergraphs (an edge can link any number of vertices) and simplicial complexes (higher-order interaction terms become fundamental units) (Bassett et al. 2014; Giusti et al. 2016). These generalizations of graphs may be critical for an accurate understanding of neuronal codes and associated computations both at micro- and macroscales (Curto et al. 2013; Reimann et al. 2017; Sizemore et al. 2017). As the field moves beyond univariate accounts to postulate more network-based hypotheses, richer network models may be required.

## Conclusion

From cellular to regional scales, neural circuitry is an interconnected system. In such a system, network modeling is a particularly useful approach for distilling interconnection patterns into tractable mathematical objects that are amenable to theory. Here we discussed the nature of network models, which share a similar networked architecture that is justified in terms of its analogies to brain structure but then differ along several dimensions: from data representations to first-principles theory, abstractions that emphasize biophysical or functional features, and different scales from elementary descriptions to coarse-grained approximations. We paid particular attention to models that have been developed to better understand the response of networked systems to perturbation enacted at a single point, at multiple points, or across extended areas or volumes of the organism. We also offered an extended discussion of generative network models that seek to identify candidate wiring mechanisms for circuit evolution or development. We suggest that network models are particularly appropriate for neural systems. Accordingly, future advances in our understanding of computation and cognition will depend on the expansion of these models in mathematical sophistication and the development of richer, network-based hypothesis of brain structure, function, and dynamics.

## Acknowledgments

We thank Blevmore Labs and Ann E. Sizemore for efforts in graphic design. We also thank D. Lydon-Staley, A. E. Sizemore, E. Cornblath, and D. Zhou for helpful comments on an earlier version of this manuscript. D. S. B. acknowledges support from the John D. and Catherine T. MacArthur Foundation, the Alfred P. Sloan Foundation, the ISI Foundation, the Paul Allen Foundation, the Army Research Laboratory (W911NF-10-2-0022), the Army Research Office (Bassett-W911NF-14-1-0679, Grafton-W911NF-16-1-0474, DCIST-W911NF-17-2-0181), the Office of Naval Research, the National Institute of Mental Health (2-R01-DC-009209-11, R01-MH112847, R01-MH107235, R21-M MH-106799), the National Institute of Child Health and Human Development (1R01-HD086888-01), National Institute of Neurological Disorders and Stroke (R01-NS099348), and the National Science Foundation (BCS-1441502, BCS-1430087, NSF PHY-1554488 and BCS-1631550). J. I. G. acknowledges support from the National Science Foundation (NSF-NCS 1533623), the National Eye Institute (R01-EY015260), and the National Institute of Mental Health (R01-MH115557). The content is solely the responsibility of the authors and does not necessarily represent the views of any of the funding agencies.

# 8

# Neuronal Morphology and Its Significance

Marcel Oberlaender

## Abstract

Since the days of Ramón y Cajal and Golgi, reconstruction of neuronal morphology has been a central element of neuroscience research. The cell body (soma) and dendrites receive and integrate synaptic input patterns from diverse neuronal ensembles. The axon, in turn, broadcasts the results of this integration process to a variety of neurons within and across brain regions. Morphological differences in the dendritic and axonal shapes are thus closely linked to a neuron's inputs, outputs, computations, and hence functions. Quantification of somatic, dendritic, and/or axonal properties by morphological reconstructions thus represents one of the major approaches to define brain areas and neuronal cell types therein. This chapter addresses some of the technical challenges involved in reconstructing neuronal morphologies and in linking morphology to other properties of the neurons, such as intrinsic physiology and synaptic connectivity. It discusses conceptual challenges involved in using morphological reconstructions for the definition of neuronal cell types, as well as for the identification of neural circuit structure and function.

## Introduction

The term "neuronal morphology" summarizes several of the structural properties of neurons. At the cellular level, somatic, dendritic, and axonal shapes— and the overlap between these post- and presynaptic neurites—determine which neurons can, in principle, be connected to each other. At the subcellular level, the shapes and density of spines and boutons along the dendrites and axons reflect the number and distributions of these contacts. Reconstruction and quantification of neuronal morphology, therefore, provide initial qualitative and quantitative insights into the structural organization of neuronal networks, and is one of the most widely used approaches in neuroscience research to delineate the borders between brain areas, to define neuronal cell types, and to identify neuronal circuits.

Reconstructing the complete morphology of individual neurons remains, however, technically challenging due to:

1. the small dimensions of the neurites, which can have diameters as thin as 100 nm;
2. the elaborate and dense projection patterns, which can reach path lengths of several centimeters even locally within a cortical area; and
3. the large volumes that are innervated by a single axon, which can span from a few cubic millimeters in cortex to the entire brain.

The vast majority of what is known about neuronal morphology originates from incomplete, partial reconstructions acquired from acute or histological brain sections. For example, one predominant approach used to reconstruct a neuron's morphology is to label it *in vitro* with biocytin via patch-clamp recording pipettes in acute brain slices of typically 300–500 μm thickness. These neurons are then typically reconstructed at the resolution limit of light microscopy (LM) using either Camera Lucida based manual tracing software, automated reconstruction routines, or combinations of both.

In addition to the issue of varying tracing accuracy across humans or different algorithms, a major caveat when reconstructing *in vitro* labeled morphologies is the truncation of neurites. Because the brain is cut before the neuron is labeled, only neurites that are contained within the brain section and that remain attached to the soma can be reconstructed. Comparison with neuron morphologies that were labeled in the intact brain (*in vivo*) revealed that depending on the slicing angle, slice thickness, and cell type, approximately 30–50% of the dendrites and more than 90% of the axon will be missing in *in vitro* reconstructions. The issue of truncation also applies to reconstructions of sparsely or densely labeled tissue with electron microscopy (EM) approaches, where the sample and imaging dimensions are typically limited to a few hundred micrometers. As a result, the number of complete neuron reconstructions remains limited and originates primarily from sparse labeling methods (e.g., cell-attached recordings *in vivo*, virus injections, genetic targeting), which are either combined with slicing the brain into consecutive histological sections or with optical clearing methods that allow imaging of large brain volumes via light sheet fluorescence microscopy (reviewed in Kleinfeld et al. 2011).

## Variations in Neuronal Morphology Define Brain Areas

At each level, neuronal morphology displays an enormous variability. These variations are often systematic and correlated with each other. For example, more than a century ago, Brodmann described the variance in the shapes and diameters of neuron somata as a function of cortical depth. These differences correlate with systematic changes in neuron densities along the vertical cortex axis (i.e., from the pial surface toward the white matter), which gave rise to the concept of cytoarchitectonic layers (Brodmann 1909). The neocortex is typically subdivided into six layers (L1–6). However, the specific laminar organization (e.g., number and/or thickness of layers) differs between cortical areas, thus providing a structural criterion to define and delineate between them (Figure 8.1a).

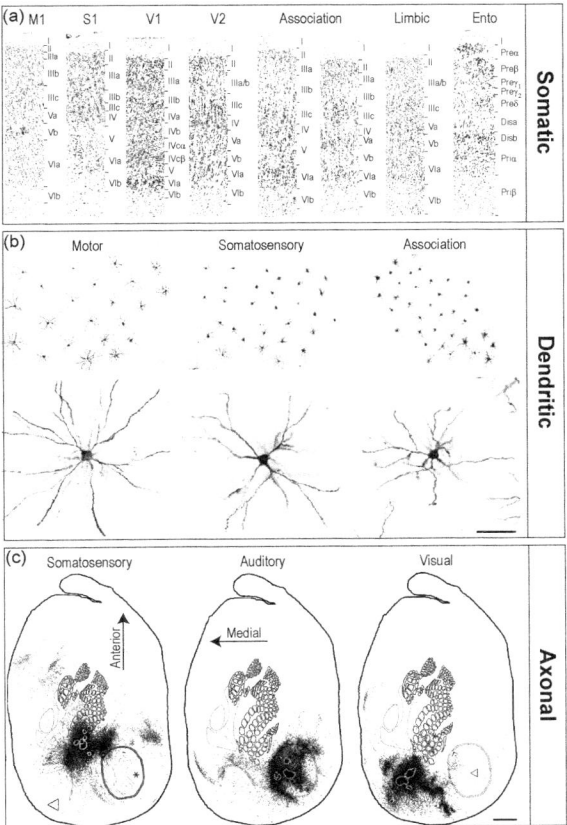

**Figure 8.1** Variations in soma, dendrite, and axon morphology define cortical areas. (a) Nissl-stained coronal sections from human cortex (left to right): primary motor cortex (M1), primary somatosensory cortex (S1), primary visual cortex (V1), secondary visual cortex (V2), association cortex of the inferior frontal gyrus and superior parietal lobule, limbic cortex of the cingulate gyrus, entorhinal cortex. Adapted from Palomero-Gallagher and Zilles (2017). (b) Tangential sections from mouse cortex. Dendritic fields of layer 3 pyramidal neurons (left to right): secondary motor cortex (M2), secondary somatosensory cortex (S2), lateral secondary visual cortex and association temporal cortex (V2L/TeA). Scale bar represents 60 μm. Adapted from Benavides-Piccione et al. (2006). (c) Distributions of intrinsic and extrinsic axons obtained from anterograde tract tracer injections into superficial layers of rat cortex (left to right): vibrissal part of rat primary somatosensory cortex (vS1; i.e., barrel cortex), auditory (A1) and visual (V1) cortex. Scale bar represents 2 mm. Adapted from Stehberg et al. (2014).

At the somatic level, the laminar differences between cortical areas extend to the level of dendrites. In marmoset monkeys, for example, dendritic fields of L3 neurons are smallest in primary visual cortex (V1), increase progressively across the hierarchy of visual areas (e.g., V2, V4), and are largest in the prefrontal cortex (Elston et al. 1999). Similar area-specific differences in dendritic fields have also been reported in other species (Figure 8.1b), such as mice (Benavides-Piccione et al. 2006). Additional dendritic features, such as spine numbers or peak densities, were also shown to vary as a function of cortical area (Elston et al. 1999). Given that dendritic length and spine density reflect the number and subcellular distributions of synaptic contacts, it is likely that such regional variations in somadendritic morphology represent structural correlates of different functional capacities (reviewed in Elston 2003).

The relationship between neuronal morphology and brain area extends to the level axons, both for intrinsic (i.e., within a brain area) as well as extrinsic (i.e., across brain areas) connections. More specifically, long-range intrinsic axons, which can travel lateral distances of multiple millimeters without entry into the white matter (WM), are thought to interconnect neurons across several of the elementary functional units of cortex (cortical columns) in an area-specific manner (Figure 8.1c). For example, horizontal intrinsic axons in the vibrissal part of rodent primary somatosensory cortex (vS1) interconnect barrel columns that represent neighboring facial whiskers within the same row along the snout (Bernardo et al. 1990). In the motor cortex of the monkey (Huntley and Jones 1991), cat (Keller 1993), and rat (Weiss and Keller 1994), horizontal intrinsic axons link regions that activate related groups of muscles. The target regions of extrinsic axons (i.e., via the WM) also depend on the cortical area in which the neurons reside but vary considerably from cell to cell even within the same cortical area (reviewed in Harris and Shepherd 2015). For example, several recent studies revealed that neurons of the same cell type and cortical area can have different *in vivo* functions, which correlate with the specific target regions of their respective long-range axons (Chen et al. 2013; Lur et al. 2016; Rojas-Piloni et al. 2017).

## Variations in Neuronal Morphology Define Cell Types

In addition to defining and differentiating between brain areas, systematic and correlated variations in soma, dendrite, and/or axon morphology are also commonly used to classify neuronal cell types within (and across) brain areas. One of the first approaches to discriminate between cell types in cortex was by classifying their soma morphology. As introduced above, variations in soma shape and size—resembling pyramids, ovoids, or spheres—correlate with vertical changes in soma densities, and hence with a neuron's layer location. Grouping neurons by layers—a widely accepted approach—provides a first-order criterion to discriminate between neuronal cell types, and countless structural, functional, and genetic studies have hence reported their results in a layer-specific manner.

Within each cortical layer, neurons display a variety of dendritic shapes and subcellular dendritic morphologies. Neurons with large, pyramid-shaped somata (pyramidal neurons, PNs) typically represent excitatory cells. In contrast, small, spherical somata typically represent inhibitory cells (INs). PNs are characterized by the presence of an apical dendrite, which projects vertically from the soma across multiple layers toward the pial surface. Both apical and basal dendrites of PNs comprise spines (Figure 8.2a), whose density distributions and shapes can vary between different dendritic compartments and across different types of PNs. Dendrites of INs are more compact and less elaborate than those of PNs; they lack an apical dendrite and have no spines. Instead, INs display bead-like swellings along their dendrites, which often represent postsynaptic structures (Figure 8.2b). However, there are several exceptions to these general rules regarding PN versus IN dendrite morphology. For example, excitatory neurons, referred to as spiny stellates in layer 4, have small spherical somata and lack an apical dendrite. Moreover, INs can have spines along their dendrites.

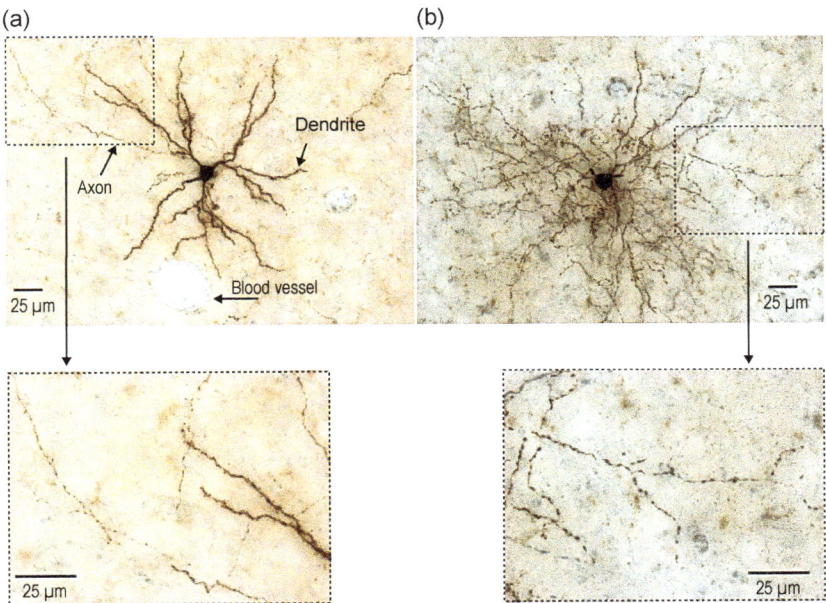

**Figure 8.2**   Soma, dendrite, and axon morphology differ between pyramidal neurons (PNs) and inhibitory cells (INs). Brightfield microscope images of histological sections that were cut tangentially through the vibrissal part of rat primary somatosensory cortex (vS1) show one *in vivo* recorded, biocytin-labeled PN (a) and one fast-spiking IN (b). From Narayanan et al. (2014).

Differences in dendrite morphology within the group of PNs provide several structural criteria to subdivide them into nine major somadendritic types (reviewed in Harris and Shepherd 2015): pyramids in superficial layer 2 (L2py), in layer 3 (L3py) and layer 4 (L4py), star-pyramids in layer 4 (L4sp), thick-tufted (L5tt) and slender-tufted pyramids in layer 5 (L5st), corticothalamic (L6ct), and corticocortical pyramids in layer 6 (L6cc), which include a group of PNs with a variety of rare morphologies, such as those that have an "inverted" apical dendrite that projects toward the white matter (Figure 8.3a). Even though somata of these different somadendritic cell types intermingle within and across layers, the local axon morphologies of PNs display cell type-specific vertical (i.e., across layers) and horizontal (i.e., across columns) projection patterns (Narayanan et al. 2015). For example, L5st PNs in rat vS1 have axonal morphologies that remain largely confined to the dimensions of a single-barrel

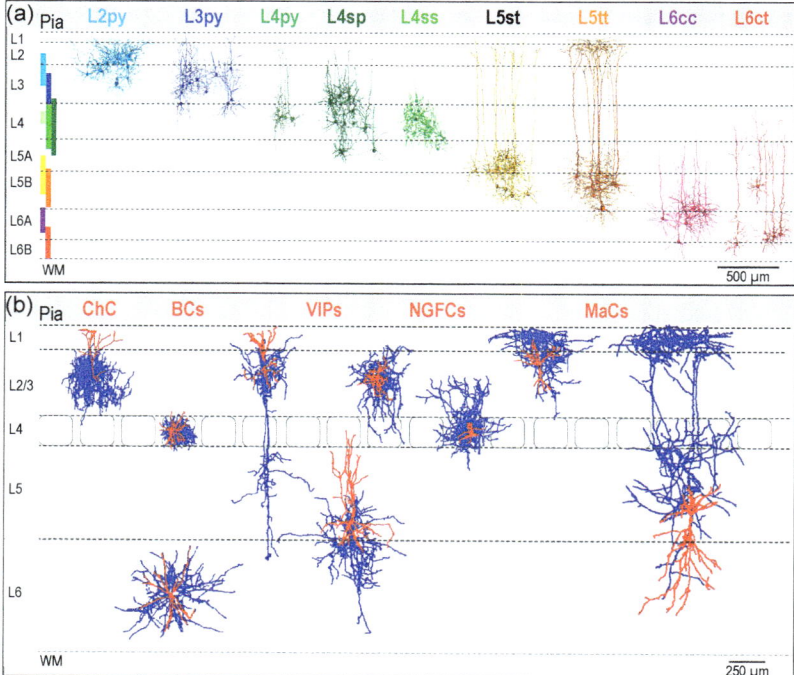

**Figure 8.3**   Variations in soma, dendrite, and axon morphology define cell types. (a) Somadendritic morphologies of the major morphological pyramidal cell types in cortex. Colored bars represent the vertical extents of the respective cell type-specific soma depth distributions in the vibrissal part of rodent primary somatosensory cortex (vS1). Adapted from Oberlaender et al. (2012). (b) Dendritic (red) and axonal (blue) reconstructions of the five major morphological inhibitory cell types in cortex. Example reconstructions from different layers of rat vS1 represent: chandelier cells (ChC), basket cells (BCs), vasoactive intestinal polypeptide cells (VIPs), neurogliaform cells (NGFCs), and Martinotti cells (MaCs). Adapted from Feldmeyer et al. (2018).

column in layer 5, and they project densely to several barrel columns in the superficial layers. In contrast, L5tt PNs have sparse and largely column-restricted axon projections to layers 2/3, but innervate multiple barrel columns in layer 5. The relationship between somadendritic cell type and intrinsic axon projection pattern extends to the long-range axonal targets of PNs (Harris and Shepherd 2015). For example, L5st PNs belong to the class of intratelencephalic neurons, which are defined by long-range extrinsic projections to other cortical (and striatal) areas. In contrast, L5tt PNs represent a class of pyramidal tract neurons defined by projections to subcortical regions.

In contrast to PNs, where soma depth location, dendrite morphology, local axon pattern, and long-range targets are closely related, different morphological types of INs are often exclusively defined by their axon projection patterns and/or specific axonal target structures (reviewed in Feldmeyer et al. 2018) (see Figure 8.3b). For example, chandelier cells (ChCs), typically found within the superficial layers (but also in layer 5), specifically innervate axon initial segments of PNs, which gives rise to their characteristic axon morphologies. In contrast, axons of Martinotti cells, typically found within layer 5, specifically project to layer 1, whereas axons of basket cells innervate somata and proximal dendrites of PNs largely within the vicinity of their own somata. However, because of the much larger variability of IN morphologies compared to those of PNs (even though INs represent less than 20% of all cortical neurons), objective criteria to define IN types and to differentiate between them remain controversial (Petilla Interneuron Nomenclature Group et al. 2008).

## Relating Neuronal Morphology to Function

The difficulty involved in reconstructing complete neurons represents just one of the challenges in trying to assess the significance of neuronal morphology necessary for understanding the basic principles of cortical circuit organization. Several additional properties are involved in determining a neuron's cellular and/or network functions, and need to be measured alongside the reconstruction of its morphology. For simplicity, these properties may be grouped into four categories: intrinsic physiology, *in vivo* activity, synaptic wiring, and genetic profile.

### Morphology versus Intrinsic Physiology

One of the standard approaches to measure the intrinsic physiological properties of individual neurons is to perform somatic, dendritic, and/or axonal whole-cell patch-clamp recordings in acute brain slices *in vitro*. By injecting currents of different shape, amplitude, and/or frequency through the recording pipette, and/or modifying the solution within which the slice is embedded, one can identify which ion channels are expressed in the different morphological compartments and in different cell types. Following the recording, the neurons are labeled and reconstructed as introduced above. Accordingly, a rich literature on the intrinsic physiological properties of morphologically identified neurons was generated and has revealed a variety of relationships between the neurons' "electrical" and morphological cell types. For example, as reviewed by Ramaswamy and Markram (2015), L5tt PNs were shown to possess a $Ca^{2+}$ channel dense region around the first bifurcation point of the apical tuft. Current injections into this $Ca^{2+}$ hot spot can trigger dendritic $Ca^{2+}$ spikes, which in turn can trigger bursts of somatic action potentials (APs)—referred to as BAC firing—when coinciding with a backpropagating AP (bAP). In contrast, L5st PNs typically lack the intrinsic (and morphological) properties to support BAC firing. As a result, L5tt and L5st PNs are often referred to as burst (BS) and regular spiking (RS) cells, respectively. However, not every BS cell in layer 5 has a thick-tufted morphology, nor does every thick-tufted PN elicit bursts (Harris and Shepherd 2015). This example thus illustrates one general caveat of *in vitro* recording/labeling: it is difficult to differentiate between biological variability and variability that is caused by the experimental approach itself. More specifically, truncation of the apical dendrites can result in false morphological classification of L5st and L5tt PNs. Truncation of the axon can transform a BS into a RS cell (Kole 2011).

## Morphology versus *In Vivo* Activity

In contrast to *in vitro* studies, which relate a neuron's intrinsic physiological properties to its morphological properties, measurements of *in vivo* activity patterns for morphologically identified neurons and cell types remain scarce. One standard approach to achieve these measurements is to perform whole-cell or cell-attached patch-clamp recordings from single neurons *in vivo* and to label the recorded neurons (e.g., with biocytin) for post hoc reconstruction. Such studies in the V1 of cat (Binzegger et al. 2004) and mouse (Vélez-Fort et al. 2014) have revealed several relationships between a neuron's morphological cell type and specific *in vivo* functions. Moreover, *in vivo* recording/labeling approaches were recently combined with injections of retrograde tracer agents, which provide additional information about the input populations (Vélez-Fort et al. 2014) and/or target structures (Rojas-Piloni et al. 2017) of the recorded and reconstructed neurons. *In vivo* recording/labeling approaches are thus capable of providing information about a neuron's soma location, dendrite morphology, local axon projection pattern, long-range axonal targets, input populations, intrinsic physiology (i.e., current injections during whole-cell *in vivo* recordings), and *in vivo* functions (e.g., during sensory stimulation). However, the number of neurons that can be recorded per animal is limited to just a very few within the same brain area. Hence, alternative approaches aim to link morphological properties to *in vivo* functions by combining population $Ca^{2+}$ imaging with EM reconstructions (Bock et al. 2011). Similar to the *in vivo* electrophysiology approaches described above, $Ca^{2+}$ imaging has also recently been combined with retrograde tracer injections (Chen et al. 2013). Thus, in principle, $Ca^{2+}$ imaging, when combined with post hoc EM reconstructions, can yield structure-function measurements similar to those from *in vivo* recording/labeling approaches, but for larger populations and with synaptic resolution.

## Morphology versus Synaptic Wiring

In addition to the dense EM approaches introduced above, several other methods are commonly used to study synaptic wiring between morphologically identified neurons. Most prominently, simultaneous patch-clamp recordings *in vitro* from multiple (up to eight) neurons are combined with post hoc reconstructions and identification of putative synaptic contact sites at the resolution limit of LM. In some cases, the putative contacts are confirmed as synapses by correlating LM and EM reconstructions. However, the impact of truncating dendrites and axons on such connectivity measurements will increase with the number of recorded neurons. Another important aspect for linking connectivity patterns and morphology to a neuron's function is the measurement of spatiotemporal synaptic input patterns during *in vivo* conditions. Recent technical advances that allow imaging of $Ca^{2+}$ hot spots (putative synapses) at several dendritic locations, while simultaneously measuring the somatic response to different sensory stimuli (Jia et al. 2010), have provided remarkable structural and functional data in a variety of sensory systems and species. Here, dendritic events were generally found to be more broadly tuned than the somatic responses, and that inputs with different stimulus preferences intermingle both spatially and temporally along the dendrites. These measurements hence provide the first direct insight into how neuronal function arises from a complex interplay of parameters at subcellular, cellular, and network scales; that is, between intrinsic physiology, dendrite morphology, synaptic wiring, and population activity.

## Morphology versus Genetic Profile

In recent years, the focus of classifying neurons has shifted from their morphological and physiological properties toward their genetic and/or molecular profiles. For example, neurons expressing three markers—the $Ca^{2+}$-binding protein parvalbumin, the neuropeptide somatostatin, and the ionotropic serotonin receptor 5HT3a—were shown to label three disjoint populations of cortical INs (Rudy et al. 2011). This discovery provided access to study and manipulate IN circuit function *in vitro* and *in vivo* via optogenetic approaches. However, several recent studies revealed that an IN's molecular marker correlates only weakly, if at all, with its morphological and electrical properties (see Tremblay et al. 2016). Revealing the significance of a molecular marker for an IN's cellular and circuit function thus remains a highly active field of research. For PNs, such a "standard set" of genetic or molecular markers has not been established so far. However, several markers exist that are capable of labeling subsets of neurons that share similar structural and/or functional properties. Moreover, gene sequencing and bioinformatics approaches applicable to morphologically and physiologically characterized neurons have recently become available. Thus far, however, these approaches have failed to reveal sets of genetic markers that, for example, correlate with a PN's long-range axonal target (Sorensen et al. 2015).

## Significance of Neuronal Morphology

To discover the structural and functional organizational principles of the nervous system, it is essential to reconstruct neuronal morphologies for the following reasons. First, a neuron's soma location, dendritic shape, and axonal projection pattern determines the pre- and postsynaptic populations to which a neuron can, in principle, be connected. Second, its morphology combined with its intrinsic properties—as defined by a variety of voltage- and ligand-gated ion channels that are expressed differently in the soma, dendrites, and axon, and which are often restricted to specific dendritic and/or axonal sub-domains—determine how a neuron integrates, transforms, and transmits synaptic input patterns. Third, its morphology and intrinsic properties combined with the specific spatiotemporal organization of synaptic input patterns—as defined by the wiring diagram and (stimulus- and state-dependent) population activity—determine how a neuron computes, for example, during sensory stimulation.

Consequently, the significance of neuronal morphology for neuronal and network functions can only be revealed once reconstructions of complete morphologies are complemented with measurements of their genetic, intrinsic, and functional properties as well as with dense reconstructions of their synaptic in- and output patterns. As discussed above, approaches that would allow us to measure all of these properties simultaneously are presently unavailable. Therefore, the major significance of morphological reconstruction for neuroscience research at the moment may be to provide a way to correlate measurements of different cellular and network properties—and at different scales—in a cell type-specific manner. Several collaborative efforts and large-scale initiatives, such as the MindScope Project of the Allen Brain Institute, aim to collect across-scale structural and functional data systematically and consistently for a particular sensory modality, such as the mouse visual system.

Arguably, one of the major challenges for such integrative approaches lies in the classification and identification of neuronal cell types, which ideally represent the canonical elements that are sufficient to describe a circuit and its functions. The large biological variability and differences in fundamental neuronal properties between species (Kalmbach et al. 2018), however, renders the definition of "meaningful" structural and functional features that can be used by classification algorithms as challenging. For example, classification of dendritic cell types is often based on a variety of custom-defined morphological (e.g., path length), topological (e.g., branch depth), and/or shape parameters (e.g., bounding box). Such differences in how morphological reconstructions are parameterized, as well as differences between the methods used to cluster the resultant branching statistics, hamper objective definitions and comparisons of morphological cell types across studies. Similarly, a standard set of intrinsic physiological features for the classification of electrical cell types has not been established. To overcome such inconsistencies, and to establish sets of "meaningful" parameters that reliably distinguish between cell types, generative models that produce synthetic morphologies (e.g., that preserve electrotonic properties of the dendrite) are currently being developed (Cuntz et al. 2010).

To reach a consensus for classification, one promising way is to generate publically available databases, such as NeuroMorpho.org which presently comprises thousands of single-neuron morphologies acquired in different species and labs. The International Neuroinformatics Coordinating Facility, Human Brain Project, Allen Institute, and others are also in the process of generating similar "big data" neuroinformatics platforms. At present, however, these large-scale efforts are based primarily on *in vitro* measurements. As discussed above, experiments in acute brain sections are hampered by severe truncation of dendrites and axons, and thus introduce variability to measurements of neuronal morphology, intrinsic properties, and synaptic connectivity which may exceed, or at the very least increase, the true biological variability. Moreover, depending on the dendritic and axonal extents of the investigated cell types, the effects of truncation may vary substantially across experiments in a largely unpredictable manner. It thus remains to be seen whether large-scale databases of *in vitro* acquired data will facilitate or hamper the identification of the elementary building blocks of neuronal circuits.

Even if the variability caused by truncation could be minimized (e.g., by *in vivo* recording/labeling), direct clustering of morphological, intrinsic physiological, and/or connectivity features may still be unsuitable for the definition of cell types. More specifically, some evidence suggests that the various cellular and network properties which can be used to parameterize neurons are related to each other (Marder and Goaillard 2006). For example, homeostatic mechanisms could compensate for morphological differences across neurons by adjusting their intrinsic properties (Figure 8.4a), so that all neurons of a particular cell type show similar functions. A cell type would thus be defined by a set of specific functional behaviors that arise from complex relationships between morphological, intrinsic physiological, and synaptic properties in high-dimensional parameter spaces (Figure 8.4b). Thus, reliable measurements of variability across neurons—and of parameter distributions in general—represent

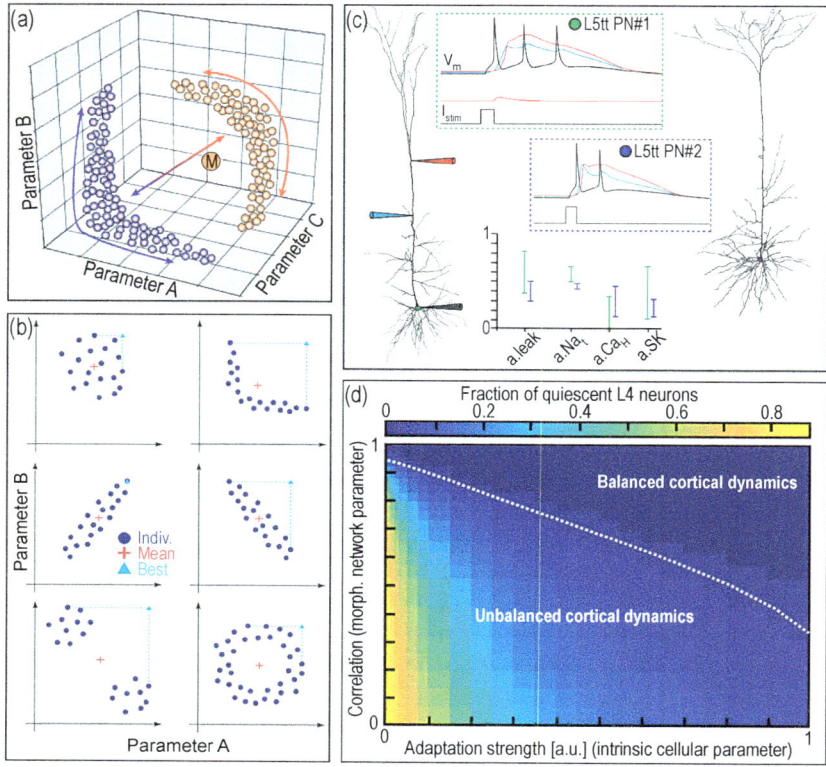

**Figure 8.4** Relationships between morphology, intrinsic physiology, and synaptic connectivity define cell type-specific functional behaviors. (a) Individual neurons that display two different types of functional behaviors are shown in purple and orange, respectively. Both populations comprise neurons with widely different values of three parameters (e.g., conductance values for three different ion channels). Neuromodulators could move neurons from one behavior to another (gradient arrow). Adapted from Marder and Goaillard (2006). (b) Example distributions of parameters for neurons that share a common behavior or set of behaviors. Adapted from Marder and Taylor (2011). (c) Morphologies of two thick-tufted (L5tt) pyramidal neurons (PNs) from the vibrissal part of rodent primary somatosensory cortex (vS1). Both L5tt PNs show qualitatively the same set of functional behaviors, for example BAC firing (dashed boxes). Morphological differences across L5tt PNs result in different active dendritic properties, as illustrated by the respective ranges of four exemplary conductance values (i.e., normalized) along the apical dendrites (a) that represent acceptable models that reproduced BAC firing of neuron #1 (blue) and #2 (green), respectively. Adapted from Hay et al. (2011). (d) Example of how relationships between morphology-related network properties (here: correlations in synaptic connectivity) and intrinsic cellular mechanisms (here: spike-frequency adaptation) can assure a circuit's proper behavior. Breakdown of balance is reflected by unrealistically sparse and temporally regular firing. Adapted from Landau et al. (2016).

an additional requirement for the definition of cell types. Alternatively, numerical simulations of biophysically detailed multi-compartmental (MC) neuron models, when combined with algorithmic approaches that allow exploration of large multidimensional parameter spaces (Marder and Taylor 2011), may also provide the general relationships between parameter distributions required to assure a specific cellular and/or circuit function.

The need for such multidimensional structure-function approaches for cell type classification can be illustrated by two examples. First, in a study by Hay et al. (2011), L5tt PNs from rat S1 were physiologically characterized in response to a set of somatic and/or dendritic current injections. These neurons were then reconstructed and converted into MC models (Figure 8.4c). The conductance values of the various ion channels along the MC model were then tuned until numerical simulations reproduced both responses to somatic and dendritic current injections. The conductance value distributions that were obtained differed substantially across L5tt PNs. Thus, despite very similar input-output behaviors, the morphological differences between L5tt PNs needed to be compensated by differences in their intrinsic properties. Second, simulations of morphologically simpler models, consisting of networks of integrate-and-fire point neurons, have recently been shown to reveal some of the general relationships between cellular and network properties that are required to assure a circuit's proper dynamics (Landau et al. 2016). Specifically, Landau et al. showed that a specific morphology-related feature, heterogeneity in incoming connectivity, has a significant qualitative impact on cortical dynamics and that the circuits' proper function depends on the interplay between connectivity structure and single-neuron intrinsic properties (Figure 8.4d).

## Conclusion

As efforts are made to elucidate the general significance of neuronal morphology and to understand the basic principles that underlie the structural and functional organization of cortical circuits, we must confront an array of issues:

- Is there a canonical circuit for the networks in different cortical areas and species?
- Is the concept that cortical columns represent elementary functional modules justified?
- Do cortical layers reflect different computational functions?
- Are there overarching principles of how noncortical processors are connected to the cortex?

In this chapter I have presented background information necessary to approach these questions. First, I introduced how systematic and correlated variations in soma, dendrite, and axon morphology allow us to define and delineate between different cortical areas and cell types. Second, I discussed some of the present technical limitations for reconstructing individual and complete neuronal morphologies. Finally, I argued that revealing the significance of neuronal morphology requires complementing neuron reconstructions with measurements of intrinsic physiology, *in vivo* activity, and (dense) synaptic wiring.

At present, experimental approaches to measure all of these properties at once are not available. Thus, consideration was given to the challenge of creating integrative approaches capable of combining data from different experimental approaches in a cell type-specific manner. In conclusion, numerical across-scale simulations provide a promising venue to explore how and to what extent the details of neuronal morphology, intrinsic physiology, and synaptic connectivity are related to each other, and how these relationships affect the dynamics and functions of cortical circuits.

# 9

# Functional Architecture
# of the Cerebral Cortex

David A. Leopold, Peter L. Strick, Danielle S. Bassett,
Randy M. Bruno, Hermann Cuntz, Kristen M. Harris,
Marcel Oberlaender, and Marcus E. Raichle

## Abstract

Recent research in the neurosciences has revealed a wealth of new information about the structural organization and physiological operation of the cerebral cortex. These details span vast spatial scales and range from the expression, arrangement, and interaction of molecular gene products at the synapse to the organization of computational networks across the whole brain. This chapter highlights recent discoveries that have laid bare important aspects of the brain's *functional architecture*. It begins by describing the dynamic and contingent arrangement of subcellular elements in synaptic connections. Amid this complexity, several common neural circuit motifs, identified across multiple species and preparations, shape the electrophysiological signaling in the cortex. It then turns to the topic of network organization, spurred by routine capacity for noninvasive MRI in humans, where interdisciplinary tools are lending new insights into large-scale principles of brain organization. Discussion follows on one of the most important aspects of brain architecture; namely, the plasticity that affords an animal flexible behavior. In closing, reflections are put forth on the nature of the brain's complexity, and how its biological details might be best captured in computational models in the future.

## Introduction

The human cerebral cortex consists of approximately 16 billion neurons (Herculano-Houzel 2009), whose integrated activity supports not only our

**Group photos (top left to bottom right)**   David Leopold, Peter Strick, Randy Bruno, Kristen Harris, Marcus Raichle, Marcel Oberlaender, Hermann Cuntz, Danielle Bassett, Marcus Raichle, Marcel Oberlaender, Peter Strick, David Leopold, Danielle Bassett, Hermann Cuntz, Marcel Oberlaender, Kristen Harris, Danielle Bassett, David Leopold, Randy Bruno, Kristen Harris, Peter Strick

higher thoughts but also our sensory perceptions, verbal communication, and complex motor actions. While the number of neurons is clearly important, it is their organization and interconnections that determine the functional principles of a working brain. From one perspective, all mammalian brains have the same basic design, which includes a layered cerebral cortex governed by highly conserved developmental constraints (Workman et al. 2013). Mammalian species differ markedly, however, in brain size, peripheral sensory adaptations, and evolved ecological specializations. These differences strongly influence the brain organization of different taxa, with one pertinent example being the relatively dense packing of neurons into the primate cerebral cortex (Herculano-Houzel 2012). The principles of cerebral cortical architecture in primates and other mammals are simultaneously manifest at multiple scales: from the synaptic microenvironment, to local circuit motifs, to large-scale brain networks measured using methods such as functional magnetic resonance imaging (fMRI). At each scale, strong genetic determinism is complemented by modification through experience, which manifests both during early development and in the adult. In recent years, neuroscientists have learned a great deal about how flexibility in function can be superimposed upon an ostensibly fixed anatomical scaffolding.

Here, we take on the task of identifying key elements of cortical architecture that shape its basic functioning, plasticity, and capacity to drive flexible behavior. This review reflects our discussions at the 27th Ernst Strüngmann Forum—the third in a series of meetings spaced out over several decades. Given the amount of research that transpired since the initial meeting (Rakic and Singer 1988), we highlight new concepts and discoveries that have emerged pertaining to the cerebral cortex, its organization, and function. We focus on recent discoveries regarding synapse formation, local and long-range functional connections, and network organization at multiple scales. Where possible, we place findings in a historical context and direct readers to recent reviews on other important features of cortical architecture: prominent laminar organization of the cortex (Palomero-Gallagher and Zilles 2017), its columnar microcircuitry (Bastos et al. 2012), and its intimate and mysterious functional relationship to other prominent brain components, such as the thalamus (Sherman 2017).

We begin by reviewing the points of articulation between neurons, including biophysical and physiological features of the dendritic microenvironment that promote certain modes of information transmission. We then investigate the structural and functional connectivity between distant areas of the cerebral cortex, which is a field of study that has come to utilize the brain's spontaneous activity. Thereafter we highlight the inherent flexibility of the cerebral cortex, from experience-dependent changes during early development to adult learning and memory. We conclude by briefly considering the importance of computational and evolutionary frameworks in shaping our future conceptions of cortical functional architecture.

## The Essence of a Neural Connection

Our current understanding of neural communication is grounded in the neuron doctrine, which is seen as the resolution of a nineteenth century debate between Ramón y Cajal and Golgi about whether neurons were interconnected through directed lines or as a broad syncytium (Bock 2013). Neurons are cells specialized to transmit information quickly through electrochemical signals that traverse a range of spatial scales. Individual neurons usually communicate through chemical cell-to-cell contacts, or synapses. The cartoon rendition of the neuron is familiar to all students of neuroscience: dendrites emerge from a cell body and a long axon makes synaptic connections with another neuron's dendrites. Unsurprisingly, the structure of real neurons is much more complex and variable than textbooks typically portray, and the physiology of synaptic connections is highly contingent on factors playing out over many spatial scales (Figure 9.1). Our knowledge about these details is growing at a rapid pace. Historically, our picture of neuronal physiology was strongly shaped by action potentials acquired in single-unit recordings. These clean and discrete pulses might suggest a brain that works by digital computation, perhaps reflecting the contemporary metaphor of the brain: the computer. However, the core of the brain's information processing is arguably its analog physiology, including the electrical, chemical, and genetic mechanisms that control the synaptic interconnections at a range of timescales.

Synaptic neuronal connections are diverse and commonly involve articulation between presynaptic axons and postsynaptic dendrites. Postsynaptic neurons integrate a massive and uneven array of axonal inputs, often stemming from diverse cell types. In the cortex, synapses onto dendritic spines have been the focus of much study, since the morphology of spines changes readily in the adult brain. These constant changes are thought to alter synaptic efficacy in the service of network plasticity and learning. Tracking the fate of individual spines can seem hopeless, given that there are estimated to be $10^{14}$ spine synapses in the human cerebral cortex (Matus 2009). Nonetheless, the principles governing their formation, retreat, or enlargement may be among the most important windows into how the adult brain remains adaptable and, in a sense, youthful: with spines, the brain's capacity for experience-dependent development seems endless.

Recent discoveries have emphasized the high specificity of synapse formation, the importance of neighboring synapses, and a number of commonly occurring synaptic and circuit "motifs" through which certain functional computations are achieved (Jiang et al. 2015). Multiple modes of plasticity are built into synaptic connections, and these adhere to complex learning rules and are subject to a wide range of cognitive and chemical contingencies. In recent years, researchers have attempted to gain a more holistic understanding of the dendritic microenvironment, its margins for plasticity, its modulation

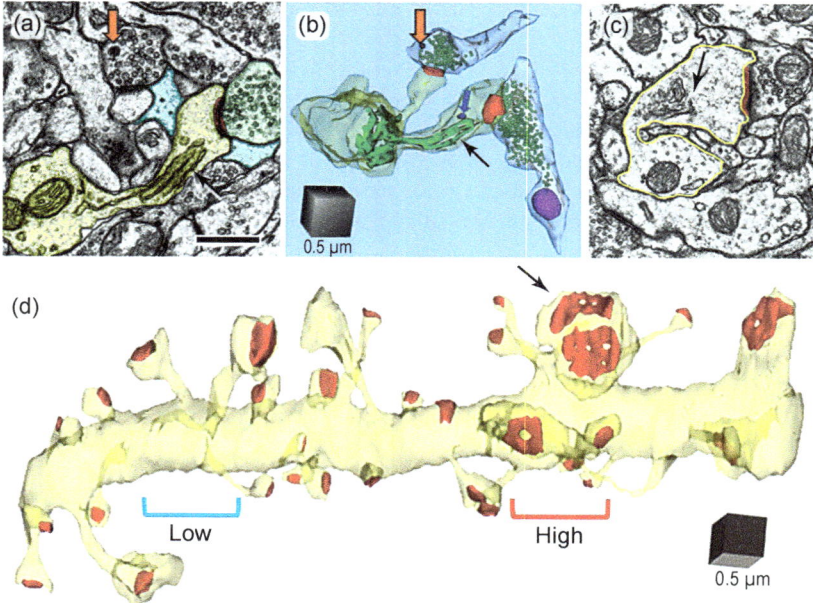

**Figure 9.1**   How do nonuniform distributions of subcellular resources in dendrites and axons influence where synaptic growth, maturation, and plasticity occur? (a) Single section electron micrograph and (b) three-dimensional reconstruction from serial section electron microscopy (3DEM) reveal extremes in the diversity of synapse size and composition between neighboring synapses on the same dendrite with different presynaptic partners. Dendrite (yellow), axon (translucent green), glia (blue), postsynaptic density (PSD, red), smooth endoplasmic reticulum forming a spine apparatus (black arrow), presynaptic dense core vesicle (orange arrow), large vesicle (blue arrow), and presynaptic mitochondrion (purple). (c) Single section through a spine with a spine apparatus (black arrow) and (d) 3DEM of the dendritic segment showing how synapses cluster even along short dendritic segments. The largest spine (arrow) along this segment contains a spine apparatus and the density of dendritic spine synapses surrounding it is high relative to other regions of the same length, where density is low. Scale bar is 0.5 µm.

by external factors, and the coordination of a dendritic tree's branches to issue action potentials in the parent cell.

## Synaptic Specificity

The basic structure of dendritic spines and synaptic densities has been long known, as early electron microscopy studies unveiled the basic structural microcomponents of neural connections (Guillery and Ralston 1964). However, what has come as a surprise in recent years is the dynamic regulation and fine-tuning of these connections. Local processes, governed by complex genetic networks and shaped by electrochemical activity patterns, continually adjust the positions, strengths, and types of synaptic connections. Ultimately, these adjustments determine, at any moment in time, how individual postsynaptic neurons integrate inputs from varied sources and ultimately issue action potentials. These transformations, in turn, define the temporally precise analog computations performed by a local patch of the cerebral cortex.

With the rise of genetic tagging of circuit elements in the mouse, researchers are moving quickly to understand the nature of synaptic specificity in a cortical column. Perhaps the best example of progress in this area pertains to the mechanistic role of various interneuron subtypes in the cortex. Inhibitory interneurons have long been recognized as important, morphologically diverse elements that serve to balance runaway pyramidal cell excitation. However, it is only in the past two decades that their molecular signatures have allowed for in-depth study. One particularly important finding has been that, unlike their pyramidal cell partners, they do not stem from a cortical origin but rather migrate tangentially into the cortical plate along multiple routes (Anderson et al. 1997). More recently, the pattern of genetically specified synaptic contacts of different interneuron subclasses has been elaborated in great detail. The spatial distribution of interneuron inputs along a cortical pyramidal cell is strongly specified by the interneuron subclass, and by extension its developmental origin and epigenetic state. Different classes of interneurons participate in blanket weak inhibition, targeted strong inhibition, and disinhibition (Kepecs and Fishell 2014). To a first approximation, these compartmentalized GABAergic synapses in the cortex are contributed by local interneurons (with some exceptions, such as the long-range GABAergic projection neurons originating in the basal forebrain). Much current work is attempting to establish when and how genetically specified interneuron subclasses find their cortical positions during early development as well as how synapse formation is regulated based on neural activity and experience (Wamsley and Fishell 2017).

By contrast, excitatory pyramidal cells receive hundreds or thousands of excitatory inputs from a varied combination of local and remote neurons. Local interconnectivity among cortical pyramidal neurons is relatively sparse, with only a tiny fraction of connections showing strong excitatory input (Lefort et al. 2009). It is important to note that neurons whose axonal and dendritic arborizations show a high degree of three-dimensional spatial overlap need not be interconnected, as the principles of pyramidal cell connectivity depend on more than spatial proximity (Mishchenko et al. 2010). While much has been learned about how cells find one another and make connections, the ultimate determinants of arborization for individual neurons or classes of neurons remain mysterious (Narayanan et al. 2017; Han et al. 2018). One fascinating observation is that growing axons appear to select their synaptic partners, in some cases, based on the projection *target* of a potential postsynaptic neuron. Such selective targeting has the interesting consequence that neurons with the same output targets tend to gather similar types of axonal inputs, and thus share their physiological response properties.

Neurons are very particular in their connections, with researchers gradually amassing a complex set of rules describing how, where, and when neural subclasses form synapses with one another in the cortical microcolumn. Let us thus take a closer look at the structure of the dendritic microenvironment, where recent experiments highlight the exquisite cellular mechanisms that enable dendrites to serve as the microscopic engines of computation and learning.

## Smart Dendrites

Dendrites, like many of the brain's elemental structures, were initially misunderstood to be simpler than they are. Dendrites were long thought to act by computing the weighted sum of proximal and distal synaptic inputs through passive electrotonic conduction. The potentials detected at the cell body then determined the digital firing of action potentials of the neuron. However, research over the past years has demonstrated that this passive and capacitive view of dendrites is inaccurate. First, cortical dendrites are replete with active currents that propagate action potentials. Dendritic action potentials are sometimes generated locally and sometimes propagated backwards from the soma. Their discovery revealed a new dimension for how neural signals are integrated. Furthermore, the precise morphology and compartmentalization of the local dendritic microenvironment can strongly affect dendritic function, with recent work showing that these aspects of dendritic structure are constantly under renovation (Bourne and Harris 2012). Over time, the invisible hand of experience-dependent learning actively remodels local spine morphology: it adjusts synaptic strength and influences the postsynaptic neuron through multiple chemical and genetic pathways. Thus dendrites are now recognized as a bed of neural computation that far exceeds what was originally envisioned through the rules of electrotonic conduction, initially conceived by Rall (London and Häusser 2005).

One important principle of dendritic organization appears to be structural and functional optimization, which goes a long way in accounting for dendritic and axonal shapes and lengths (Chklovskii 2004). Dendrites optimize the amount of resources, such as cable length, and optimally enforce short conduction times and efficient current transfer from synaptic signals toward the soma (Cuntz et al. 2010). The close relationship between anatomical features and principles of connectivity serves as the basis for a large number of compartment models that accurately account for neuronal electrophysiology (Hines et al. 2004).

Electron microscopy has offered deeper insights into the complex cell biology of dendrite remodeling (Bourne and Harris 2012). Creating spines, selecting axonal partners, and adjusting synaptic strengths all require a systematic redistribution of local subcellular resources, including plasma membrane, structural and metabolic molecules, and cellular organelles such as mitochondria. Ultimately, it is through the parallel microscale adjustment of these elements within trillions of subcellular microcosms that the brain is able continually to tune and update its analog computations to support flexible cognitive and executive functions.

One surprise from recent years is that connections on the distal dendrites can be just as effective in driving the postsynaptic cell as those on proximal dendrites, contrary to conventional wisdom (Bromer et al. 2018). This may be for the simple reason that dendrites can actively adjust the strength of a synapse in any location, easily overriding the natural biases due to electrotonic conduction in a canonical dendrite model. In fact, investigations into this matter using electron microscopy (EM) suggest that synapses at distal dendrites are systematically larger than those near the soma. A look at a dendritic segment through EM reveals an uneven distribution of synapses of all sizes, which again is thought to reflect the specificity of synaptic connections (Figure 9.1). However, despite the nonuniformity of cellular resources, systematic investigation reveals a tight relationship between the size of a postsynaptic surface and the number of presynaptic vesicles and the presence or absence of a presynaptic mitochondrion (Figures 9.1a, b). The rules governing synaptic modification are not well understood, and those rules that have been well characterized pertain only to a subset of synapses. In the hippocampus, it has been observed that approximately 5% of spines are eligible to undergo changes in their synaptic size over time. These regional differences in plasticity speak further to the specificity of interconnections and may become important as we learn more about the learning principles that govern changes in circuit operation.

Finally, the spines themselves, which are abundant on many cortical pyramidal cells, are also highly structured, varied, and subject to morphological change (Bailey et al. 2015). Some large spines contain a so-called *spine apparatus*—an organelle involved in calcium regulation, protein and lipid trafficking, and posttranslational modification of proteins. Small spines lack this apparatus and have fewer of these resources. Some small or large spines contain a presynaptic dense core vesicle known to transport active zone proteins and vesicles between synapses. It is also important to point out that while spines have some degree of independence, the size/resource principle can extend beyond individual spines. For instance, on a given dendrite, the density of synapses is higher surrounding a large spine containing a spine apparatus (Figure 9.1c), compared to other regions where there is no proximal spine apparatus (Figure 9.1d). Future investigation should reveal the extent to which local subcellular resource allocation is a general principle that determines where synapses form, stabilize, and undergo plasticity across dendritic arbors, cell types, and brain networks. Ultimately, refined markers are needed to determine how synapses, when activated, actively redistribute resources during behavior and learning.

**Circuit Motifs**

With the collection of myriad observations about individual synapses, dendritic environments, and circuit contexts, the scientific community has come to discern recurring patterns or "motifs" which present themselves commonly. These motifs often refer to patterns of connectivity that specify certain physiological computations and, in some cases, extend to larger principles of cortical architecture and operation. These advances have been facilitated greatly, albeit not exclusively, through the advent of paired recording studies *in vitro* and transgenic techniques for targeting specific cell types, the latter driven primarily by the genetic tractability of circuits in the mouse. A now classic example, found throughout diverse parts of the nervous system, is the strong feedforward inhibitory circuit, which has been studied extensively *in vitro, in vivo*, and *in silico* (reviewed in Bruno 2011). The key ingredients of this motif are that a group of presynaptic neurons excite a downstream population of interconnected excitatory neurons and inhibitory neurons but provide greater drive to the inhibitory population (Figure 9.2a). Strong disynaptic inhibition then favors the propagation of signals encoded in the synchrony, rather than the absolute firing rates, of the presynaptic neurons. Another example that has gained much attention in recent years is a disinhibitory motif whereby excitatory synaptic input or nicotinic modulation of vasoactive intestinal peptide (VIP) inhibitory cells is effectively able to activate layer 2/3 pyramidal neurons via suppression of somatostatin inhibitory cells (Figure 9.2b). This motif has now been implicated in state-dependent modulation of sensation as well as in learning (Letzkus et al. 2011; Lee et al. 2013b; Pfeffer et al. 2013; Kepecs and Fishell 2014).

Consistent architectural features of columnar circuitry in mammalian neocortex have also become increasingly clear. These have been examined most closely in rodent sensory cortex (Figure 9.2c) but several aspects generalize across species and cortical areas. For instance, primary thalamic relay nuclei of at least the visual, somatosensory, auditory, and motor systems bifurcate to arborize in the middle (largely intracortical) layers and more sparsely at the border of layers 5 and 6 (both intracortically and subcortically projecting layers) in all mammals. This may allow some functional independence of the upper versus deep layers (Constantinople and Bruno 2013; Pluta et al. 2015). In contrast, secondary thalamic nuclei mainly innervate layer 5A and layer 1, which is also a target of other cortical regions and may exploit dendritic nonlinearity to enable top-down control (Larkum 2013). Layer 6 provides corticothalamic feedback but also targets excitatory and inhibitory cells of other layers and has been suggested as an important means to control circuit gain (Olsen et al. 2012; Vélez-Fort et al. 2014). Pyramidal neurons in layers 2/3 and 5 have extensive interconnections (omitted from Figure 9.2c for clarity). The degree to which these excitatory networks comprise motifs that depend on more than first-order connectivity statistics is a major area of active research.

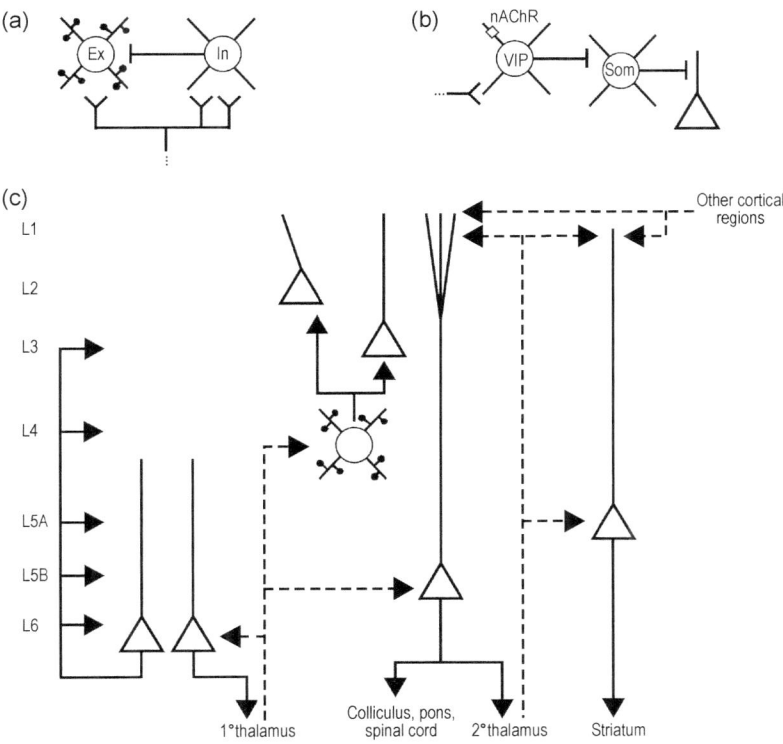

**Figure 9.2** Examples of cortical circuit motifs. (a) Schematic of *feedforward inhibition*, where incoming information impinges in parallel on excitatory (Ex) neurons and local interneurons inhibiting the excitatory neurons (In). (b) Schematic of *disinhibition*, where excitatory input or neuromodulation stimulates one family of inhibitory interneurons (VIP), which inhibits another family of inhibitory interneurons (Som), synapsing on a pyramidal neuron, altogether resulting in the excitation of the pyramidal neuron. (c) Canonical patterns of input and output as well as arborization to different cortical layers.

## Summary

The connections among neurons are at the heart of understanding information processing in the cerebral cortex. The twentieth-century metaphor of a *wiring diagram* fails to take into account the remarkable complexity surrounding synapses, dendrites, and functional specificity among genetically specified cells comprising circuit motifs. In some ways, synaptic microenvironments are more like living ecosystems, in which a panoply of neurites, organelles, and genetic instructions cooperate and sometimes compete for resources. At the same time, an understanding of the cerebral cortex cannot rely solely on microscopic structure and function, since much of the brain's architecture hinges on evolved cortical areas, long-range connections, and interplay with more primitive subcortical structures. Together, their operation as a coordinated, single entity is at the heart of brain function. We next discuss the concept of large-scale functional connections, specifically how they are investigated and summarized as organized networks distributed across the cerebral cortex.

## Functional Connections in a Restless Brain

Investigating the large-scale organization of the brain involves a different set of questions and tools, and dates back to the nineteenth century. Most of the early neuroanatomists assessed cortical organization through postmortem methods, revealing histological subdivisions (Brodmann 1909), gross fiber bundles (Curran 1909), and specific patterns of fiber degeneration following a lesion (Nauta and Gygax 1954). Through painstaking work, a portrait of the brain's anatomical connectional skeleton gradually took form and, for the cortex, is perhaps best summarized in the diagrams constructed by Felleman and Van Essen (1991). These diagrams put forth a hierarchy of cortical areas based in part on laminar patterns of inter-areal projections.

A handful of early physiological studies also appropriated the brain's signaling capacity to study its large-scale organization. For example, in chemical neuronography, small amounts of the neuroactive agent strychnine were applied to a given cortical site in an experimental animal instrumented with large electrocorticography arrays. The depolarizing action of the strychnine led to voltage deflections in a subset of cortical surface electrodes, thus revealing which areas received axonal connections from the stimulated site (Pribram and MacLean 1953). Chemical stimulation was gradually replaced by "electroanatomy," where the effects of electrical stimulation at one location were assessed at other locations across the brain (Miller and Bloomfield 1983). Together with the systematic investigation of brain circuitry through lesions and electrophysiological recordings, the concept of *functional anatomy* gradually emerged, placing emphasis on the large-scale organizational principles of networks in the brain, and particularly the cerebral cortex.

### Structured Spontaneous Activity

An unexpectedly fruitful source of neural signals with which to study brain organization has been spontaneous activity. Traditionally viewed as a nuisance background signal, little attention was paid to the spatial organization of spontaneous activity until the 1990s. Then, two different brain imaging methods abruptly increased neuroscientists' respect for this ongoing background activity: brain imaging in anesthetized animals and in awake humans. Optical imaging methods in the anesthetized cat showed that spatial patterns of spontaneous activity in the visual cortex followed the pattern of orientation preferences present in the local functional architecture (Arieli et al. 1995; Kenet et al. 2003). Around the same time, human fMRI studies demonstrated that spontaneous hemodynamic fluctuations in subjects, in the absence of any task, showed correlated activity within established functional networks (Biswal et al. 1995). The brain's ongoing signals, it appeared, could be harnessed as a tool for studying the layout of its functional networks. In the two decades that followed, this approach contributed significantly to, and in some ways even came to dominate, the study of the human brain.

Analyzing spontaneous activity forces researchers to depart from conventional experimental paradigms, in which brain responses are typically locked in time to stimuli or actions. During the resting state, brain organization is characterized in terms of the internal statistical dependencies of neurons or voxels at different spatial positions. In the simplest case, this involves computing the temporal correlation of a signal measured at one location with all other simultaneously recorded locations, rendering a brain-wide map. Tools to formalize terms and concepts related to neural interactions were initially developed in the context of single unit electrophysiology (Gerstein and Aertsen 1985). This formalism, summarized and expanded by Friston et al. (1995), set the stage for thousands of future neuroimagers to study what is now termed *functional connectivity*. Broadly defined, functional connectivity is the statistical relationship between the dynamic neural activity measured in two or more parts of the brain. This statistic is sometimes computed between pairs of points, but can also be evaluated for many areas in parallel using data-driven methods, such as independent component analysis (Smith et al. 2013). It is important to note that the relationship between functional and anatomical connectivity is complex and often underdetermined, particularly when the functional signal is assessed through an indirect measure such as blood-based hemodynamic responses. Nonetheless, the emergence of functional connectivity in the fMRI field has revolutionized the study of the human brain, by first establishing the basic correlations between related areas and then offering a new way to visualize and study brain networks. These methods and descriptions currently play a fundamental role in research into the human brain, including its dysfunction in psychiatric and neurological disorders.

A fascinating aspect of spontaneous activity, one that has drawn additional attention, is its potential *effects* on normal brain operation. At the microscopic level, ongoing activity provokes neurons to vary their responses from trial to trial. In an early observation by Bishop (1932), electrical stimulation of the optic nerve led to neural responses in the visual cortex that varied with each stimulation, an effect attributed at the time to the state of the cortex. In more recent studies in experimental animals, spatially coherent waves of ongoing activity have been shown to explain a large proportion of the response variance, even in primary sensory areas (Arieli et al. 1995; Fukushima et al. 2012).

Within the domain of human fMRI, ongoing fluctuations have been shown to alter task-based responses and directly impact perception and behavior. For example, subjects detecting faint visual stimuli are likely to be fooled into a false percept on trials in which fMRI activity in the visual cortex is high in the absence of a stimulus (Ress et al. 2000). Similarly, the reaction time of button presses is shortened during trials in which activity is high in sensorimotor areas. Large-scale fluctuations are sometimes directed by expectations or the structure of a specific task. For example, Sirotin and Das (2009) demonstrated a robust task-entrained change in hemodynamic signals in the primary visual cortex amid an unchanging visual stimulus. These results demonstrate that hemodynamic responses in sensory cortical areas can be subject to cognitive influences, such as the expectation of a stimulus. Such anticipatory modulation might be directed through long-range connections from the basal forebrain (Turchi et al. 2017), the adrenergic system (Reimer et al. 2016), the frontal cortex (Noudoost and Moore 2011), or the amygdala (Hadj-Bouziane et al. 2012), all of which have the capacity to alter responses of neurons to sensory stimuli.

Neural activity and hemodynamic fluctuations have a notoriously complex relationship (Logothetis 2008), part of which can be seen in relationship to cellular metabolism. One straightforward explanation of neurovascular coupling is that neural responses spend local energy; this, in turn, causes local metabolic increases, which summons more regional blood perfusion (Magistretti 2000; Raichle and Gusnard 2002). There are many known examples, however, in which this linear explanation fails, particularly since fluctuations can stem from physiological signals other than from neurons, including metabolic changes (Goldbeter 1996) and even hemodynamic changes themselves (Moore and Cao 2008). These multiple levels of biological complexity pose significant challenges for pinpointing the neural processes that underlie the commonly observed whole-brain correlation patterns in humans, and for understanding the causal chain by which ongoing fluctuations might influence functional responses and behavior.

Nonetheless, through the mapping of fMRI temporal correlations, the resting human brain offers an array of at least a dozen interleaved networks, many of which are straightforward to identify in the majority of subjects and appear to be in a mature form in early childhood (Damoiseaux et al. 2006; Thornburgh et al. 2017). While less studied in animals, the basic features of many of the resting-state networks appear similar (Hutchison and Everling 2012; Belcher et al. 2013). These networks have provided a new and extremely useful approach to study brain organization and physiological processes in healthy subjects and patients. For example, it is now possible to use spontaneous signals to establish the functional layout of the brain, and to use this information to define regions of interest for analysis in subsequent fMRI experiments. It is also possible to compare the functional integrity of such networks between patient groups and control subjects through straightforward resting-state scans collected in just a few minutes.

## Conceptualizing and Analyzing Neural Networks

A network, broadly defined, is a complex set of interacting elements. Above we highlighted the whole-brain networks identified using fMRI. However, neuroscientists also speak of networks defined by the interactions among genes, cell types, interconnected areas, or even individual animals. At some level, each network is embedded within a superordinate network that may be impacted by its perturbation. In terms used by network theoreticians, if a given node is disrupted (e.g., by suppressing activity of a cell type within a circuit) or a given edge removed (e.g., by suppressing one particular set of connections), other network components are likely to be affected. Such "diaschisis" or "off-target effects" present a potential pitfall for scientists attempting to draw cause-and-effect conclusions through manipulations of a biological system. The remedy might be for a researcher to focus on a separate perturbation of multiple nodes and edges, perhaps under multiple conditions. To approach the problem this way, however, one needs a considerable understanding of the network structure, and this is often only achievable through an integrated computational framework. At the same time, well-developed network approaches promise to facilitate unity *across* levels of analyses, not just among components at one level. The feasibility of studies which cross levels (genetic, cellular, circuit, system, behavior) has increased tremendously, and much effort is currently being directed toward network models that link two or more levels.

As mentioned above, functional connections are sometimes characterized by their pairwise interactions, primarily because the computational methods involved are straightforward. However, interactions in real brains are much more complex, may be more difficult to assess, and are ultimately tuned at a systems level to achieve certain behaviors. Encapsulating and analyzing these dynamics is of great interest as we attempt to integrate isolated descriptions of anatomical connections or physiological measurements into principles of brain function. As in many fields of biology, fitting the components of a complex system into a rigorous mathematical framework is an immense challenge. Nonetheless, this approach has already led to fruitful insights into modes of cortical dynamics.

Because whole-brain wiring diagrams capture only particular aspects of brain connectivity, quantitative models of brain architecture have gained popularity. Quantitative models require a mathematical language capable of capturing different types of interaction in a highly interconnected network. Built on fundamental mathematics in the form of graph theory and fundamental physics in the form of statistical mechanics, network science provides exactly such a language (Albert and Barabasi 2002) (Figure 9.3). In its simplest form, network science can be used to model intricate connectivity patterns as graphs, where computational units (neurons, ensembles, cortical columns, brain areas) are represented as nodes, and connections between them (functional relations or physical links) as edges (Bassett and Sporns 2017). By modeling the system

(a)   ## Directed connectivity matrix (213 x 213 regions)

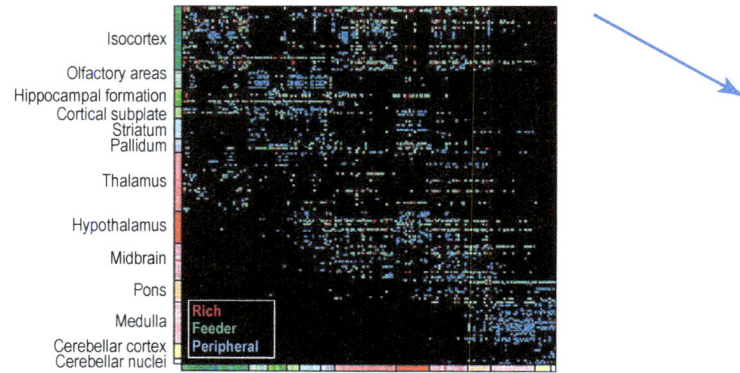

(b)   ## Gene expression (213 regions x 17,642 genes)

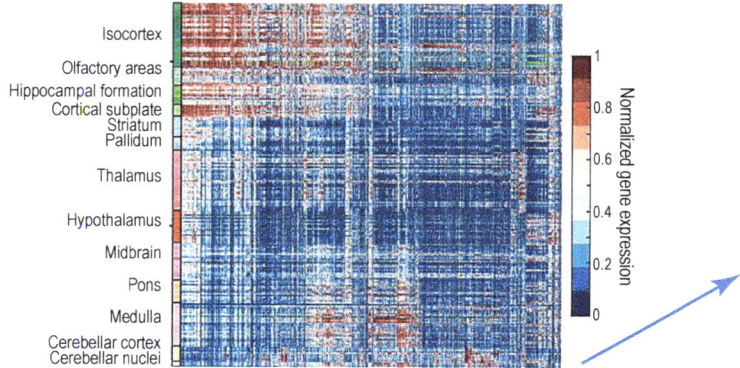

**Figure 9.3**  Relations among anatomical connectivity and gene co-expression networks. (a) Matrix of anatomical connections among 213 mouse brain regions: regions (nodes) with more than 44 distinct connections were considered hubs, and connections were classified as hub→hub (rich), hub→nonhub (feeder) or nonhub→nonhub (peripheral). (b) Normalized expression levels of 17,642 genes across 213 brain regions: genes with highly correlated expression profiles are placed near each other.

as a graph, one can apply computational tools to characterize quantitatively the architecture of the graph with various metrics, and then compare those metrics across measurement modalities, spatial scales, temporal scales, individuals, and species (van den Heuvel et al. 2016).

Several concepts from network science have proven useful in our understanding of neural systems. Locally, processing units tend to display strong clustering with neighboring regions, and these clusters combine to create

(c)

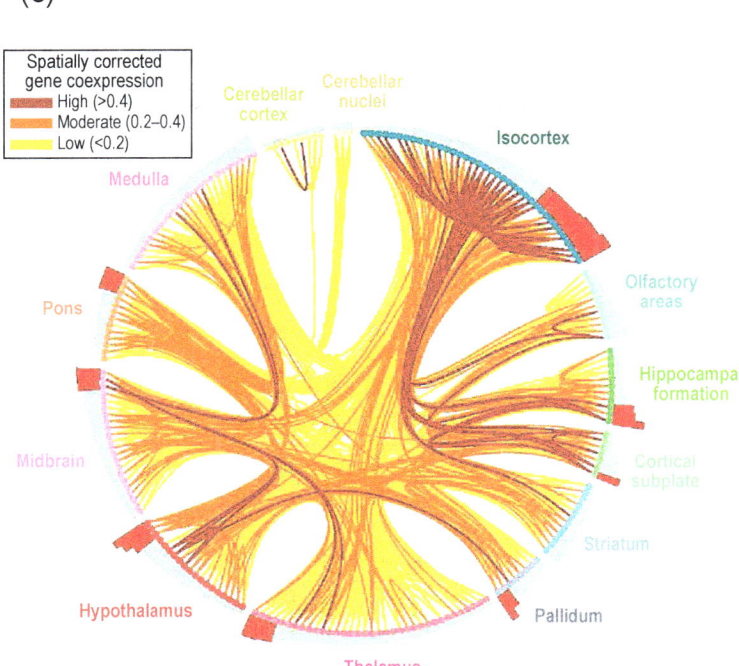

**Figure 9.3 (continued)** (c) Brain regions have been arranged around a circle, ordered by number of connections (bars) in each anatomical subdivision. Hubs are marked by red bars. The connection diagram traces anatomical connections between pairs of brain regions, color-coded by the corresponding gene co-expression value, after applying a correction for spatial distance. Statistical analysis revealed strongest gene co-expression among pairs of regions linked by reciprocal connections (as compared with unidirectional or unconnected pairs), as well as for rich connections linking hubs (as compared with feeder and peripheral connections). Genes driving correlations in expression in connections involving hub regions are functionally enriched in oxidative energy metabolism. Connectivity data derived from Bassett and Sporns (2017).

modules (Sporns and Betzel 2016). Each module is composed of a set of nodes that are more densely interconnected to one another than they are to nodes in other areas of the network. Complementing the local clustering, strong long-distance connections exist to link diverse areas of the network and enhance the complexity of functional dynamics (Betzel and Bassett 2018). The combination of local clustering and a few long-distance connections produces a small-world network topology, which intuitively can support segregated processing of information in combination with information transmission to spatially disparate units (Bassett and Bullmore 2017). Within this modular, small-world topology, one also observes network hubs—nodes that display an unexpectedly high number of edges—that are often connected to one another, and which are thought to be capable of exerting a particularly salient influence on the system (van den Heuvel and Sporns 2013). Recent work bridging network neuroscience and control theory has identified additional node types thought to be capable of enacting diverse control strategies, altering system dynamics to support cognitive function (Gu et al. 2015; Kim et al. 2018). Other complementary work that bridges network neuroscience and applied algebraic topology has identified additional network motifs that capture higher-order interactions, which may also be particularly important for neural computations (Giusti et al. 2016). These and related efforts have demonstrated promise in understanding (and quantitatively characterizing) alterations in network organization that accompany neurological disease and psychiatric disorders (Stam 2014; Fornito et al. 2017).

One of the advances over the past three decades has been in how we think about neural networks—progress that has resulted, in part, from improvements in computational methods and speed. What has changed at a conceptual level is the linearity, or perhaps seriality, in thinking about brain operation. This change is manifest at the level of the synapse, with highly selective, multidirectional, and dynamic analog interactions in the neuropil. This has also altered how we look at whole-brain networks more broadly, characterizing them not within the framework of a climbing hierarchy but rather as a syncytium of interconnected areas. There may be some historical irony here, in that the emerging network conceptualization of the brain may resonate somewhat more with Golgi's original syncytium view of the brain's organization than Cajal's revered neuron doctrine.

## The Far Reach of Cerebral Control

The exquisite spatial organization of spontaneous signals in the cortex came as a surprise to many systems neuroscientists. As described above, the discovery was made only after stepping away from conventional paradigms designed to tap into the presumed sensory, cognitive, and motor functions of the cerebral cortex. In some ways, this finding fits with the long-known fact that much of the brain's activity is concerned with internal regulation and homeostasis rather than interaction with the external environment. Beyond homeostasis, internal signaling can also shape a range of behaviors by adjusting state parameters that require action. For example, small groups of cells in the hypothalamus and elsewhere act through the endocrine and descending autonomic systems, utilizing bidirectional communication with visceral organs in the regulation of feeding, sexual behavior, and other actions critical for survival.

Subcortical brain structures have been shown to regulate aspects of immune functions (Wrona 2006), body metabolism (Morton et al. 2006), and perhaps the gut microbiome (Foster et al. 2017).

The surprising thing that we have learned in recent years is that the *cerebral cortex* has a hand in controlling autonomic and visceral function. The polysynaptic cortical control over these structures was determined using injections of rabies virus into the end organs. Rabies is a retrograde virus that crosses synaptic connections and can thus be used to identify higher-order upstream neurons, including those in the cerebral cortex (Dum and Strick 2013).

Several studies have demonstrated that restricted regions of the motor cortex and a handful of other cortical areas hold reign over stations in the peripheral nervous system as well as over visceral organs (e.g., kidney, adrenal medulla, stomach, heart) and likely many other corporeal structures (Levinthal and Strick 2012; Dum et al. 2016). After several synaptic crossings, typically four, through sympathetic or parasympathetic chains, the retrogradely transported rabies virus reaches layer 5 neurons of highly specific and circumscribed cortical regions. In the motor cortex, these regions occupy portions of the homunculus map of the body that are commensurate with their possible roles. For example, one of the sites projecting to the adrenal medulla is present in the face area of the motor cortex, which may be related to the regulation of sympathetic responses elicited in concert with facial emotions. These findings offer an entirely new perspective on cortical anatomy and may have potential implications for clinical work, where it is known, for example, that stimulation or transection of the vagus nerve can relieve symptoms that might originate from cortical dysfunction.

## Summary

Over the last three decades, the appropriation of spontaneous functional signals in the study of brain organization has opened new doors for the study of cortical organization, particularly in humans. It has also launched multidisciplinary approaches for the analysis of cortical activity, gaining insights from disciplines accustomed to making sense of complex interactions in large-scale networks. At the same time, neuroscientists remain aware that the cerebral cortex is not a computer in isolation, but is integrated within a biological system that includes the body itself, and can exert descending control over endocrine, immune, and even visceral functions.

## Design Principles Promoting Flexible Behavior

The capacity to learn, adjust, and flexibly direct behavior constitutes a very important architectural design consideration of the cerebral cortex. These capacities derive from a multiplicity of mechanisms that are manifest simultaneously at multiple scales, from the synapse to the system. Many of the mechanisms for neural plasticity found in the adult resemble those that were involved in the initial building of the brain and may actually extend developmental windows that permit lifelong modification of neural circuits.

**Essential Early-Life Training of the Cerebral Cortex**

The cerebral cortex has a protracted period of growth and maturation compared to many brain structures outside the telecephalon, which reflects one of its fundamental design principles: it needs training. The genetic specification of cortical development is staggering and strongly conserved across mammals (Workman et al. 2013), with the most prominent difference related to factors such as brain size and peripheral sensory specialization. Built into that genetic program, however, are explicit mechanisms for the postnatal refinement of cortical connections guided by activity and experience. These developmental steps are critical if we are to understand how the cerebral cortex of a particular mammal (e.g., a human) takes form. Early-life plasticity, through typical sensory experiences, ecological constraints, and parental relationships, has the effect of shaping species-specific sensory imputs, cortical connections, and processing domains.

Interestingly, some of the relevant behaviors that drive cortical training originate not in the cortex itself, but rather in the early-developing control centers of behavior which reside in the hypothalamus and midbrain (Swanson 2000). Thus, early-life plasticity may stem from one part of the brain training another, with innately programmed subcortical structures driving experiences needed for the normal development and maturation of the cerebral cortex. Among primates, whose extended childhood offers much time for experience to mold the adult patterns of cortical connections, experience-dependent learning is particularly obvious in the domain of complex social interaction.

The postnatal shaping of the cerebral cortex involves multiple mechanisms, and builds in an initial exaggeration, or exuberance, of axonal projections followed by a gradual pruning and restriction to their adult target locations. One measurable example of this exuberance in primates is the initial overproduction of interhemispheric fibers passing through the corpus callosum. In rhesus monkeys, for example, less than one third of the interhemispheric fibers present at birth persist into adulthood (LaMantia and Rakic 1990). For other corticocortical connections, less is known about the principles that underlie the elimination of axons and synapses during early life, particularly in primates. Recent work, however, has tracked the age-dependent expression of other developmental markers; for example, those related to myelination and structural molecules such as neurofilament protein in the primate visual cortex (Mundinano et al. 2015). These studies illustrate that the dorsal, parietal areas mature more quickly than the ventral, temporal areas, and that early maturation of the cerebral cortex may be driven by a transient thalamic connection that is present only during early development (Mundinano et al. 2018).

The important message that we wish to stress is that the cerebral cortex is, from the beginning, critically shaped by early sensory experiences and behaviors. Many of these behaviors are primitive or innate, driven by specialized subcortical circuits. Subject to the sensory and social consequences of these innate actions, the cerebral cortex prunes and refines its connections and steadily takes control over many overt and internal behaviors. The cortex learns to interpret complex sensory signals, to establish contingencies, to withhold reactions and reflexes, and to plan goal-oriented action sequences critical for survival. This learning is an essential element of cerebral cortex design and may ultimately be as important as the adult wiring diagram for understanding its core principles.

## Learning in the Adult: Plasticity of Synapses and Systems

Following the extreme plasticity evident during development, the adult brain retains a critical ability to alter itself in response to the environment; this serves as the basis for learning and memory. Above, we provided an overview of the complex and interacting cellular substructures involved in the tuning of neural connections. Over the last decade, researchers have discovered myriad modes of synaptic regulation stemming from mechanisms that govern changes to the chemical, genetic, and structural composition of synaptic connections (Alberini 2009; Holtmaat and Svoboda 2009; Bailey et al. 2015).

At one level of description, synapses can be seen as independent actors, regulating their potency independently of their neighbors. The phenomenon of long-term potentiation has long been known (Bliss and Lomo 1973), and the list of factors regulating the strength of individual synapses is ever growing (Malenka and Nicoll 1999). At the same time, a broader view of the synaptic microcosm, highlighted in an earlier section, indicates that efficacy of a given synapse is highly dependent on extrasynaptic factors. In other words, altering the morphology or synaptic strength at one location can affect the contribution of neighboring synapses. This partial dependency of nearby sites on the dendritic shaft, together with other active features of dendritic physiology, provides many degrees of freedom for fine-tuning circuit plasticity. While learning principles are still being discovered, recent work suggests that even extrinsic cells, other than presynaptic and postsynaptic neurons, play important roles in shaping the dendritic microenvironment. These cells range from microglia, which contribute to synaptic modification by actively and aggressively removing spines, to particular subtypes of inhibitory interneurons, whose level of input may be important to shift neurons into a mode that is receptive to synaptic plasticity (Bavelier et al. 2010).

Under some conditions, opening the window to synaptic modification can resemble a local reversion to a more immature state of brain development. While this is a somewhat new field of study, the genes thought to be involved in this process are sometimes called *neotenous*, referring to their prominent role in early life. Such developmental genes have been found to co-localize with other markers of brain activity, such as cortical areas undergoing aerobic glycolysis (Goyal et al. 2014). As we learn more about the regulation of plasticity through physiological, chemical, and genetic mechanisms, opportunities may arrive to improve plasticity for a range of cognitive and neurological disorders.

At a more holistic level, the cerebral cortex participates in multiple systems that appear designed to adapt behavior through various types of learning. This also involves cortical connections to external brain structures. Although this is a very broad topic, brief mention is warranted here since so much of the cerebral cortex participates in these adaptive systems. The two most prominent regions working with the cortex to facilitate behavioral adaptation are the basal ganglia and cerebellum (Bostan and Strick 2018). These structures are anatomically interconnected to form an integrated network capable of adaptation over different timescales and under different contingencies. Often overlooked, this network is topographically organized so that motor, cognitive, and affective territories at each node are interconnected with the corresponding territory of another node. During a particular task, the interlinked nodes are coactivated, and during learning there is an orderly shift in the progression of learning between the different structures as the performance level changes. Within each structure, activation in cognitive territories often predominates when a task is first performed. Thereafter, learning is expressed in motor territories, where changes emerge simultaneously with improvements in motor performance. Figure 9.4 illustrates the temporal evolution of the involvement of different networks in behavior. It is typical of the changing involvement of corticocortical networks (e.g., default mode network, dorsal attention network) in various tasks. The conserved connectivity of the cerebral cortex with external elements such as the basal ganglia and cerebellum allow the adult brain of mammals to learn and adapt their behaviors. This general mammalian pattern has been expanded in primates and is particularly prominent in the massive human brain, conferring a remarkable level of flexibility in adult behavior. Unlike any other animal, a human being can participate in learned activities as varied as acrobatics, music, politics, and typing on a keyboard. This flexibility stems from the multiscale organization of learning mechanisms in the brain, from plasticity in the dendritic microenvironment to whole-brain circuits specialized for learning. Through its interconnection with areas such as the basal ganglia and cerebellum, the cerebral cortex learns to control myriad aspects of behavior, conferring a centrality to our actions and thoughts, the expression of which, through consciousness, is one of the most fascinating and elusive aspects of brain science.

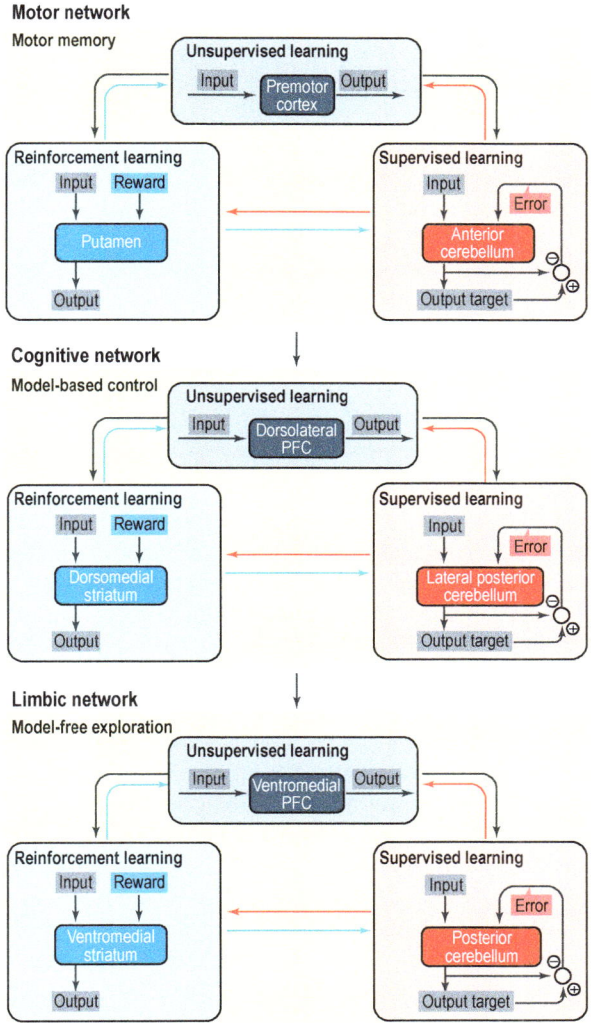

**Figure 9.4** Functionally related cortical, basal ganglia, and cerebellar sites within interconnected networks participate in progressive stages of action planning. On the basis of these results, learning through exploration involves a limbic network, including the ventromedial prefrontal cortex (PFC), ventromedial striatum, and posterior cerebellum. Model-based learning involves an associative (cognitive) network, including the dorsolateral PFC, dorsomedial striatum, and lateral posterior cerebellum. Performance based on motor memory involves a motor network, including the supplementary motor areas, putamen, and anterior cerebellum. The authors' imaging data suggest that as learning progresses, the sites of activation shift in a topographically organized fashion. Our interpretation of these data is that each stage of the learning process involves a different set of interconnected basal ganglia, cerebellar, and cerebral cortical regions. Reprinted with permission from Bostan and Strick (2018).

## Summary

Without the capacity to learn and adapt its behavior, a brain would be unsuccessful in a complex and contingent world. Many if not most interactions
among animals, including predation, foraging, social interactions, and navigation, require continual learning and modification of behavior. The cerebral
cortex is at the heart of this flexibility. In early life, its neural circuits undergo
major experience-dependent changes as they tune themselves to the basic
statistics and behavioral requirements of the environment. In adulthood, the
brain retains the capacity to learn and adapt, facilitated both by the capacity
to reconfigure cortical synaptic connections, as well as the utilization of cortical connections to subcortical structures specialized to support behavioral
flexibility.

## Conclusions

The functional architecture of the cerebral cortex is a formidable topic and
its details could fill multiple volumes. In this chapter we have highlighted
three areas that have seen particular conceptual advancement over the past
decades: the synaptic microenvironment, large-scale cortical networks, and
plasticity and adaptation in cortical circuits. Many other important features of cortical architecture include the prominent laminar organization
of cells in cortical microcircuits, the growth and initial wiring of the brain,
the functional consequences of its columnar organization, and its principles
of connectivity with subcortical structures, such as the thalamus, striatum,
claustrum, and superior colliculus. A growing body of exciting findings has
investigated functional specificity through genetically modified mice, as well
as specificity of interneurons and circuit motifs. Details from this burgeoning
field will continue to shape our understanding of functional architecture at
both microscopic and macroscopic scales. One of the great challenges in the
study of the brain is to synthesize a large number of details into principles
for understanding. While many details are known about the cerebral cortex,
our level of understanding about its overarching architectural and functional
principles remains, arguably, primitive. For the optimist, this is a situation of
great opportunity, where any scientist who is able to identify and integrate
the most relevant of these details will reap the benefits of fundamentally new
insights into brain function.

There is no denying that remarkable conceptual progress has been made
over the last decades. Reflecting back to the first meeting in this series in
1987, the community was largely familiar with the basic connectional anatomy of the cerebral cortex. Cortical anatomists and physiologists pictured the
brain as a step-by-step sequence of hierarchical computations, in some ways
captured by the newly available information codified in the VLSI diagram of

the cerebral cortex (Felleman and Van Essen 1991). There was also the general awareness that perhaps superimposed on that hierarchy were two distinct streams for visual information processing: one concerned with objects, the other with locations (Ungerleider and Mishkin 1982). However, there was not yet a community of brain imagers, outside of a few specialists in positron emission tomography. At the top of the cortical hierarchy sat the hippocampus alone, which, perhaps coincidentally, was the target of study for most synaptic physiologists.

Our current shared understanding of the cerebral cortex has changed considerably, particularly in reference to the richness of synaptic interconnections and the large-scale organization of networks. The view of a serial processing hierarchy has been complemented by the concept of interacting and dynamic functional networks. As is often the case, the scientific conquests have had the effect of raising new questions at a rate that seems to exceed our steps forward in understanding. The staggering biological complexity of the brain is not for the weak willed, and neuroscientists struggle to gain traction on basic architectural and functional principles.

One critical element for the future is almost certainly the continued development and refinement of computational and theoretical tools. Within the realm of computation, there is a spectrum of detail built into different models, ranging from realistic, elemental models of the brain, to mathematical encapsulations of brain function, to deep learning approaches that only marginally reflect brain operation. An important question for the future is how closely to tune modeling efforts to the empirical details of brain anatomy and physiology as they are continually revealed. Perhaps even more important is the philosophical question of where the essence of brain function, cognition, and behavior lies. Is there a primacy of computational descriptions, as one might conclude from visionary thinkers such as David Marr (1982)? Or does focusing on computation while ignoring biological details come at a cost that is too heavy to bear? Theorists will have to grapple with these questions as both computation power and the capacity to collect empirical data about the brain accelerate.

A complementary perspective on cognition and behavior is that structural and functional principles of the brain can only really be understood through genetics, development, and evolution. From this perspective, the human brain is a fundamentally composite structure, with evolved layers of control subsuming and overriding more primitive control systems. A truly mechanistic understanding will therefore depend on disentangling ancestral versus derived features of the brain, including its multiscale structural and functional details, its layers of genetic and molecular control, its growth and refinement during embryonic and postnatal development, and much more. While this pursuit will also draw upon computational tools, it pushes neuroscientists to embrace the natural biological complexity that has continually shaped the human brain, and its massive and powerful cerebral cortex, over hundreds of millions of years.

## Acknowledgment

This research was supported in part by the Intramural Research Program of the NIMH (ZIA MH002838).

# Functional Properties of Circuits, Cellular Population, and Areal Level

# 10

# Cortical Dynamics

Wolf Singer

## Abstract

A hallmark of cortical organization is the coexistence of serial feedforward with re-entrant processing. The latter is based on feedback projections from higher to lower processing levels and massive reciprocal excitatory projections which link neurons located within the same cortical areas as well as cortical areas occupying the same level in the processing hierarchy. These reentrant connections, together with local negative feedback loops, give rise to exceedingly complex dynamics that are characterized by oscillations in a broad range of frequencies, synchronization of discharges, and cross-frequency coupling. Evidence is reviewed which suggests that these dynamic properties support specific computations: the flexible binding of distributed neurons into functionally coherent assemblies, the attention-dependent selection of sensory signals, the conversion of semantic relations into temporal relations, the comparison of stored priors with sensory evidence, the selective routing of signals in densely interconnected networks, the definition of relations in the context of learning, and the dynamic formation of functional networks. Arguments challenging a functional role of oscillations and synchrony, due to their volatile nature, are discussed in relation to recent evidence that highlights the advantages of volatility.

## The Encoding of Relations

Organisms have evolved cognitive functions that enable them to construct an internal model of the world. This model permits interpretation of sparse and noisy sensory information, the generation of predictions, and the planning of well-adapted responses. To generate and exploit such models, organisms have developed efficient mechanisms to extract relevant features selectively from the plethora of available physical and chemical signals, to detect consistent spatial and temporal relations between these features, to memorize the corresponding relational constructs, to compare sensory evidence with stored knowledge, to evaluate the behavioral relevance of the actual conditions, and to select the appropriate response. In this chapter, I focus on mechanisms that underlie the detection and encoding of relations, as relations define the virtually infinite space populated by cognitive objects that can be generated by combining a limited set of elementary features. With a limited number of symbols, for instance, alphabets sufficed throughout history to compose vast bodies of literature due to the richness and the combinatorial power of relational codes.

One common strategy for the detection of relations and their encoding in neuronal responses, realized in virtually all neuronal systems, is based on convergent feedforward circuits. Neurons tuned to respond to particular, frequently co-occurring, and hence related features of the environment are selectively connected to higher-order target cells. By adjusting the gain of these converging connections and the threshold of the target cell, it is assured that the latter responds preferentially to only a particular conjunction of features (Barlow 1972). In this way consistent relations among features become represented by the activity of conjunction-specific neurons. Iterating this strategy across multiple layers in hierarchically structured feedforward architectures then leads to the representation of complex relational constructs by conjunction-specific neurons of higher order. This basic principle for the evaluation and encoding of relational constructs has been realized independently during evolution in the nervous systems of different phyla (molluscs, insects, vertebrates) and reached the highest degree of sophistication in the hierarchical arrangement of processing levels in the cerebral cortex of mammals and, in particular, primates. It has also been implemented in numerous versions of artificial neural networks (Rosenblatt 1958; Hopfield 1987; DiCarlo and Cox 2007; LeCun et al. 2015). The highly successful, recent developments in the field of "deep learning" (LeCun et al. 2015) capitalize on the scaling of this principle in large multilayer architectures.

There are, however, marked differences between these artificial neuronal networks and the processing architectures found in the nervous systems of highly evolved species, suggesting that these exploit additional principles of information processing. This is particularly true for brains endowed with cortical structures, such as the hyperstriatum in birds and the hippocampus and neocortex in higher vertebrates. These structures possess feedback connections between processing levels; in addition, neurons within the same layer are reciprocally coupled by myriads of recurrent lateral connections. These reentry connections are missing in most feedforward artificial systems, but in natural brains they are more abundant than feedforward connections (Markov et al. 2014; Bastos et al. 2015). Differences between natural and artificial systems exist also with respect to the learning mechanisms. In technical systems, the supervised adjustment of the synaptic gain of feedforward connections is achieved by the so-called "backpropagation algorithm," which is biologically implausible and differs from the various learning mechanisms implemented in natural brains. This latter difference is not crucial, because the results of training procedures are ultimately similar; however, the differences in processing architectures are consequential as they permit computations that go beyond those realizable in feedforward architectures. The latter possess no short-term memory and therefore have difficulties to process and classify temporal relations. Recurrent networks, by contrast, exhibit fading memory and hysteresis because of the nonlinear dynamics that evolves on the backbone of reverberating circuits (see below). Although it is argued that any multilayered feedforward network can be unrolled so as to function like a recurrent network, the number of layers required to cope with behaviorally relevant temporal relations (e.g., in speech processing) is prohibitive. Recent designs of artificial systems, therefore, incorporate recurrent networks that realize a function characterized as long short-term memory (Hochreiter and Schmidhuber 1997; Silver et al. 2017; Banino et al. 2018).

Recurrent networks are not only better suited for the processing of temporal relations, they also offer complementary solutions for the encoding of spatial relations. As proposed by Donald Hebb (1949), consistent relations among features can be encoded by forming functionally coherent assemblies that bind individual, feature-coding neurons together. Neurons coding for features that typically co-occur, and thereby define a particular cognitive object, get bound together into an assembly which, as a whole, represents a concrete perceptual object or a more abstract cognitive content (e.g., category, concept, action plan). In this case, the *binding* of specific features is not achieved through the convergence of feedforward connections onto conjunction-specific neurons but by reciprocal connections between feature-selective neurons. These connections are strengthened by correlation-dependent synaptic plasticity mechanisms (Hebbian synapses, see below) and by selectively increasing the mutual interactions between nodes encoding a related feature, which enhance the vigor and/or coherence of the responses of the respective nodes. In this way, consistent relations among features characteristic for perceptual objects or, at higher processing levels, among more abstract contents, are translated into the weight distributions of reciprocal connections among the nodes of the network. Accordingly, the information about the presence of a particular constellation of features is not represented by the discharge of a single conjunction-specific neuron but by the amplified or more coherent responses of a distributed assembly of neurons. If a particular feature constellation matches the weight distributions of the recurrent connections linking the respective feature-sensitive nodes, the assembly will self-amplify its responses through reverberation and then assume the same function as a conjunction-specific neuron, except that now the coding unit is an assembly of cells.

Both relation-encoding strategies have advantages and disadvantages, and evolution has apparently opted for a combination of the two. Feedforward architectures are well suited to evaluate relations between simultaneously present features (e.g., spatial relations), and they allow for fast processing because they rely exclusively on a series of simple summation and thresholding operations. Moreover, they are easy to implement and are robust because interactions are essentially linear, well-controllable, and show no runaway dynamics. However, feedforward architectures are less apt to handle relations among temporally segregated events because they lack memory functions. In addition, they are costly in terms of hardware requirements. All information about the statistical contingencies of features must be stored in the weights of the feedforward connections; because the dynamic range of neurons limits the number of converging driving connections, processing hierarchies require a large number of levels to cope with the combinatorial complexity of possible feature constellations. Since the number of required output units scales linearly with the number of relations that can be analyzed and encoded (combinatorial explosion), biological systems that rely exclusively on feedforward architectures can only afford representation of a limited number of behaviorally relevant relational constructs. Finally, feedforward architectures cannot easily cope with entirely new constellations of features because they lack the associative capacities of recurrent networks.

By contrast, assemblies of recurrently coupled, mutually interacting neurons can cope very well with the encoding of temporal relations (sequences) because such networks exhibit fading memory due to reverberation. Assembly codes are also much less costly in terms of hardware requirements, because individual feature-specific neurons can be recombined flexibly to yield a very large number of different assemblies, each representing a different cognitive content (combinatorial code). In addition, coding space increases dramatically because information about the statistical contingencies of features can be stored in the synaptic weights of feedforward connections as well as in the weights of the recurrent and feedback connections. Finally, the encoding of entirely new or the completion of incomplete relational constructs is facilitated by the nonlinear dynamics of recurrently coupled networks; their dynamics allows for self-organization and hence the completion of patterns and the generation of novel associations (generative creativity).

These advantages, however, come with a price. Processing may be slower than in purely feedforward architectures because assembly formation depends on time-consuming self-organizing processes based on the nonlinear dynamics of recurrent networks that exhibit reverberation, hysteresis, and attractor dynamics. Moreover, implementation of this distributed combinatorial coding strategy is not trivial; additional mechanisms are required to assure stability of network dynamics. To exploit the advantages of recurrent networks fully, their dynamical range needs to be well controlled because

- fast formation of assemblies requires a delicately regulated level of resting activity, and
- if global excitation drops below a critical level, recurrent networks may cease to operate or, if a critical level of excitation is reached, engage in runaway dynamics and become epileptic.

For cortical structures, this problem is taken care of by a number of cooperating, self-regulating mechanisms (involving inhibitory interneurons, excitation–inhibition balance, and ascending modulatory systems) that keep the network within a narrow working range just below criticality, beyond which it would enter into a chaotic regime.

Another particularly challenging problem, known as the superposition catastrophe, involves the segregation of simultaneously active assemblies, in particular if they comprise spatially interleaved neurons or have to share some of the feature-selective neurons. If assemblies were solely distinguished, as proposed by Hebb, by enhanced activity of the constituting neurons, it becomes difficult to distinguish which of the more active neurons actually belong to which assembly (the "binding problem"). The option to multiplex in time coexisting rate-coded assemblies is also problematic because readout of enhanced discharge rate requires temporal integration in downstream structures. Given the low discharge rate of cortical neurons, it might take several hundreds of milliseconds before the more active members of assemblies become distinguishable from less active neurons, and hence multiplexing is achievable only on a slow timescale. Moreover the inability to configure temporally overlapping representations would jeopardize the associative capacities of assembly coding. It has thus been proposed that the salience of the responses of neurons temporarily bound into assemblies should not be enhanced solely by an increase of their discharge rate but by the precise synchronization of their action potentials (Gray et al. 1989; Gray and Singer 1989; Singer 1999). There is ample evidence that synchronous inputs are particularly effective in driving neurons above threshold. Thus, activation of target cells at the subsequent processing stage can be assured by increasing either the rate or the synchronicity of discharges in the afferents converging from the respective lower level. The advantage of increasing salience by synchronization is that integration intervals for synchronous inputs are very short, thus allowing for instantaneous detection of enhanced salience. Hence, information about the relatedness of responses can be read out very rapidly. *In extremis*, single spikes can be labeled as salient and belonging to a particular assembly if synchronized with a precision in the millisecond range. Thus, assemblies defined by synchrony rather than rate increases can be multiplexed at a much faster rate than rate-coded assemblies without becoming confounded.

At this point, one might argue that the readout of assemblies again requires conjunction-specific neurons that convert relational information into discharge rates and that assembly coding offers no advantages. This argument would probably hold if the output of the system would solely be conjunction-specific grandmother cells at the top of the sensory processing streams, or individual command neurons at the top of the inverse hierarchy of motor output. However, this is not the case. As far as we understand the system's coding strategy, information remains distributed over many nodes: from the very early sensory structures all the way to the motor output. Thus, there is no bottleneck requiring condensation of relational information in the discharges of individual cells (grandmother cells), and hence assembly coding can be maintained throughout the whole processing stream. Assemblies formed at a lower level can ignite corresponding assemblies at the respective next processing stage. The ability to cope with the "combinatorial explosion" (i.e., the virtually infinite number of possible feature constellations) is one of the advantages of assembly coding: it permits hardware-efficient, flexible binding of features represented by the reciprocally coupled network nodes at all levels of processing. However, the provision of an anatomical backbone for assembly coding is not the only advantage of recurrent networks. In addition to the ability to cope with the encoding of temporal relations mentioned above, it is likely that the complex dynamics which evolve in recurrent networks provide additional computational options. These, as well as more complex dynamic properties of cortical networks, are considered later in the chapter.

## Temporal Relations as Code for Semantic Relations

Consistently (i.e., frequently and stereotypically) occurring temporal relations between real world events signal relatedness. Simultaneously occurring events usually have a common cause or are interdependent because of interactions. If one event consistently precedes the other, the first is likely the cause of the latter; if there are no temporal correlations between the events, they are most likely unrelated. Nervous systems exploit this fact in the detection and encoding of relations. Accordingly, learning rules adopted by evolution are sensitive to such temporal relations, thereby permitting the generation of internal models of relational constructs that have considerable predictive power—a likely reason for the striking conservation of the mechanisms supporting use-dependent modifications of synaptic transmission. Without exception, the associative mechanisms of synaptic plasticity evaluate correlations with a precision in the millisecond range. For the evaluation of correlations over longer time spans, required for the detection and encoding of contingencies separated by long intervals, additional mechanisms have been implemented. These involve memory functions at different timescales, ranging from fading memory in recurrent networks over short- to long-term memory mechanisms in devoted structures of the brain.

Expressing semantic relations in temporal relations does not, of course, exclude the common encoding of relations in the responses of conjunction-specific neurons (labeled line codes). However, unless the connections converging on such neurons are genetically determined, the input connections to the future conjunction-specific neurons have to be selected by experience. This selection, in turn, is again based on the time-sensitive mechanisms of synaptic plasticity, both during development and learning. Thus, the implementation of nongenetically specified, conjunction-specific "binding" neurons also requires that semantic relations become recoded in temporal correlations.

The exquisite sensitivity for temporal relations of associative synaptic plasticity mechanisms constrains the strategies used by cortical networks for the encoding of relations. Therefore, the respective mechanisms will now be briefly reviewed.

## Time-Sensitive, Associative Synaptic Modification Rules in a Nutshell

Interestingly, the mechanisms that support activity-dependent shaping of neuronal architectures during development and those that mediate use-dependent long-term modifications of synaptic gain, thought to underlie learning in the adult, share numerous similarities (for a review, see Singer 2018b). Major differences are that during development, functionally weakened synaptic connections eventually get physically and irreversibly removed while the pool of connections available for selection is permanently replenished through newly formed connections. The initial steps, however, that serve the evaluation of temporal relations are based on very similar molecular mechanisms and correspond to the rules proposed by Hebb (1949) for the strengthening of interactions between neurons that exhibit correlated activity: neurons wire together if they fire together. Hence statistical contingencies get translated into coupling strength. The first experimental confirmation of Hebb's hypothesis was the seminal discovery by Bliss and Lomo (1973) that tetanic stimulation of excitatory pathways in the hippocampus causes a long-term potentiation (LTP) of synaptic transmission. Subsequent studies demonstrated that LTP induction required a critical level of postsynaptic depolarization. If postsynaptic cells are prevented from responding to excitatory input by concomitant inhibition or hyperpolarizing current injection, modifications either do not occur or change to long-term depression (LTD) (Artola et al. 1990). With the advent of Ca imaging it became clear that the initial trigger for both LTP and LTD is a surge of calcium in the postsynaptic dendrites and that the polarity of the modifications depends on the rate of rise, amplitude, and sources of this Ca increase. Fast and strong increases lead to LTP, whereas slow and smaller increases trigger LTD (Bröcher et al. 1992; Hansel et al. 1996, 1997). Accordingly, both modifications can be obtained by raising intracellular Ca concentrations through the liberation of caged Ca in a concentration-dependent manner (Neveu and Zucker 1996). Moreover, the source of the Ca increase is of importance. Calcium entering through N-methyl-D-aspartate (NMDA) receptor-associated channels favors the induction of LTP, whereas Ca entering solely through voltage-dependent Ca channels is more likely to trigger LTD. The complex molecular cascades leading to these changes in synaptic gain have been thoroughly studied, and evidence indicates that both pre- and postsynaptic modifications are involved (for a review, see Morishita et al. 2005).

The evidence that both experience-dependent circuit modifications during development as well as use-dependent changes of synaptic efficiency in the adult depend on correlations between pre- and postsynaptic activation found a mechanistic explanation when it was discovered that both involve activation of NMDA receptors. These function as coincidence detectors because they are permeable for Ca ions only if glutamate is bound to the receptor while the postsynaptic cell is sufficiently depolarized to remove the magnesium block (Nowak et al. 1984; Artola and Singer 1987; Kleinschmidt et al. 1987; Bear et al. 1990; for a review, see Collingridge and Singer 1990). Since the level of depolarization of the postsynaptic membrane does not only depend on the activity of the local excitatory synapses, but also on all the other excitatory and inhibitory inputs, this mechanism also accounts for the cooperativity that characterizes use-dependent synaptic modifications. Even weak inputs can increase their gain if they are active in synchrony with other nearby excitatory inputs that contribute to depolarization and the removal of the magnesium block. With the advent of two-photon imaging technology it became possible to demonstrate *in vivo* that contingent activation of weak inputs converging onto the same dendritic branch could induce sufficient depolarization to activate regenerative dendritic responses (Na and Ca spikes) and to induce LTP (Grienberger et al. 2015). Conversely, concomitant activation of inhibitory inputs can prevent even strongly activated inputs from depolarizing the postsynaptic dendrite above LTP threshold. In this case, presynaptic activity that would normally induce LTP may either induce LTD or induce no change at all (Artola et al. 1990).

The discovery that spikes can backpropagate into dendrites and contribute to the postsynaptic depolarization that gates synaptic plasticity revealed yet another aspect of time-sensitive plasticity (Markram et al. 1997; Bi and Poo 1998; Stuart and Häusser 2001). Varying the timing between a single excitatory postsynaptic potential (EPSP) and the backpropagating spike showed that small changes in the *temporal relations* have a massive impact on synaptic modifications. When the EPSP precedes the backpropagating action potential by less than 50 ms, the synapse potentiates and the strength of potentiation increases with decreasing delay. However, once the EPSP occurs after the backpropagating spike, there is a sharp transition toward LTD. The underlying mechanism is the same as detailed above. If the backpropagating spike occurs shortly *before* the EPSP, it can contribute to lifting the Mg block, thus allowing LTP to occur; if it arrives *after* the EPSP, the repolarizing currents prevent NMDA receptor activation, and LTD is the likely result. This special case of a use-dependent synaptic modification, known as spike timing-dependent plasticity (STDP), has an important implication: use-dependent synaptic modifications are not only sensitive to the coherence of converging activity but also to causal relations. The gain of excitatory connections increases if their activity can be causally related to the activation of the postsynaptic neuron and weakens when this is not the case.

These empirical results have been formalized in rules addressed as the BCM (Bienenstock et al. 1982), ABS (Artola et al. 1990), and STDP (Markram et al. 1997; Bi and Poo 1998) rules. In conclusion, the net effects of these use-dependent synaptic modifications of excitatory connections are

- a strengthening of (reciprocal) connections among pairs of cells that are frequently activated in temporal contiguity or when one cell successfully drives its target neuron,
- a strengthening of the gain of converging inputs that are frequently active in temporal contiguity,
- a weakening of connections among pairs of cells whose activity is un- or anti-correlated or when one cell discharges shortly after its target neuron,
- a weakening of inputs active in contiguity with inhibition of the post-synaptic cell, and
- a weakening of connections that are inactive while the postsynaptic cell is strongly activated by other inputs (heterosynaptic depression).

Thus, the crucial variable that determines the occurrence and polarity of synaptic gain changes is the *temporal relation* (contiguity) between discharges in converging presynaptic inputs and/or between the discharges of presynaptic afferents and the depolarization of the postsynaptic neuron. These mechanisms evaluate correlation patterns with a precision in the range of tens of milliseconds and, in the case of STDP, even in the millisecond range. Moreover, the dependence of synaptic modifications on cooperativity between pre- and postsynaptic activity predicts that synchronous oscillations should provide a particularly favorable condition for the induction of use-dependent synaptic modification. This prediction is supported by the finding that use-dependent modification of orientation maps in the visual cortex is facilitated by entrainment of cortical networks in synchronous gamma oscillations (Galuske et al. 2019).

However, a learning mechanism has recently been discovered (Bittner et al. 2017) that operates on much longer timescales and is associative but does not seem to evaluate causal relations. If a postsynaptic neuron is sufficiently depolarized to generate a plateau potential—an active, $Ca^{2+}$-dependent dendritic response—then EPSPs arriving before *or* after this plateau potential undergo potentiation. This potentiation is maximal for EPSPs arriving in temporal contiguity with the plateau potential and decays with increasing temporal distance from the plateau potential. Thus, this mechanism is also sensitive to temporal coherence but allows for a gradation of potentiation as function of the temporal offset between the postsynaptic "reward" signal and incoming EPSPs. Obviously, similar to STDP, this mechanism is ideally suited for the learning of sequences as it converts sequence order into graded changes of synaptic efficiency. At first sight this mechanism seems to relax the constraint of precise timing relations between pre- and postsynaptic events. However, if information about sequence order were to be stored by the gradation of synaptic weights, the temporal relations between input sequences and the postsynaptic event, the plateau potential, need again to be precise and reproducible. Precise temporal sequences in population responses of cortical neurons are commonly observed, and there is evidence from the visual cortex that the sequence order contains as much stimulus-specific information as the firing rate of individual neurons. When stimulated with a moving grating, cortical neurons discharge in a fixed sequence, as revealed by systematic offsets of cross-correlation peaks in the range of $\pm$ 10 ms. The order of the neurons in this sequence changes with stimulus orientation and allows determination of stimulus orientation and direction of motion with the same precision as measurements of discharge rates (Havenith et al. 2011).

Use-dependent gain changes have also been described for inhibitory connections onto excitatory neurons (I/E) and connections among inhibitory neurons (I/I). These modifications are also sensitive to the relative timing of pre- and postsynaptic activity but the modification rules appear to be more heterogeneous than for E/E connections, which matches the large diversity of interneuron types. The database for I/E and I/I plasticity is comparatively sparse but both Hebbian and anti-Hebbian modifications have been observed (Moore et al. 2010).

In conclusion, the time-sensitive mechanisms of synaptic plasticity, especially those identified for E/E connections, are well suited to convert information about the relatedness of events encoded in the precise timing relations of neuronal discharges in lasting changes of functional architectures.

## Stimulus-Locked versus Internally Generated Temporal Relations

The fact that the learning rules are exquisitely sensitive to the precise temporal relations between individual discharges of connected neurons has far-reaching implications for the way nervous systems capture and encode information about relations. First, it has to be assured, if temporal relations between external events are to be evaluated, that the timing relations between events in the environment are reliably encoded in spike timing to permit learning of correct associations. This requirement is met in all sensory modalities by the implementation of transmission chains, commonly referred to as "phasic systems," which operate with high temporal resolution and accuracy. *In vivo* recordings from higher visual areas as well as the auditory and the somatosensory cortex have revealed that the discharge latencies of individual neurons signal the temporal structure of stimuli with extreme precision in the millisecond range. This proves that precise timing of discharges can be preserved despite numerous intervening synaptic transmission steps (Buracas et al. 1998; Reinagel and Reid 2002). Simulation studies, partly based on the concept of synfire chains proposed by Moshe Abeles (1991), confirmed that conventional integrate-and-fire neurons are capable of transmitting temporal information with the required precision (Mainen and Sejnowski 1995; Diesmann et al. 1999).

Additional mechanisms are required, however, when selective associations have to be established between neuronal responses that lack precise temporal structure. Such binding operations are most likely required for the association of sensory responses to stimuli lacking a temporal dimension or for the association of internally generated activity. A parsimonious solution would be to implement intrinsic mechanisms that impose temporal structure on neuronal responses that satisfies the contingency requirements of the classical plasticity rules and to utilize the existing plasticity mechanisms also for the association and segregation of signals that initially lack temporal structure.

## Mechanisms for the Generation of Temporally Structured Activity

Neuronal mechanisms capable of generating temporally structured activity are diverse, abundant, and evolutionarily ancient. A common and highly conserved strategy to generate temporally structured activity is the oscillatory patterning of activity, the basic principle of parsing time used in virtually all clocks. Neuronal networks have a high propensity to engage in oscillatory activity. These oscillations cover a broad frequency range, from below 0.1 to more than 200 Hz, and they tend to occur in typical frequency bands that are characteristic for particular brain structures and brain states. As reviewed recently (Buzsáki et al. 2013), these frequency bands are surprisingly well conserved across different species and even across different phyla. This suggests that they reflect some basic dynamics of nerve cells and/or circuits and are adapted to serve particular cognitive and/or executive functions. Spectral decomposition of global measures of brain activity, such as EEG or MEG or local field potential (LFP) recordings, usually reveals a continuous distribution of oscillation frequencies, the power of the oscillations decreasing with increasing frequency (the 1/f rule). To be addressed as such, an oscillatory process should appear in the power spectrum as a narrow-band "bump" of increased power. This distinction is important because with certain recording techniques (e.g., electrocorticography) one usually observes a task-associated broadband increase of power in the high-frequency range (from ~ 80–120 Hz) that is often addressed as "enhanced activity in the gamma-frequency range." This broadband activity reflects synaptic currents and action potentials and is a good measure of neuronal activity but it must not be confounded with narrow-band gamma oscillations.

Common to all oscillatory processes is that the neurons engaged in an oscillation undergo periodic changes of excitability. Phases of increased excitability, often associated with action-potential generation, alter with phases of low excitability (Fries et al. 2007). The mechanisms causing these cyclic changes in excitability are heterogeneous. In certain cells, often addressed as pacemaker neurons or clock cells, these cyclic oscillations of excitability are caused by interactions among voltage-gated ion channels that have antagonistic effects on the membrane potential (Heyer and Lux 1976). Here oscillation frequency depends on channel kinetics, membrane time constants, and driving forces. Individual pacemaker neurons usually operate in a characteristic frequency band and are typically found in pattern generator circuits that control rhythmic motion. Swimming, locomotion, respiration, heartbeat, and peristaltic movements are prime examples (Marder and Buchner 2001; Grillner 2006). However, cells with such properties are also found in structures traditionally not considered to be involved in pattern generation, such as the visual cortex. The chattering cells that engage in gamma oscillations when depolarized by current injection are one example (Gray and McCormick 1996). Recently it was discovered that such cells have narrow spikes and constitute a large fraction of the excitatory neurons in monkey V1 (Vinck, pers. comm.). Finally, even standard neurons can be considered as (relaxation) oscillators because action-potential firing is followed by refractory periods imposing cyclic alterations of excitability.

Another prominent mechanism for the generation of oscillations are circuit motifs giving rise to an oscillatory patterning of responses, the most common involving negative feedback loops, also addressed as recurrent inhibition. Excitatory neurons drive inhibitory neurons that inhibit the very same excitatory cells and these reciprocal antagonistic interactions naturally lead to an oscillatory patterning of the responses of both cell populations, the discharges of the inhibitory cells lagging slightly behind the discharges of the excitatory neurons (Whittington et al. 2000; Börgers and Kopell 2008; Buzsáki and Wang 2012). Here the frequency and regularity of the oscillations depend on a host of variables such as (a) the time constants of EPSPs, IPSPs, and dendritic integration, (b) the conduction delays of the feedback loops, (c) the excitatory drive, and (d) the embedding of the oscillator circuits in the network. Circuits tend to have their characteristic preferred oscillation frequency that can, however, be modulated over wide ranges by excitatory drive (Lowet et al. 2017).

Prominent examples are the septohippocampal circuits which generate the theta rhythm (Buzsáki 2006), the thalamocortical interactions responsible for the alpha rhythm (Steriade et al. 1993), and the cortical microcircuits which generate the beta and gamma oscillations known as ING and PING circuits (Kopell et al. 2000; Börgers and Kopell 2008; for a review, see Buzsáki et al. 2013). The excitability of cells participating in such oscillating circuits also undergoes a periodic modulation. At the peak of increased excitability, incoming EPSPs have a high probability to summate effectively and to generate action potentials while they are barely effective when arriving during the subsequent phase of enhanced inhibition. In this phase, EPSPs cannot summate effectively because of shunting inhibition and are less likely to reach firing threshold because of hyperpolarization. Thus, when cells oscillate, irrespective of whether the oscillations are due to pacemaker currents or circuit interactions, their ability to relay signals is modulated periodically, whereby the duration of the windows of opportunity decreases with oscillation frequency.

In conclusion, nervous systems are endowed with diverse and highly conserved mechanisms that are capable of imposing temporal structure on neuronal activity. How these temporal patterns can be used to establish precise temporal relations between the discharges of different neurons will be discussed in the following section.

## Mechanisms for the Establishment of Temporal Relations

Consistent temporal relations among the discharges of pairs of neurons are commonly addressed as correlations. These can have several causes, some of which are trivial. Discharges can become correlated by external events. This is a frequent phenomenon, addressed as stimulus locking, and can be distinguished from internally generated correlations by trial shuffling (shift predictor). Simple causes for internally generated temporal correlations are (a) common input from bifurcating axons, in which case one observes sharp and single peaks in the correlogram with close to zero phase lag; (b) a direct connection from a feeder cell to the target cell, in which case the correlogram also exhibits a single sharp peak, but now with a consistent offset; (c) reciprocal E/E connections among excitatory neurons, in which case correlograms tend to have broader peaks centered around zero; and (d) common fluctuations of global excitability, usually caused by descending or ascending modulatory systems, that regulate excitability in a global and state-dependent way. In this case one observes joint fluctuations of discharge rates that lead to very broad peaks in the correlograms. These coordinated rate fluctuations are addressed as "noise correlations." The term suggests that they are considered detrimental for information transmission, in particular for population coding (Averbeck et al. 2006). The controversies regarding the consequences of noise correlations will be discussed in detail later in this review.

A particularly versatile mechanism for the dynamic generation of temporal relations exploits the nonlinear interactions among *coupled oscillators* and therefore warrants an in-depth discussion.

## The Establishment of Temporal Relations
## in Coupled Oscillator Networks

An important feature of oscillatory circuits is their propensity to resonate and be entrainable by periodically modulated inputs. As observed as early as 1665 by Christiaan Huygens, a Dutch watchmaker, very weak interactions suffice to synchronize coupled oscillators if their preferred frequencies are similar. Huygens noticed that pendulum clocks synchronized their beat when fixed to the same timber due to the weak interactions caused by mechanical coupling. If the difference between preferred frequencies increases, stronger coupling is required to assure synchrony with stable phase locking, and if the frequency difference increases beyond a critical point, synchronization becomes unstable. Phase offset gradually increases and this may lead to intermittent phase resetting or a complete breakdown of synchrony. These complex and highly nonlinear relations have been analyzed in numerous theoretical studies (e.g., Winfree 1967; Aronson et al. 1990; Kuramoto 1990) and are summarized in the so-called Arnold tongue regime (Glass and Sun 1994). A graphical representation of synchronization behavior relating the difference in preferred frequency to increasing coupling strength leads to a "tongue"-shaped surface of possible synchronization regimes: the Arnold tongue.

Such reciprocal coupling between oscillatory circuits is a common motif in recurrently coupled neuronal networks and the cerebral cortex is a prime example. Thus, the Arnold tongue formalism can be applied to describe the relation between coupling strength and synchronization probability. Applied to the cerebral cortex, this predicts that the probability of two coupled columns synchronizing should increase with the strength of their reciprocal coupling. Experimental evidence from the visual cortex indicates that this is indeed the case. The network of the tangential intracortical connections that reciprocally couple neurons located in different functional columns is anisotropic. Columns responding to features that have a high probability to co-occur in natural scenes are more strongly coupled than columns tuned to features that are rarely contiguous (Gilbert and Wiesel 1989; Bosking et al. 1997; Stettler et al. 2002; Pecka et al. 2014). This selectivity of coupling is to a large extent due to experience-dependent pruning processes (Singer and Tretter 1976; Löwel and Singer 1992; Smith et al. 2015) whereby the statistical contingencies of features in the outer world are translated into the functional architecture of the network of intracortical recurrent connections (Iacaruso et al. 2017).

In addition to coupling strength and preferred oscillation frequency, synchronization probability also depends critically on the conduction velocity of the coupling connections that varies over a wide range in neuronal networks. If conduction delays exceed a critical value, synchronization breaks down and interactions may lead to drifting phase behavior or a complete shutdown of the oscillations of one or all of the coupled oscillators (Aronson et al. 1990; Niebur et al. 1991; Reddy et al. 1998; Vicente et al. 2008; for a review, see Pajevic et al. 2014). Thus, synchronization by reciprocal coupling can only be achieved over larger distances if coupling connections are fast conducting or if oscillation frequency is reduced. This agrees with the evidence that, in general, long-distance synchronization is more common in the beta and alpha bands than in the gamma band (Buzsáki et al. 2013). However, there is evidence that high-frequency gamma oscillations can become synchronized between the two hemispheres and that this synchronization is mediated by the reciprocal callosal connections rather than a common synchronizing input from subcortical projections (Engel et al. 1991). Thus, even high-frequency oscillations can be synchronized over fairly large distances. Gamma oscillations can also synchronize between distant cortical areas of the same hemisphere (Buschman and Miller 2007; Gregoriou et al. 2009), and analysis of phase lags and Granger causality suggests that this long-range synchronization results from direct (reciprocal) interactions. In this case, however, it cannot be excluded that the respective distant oscillators are synchronized in addition by common oscillatory input from a third source, which could either be another cortical area or a subcortical projection system (e.g., Saalmann et al. 2012).

Still another possibility to coordinate high-frequency oscillations is phase locking to a slower oscillatory process. Evidence that fast oscillations can be coupled to slow oscillations is available and addressed as cross-frequency coupling (for review and critical discussion, see Palva et al. 2005; Canolty et al. 2006; Montgomery et al. 2008; Belluscio et al. 2012; Aru et al. 2015). The most commonly observed form of coupling is that the amplitude (power) of the fast oscillations is modulated periodically by the phase of the slow oscillation. However, it is not yet clear whether this coupling can actually assure phase synchronization of concomitantly modulated fast oscillators, as this would require phase locking or coordinated phase resetting of the fast oscillations rather than solely amplitude modulation.

Finally, an important variable determining synchronization probability is the connectivity motif of the coupling connections. It matters whether the coupling connections are excitatory, inhibitory or both, whether they impinge only on excitatory or inhibitory elements of the (oscillatory) circuits or on both, and whether more than two oscillators are coupled with one another (Vicente et al. 2008; Pérez et al. 2011). As the most frequent motif for the coupling of oscillatory circuits, anatomical evidence suggests reciprocal interactions between the inhibitory elements of oscillatory circuits (I-I connections) and excitatory connections that impinge both on the excitatory and inhibitory elements (E-E/E-I connections). Prominent examples are the septohippocampal circuits supporting the theta rhythm (Buzsáki 2006), the thalamocortical interactions responsible for the alpha rhythm (Steriade et al. 1993), and the cortical microcircuits generating the gamma oscillations, known as ING and PING circuits (Kopell et al. 2000; Börgers and Kopell 2008; Veit et al. 2017; for a review, see Buzsáki et al. 2013). In general, long-range interactions are mediated by glutamatergic projections but recent tracing studies indicate that inhibitory projections also span large distances. Examples are putative GABAergic neurons in the cerebral cortex that project across the corpus callosum (Buhl and Singer 1989), inhibitory cells in the septum and entorhinal cortex that innervate the hippocampus, and GABAergic cells in the basal forebrain that project to the cerebral cortex (for a review, see Caputi et al. 2013). These GABAergic long-range projections tend to innervate selectively inhibitory neurons in the respective target structures and are therefore in a good position to influence, via I-I interactions, the respective oscillatory circuits.

In conclusion, the probability that reciprocally coupled oscillators synchronize, irrespective of whether they consist of pacemaker neurons or oscillatory microcircuits, increases with the similarity of their preferred oscillation frequencies, the strength of mutual coupling, and the conduction velocity of the reciprocal connections. In addition, state variables such as arousal, expectation, and readiness to act play an important role. They modulate the excitability of the nodes and thereby the entrainability of the coupled oscillator networks. An example is the cholinergic facilitation of induced gamma oscillations (Herculano-Houzel et al. 1999; Rodriguez et al. 2004).

So far only pairwise interactions have been considered but in reciprocally coupled networks, such as the cerebral cortex, interactions among multiple nodes lead to much more complex temporal patterns. These have been identified by multisite recordings and appear as stereotyped sequences of discharges distributed across nodes (songs, synfire chains) (Abeles 1991; Diesmann et al. 1999; Yuste et al. 2005; Gansel and Singer 2012) and traveling waves (Ermentrout and Kleinfeld 2001).

Altogether these mechanisms provide ample opportunities to impose temporal structure on neuronal activity and to establish precise temporal relations among discharges. These relations can then be converted by the established mechanisms of time-sensitive synaptic plasticity into modifications of functional architectures, whereby the selectivity with which the gain of synaptic connections between oscillating nodes can be modified depends critically on the oscillation frequency. For fast oscillations, for example, in the gamma range, the window of opportunity for the strengthening of synapses (i.e., the phase of heightened excitability) is very short, in the range of 10 ms or less (Wespatat et al. 2004), whereas for theta oscillations it is in the range of 100 ms. Thus, the faster the oscillations are, the greater the temporal selectivity is with which synchronous firing can be converted into lasting gain changes of synaptic connections.

In conclusion, the oscillatory patterning of neuronal responses and their synchronization can endow neuronal responses with the temporal structure required to make use of the established time-sensitive mechanisms of synaptic plasticity. This permits the selective association of neuronal populations whose responses are not time locked to stimuli but generated internally (e.g., during imagery or recall of memories). Evidence that this option is likely exploited has been provided by investigations on memory consolidation in human subjects (Miltner et al. 1999; Axmacher et al. 2008; Fell et al. 2011) and animals (for additional references, see Yamamoto et al. 2014; Singer 2017).

## The Role of Oscillations and Synchrony in Information Processing and Coding

An undisputed function of oscillations in neuronal systems is the generation of rhythmic movements (see above). Furthermore, the cyclic changes of excitability associated with oscillations play a crucial role in the gating of signal transmission and the routing of activity. Thus, transmission of sensory signals is modulated as a function of their timing relative to the phase of self-generated or stimulus-evoked oscillations (for a review, see Lakatos et al. 2008, 2013; Van Rullen 2016), and this modulation likely plays a role in the discontinuous sampling of information (Sergent et al. 2005; Landau and Fries 2012; Ni et al. 2017). Engaging in an oscillation can also increase the saliency of neuronal signals because it often results in burst firing (Gray and McCormick 1996). Bursts, in turn, enhance the impact of excitatory input on target cells without increasing average discharge rate, especially in sparsely connected networks (Larkum 2013).

Particularly consequential functions emerge once several nodes of a network engage in oscillations and synchronize. The high propensity of coupled oscillators to synchronize (see above) can be exploited to establish precise and highly specific temporal relations between the discharges of the neurons participating in the respective oscillatory circuits. These temporal relations can range from zero phase lag synchrony, in which case the engaged neurons discharge simultaneously, to synchronization with various phase lags, in which case one obtains precisely timed *discharge sequences* distributed across populations of neurons (Havenith et al. 2011). This option is exploited for a host of executive functions. The control of bird song, speech, and composite movements are but a few examples (Marder and Buchner 2001; Suthers and Margoliash 2002; Grillner 2006).

It is likely, however, that these dynamics may also play a role in cognitive processes. Provided that information can be encoded in and read out from temporal patterns in the discharges of distributed neurons, for which there is evidence (Masquelier et al. 2009), coding space can be considerably enlarged by exploiting an additional dimension of fine-grained temporal relations.

## The Role of Oscillations and Synchrony in the Dynamic Gating of Neuronal Interactions

Another important function supported by the synchronization of nodes is the dynamic gating of interactions which permits flexible routing of signals across the backbone of fixed anatomical connections without requiring synaptic gain changes. If two anatomically coupled nodes engage in oscillatory activity, phase and frequency shifting can be used to selectively and dynamically modulate the strength of functional coupling without changing the gain of synapses. If phase relations are adjusted such that the EPSPs of one node arrive at the other during its phase of heightened excitability, transmission is facilitated, whereas it is blocked when phase is shifted by 180° (Fries 2005, 2009). Direct experimental evidence for such a phase-dependent gating mechanism has been obtained in the visual system (Womelsdorf et al. 2007; Bosman et al. 2009; Bastos et al. 2015).

This dynamic gating mechanism has been shown to serve multiple functions. Recent evidence indicates that it is exploited for attention-dependent signal selection. If two oscillatory signals compete for transmission in an upstream target area, the signal wins that succeeds to phase lock with oscillations in the target area (for a review, see Maris et al. 2016).

Synchronization has also been shown to support dynamic configuration of functional networks. Accordingly, measures of coherence among oscillatory signals such as phase locking indices, paired phase consistency, Granger causality, and transfer entropy are commonly applied to identify functional networks. As expected, these measures reveal networks composed of nodes linked by strong anatomical connections because, as discussed above, these nodes have a higher probability to engage in synchronous activity than weakly coupled nodes. An example of such preconfigured functional subsystems is the default network or the language network. However, coherence analysis also revealed networks that get configured on the fly in a context- and task-dependent way.

Cats trained to perform a motor response to the change of a visual stimulus synchronized, in anticipation, those cortical areas that needed to be engaged to accomplish the task. In response to an auditory cue announcing the next trial, neurons in the visual, parietal, somatosensory, and motor cortex of the hemisphere controlling the motor response engaged in beta oscillations that were synchronized with close to zero phase lag *before* the visual stimulus appeared (Roelfsema et al. 1997). Thus, communication among processing stages required to accomplish the task is likely to be facilitated in anticipation by entrainment in coherent oscillations. By enhancing coherence of oscillatory activity, similar task-dependent formation of functional networks has been observed for diverse executive and cognitive functions that require dynamic coordination of distributed processing nodes (Buschman and Miller 2007; Siegel et al. 2008; Gregoriou et al. 2009; Grothe et al. 2012; Salazar et al. 2012; Dotson et al. 2014; for a review, see von der Malsburg et al. 2010).

Finally, as proposed following the discovery of synchronized gamma oscillations in the visual cortex, this gating mechanism also supports the dynamic binding of neurons driven by perceptually groupable (semantically related) features into functionally coherent assemblies (distributed coding). As reviewed above, the architecture and coupling strength of intercolumnar horizontal connections reflects the statistical contingencies of feature constellations. Neurons tuned to features that have a high probability to co-occur, and hence to be related, are more strongly coupled than neurons coding for unrelated features. Consequently, and as shown in numerous studies, neurons responding to related features—those that tend to be bound perceptually according to the well-established Gestalt rules for perceptual grouping and scene segmentation—have a high probability to synchronize their responses if their preferred stimuli are presented together. Thus, synchronization probability reflects the Gestalt rules for perceptual binding. Nearby columns responding to closely spaced contours are all strongly coupled, which corresponds to the Gestalt rule of vicinity. More widely separated columns exhibit strong coupling only if they respond to the segments of continuous elongated contours or contours with similar orientation, especially when these are aligned collinearly and/or move with the same speed in the same direction. This reflects the Gestalt criteria of continuity, similarity, collinearity, and common fate. As shown in numerous experimental and simulation studies, neurons in these strongly coupled columns have a high probability of synchronizing their oscillatory discharges with close to zero phase lag when co-activated by their preferred features (Gray et al. 1989; Gray and Singer 1989; König and Schillen 1993; Schillen and König 1994; Kohn and Smith 2005; for reviews, see Singer 1999 and Uhlhaas et al. 2009).

Thus, temporal contiguity of discharges appears again to serve as a code of relatedness. However, in this case it is not used for developmental pruning of connections or the association of neuronal assemblies in the context of learning; it serves to define relations during the processing of visual patterns as proposed by the "binding by synchrony" hypothesis.

In conclusion, precise timing relations among neuronal discharges are used during development and learning to translate consistent correlations among features in lasting changes of coupling strength, and these anisotropies in coupling, in turn, determine the probability of synchronization once such "trained" networks are presented with sensory stimuli. Thus, synchronization is used in an instructive way during development and learning to store "knowledge" about statistical contingencies in the functional architecture of networks, and it is then used again for the readout of stored knowledge during information processing.

This gating of neuronal interactions by synchronization of oscillatory activity likely plays a more important role than is suggested by the firing statistics of neurons considered in isolation. The reason for this is that neurons participate considerably more often in oscillatory processes than is suggested by extracellular recordings of discharges. Even when the population of interacting neurons engages in regular oscillations—as indicated by oscillatory field potential fluctuations and periodic fluctuations of the cells' membrane potential—the discharges of individual cells may still exhibit Poisson statistics because of cycle skipping (Zeitler et al. 2008; Palmigiano et al. 2017).

While there is common agreement that precise correlations among neuronal discharges play a crucial role in synaptic plasticity during developmental pruning and learning, the proposal that such correlations also play a role in signal processing has given rise to substantial controversies. Thus a brief summary of the most pertinent arguments for and against a functional role of synchrony in signal processing seems warranted when discussing the role of cortical dynamics.

## Counterarguments

It has been argued that synchronized oscillations cannot have a functional role because they are too volatile and nonstationary to support reliable processing. To appreciate this criticism, it is necessary to consider the constraints of the commonly applied methods for the detection of oscillations and synchrony, such as Fourier and wavelet analysis, auto- and cross-correlation techniques, or various coherence measures. These techniques are applicable only if oscillation frequencies are reasonably stable and if oscillatory episodes last for a substantial number of cycles. Even for the analysis of the high-frequency gamma oscillations, measurement windows typically range from 200 ms to 1 s. As a consequence, most studies investigating oscillations and synchrony have been performed under conditions that favor the occurrence of sustained, frequency stable oscillations. Typical cases are the induction of gamma oscillations in the visual cortex with drifting grating stimuli or isolated contours. These studies generated a host of interesting insights into the dependence of oscillations and synchrony on stimulus features, central states, attention, and behavioral goals. However, because of the applied stimulation and analysis techniques, these studies also nurtured the notion that functionally relevant synchronization can only be established under very restricted stimulation conditions and that it is absent or too spurious to serve a function under more natural conditions, especially when information processing has to occur on fast timescales, as is the case during visual exploration. Primates perform on average four saccades per second, which implies that new visual information is sampled every 250 ms (Maldonado et al. 2008; Ito et al. 2011), and psychophysical evidence indicates that familiar scenes can indeed be segmented and recognized within such short intervals. Thus, it has been argued that there is not enough time to rely on synchronization of sustained oscillations for feature binding and the formation of object-specific neuronal assemblies. Even more problematic is that synchronous oscillations, when assessed with conventional Fourier or wavelet analyses, decrease in amplitude or become undetectable when complex stimuli are presented, such as cluttered scenes or images that lack clear high-contrast boundaries (Lima et al. 2010). Moreover, it has been argued that the dependence of oscillation frequency and power on contrast, motion speed, eccentricity, and complexity of stimuli is incompatible with the idea that spike synchronization can be used to encode semantic relations, as postulated in the binding by synchrony hypothesis (Singer 1993; Singer and Gray 1995), and to gate communication by coherence, as postulated by the CTC hypothesis (Fries 2005). These arguments have been made explicit by Atallah and Scanziani (2009), Burns et al. (2010, 2011), Ray and Maunsell (2010), Jia et al. (2013a, b); for reviews, see Ray and Maunsell (2015) and Palmigiano et al. (2017).

Before reviewing more recent evidence that counters these arguments, it is worth noting that the constraint of processing speed also poses a problem for coding strategies that rely on joint rate increases for the definition of relations, because discharge rates of cortical neurons are low and can carry only little information when integrated over short intervals. This problem could, in principle, be solved by averaging over large populations of cells. However, considerations on sparse coding and noise correlation make this solution appear suboptimal as well (Averbeck et al. 2006). For these very reasons, it has been proposed that encoding relatedness by spike synchronization, rather than by joint rate increases, accommodates the superposition problem inherent in assembly coding (see above) as well as the constraint of processing speed (Singer 1999; Van Rullen et al. 2001, 2005; for a review, see Korndörfer et al. 2017).

## Supporting Evidence

Recent experimental studies that combine multisite recordings with time-resolved analysis of coherent oscillations and spike synchronization in awake, behaviorally trained monkeys have shown that high-frequency oscillations and their synchronization also occur under natural conditions. In addition, they have confirmed the volatile features of synchronization phenomena: nonstationarity, frequency variability, short duration, rapid phase shifts, and fast formation and dissolution of coherent states. These studies also revealed, however, that the brief bouts of coherent activity contain information about the contents of working memory (Lundqvist et al. 2016), the communication between cortical areas (Siegel et al. 2008), the dynamic formation of functional networks (Buschman and Miller 2007), and the direction of information flow (Bastos et al. 2015; Lowet et al. 2017).

The indications that even short periods of synchronization are functionally relevant and informative are in perfect agreement with the dynamics of simulated recurrently coupled networks of spiking neurons. Korndörfer et al. (2017) recently demonstrated that neurons in reciprocally coupled networks engage very rapidly in synchronous discharges when activated by structured input, and that the synchronization probability is determined by the strength of coupling. This confirms the experimental evidence that neurons responding to features that need to be bound together for perceptual grouping and scene segmentation synchronize their responses if activated by patterns containing groupable features (Gray et al. 1989). In the study by Korndörfer et al. (2017), no explicit oscillatory properties of the nodes were implemented, but the spiking neurons had the usual refractory period and hence shared features of relaxation oscillators. The additional implementation of inhibition influenced network dynamics, as expected, but did not interfere with the fast synchronization of spikes.

Another comprehensive simulation study by Palmigiano et al. (2017) investigated the effect of synchronization on information transfer in coupled oscillator networks. They simulated *delay-coupled* recurrent networks with spiking excitatory and inhibitory neurons that shared essential connectivity motifs of supragranular cortical layers. In this study, two regimes of network dynamics were examined. One, addressed as the transient synchrony regime, was characterized by low coherence, as is typically observed for resting-state activity in awake animals or when cortical networks are challenged with complex stimuli. Spike statistics were stochastic but field potentials computed from averaged membrane potential fluctuations exhibited bursts of synchronous gamma oscillations which only lasted a few cycles. When the strength of recurrent inhibition was increased, together with an increase in background drive, oscillation frequency and discharge rate remained in the same range, but oscillations became more sustained and synchrony more robust. This second regime resembles *in vivo* conditions when responses are evoked with regularly structured, high-contrast stimuli that induce strong excitation and inhibition such as drifting gratings. The interesting and unexpected outcome of measurements of information transfer in this delay-coupled oscillatory network was that phase-dependent gating of information transfer was as effective and even more flexible in the transient synchrony regime than in the regime exhibiting more sustained oscillations and synchrony. The short duration of the oscillatory bursts allowed for fast opening and closing of transmission channels by frequency tracking and the direction of information flow rapidly reversed if phase relations changed. Palmigiano et al. concluded that "features that at first sight appear to be noncompliant with information routing may actually provide the brain with a particularly flexible routing mechanism." This conclusion received strong support from the recent experimental study by Lowet et al. (2017), who investigated the dynamics of synchronized gamma oscillations in V1 of awake monkeys using massive parallel recordings. The observed dynamics resemble in great detail those reproduced by the simulated network and confirm essential predictions of the hypothesis that supragranular layers of the cerebral cortex can be considered as a delay-coupled recurrent oscillator network. In this study, all predictions derived from the Arnold tongue formalism could be confirmed experimentally.

Taken together, both the results of electrophysiological experiments and simulation studies indicate that synchronization of spike discharges can be fast enough to serve feature binding/perceptual grouping within the short intersaccadic fixation intervals (see also Lowet et al. 2016). Moreover, the transient and variable nature of synchronized gamma oscillations, characteristic of free viewing conditions and the processing of complex scenes, does not compromise the many other functions assigned to synchronization. Rather, the fast fluctuations between synchronized and uncorrelated states are advantageous for the rapid and flexible definition of relations between distributed neuronal responses, as is required for attention-dependent input selection, the flexible and selective routing of signals on the backbone of the fixed connectome, the task-dependent formation of functional networks, and the flexible binding of distributed feature detectors into functionally coherent assemblies. Recent experimental results obtained in awake monkeys by Lowet et al. (2016), Bosman et al. (2009) on the effect of microsaccades, and Brunet et al. (2015) on oscillations during free viewing of natural scenes are fully compatible with this view.

## Dynamics beyond Oscillations and Synchrony

The nonlinear interactions within delay-coupled oscillator networks give rise to very complex dynamics that provide a very high-dimensional space that, in principle, can be exploited for flexible and efficient computation. In the following, some of these options are discussed and substantiated with recently obtained experimental evidence. I argue that such networks are particularly suited to cope with a number of hitherto poorly understood functions: the encoding of temporal sequences, the storage of vast amounts of information about the environment in the networks of sensory cortices, the ultrafast retrieval of information in processes requiring comparison between input signals and stored knowledge, and the fast and effective classification of complex spatiotemporal input patterns.

Theories of perception, formulated more than a hundred years ago (von Helmholtz 1896), and a plethora of experimental evidence indicate that the brain interprets sparse and impoverished input signals on the basis of an internal model of the world. The information provided by this model is used to reduce redundancy, to facilitate segregation of figures from background, to bind signals evoked by features constituting a perceptual object, and to enable classification and identification. Given the daunting complexity of the visual world, the store containing such an elaborate model must have an immense capacity to accommodate the vast number of statistical contingencies required for the interpretation of ever-changing sensory input patterns. Moreover, this massive amount of prior knowledge needs to be arranged in a configuration that renders it accessible within fractions of a second to meet the constraint of known processing speed.

The proposal is that these requirements can be met if encoding, storage, and processing of information take place in the high-dimensional state space provided by complex systems with nonlinear dynamics.

As reviewed above, information about contingencies in the outer world is stored in the synaptic weight distributions of the tangential intracortical connections that reciprocally couple feature selective nodes. In low visual areas, the nodes are selective for elementary features, such as the orientation of contour borders, whereas in higher visual areas, the nodes code for increasingly complex constellations of elementary features. Accordingly, one expects that the weight distributions of the tangential coupling connections in these areas reflect contingencies of higher order. However, because multisite recordings and correlation analysis have not been performed in higher visual areas, we know little about the interactions between network nodes in higher visual areas.

## Resting- and Stimulus-Induced States

The brain's spontaneous activity is constrained by its functional architecture. Hence, the dynamics of resting activity must reflect the weight distributions of the structured network that harbors the entirety of latent internal priors. This predicts that resting activity is high dimensional and represents a vast but constrained manifold inside the universe of all theoretically possible dynamical states. If input signals become available, they are likely to trigger a cascade of effects: they drive in a graded way the feature-sensitive nodes, thereby constraining the network dynamics. If the evidence provided by the input patterns matches well the priors stored in the network architecture, the network dynamics will collapse to a specific substate that provides the best match with the corresponding sensory evidence. Such a substate is expected to have a lower dimensionality than the resting activity, to exhibit specific correlation structures, and to be metastable due to reverberation among nodes supporting the respective substate. Because these processes occur within a very high-dimensional state space, substates induced by different input patterns are likely to be well segregated and therefore easy to classify. They can then serve either as input to the next cortical processing stage, where higher-order priors are implemented, permitting emergence of more abstract interpretations, or they can be classified by local readout units that directly feed into executive centers.

## Analogies from Computational Studies

In a much simplified version, the nonlinear dynamics characteristic of recurrent networks are exploited for computation in certain AI systems, the respective strategies being addressed as "echo state, reservoir, or liquid computing" (Buonomano and Maass 2009; Lukoševičius and Jaeger 2009; D'Huys et al. 2012). The following Gedanken experiment illustrates the principle.

Objects impact at different intervals and locations in a pond of water and generate propagating waves whose parameters reflect the size, impact speed, and location of the object. The wave patterns fade with a time constant determined by the viscosity of the liquid, interfere with one another, and create a complex dynamic state. This state can be analyzed by measuring, at several locations in the pond, the amplitude, frequency, and phase of the respective oscillations, and from these variables a trained classifier can subsequently reconstruct the exact sequence and nature of the impacting "stimuli."

Fernando and Sojakka (2003) put these ideas into practice using an actual bucket of water and showed that the interference between waves on the water surface allowed a simple perceptron to solve the XOR problem. The advantages of this computational strategy are as follows:

1. Low-dimensional stimulus events are projected into a high-dimensional state space, where nonlinearly separable stimuli become linearly separable.
2. The high dimensionality of the state space can allow for the mapping of more complicated output functions (like the XOR) through simple classifiers.
3. Information about sequentially presented stimuli persists for some time in the medium (fading memory). As such, information about multiple stimuli can be integrated over time, allowing for the mapping of complicated temporal functions.

These properties make artificial recurrent networks extremely effective for complex sequence processing as demonstrated by a number of applications: grammar learning, automatic driving systems, generation of handwritten text, image captioning, and so forth. Recent simulation studies have actually shown that performance of an artificial recurrent network is substantially improved if the recurrent connections are made adaptive and can "learn" about the feature contingencies of the processed patterns (Lazar et al. 2009; Hartmann et al. 2015). Interestingly, the local plasticity mechanisms used for learning and adaptation in these models have lasting consequences on the spatiotemporal structure of spontaneous activity. Changes in the structure of spontaneous activity are consistent with the statistical properties of the previously observed stimuli and have functional consequences for stimulus disambiguation and interpretation (Hartmann et al. 2015).

## Experimental Evidence

The structural similarities between these artificial recurrent networks and supragranular layers in the cerebral cortex suggest that some of the computational strategies applied in reservoir computing might actually be used by the cerebral cortex. Experimental studies testing this hypothesis are still rare and have become possible only through the advent of massive parallel recordings from the network nodes. So far, however, the few predictions that have been subject to experimental testing appear to be confirmed. The covariance structure of resting activity does indeed reflect the anisotropic layout of the tangential reciprocal connections (Bosking et al. 1997; Fries et al. 2001a; Kenet et al. 2003), is modified by learning (Lewis et al. 2009; Kundu et al. 2013), and reveals hallmarks of an internal model of the environment (Berkes et al.

2011). Evidence is also available that cortical computations exploit the high-dimensional dynamics of recurrent networks to encode information about stimulus sequences. Responses to successively presented visual stimuli (letters and numbers) were recorded with matrix electrodes simultaneously from a random sample of ~ 60 neurons in cat primary visual cortex; linear classifiers were trained on short segments (5–100 ms) of the activity vectors of a training set of responses; these classifiers were then used to identify the nature of the presented stimuli in a test set (Nikolic et al. 2009). These experiments revealed the following:

1. Information about a particular stimulus persists in the activity of the network for up to a second after the end of the stimulus (fading memory).
2. Information about sequentially presented stimuli superimposes so that two subsequent stimuli can be correctly classified some time after the end of the second stimulus.
3. Information about stimulus identity is distributed across many neurons (> 30) and encoded in the correlation structure of the responses.

Evidence has also been obtained that sensory signals matching the priors stored in the network cause a collapse of high-dimensional network dynamics into metastable subregions of the state space that are stimulus specific and distinguished by enhanced coherence (covariance, synchrony), reduced variability, and lower dimensionality (Churchland et al. 2010; Banyai et al. 2018; Klein et al., in prep.; Lazar et al., in prep.). Stimuli that do not match the stored priors (e.g., "unnatural" stimuli) evoke substates that are less coherent, less stimulus specific (Banyai et al. 2018), and more difficult to classify (Lazar et al., pers. comm.). This could explain why grating stimuli elicit particularly strong synchronized gamma oscillations. A grating matches several priors (Gestalt criteria, grouping criteria), namely those of continuity, collinearity, similarity (in this case in the orientation domain), regularity, and (if the grating drifts) common fate. As discussed by Vinck and Bosman (2016), stimuli that match predictions stored in the cortical network appear to give rise to particularly well-synchronized gamma oscillations. Evidence also indicates that the cortical network "learns" about stimulus statistics and exploits this knowledge to optimally segregate the representations of different stimuli. Repeated presentation of stimuli has been shown to cause unsupervised changes in the network with the consequence that familiar stimuli evoke substates that are better classifiable than those evoked by less familiar stimuli, because they are better segregated in high-dimensional dynamic space (Lazar et al., in prep.). These experience-dependent modifications of the network have an impact also on the structure of resting-state activity: the vectors specific for highly familiar stimuli are spontaneously replayed (Lazar et al., pers. comm.).

There are also indications that top-down mechanisms related to attention and expectancy constrain the dynamic space in anticipation of having to respond to the cued stimulus. This top-down effect is associated with a massive enhancement of gamma oscillations induced by the attended (Fries et al. 2001b) or expected stimulus (Lima et al. 2011). Also, one observes a ramp up of the power of gamma oscillations and a change in dimensionality of the network dynamics following presentation of a cue that instructs the animal about the sequence of future events (Klein et al., in prep.). These top-down influences change the correlation structure of the activity vectors and could contribute to speeding up the formation of specific substates once sensory evidence becomes available.

## Concluding Remarks

Despite considerable effort, there is still no unifying theory of cortical processing. As a result, numerous experimentally identified phenomena lack a cohesive theoretical framework. This is particularly true for the dynamic phenomena reviewed here, because they cannot easily be accommodated in the prevailing concepts which emphasize serial feedforward processing and labeled line codes. However, with its preponderance of reciprocal connections and the rich dynamics that result from these reciprocal interactions, the cortical connectome suggests that additional processing strategies are implemented. Here I proposed concepts that assign specific functions to oscillations, synchrony, and the more complex dynamics that emerge from a delay-coupled recurrent network. These concepts are fully compatible with the robust evidence for the encoding of relations by anatomical convergence (labeled line codes) but complement this mechanism through a scenario in which the precise temporal relations among the discharges of coupled neurons serve as complementary code for the definition of semantic relations, both in signal processing and learning. In addition, I introduced a computational strategy that capitalizes on the high-dimensional coding space offered by reciprocally coupled networks. In this conceptual framework, information is distributed and encoded in the discharge rate of individual nodes (labeled lines) as well as, to a substantial degree, in the precise temporal relations among the discharge sequences of distributed nodes. The core of the hypothesis is that the dynamic interactions within delay-coupled recurrent oscillator networks (a) endow responses with the precise temporal structure required for the encoding and learning of semantic relations, (b) exhibit complex, high-dimensional correlation structures that reflect the weight distributions of the coupling connections and serve as an internal model, and (c) permit fast convergence toward stimulus-specific substates that are easy to classify because they occupy well-segregated loci in the high-dimensional state space. Analysis of the correlation structure of these high-dimensional response vectors is still at the very beginning. However, methods are now available for massive parallel recording from large numbers of network nodes in behaving

animals. It is to be expected, therefore, that many of the predictions deriv-
able from this novel concept will be amenable to experimental testing in the
near future.

# 11

# Computational Elements
# of Circuits

Johannes Leugering, Pascal Nieters, and Gordon Pipa

## Abstract

Information processing in the brain is implemented across several temporal and spatial scales by populations of neurons. This chapter addresses how single neurons, small network motifs, and larger networks, in which emergent dynamics are largely shaped by the connectivity of the system, contribute to this processing of information. Computation is defined as a semantic mapping; that is, it is the process by which representations of external (e.g., stimulus-driven) or internal (e.g., memories) information change. A feature specific to neuronal computation is that mappings are mostly local, constrained by connectivity patterns between neurons. This implies that complex mappings from local information onto representations that are highly relational and abstracted, and which rely on information between distant parts of the system, require mechanisms that can bridge, bind, and integrate pieces of information across large scales. An overview of this process in the nervous system is delineated: Local information processing is described at the level of individual neurons and small motifs. Emergent phenomena are addressed that implement information processing across large recurrent neuronal populations. Finally, an omnipresent but mostly ignored feature of neuronal systems, delay-coupled computation, is described.

## Information Processing in Single Neurons and Populations

An understanding of how information is processed in neural systems begins with a consideration of how an individual neuron perceives and processes information, before extending this scope gradually to larger systems. Our goal in this chapter is to present a concise, abstract view of computation in neural systems, understood to be key to a meaningful change in the representation of information. In the interest of brevity, the biological complexity of neurons and networks (e.g., the role of specific ion channels or the potential influence of glia cells and neuromodulators) will not per se be addressed.

## A Stochastic Process Linear-Nonlinear Neuron Model

From the perspective of the linear-nonlinear (LN) model, a neuron is a computational unit that receives a multivariate time-varying input signal through its synaptic inputs and generates a univariate time-varying output signal. This mapping from input to output signals is near instantaneous (at least time-invariant), as the neuron itself is assumed to have, at most, a very limited internal memory[1] and be subject to noise.

In the mathematical framework of stochastic processes, a neuron can thus be concisely described as a nonlinear, causal, time-invariant operator that maps a multivariate stochastic process onto a univariate stochastic process. We make several simplifying assumptions that result in a convenient class of neuron models (Ostojic and Brunel 2011; see also Figure 11.1):

- The neuron's operation can be modeled as a leaky integrator or, even simpler, an instantaneous input-output mapping.
- It is composed of a linear operator, which reduces the multivariate input arriving at different synapses along the dendritic tree to a univariate input to the neuron's soma, followed by a nonlinear transformation.
- The linear operation is parameterized by synaptic weights, which can be positive or negative.
- The nonlinear transformation, which we refer to as the activation function or just nonlinearity, is a monotonically increasing, (locally) differentiable and bounded function.

While the activation could be further used in a spike generation process as an instantaneous firing rate, we treat it here as the neuron's continuous state or output. Each neuron in a population independently processes its own input (which may be correlated to other neurons' inputs), and its state provides one component of the entire population's multivariate state. The computation carried out by a population of neurons, mapping a multivariate input signal onto a multivariate state, must thus arise component-wise from the computations realized in the individual neurons. Each neuron, however, is limited to those operations which can be realized by a LN model under the above constraints.

To better understand the capabilities and limitations of this class of models, it helps to analyze them from a machine learning perspective, where such models commonly appear under different guises and names.

---

[1] The exception to this rule is found in plasticity mechanisms, which we assume to operate on a much slower, separate timescale than that of the output signal, and thus they can be treated as virtually constant in this context. The commonly made assumption of near instantaneous operation of the neuron further presumes that slower active dendritic processes do not substantially contribute to computation, which can be called into question and may turn out to be an overly simplistic perspective.

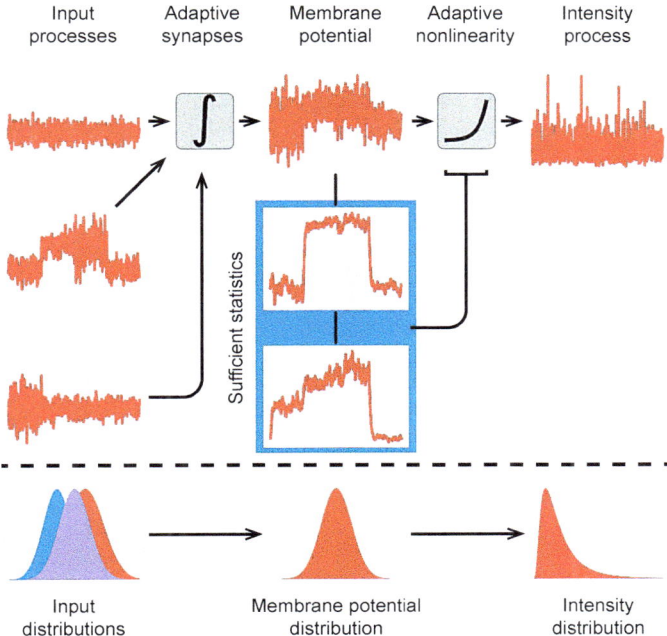

**Figure 11.1** A model neuron receives a linear combination of multiple time-varying stochastic processes that are scaled by adaptive, synaptic weights and integrated into the neuron's membrane potential. By sensing some sufficient statistics of the membrane potential, the neuron's nonlinearity can be adjusted to achieve an activation (or intensity) with desirable statistical properties. Assuming stationarity of the input processes, the neuron's nonlinearity can be determined by the desired mapping from the univariate membrane potential distribution to an intensity distribution. Adapted from Leugering and Pipa (2018).

### LN Models in Machine Learning

Using a Heaviside function for the nonlinearity, LN models appear in machine learning in the form of linear hard-margin classifiers, such as the classical perceptron (Rosenblatt 1958), linear support vector machines (Hearst et al. 1998), or depth-one decision trees (so-called "decision stumps"; Criminisi et al. 2012). With continuous nonlinearities, such as the logistic function, these models can be used as soft-margin classifiers and regressors, as in generalized linear models (GLMs) (McCullagh and Nelder 1989), where the nonlinearity is used to relate a linear combination of input features to the expected value of the (task-specific) label associated with the data. To improve performance, multiple instances of such models can be combined laterally to form an ensemble, used in a boosting procedure or stacked hierarchically, like the layers of an artificial neural network (Hopfield 1988) or the levels of a decision tree.

Computation in this context simply refers to the ability of the model to encode specific task-relevant information about its inputs into its output. The same claim has been made for individual biological neurons, as well as whole layers of neurons in deep networks under the "information bottleneck" principle. The deceptively simple argument is that each neuron (or each layer of a network, respectively) is presented with a high-dimensional input signal that carries task-relevant, as well as irrelevant, information, and, in a noisy environment with limited capacity to transmit information, ought to transform it into an informative low-dimensional output signal (Becker 1996).

**Supervised Learning**

In a supervised setting, where the desired output of the model is known at all times, the extraction and transmission of task-relevant information with simultaneous suppression of task-irrelevant "noise" represents a form of lossy compression. In multilayer networks, backpropagation can provide a supervised error signal for each layer and ultimately each neuron, thus allowing it to locally solve a lossy compression problem, which has been hypothesized as the theoretical mechanism underlying the surprising success of deep neural networks (Shwartz-Ziv and Tishby 2017).

**Unsupervised Learning**

The concept and potential mechanisms of error backpropagation in biological neural networks, however, are controversial, and the existence of supervised target signals may be called into question altogether. In the absence of supervision, the information bottleneck principle can be restated as the objective for each neuron to simply encode its inputs into its output in the most informative way possible, since it cannot distinguish task-relevant from irrelevant information.

The information encoding of the output signal is reflected in the firing statistics, with heavy-tailed firing rate distributions corresponding to sparse spiking codes, and narrowly peaked distributions corresponding to tonic firing or bursting codes. By driving synaptic plasticity, this can, in turn, shape the topology of synaptic connections and lead to the formation of specific motifs, thus allowing a population of neurons to implement task-relevant computation without supervision.

Under the biological constraints imposed on the neuron (e.g., bounded firing rates, energy limitations), the mutual information between input and output is bounded by the entropy attainable by the output distribution. A common objective is thus for the neuron to enforce a maximum entropy distribution of its outputs by appropriately adjusting its nonlinearity, while simultaneously tuning its synaptic connection weights to project the multidimensional input signal onto the most informative subspace. Equivalently, for a population of neurons, the information bottleneck objective is to realize a maximum entropy joint distribution, such that each marginal distribution of an individual neuron's output satisfies the biological constraints.

### Unsupervised Learning Application: Independent Component Analysis

As it turns out, this objective fully determines a unique optimal choice of nonlinearity for a given family of input distributions and a desired output distribution. It also implies that the linear subspaces selected by the neurons' respective synaptic input weights should correspond to the main independent components. Consequently, this problem is also referred to as independent component analysis (ICA), a generalization of principle component analysis which can no longer be solved by linear methods (Hyvärinen and Oja 1998; Triesch 2007).

This intuition transfers seamlessly to a framework of stochastic processes (Leugering and Pipa 2018), where a population is tasked with mapping its (stationary) multivariate input process onto a multivariate output process, with a joint distribution composed of independent components with given marginal distributions. By factoring the population's joint distribution into its marginal distributions and a copula function, it becomes apparent that this objective can be achieved through the interaction of two distinct mechanisms:

1. The copula function captures all of the dependency structure present in the joint distribution and depends only on the choice of synaptic input weights of the population; thus it can be adjusted by synaptic plasticity.
2. The marginal distribution of each neuron's output can be enforced purely by an appropriate choice of nonlinearity; thus it can be adjusted by intrinsic plasticity.

Since all of the information required to solve the ICA problem is available locally to the neurons or their synapses, it can be solved by the LN model discussed above using simple, biologically plausible mechanisms of intrinsic and synaptic plasticity in a time-continuous, noisy setting.

Using motifs of several laterally inhibiting neurons, different independent components can be found, leading to a highly informative, multivariate output signal. As shown in Figure 11.2, such a structure can be used to learn, in an unsupervised fashion, to classify MNIST images with just a handful of neurons. For an in-depth discussion of this result, see Leugering and Pipa (2018).

### Computation in Networks Using Emergent Properties

The cerebral cortex is a highly distributed system with reciprocal connections that shape neuronal activity through self-organizing and that can create

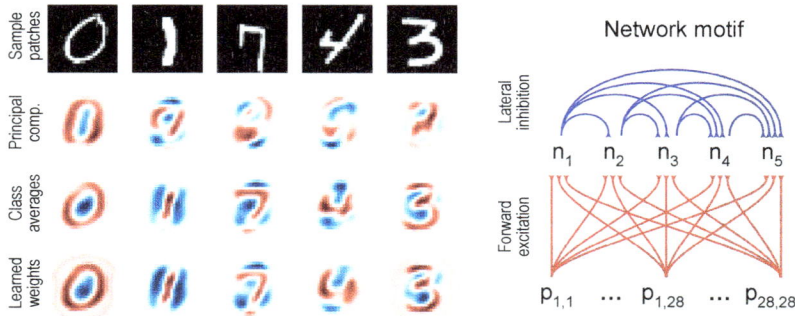

**Figure 11.2**   A small motif of five neurons receives feedforward excitation from 28 × 28 neurons, representing pixels of visual inputs. Images from the MNIST database are presented successively, while the synaptic weights and each neuron's nonlinearity are adjusted by local synaptic and intrinsic plasticity, respectively. Only the combination of both learning mechanisms leads to the (unsupervised) discovery of independent components in the input space, corresponding to the average input for each class. Lateral inhibition learns to decorrelate the neurons and ensures that different components are discovered, reducing redundancy and thus maximizing the information content of the motif's output. Results from Leugering and Pipa (2018).

coherent states able to encode representations of sensory objects, decisions, and programs for motor acts (Uhlhaas et al. 2009). The topology of the connectivity shares properties with small world networks having no singular center where all information converges (Gerhard et al. 2011). This raises questions of how the numerous computations on the level of single neurons are coordinated and bound together to give rise to coherent percepts and actions, and how relations between simultaneously represented contents can be encoded. One option is that neuronal synchrony can implement both features. Several mechanisms have been discussed—for instance mediated by inhibitory synapses, enhanced via gap junctions, induced by motifs of neuronal connectivity (Vicente et al. 2008; Pérez et al. 2011; Messé et al. 2018)—that can induce neuronal synchronous firing even despite long conduction delays. However, one of the central challenges that has not been sufficiently addressed is that the mechanism needs to enable the neurons to synchronize and desynchronize in a stimulus-specific fashion, and thereby to encode relationships. Noise-induced coherence is one such mechanism that was recently demonstrated to produce fast, stimulus-specific, and biologically plausible synchronization patterns.

First discussed in complex and excitable systems (Pikovsky and Kurths 1997), noise-induced coherence is a process that can structure and synchronize the activity of the system based on noisy or even unstructured input. The nature of noise-induced coherence is that the complex system (the dynamical elements, e.g., neurons, together with the network topology) defines patterns that exhibit enhanced coherence if the system is driven by a corresponding motif, neuron-specific optimal amplitude of unstructured noise. In other words, and in respect to neuronal networks, noise translates a pattern of neuron-specific firing rates into patterns of coherent and synchronized population responses (i.e., translation of a neuron-specific rate code to a population-based sync code). Importantly, this translation is network specific, which opens the possibility that the expression of synchronous events is not only driven by the stimulus-specific rate pattern but also by the network, and its structure is shaped by neuronal plasticity.

## Transformation of Spike Rate Coding to Coherent Population Codes via Noise-Induced Coherence

To illustrate the mechanism and encoding based on noise-induced coherence, let us consider an example for the visual cortex V1. In general, it is known that network structure of cortical networks is at least partially shaped by the experience of past activation mediated by neuronal plasticity. For V1, this implies that the connection strength horizontal connections in V1 reflect the aggregate statistics of natural visual scenes (Onat et al. 2013); that is, V1 cells with nearby receptive fields are preferentially connected, and specifically when they select for similar visual stimuli. Figure 11.3a shows a network simplified to such a V1 prototypic connectivity pattern. The system receives stimulus-specific input described by neuron-specific retinal coordinates which match their cortical position (retinotopy) and have a particular angle (orientation tuning) presynaptic spike rates (i.e., uncorrelated and rate-modulated Poisson firing). To illustrate the effect of noise-induced coherence, we use two kinds of stimuli: one that is open and composed of two shorts blocks, and one that is closed and composed of a longer bar. Given the retinotopic mapping, this implies that the activation pattern, in comparison to the underlying network, results in different shortest path lengths between stimulus-driven cells. Only few of these cells will have direct connections, since horizontal connections preferably connect cells with nearby receptive fields. More generally, the network connectivity implies a metric for possible stimulus patterns. Given this metric, for V1, the shortest path between any two responding cells will likely be longer, on average, for a scattered stimulus than for a more compact stimulus. As a result, the same cells which receive a presynaptic input pattern matching the connectivity of the network (here, cells that are part of a continuous patch) exhibit stronger noise-induced coherence than others. Such mechanisms can be generalized to more complex encoding schemes, depending on the connectivity patterns of the network. For example, the well-known orientation tuning of cells in V1, in combination, and network motifs described by preferred connectivity across cells with similar orientation will result in enhanced coherence of neurons that encode chains of shorter line segments (see Figure 11.4). In general, noise-induced coherence is a mechanism that can measure the similarity between the network connectivity and the stimulus-induced spike rate pattern (Korndörfer et

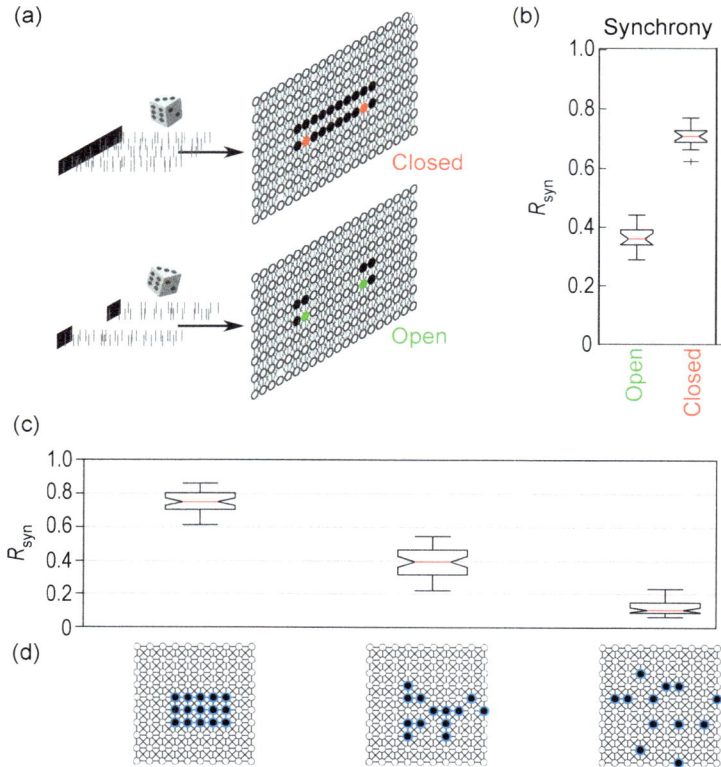

**Figure 11.3** (a) Noise-induced coherence for two alternative presynaptic stimuli (red, a closed line; green, an open line segment). Coherence is measured between the green- or red-highlighted neurons. The only difference is the context given by the stimulus drive, which itself is composed of unstructured Poisson noise. The network topology is defined by nearest neighbor connection matches, and stimuli are matched using retinotopic mapping. (b) Synchrony measure of the pairs of neurons shown in (a) and for the two stimulus conditions. Coherence is higher for the compact closed line, since the shortest path length between stimulus-driven neurons is smaller for the closed contour. (d) This feature is generalized to the amount of scattering of a stimulus; that is, the greater the scattering, the larger the shortest path length between neurons, given the metric of the underlying network connectivity. The resulting stimulus-induced coherence (c) is the largest for the most compact, and the lowest, for the most scattered stimulus. Adapted from Korndörfer et al. (2017).

al. 2017). It can therefore be a measure of how well the stimulus matches a prior learned by neuronal plasticity and encoded in the network's connectivity. Here the stimulus-induced spike rate reflects a classical labeled line code. Thus, spike synchrony generated by noise-induced coherence carries synergistic information that reflects to which degree the current stimulus encoded by the spike rate is expected, given past stimulus experiences. Such a signal could be used early after input onset in a feedforward fashion, for instance,

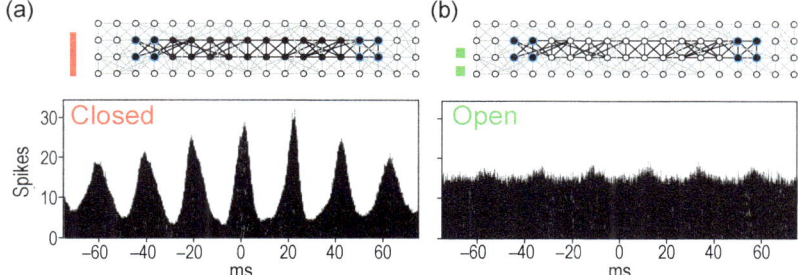

**Figure 11.4** Average cross correlation of noise-induced coherence between two sets of neurons marked in blue for two different stimulus conditions: (a) closed line segment and (b) open line segment. The noise-induced coherence is stronger for the closed, compared to the open, condition. In the original publication (Korndörfer et al. 2017), it is shown that this increased coherence can be decoded as closed contour as early as a few spikes after stimulus onset (70 ms).

to guide attention toward stimuli composed of plausible parts. In contrast to many other types of synchronization, it also does not require, but can be improved by, inhibitory cells (Korndörfer et al. 2017), and it produces firing patterns that closely resemble *in vivo* recorded patterns (e.g., Gray et al. 1989; Uhlhaas et al. 2009).

## Reservoir Computing

Most mechanisms discussed over the past decades for neuronal information processing require highly structured networks, specific types of dynamical processes, and very specific encoding schemes of information (e.g., rate code versus population spike codes). A frequently used feature of computational models is that they rely on attractor dynamics, which can be trained to implement specific computational features, such as associated memory in Hopfield networks (Hopfield 1982) or the winner-takes-all mechanism (Maass 2006) for decision making, for example.

Like the ICA network discussed above, all of these computational models implement a clearly defined information processing principle and rely on a very specific type of implementation, in terms of connectivity and dynamical elements. This is a strong advantage, since it allows us to study principle and well-defined behavior, and to reduce the computation to a minimal set of required properties. At the same time, this reductionism also renders the models biologically implausible, since biological systems are subject to noise on pretty much any property, such that neuronal networks are mostly random with some statistical preferences for certain motifs, and neurons are diverse in type and morphology.

Therefore, a strikingly different model for neuronal computation is *reservoir computing*, originally introduced as liquid state machines by Maass et al. (2002) or echo-state networks by Jaeger and Haas (2004). In contrast to most other computational principles, the recurrent network of a reservoir computer can be unstructured and random. This surprising property results from the simple insight that the distance between random mappings of states is growing fast, with increasing dimensionality of the mapping. In other words, implementing a certain computation does not require a dedicated network with specific connectivity tailored for the given task but, in principle, only a random network that implements a sufficiently high-dimensional random mapping. In the field of machine learning, this is known as feature expansion or kernel machines (Schölkopf and Smola 2002). Further, reservoir computing makes explicit use of the recurrence of neuronal networks to maintain an echo (i.e., memory capacity) of past inputs. This echo is mediated by reverberating activity, generated by the recurrent connectivity. Together, feature expansion and memory of the system can render a reservoir computer a universal computer (Buonomano and Maass 2009). The only task-specific element in reservoir computing is a task-specific mapping that can be learned by supervised, semi-supervised (Toutounji and Pipa 2014), or reinforcement learning algorithms (Aswolinskiy and Pipa 2015).

The remarkable insight of reservoir computing is that random recurrent networks can implement, in principle, any kind of computation if the networks are sufficiently complex. From a biological point of view, this implies that initially unstructured networks can bootstrap themselves, based on neuronal plasticity, to improve performance. Importantly, it can operate initially even without any structure.

## Computation in Delay-Coupled Systems

When describing computation in the nervous system from the perspective of abstract single neurons or recurrent networks which show emergent behavior as a collective, a simplifying assumption is often made: interactions between neurons are instantaneous and not delayed. This is simply because delays in differential equations complicate the analysis of such systems significantly, and deriving theoretical results is a lot harder.

In biophysical reality, however, the brain is a network of nodes and wires that must be subject to transmission delays. For instance, conduction delays of tens of milliseconds occur in axonal transmission of spikes (Ringo et al. 1994). Interspike intervals, indicative of the timescales on which neurons compute outputs, have been found on the same scale in the motor system (Calvin and Stevens 1968) or in retinal ganglion cells (Levine and Shefner 1977). It is thus clear that delays play a role in the dynamics and computational properties of neural networks. A long-established example, where this role is well understood, is audio processing: transmission delays on delay lines are used to distinguish left ear input from right ear input, and interpolate the location of a sound source (London and Häusser 2005).

In cortical structures, network motifs or microcircuits have been found that circumvent transmission delays and lead to zero time lag synchronization (Vicente et al. 2008). On the other hand, in microcircuits where transmission delays are modeled, only very specific topologies allow for coherent spiking activity, which delays control phase differences between oscillatory neurons (Pérez et al. 2011). So, locally, transmission delays control phase transitions between in-phase and out-of-phase response, whereas, globally, axonal delays can stabilize coherent response-important phenomena in neural computation.

Even still, these examples only describe how delays can negatively impact behavior of microcircuits or stabilize existing behavior. Future work should investigate the degree to which the added complexity of delay-coupled systems can be exploited for computation.

## A Single Node with Delayed Feedback

Stabilizing emergent phenomena may not be the only mechanism by which delays can aid computation in the nervous system. Instead, the benefit of delayed interactions can be illustrated theoretically by examining a single computational node with delayed feedback. This very simple setup is described by a delay differential equation:

$$dx(t) = f\big(x(t), \, x(t - \tau)\big) dt. \tag{11.1}$$

The equation can be solved by a trick known as the method of steps (Guo and Wu 2013), which is both intuitive and illustrative of the complexity of delayed interactions: Assume that the solution to Equation 11.1 on some interval, $[t_0 - \tau, t_0]$, is known and denote that solution $\phi_0$. For the subsequent overlapping interval, $[t_0, t_0 + \tau]$, Equation 11.1 can then be rewritten as

$$dx(t) = f\big(x(t), \, \phi_0(t - \tau)\big) dt, \tag{11.2}$$

since for all $t \in [t_0, t_0 + \tau]$, it holds that $t - \tau \in [t_0 - \tau, t_0]$, where $\phi_0$ is the solution. This is now an ordinary differential equation and can be solved using traditional methods. However, the starting condition for the new solution is now a tuple $(\phi_0, \phi_0(t_0))$ of a function, and the function is evaluated at $t_0$. Further, the solution on the interval $[t_0, t_0 + \tau]$ is again a function; let that function be $\phi_1$. In the method of steps, this procedure is iterated with this new starting value for the next interval of length $\tau$. In general, if $t_i = t_0 + i\tau$, then $\phi_i$ is the solution on the interval $[t_{i-1}, t_i]$.

Even though Equation 11.1 is a differential equation of a single, scalar variable, solving it involves mapping functions onto functions for each $\tau$ interval or cycle (Figure 11.5a) and is therefore infinitely dimensional. Delay differential equations are a subclass of partial differential equations whose state is described by functions, instead of finite-dimensional state vectors.

By introducing one simple, delayed feedback to a dynamical system, we elevate the complexity from one to infinitely many dimensions. This complexity

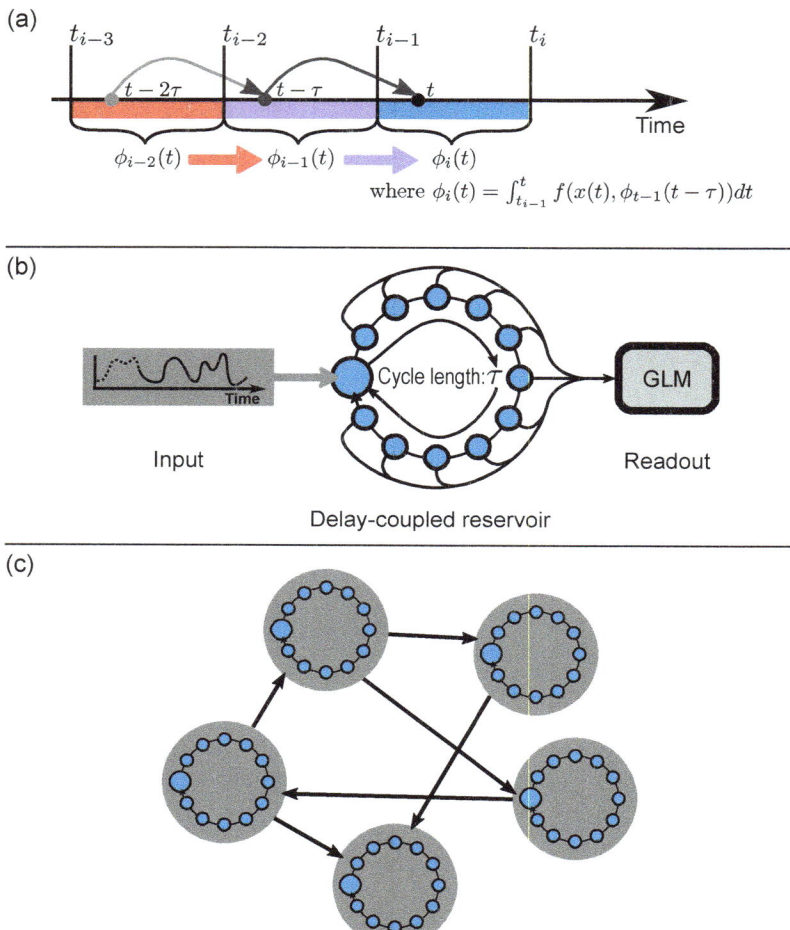

**Figure 11.5** (a) Schema of solving delay differential equations with the method of steps. Functional solutions $\phi_i$ are mapped onto solutions $\phi_{i+1}$ via an integral over the original delay differential equations, where the delay dependency is replaced by a dependency on the last solution. (b) A delay-coupled reservoir utilizes the complexity of delay differential equations for computation by creating an input-driven dynamical system and feeding sampled activity during one $\tau$ cycle into a GLM readout trained to solve a specific task. (c) Networks of delay-coupled nodes can be understood as small recurrent systems inside a larger recurrent system. This model may be used to model the complexity of recurrence and delay coupling at the same time.

not only makes solving the mathematical problem more difficult, it also leads to dynamics that can be used for computation and inference.

## The Delay-Coupled Reservoir

Introduced by Appeltant et al. (2011), the delay-coupled reservoir is a system described by a delay differential equation, such as Equation 11.1, but driven by an input $u(t)$:

$$dx(t) = -ax(t) + g(x(t-\tau), u(t)), \qquad (11.3)$$

where $g$ is a nonlinear function. The key insight from this work is that the activity in/of this simple one-node recurrent system can be sampled $N$-times during one delay cycle of length $\tau$, and this sampled activity can be treated as the $N$-dimensional of a reservoir computer with $N$ nodes. The input $u(t)$ is adapted to also change on the timescale of one $\tau$-cycle, such that each $\tau$-cycle associates one $N$-dimensional vector of activations with one input value. Following the reservoir computing procedure, this activation vector can then be used in a linear readout to learn a time-invariant, fading-memory function on the input (Figure 11.2).

The on-the-surface simplicity of delay differential equations leads to straightforward hardware implementations, where some nonlinear element is driven by input and self-coupled via a delay line. These simple building blocks have led to implementation based on standard electronic building blocks, but they also allow for the exploration of new computing devices, as in using delay-coupled lasers and photonics (Larger et al. 2012).

The hidden complexity of the system, however, allows it to be used in time-series forecasting, speech recognition, and even volatility prediction for financial markets (Appeltant et al. 2011; Grigoryeva et al. 2014).

This complexity, and the process of obtaining a vector of activity, can also be looked at theoretically using the method of steps. The iterative solution of the ordinary differential equation for subsequent $\tau$-cycles, or intervals of length $\tau$, can then be approximated analytically and written as a vector update equation for the $N$-relevant sample points directly (Schumacher et al. 2013). Thus, for computation within a reservoir computing setup, the infinite dimensionality of space of solutions to the delay differential equation reduces to $N$ dimensions, a free parameter of the model. One can therefore profit from the potentially infinite dimensionality of a functional state. In practice, with a chosen decay rate $\alpha$ and a specific nonlinearity $g$, choosing arbitrarily large $N$ does not benefit specific machine learning tasks above a task-dependent soft threshold. Nevertheless, researchers have seen benefits in expanding the dynamics of this simple, nonlinear, and delayed feedback-coupled node into an $N = 50$ up to an $N = 800$ dimensional state vector, as input to the linear regressor.

Instead of sampling the activity of the delay-coupled reservoir at $N$ evenly spaced points, one can optimize the placement of these readout points. Here, it is useful to treat the $N$ sampling points as a network of virtual nodes with a very particular connectivity structure: a lower diagonal exponential decay matrix. The distance from one sampling point, or virtual node, to the next can then be fine-tuned according to a homeostatic plasticity rule. It presupposes that good spatiotemporal computational performance is achieved when different virtual nodes are both sensitive to their inputs and as diverse as possible (Toutounji et al. 2015). The experiments in the study show that this rule does indeed lead to increased performance. From the point of view of the readout, the result permits a crude biological interpretation: a linear-nonlinear output neuron optimizes the locations along an axon, where it "reads" the activity of another neuron with complex time-dependent dynamics. Clearly, this interpretation is somewhat bold, but it highlights the potential of future research that uses delayed feedback models to encode and then decode information in temporal dynamics.

The delay-coupled reservoir can also serve as a model system to investigate how two different delays might interact. In a previous study, Nieters et al. (2017) highlighted strange dependencies that arise if Equation 11.3 is expanded to

$$dx(t) = -ax(t) + g\big(x(t - \tau_1), x(t - \tau_2), u(t)\big). \tag{11.4}$$

Delays that are close to simple rational, or even integer multiples of each other, lead to a poorly performing reservoir computer—how close is too close is controlled by the decay rate $\alpha$ of the exponential decay in the system. A too strong dependency of a sampling point onto its own history—the effect of choosing the $\tau_2 = 2\tau_1$—is detrimental. This delicate sensitivity to the choice of a second delayed feedback is reminiscent of the sensitivity to different delays in microcircuits mentioned earlier but is, of course, also an artifact of the discretized system used to model the activity at $N$ sampling point with an analytic approximation and discretization. Future work must focus on a more realistic setting, where delays are distributed and continuous to investigate whether sharp transitions between well- and badly performing models also occur.

The takeaway from previous investigations into delay-coupled computation is that the added complexity can induce a complex temporal dynamics readout by an appropriate readout mechanism, which can benefit computation significantly. Reservoir computing is a compatible concept that embeds models, such as the delay-coupled reservoir, in the context of neural computation. More work is needed to connect the observed effects of delay coupling more closely with biological reality. Studies also highlight how complex neural networks may actually be that are subject to multiple delayed interaction effects. A possible perspective to study such systems abstractly is to connect single nodes with distributed delays recurrently in a reservoir, in the sense of a classical recurrent network. In such a network of networks, each node itself can be regarded as a simple recurrent system (Figure 11.3).

## Discussion

In this chapter, we have discussed several computational principles at the level of individual neurons and networks of neurons, and addressed the implications of delayed communication. These principles ranged from specifically tuning single neurons to implement well-defined computational tasks (i.e., independent component analysis) to reservoir computing to implement computing based on randomly connected networks and random feature expansion. This diversity and wide range of functions can be viewed as either an overwhelming complexity that might just hide a key underlying unifying principle not yet uncovered, or a rich diversity used by the evolution as a large reservoir of tools and tricks to implement efficient computational circuits. If the latter is true, then the simple question of which computational principle do we discard is not sufficient. Instead, we need to address efficiency in terms of performance and the use of resources, robustness to noise and structural changes, and generalizability of the computational principles for different tasks. The ultimate question, however, remains essentially open: How does the cortex, or the brain, compute information?

# 12

# Coding in Large-Scale Cortical Populations

Jason N. MacLean and Nicholas G. Hatsopoulos

## Abstract

Theories of information coding in cortical populations have been put forth for many years, but only recently have experimental methods become available to permit simultaneous recordings from hundreds of neurons, thus allowing these theories to be tested. This chapter discusses some of the more prominent theories and argues that they fall along a spectrum of coding schemes, ranging from population codes that are built up from single-neuron tuning functions to codes that emerge from the collective dynamics of cortical populations. At the extremes, these theories are incompatible: one relies on single neurons whereas the other ingrains coarse neuronal activity into low-dimensional trajectories that summarize the covariance of activity across multiple neurons. It is proposed that both can be reconciled using a hierarchical coding scheme where relevant information is represented at the level of large-scale spatiotemporal patterns, and both individual neurons and the temporal interrelationships convey information. Antecedents to this contemporary theory can be seen in Donald Hebb's assembly phase sequences (Hebb 1949): information is encoded at the single-neuron level in terms of tuning functions, but spatiotemporal patterning of individual neurons provides context to interpret the population code fully. Moreover, the encoding perspective proposed here explicitly incorporates the synaptic implementation of the code, thus strengthening the postulate.

## Introduction

Thanks to the ongoing development of technologies, such as multielectrode arrays and optical imaging, it has become increasingly routine to record simultaneously from hundreds, and soon even thousands, of neurons in awake-behaving animals. From this perspective, we begin by arguing *why* the cortex needs to use a population scheme to encode sensory, cognitive, and motor information. We then proceed to describe different population schemes and argue that it is necessary to understand *how* as well as *what* is encoded in the population to fully delineate the nature of the cortical population code: How is the code implemented mechanistically from a cellular and synaptic perspective? What is being represented, and for which purpose, at any moment in time by the active cortical neuronal population? From an encoding perspective, the issues of *how, what,* and *for which purpose* must be established to permit a complete theory. From a decoding perspective, it is possible to black box the *how* and *for which purpose* and instead establish only *what* is being encoded.

Population coding schemes can be considered on a spectrum of neural representations where the extremes are seemingly mutually exclusive. At one end of the spectrum, population codes are built up from the sum of activity of a collection of single neurons, which we term "single neuron level" (SNL) codes. SNL codes assume that sensory, cognitive, or motor information resides at the single-neuron level in the form of tuning curves, for example. Tuning curves represent the trial-averaged response of a single neuron and reflect the fact that a neuron is more likely to spike when a specific statistical feature is present in a stimulus, when a percept, thought, or decision is generated, or preceding a specific motoric feature. The population, in turn, codes information by aggregating or combining information across neurons. A pooling code is perhaps the simplest example of an SNL code in which information is represented as a weighted average of activities of single neurons. At the other end of the spectrum, information is encoded at the neural ensemble level (NEL). One example of an NEL code assumes that single neurons in isolation are uninformative but collectively they serve as cogs in a multineuronal machine whose spatiotemporal dynamics carry information. The characterization of motor cortex as a dynamical system put forth by Shenoy and Churchland exemplifies such a code (Churchland et al. 2012; Shenoy et al. 2013). However, we subscribe to the viewpoint that the cortex employs NEL codes that do not ignore single neurons and their tuning properties while simultaneously employing higher-order statistical relationships among neurons as a coding element. Changes in both rate and spiking correlations, in response to sensory stimuli or in relation to motor output, are generally linked (de la Rocha et al. 2007; see however, Biederlack et al. 2006). To understand what we mean by this perspective, consider the analogy of our decimal system for representing numbers. Single digits, such as "1" and "2," have distinct meanings and, therefore, code information. However, "12" refers to something quite different than the two isolated digits that comprise it and yet the ensemble code depends on the meanings of the individual digits. This suggests a hierarchical coding scheme where information is carried at multiple levels of a hierarchy.

## Why Use a Population Coding Scheme?

At the cortical level, it is well-established that single neurons are highly variable from moment to moment and trial to trial, even under well-controlled conditions (Heggelund and Albus 1978; Tolhurst et al. 1983; Vogels et al. 1989; Britten et al. 1993; Arieli et al. 1996; Lin et al. 2015). Moreover, neurons are interconnected by large numbers of dynamic and generally weak synapses and consequently must integrate multiple inputs to spike. It is likely that cortical codes depend on the collective activity of large groups of neurons. Fortunately, we now have the technological ability to experimentally evaluate population coding schemes. As we discuss below, it remains unclear what the numerical size and spatial extent of an encoding population is beyond the fact that it is more than a single neuron.

## SNL Codes

The population vector, one of the best known SNL codes, has proven effective in decoding reach direction from a group of motor cortical (M1) neurons (Georgopoulos et al. 1986). By fitting directional tuning curves of individual M1 neurons to cosine functions, the preferred directions of these cells are computed and represented as two- or three-dimensional vectors. The population vector is then calculated as a weighted vector sum of the preferred directions, each weighted by its firing rate. This approach can be extended to incorporate the sum of population vectors (Gilbert and Wiesel 1990; Vogels 1990). Although originally applied to sequentially recorded neurons, this approach has been extended to simultaneously recorded neurons with good decoding success (Taylor et al. 2002). Population vector decoding has been applied to a variety of systems, including the visual system, to estimate the orientation of faces (Oram et al. 1998) and, in the auditory system, to localize a sound source based on intra-aural time differences (Fitzpatrick et al. 1997).

In general, the population vector decoder does not provide an optimal estimator in the sense that it does not always minimize the variance of the estimator (Deneve et al. 1999); this can result in poor decoding performance (Montijn et al. 2016). Other SNL population coding schemes have been shown to be optimal under certain assumptions, including the optimal linear estimator (OLE), the indirect OLE, and maximum likelihood estimation (Salinas and Abbott 1994; Deneve et al. 1999; Wang et al. 2007b). The indirect OLE approach has been extended for real-time, closed-loop neural prosthesis control of a 10-dimensional robot in a paralyzed patient (Wodlinger et al. 2015). In this application, single-neuron tuning models included not only reach direction but also wrist orientation and grasp velocities. Maximum likelihood estimation can also be implemented using a recurrent network model under certain assumptions (Deneve et al. 1999).

Despite the general success of the population vector and other near-optimal estimators, an important assumption underlies these population decoding schemes: the firing rates of the population are assumed to be statistically independent, conditioned on the stimulus, decision, or movement. It is this assumption, in fact, which provides the population vector scheme with its ability to reduce noise by averaging over neurons, which in turn makes it particularly effective for decoding (Zohary et al. 1994). However, trial-to-trial correlations in spike counts that are independent of the specific stimulus or movement (i.e., noise correlations) have been documented in many sensory and motor cortical areas and are often considered to be a nuisance for SNL population codes because the variance of a population mean estimator will not decrease as 1/N, where N is the number of neurons in the population. More generally, however, noise correlations do not always reduce population-level information and, in fact, can improve stimulus discrimination in population activity (Poort and Roelfsema 2009). Theoretical work by the Pouget group has identified a specific kind of noise correlation (differential correlations) that is proportional to the product of the derivative of the tuning curves and impacts the information capacity of SNL population codes (Moreno-Bote et al. 2014). The assumption of independence makes it difficult to extend this framework to an encoding model that incorporates synaptic and cellular mechanisms, since neurons are interconnected and collectively drive and support activity in the active population; that is, they are not independent (Renart et al. 2010). However, the utility of SNL codes makes clear that single neurons do indeed represent information. We suggest that this fact should be incorporated into any cortical population coding scheme.

## NEL Codes

Regardless of the cortical area recorded or the recording method, researchers have found evidence of structured spatiotemporal activity in all cases: this is consistent with the Hebbian assembly phase sequence, which postulates that across trials, functionally related assemblies or ensembles of neurons reliably propagate spikes from one ensemble to another (Hebb 1949; Abeles and Gerstein 1988; Villa et al. 1999; Beggs and Plenz 2004; Gourevitch and Eggermont 2010; Bathellier et al. 2012; Gansel and Singer 2012; Palm et al. 2014; Peters et al. 2014; Reyes-Puerta et al. 2015). This coding scheme is made more attractive by the fact that it easily allows for inclusion of cell identity and is consequently compatible with cellular and synaptic mechanisms (Kruskal et al. 2013).

### Correlation Codes

While it is widely recognized that cortical neurons exhibit correlated firing at different timescales and constrain population activity patterns, we postulate that these correlations code information.

*Synchrony*

Synchronous firing of neuron pairs at the millisecond timescale have been shown to signal a variety of different sensory and motor states. Pairs of oscillating neurons in visual cortex have been shown to synchronize when stimulus features form a single, coherent visual stimulus (Gray et al. 1989). This and other experiments have been used to support the view that synchronization serves to bind stimulus features that are perceptually related (Singer 2018a). In the motor domain, synchrony has been shown to encode movement direction and expected visual cues for initiating movement in primary motor cortex (Riehle et al. 1997; Hatsopoulos et al. 1998; Kilavik et al. 2009; Denker et al. 2011). Moreover, larger groups of synchronous M1 neurons have been shown to signal task epochs and behavioral conditions in an instructed delay reach-to-grasp task (Torre et al. 2016). By identifying so-called cortico-motoneuronal cells that make direct synaptic connections with motor neurons in the spinal cord via spike-triggered electromyography recordings, it has been found that cortico-motoneuronal pairs which share similar muscle fields tend to be more synchronized than pairs that have different projection fields (Jackson et al. 2003). Moreover, modeling work suggests that inputs to motor neurons can generate substantially more force output when they are synchronized (Baker et al. 1999). Therefore, cortical synchrony may provide a mechanism for directly affecting behavioral output.

*Noise Correlations*

Despite their potential detrimental effect on SNL population codes, noise correlations have been shown to signal different behavioral states. For example, spike count correlations in frontal eye fields have been shown to vary dynamically in distinct ways for different saccadic eye movements, despite the fact the neurons' firing rates do not modulate (Vaadia et al. 1995). Spike count correlations between pairs of M1 neurons have been shown to increase when two-element movement sequences were preplanned, compared to when they were planned one at a time, even when the firing rate modulations of the neurons did not differ under the two conditions (Hatsopoulos et al. 2003). This suggests that task complexity impacts the correlational structure.

## Latent Variable Codes

A latent variable code is a form of NEL code that assumes there are latent variables that are not explicitly observed but inferred from spiking data (i.e., the manifest variables). Principal components analysis and factor analysis are examples of continuous latent variable methods that have been used extensively to characterize large-scale cortical recordings (Mazor and Laurent 2005; Yu et al. 2009a). Unlike pooling codes, these methods depend on the covariance of neuronal activity to find subspaces in which most of the variance of activity resides. It is postulated that population activity resides in a much smaller subspace than what would be theoretically possible because of (a) the interconnections between neurons, (b) the minimal coding requirements for the task at hand (see below), or (c) due to common input. Regardless, these subspaces have been demonstrated to enable decoding of cortical population activity and have also provided insights into how cortex represents information. For example, animals have difficulty learning to control brain–machine interfaces when this requires generating population activity outside of these intrinsic subspaces (Sadtler et al. 2014). Moreover, a recent study has shown that choice behavior in a visual discrimination task depends on population activity that resides in a principal component subspace, even though it is suboptimal (Ni et al. 2018).

Besides serving to reduce the dimensionality of high-dimensional neural data, continuous latent variables reveal a dynamic structure in the active population that is not evident at the single-neuron level. In an elegant set of studies, Shenoy and colleagues have characterized motor and premotor ensembles as neural trajectories that exhibit relatively simple rotational dynamics during movement execution in a reduced subspace (Churchland et al. 2012; Shenoy et al. 2013). Different movements (e.g., movements to different goals) correspond to different initial conditions within the movement subspace, but all movements share these underlying rotation dynamics. Moreover, movement planning or preparation resides in an orthogonal subspace to the space that corresponds to the muscle activation and movement (Kaufman et al. 2014). According to Kaufmann et al., single neurons serve only as the building blocks of a multineuronal dynamical system that traverses different portions of state space. If anything is coded, it is the location or dynamics of the population state that represents planning or movement of a particular type. From an encoding perspective, the informative nature of structured population dynamics indicates that the spatiotemporal interactions between neurons, and the corresponding cellular and synaptic mechanisms that generate this structure, will be fundamental to any unifying theory of cortical population coding.

Hidden Markov models (HMMs) are examples of discrete latent codes that have been used to characterize cortical population dynamics as a sequence of discrete "hidden" states which account for shared variability across the population (Radons et al. 1994; Abeles et al. 1995; Seidemann et al. 1996; Kemere et al. 2008). The probability of a state transition is based solely on the current state of the system. Moreover, each hidden state has an associated probability distribution for observed spiking responses across the population. As compared to populations of poststimulus time histograms, coupling hidden states can more accurately predict the spiking statistics of population activity, particularly when activity is not time-locked to an external stimulus (Abeles et al. 1995). Moreover, each state is associated with distinct pairwise neural correlations, often reflecting different behavioral states. Recently we applied a HMM to simultaneously recorded spiking data from primary motor cortex while monkeys engaged in reaching movements and found that population activity appears to transition among a small set of states (Kadmon Harpaz et al. 2018). More importantly, we discovered that state transitions correspond to velocity extrema of reaching, such that a given state corresponds to either an accelerative or decelerative phase of reaching in a particular direction. Simulations using single-neuron tuning models of direction and speed applied to the reaching movements could not replicate the findings of our HMM. This implies that motor cortical population dynamics may be more accurately characterized as transitions among discrete hidden states that code for accelerative/decelerative movements in particular directions.

## Spatiotemporal Patterning

Most NEL coding schemes consider information encoded in the pattern of activity across a population but largely ignore the spatial layout of neurons within the population, despite the strong anisotropy of neurons and their connection likelihood. Studies using multielectrode arrays and voltage-sensitive dyes, however, have documented propagating wave activity consistent with spatiotemporal codes within the visual (Arieli et al. 1995; Prechtl et al. 1997; Roland et al. 2006; Xu et al. 2007), somatosensory and sensorimotor (Petersen et al. 2003; Ferezou et al. 2007), auditory (Song et al. 2006; Witte et al. 2007), and motor cortices (Rubino et al. 2006; Takahashi et al. 2015) as well as hippocampus (Lubenov and Siapas 2009; Patel et al. 2012; Zhang and Jacobs 2015). In the CA1 region of hippocampus, local field potential (LFP) waves propagate along the septotemporal axis, mediated by theta oscillations, and may serve to encode not only the present location of the animal but also the past and future locations (Lubenov and Siapas 2009). In primary and premotor motor cortices, LFP waves in beta oscillations propagate along the rostrocaudal and medial-lateral axes of the cortical surface, respectively. The propagating direction and speed do not vary with movement direction and thus do not appear to code for movement parameters. However, in an instructed delay paradigm, where a visual target of an upcoming movement is presented, visually evoked waves do encode upcoming movement direction in the amplitude and timing of these evoked potentials (Rubino et al. 2006). Moreover, the sequential firing of pairs of M1 neurons carries more directional information when the two neurons are oriented along the LFP wave-propagating axis (Takahashi et al. 2015).

We have documented another spatiotemporal pattern of activity in motor cortex that may be important for initiating movement (Best et al. 2017; Balasubramanian et al. 2019). Voluntary movement initiation involves the modulation of large populations of M1 neurons around movement onset. Despite knowledge of the temporal dynamics of cortical ensembles that lead to movement, the spatial structure of these dynamics, across the cortical sheet, have been largely ignored. We have shown (Best et al. 2017) that the timing in attenuation of the beta frequency oscillation amplitude, a neural correlate of corticospinal excitability (Pfurtscheller and Lopes da Silva 1999), forms a spatial gradient across motor cortex prior to movement onset with a defined beta attenuation orientation from earlier to later attenuation times. We have also shown that a similar propagating pattern is evident in the modulation times of populations of M1 neurons. Interestingly, even though M1 neurons modulate activity during movement preparation well before movement onset, these modulation times do not exhibit spatial structure, suggesting that spatiotemporal structure in modulation may be necessary to trigger movement. It should be emphasized that such spatiotemporal patterns do not lessen the possibility that single neurons also carry information. Rather, we argue that cortex may be using a hierarchical coding scheme: large-scale patterning signals certain global aspects of behavior while single neurons code for more specific aspects or, alternatively, large-scale patterning provides context for single-neuron codes which together form a population code.

## How Would an Ensemble Code Work?

To truly understand the code, we must eventually synthesize it with the cellular and synaptic mechanisms that implement the code. Neocortical microcircuitry includes the neurons and synaptic connections within volumes of ~500 μm × 500 μm × 1000 μm. Individual neurons are highly interconnected and connection likelihood is biased toward neighboring neurons (Song et al. 2005; Ko et al. 2011; Perin et al. 2011). Moreover, in primary sensory cortices, only ~5% of connections that a neuron receives arise from ascending inputs (Peters and Payne 1993; Douglas and Martin 2004), and a comparable portion originates from distal cortical regions (Budd 1998). The microcircuit, then, represents the scale over which (a) most excitatory and inhibitory interactions take place, and (b) synaptic connections strengthen or weaken according to the relative spike timing between pre- and postsynaptic neurons (Kruskal et al. 2013). This is also the scale at which Hebbian learning occurs. Consequently, microcircuits form the building blocks from which the cortical population code is built. Imaging approaches permit researchers to densely sample neurons allowing microcircuit dynamics to be more directly linked to synaptic mechanisms (Ko et al. 2011; Chambers and MacLean 2015).

Individual synaptic connections are weak, ranging from 0.2–1.0 mV (Holmgren et al. 2003), and patterns of spiking are complex and variable, which make the mapping between structure and functional dynamics far from straightforward. However, a small set of correlations in population activity is indicative of synapses that are actively involved in the recruitment of postsynaptic neurons; namely, those synaptic inputs that occur at just the right time to drive the postsynaptic neuron to threshold (Ko et al. 2011; Chambers and MacLean 2015, 2016). We found that these "recruitment" synapses are directly linked to specific higher-order motifs in population correlational structure. To make this observation, we accounted for the nonlinear integrative properties of neurons by using a combination of spiking neuronal network models and experimental measurements (Chambers and MacLean 2016). Fan-in triangles, where two input neurons are themselves connected, coordinate the timing of presynaptic inputs during ongoing activity to facilitate postsynaptic spiking. Interplay between higher-order synaptic connectivity and the integrative properties of neurons constrains the structure of network dynamics and shapes the routing of information in neocortex (Chambers and MacLean 2016).

We have also linked higher-order correlational structure to the activity of neurons in mouse visual cortex by imaging microcircuit responses to visual input in awake, head-fixed, ambulating mice. We found that trial-averaged tuning properties of neurons explain only a small fraction of the single-trial activity of neurons, similar to other studies (Reimer et al. 2014; Montijn et al. 2016). By summarizing the dynamics as a functional network, we are able to use the neighbors of a neuron, necessarily including both tuned and untuned neurons, to predict individual neuron activity on a single-trial basis. Perception and behavior take place in real time, after all, so it is necessary that any population encoding model of stimulus representations in cortex encompass single-trial responses. Moreover, again a specific triplet motif maximized predictions of single-trial responses (Dechery and MacLean 2017). Consistent with these results from visual cortex, Meshulam et al. (2017) found that they were able to best predict single-neuron activity in hippocampus by employing a maximum entropy model that incorporated the state of all of the recorded neurons, including neurons that have well-defined place fields as well as those which do not. Together, these studies indicate that the collective behavior of neurons, both tuned and untuned, are necessary to predict single-trial neuronal responses and argue strongly in favor of an NEL coding framework.

At any given moment in time, it is unclear how many neurons are necessary to encode a stimulus or motor output. Using connectivity estimates, synaptic strength estimates, as well as membrane potentials and conductance states of individual neurons, a lower bound estimate of approximately 100 presynaptic neurons has been postulated (Ainsworth et al. 2012). As described above, low-dimensional summaries of population dynamics are a very effective means to decode population activity (Briggman et al. 2005; Churchland et al. 2007; Harvey et al. 2009). Consistently, a number of studies have found that the number of observed spatiotemporal neuronal activity patterns are limited (MacLean et al. 2005; Luczak et al. 2007, 2015; Luczak and Maclean 2012). Moreover it appears that baseline connectivity also constrains feasible dynamics during learning (Shenoy et al. 2013; Sadtler et al. 2014; Shenoy and Carmena 2014). All of these data suggest that population dynamics occupy a much lower-dimensional space than the number of neurons recorded, which indicates diminishing returns as the number of recorded neurons increases into the thousands. Two recent studies, however, provide compelling arguments that the number of neurons, and more loosely the number of dimensions, necessary to encode a motor output or sensory stimulus likely depends on what is being encoded. The neural task complexity theory (NTC) posits that the complexity of the task that an animal is performing combined with the smoothness of the recorded neural trajectories determine the size of the population code (Gao et al. 2017). Further, NTC theory maps the behavioral parameters within neural dynamics and predicts that this volume will be small when tasks are simple and trajectories are smooth (i.e., the covariance of neuronal activity is well captured in the recorded population). NTC theory predicts that the size of the coding pool and, relatedly, the number of dimensions necessary to summarize population dynamics scale with the nature or difficulty of the task or stimulus.

# 13

# Functional Properties of Circuits, Cellular Populations, and Areas

Kenneth D. Harris, Jennifer M. Groh, James DiCarlo,
Pascal Fries, Matthias Kaschube, Gilles Laurent,
Jason N. MacLean, David A. McCormick,
Gordon Pipa, John H. Reynolds, Andrew B. Schwartz,
Terrence J. Sejnowski, Wolf Singer, and Martin Vinck

## Abstract

A central goal of systems neuroscience is to understand how the brain represents and processes information to guide behavior (broadly defined as encompassing perception, cognition, and observable outcomes of those mental states through action). These concepts have been central to research in this field for at least sixty years, and research efforts have taken a variety of approaches. At this Forum, our discussions focused on what is meant by "functional" and "inter-areal," what new concepts have emerged over the last several decades, and how we need to update and refresh these concepts and approaches for the coming decade.

In this chapter, we consider some of the historical conceptual frameworks that have shaped consideration of neural coding and brain function, with an eye toward what aspects have held up well, what aspects need to be revised, and what new concepts may foster future work.

Conceptual frameworks need to be revised periodically lest they become counterproductive and actually blind us to the significance of novel discoveries. Take, for example, hippocampal place cells: their accidental discovery led to the generation of new conceptual frameworks linking phenomena (e.g., memory, spatial navigation, and

---

**Group photos (top left to bottom right)** Kenneth Harris, Jennifer Groh, James DiCarlo, Wolf Singer, Jason MacLean, Andrew Schwartz, David McCormick, Pascal Fries, Terry Sejnowski, Kenneth Harris, John Reynolds, Matthias Kaschube, Jennifer Groh, Wolf Singer, Gilles Laurent, James DiCarlo, Gordon Pipa, Pascal Fries, Terry Sejnowski, Gilles Laurent, Martin Vinck

sleep) that previously seemed disparate, revealing unimagined mechanistic connections. Progress in scientific understanding requires an iterative loop from experiment to
model/theory and back. Without such periodic reassessment, fields of scientific inquiry
risk becoming bogged down by the propagation of outdated frameworks, often across
multiple generations of researchers. This not only limits the impact of the truly new and
unexpected, it hinders the pace of progress.

## Outline and Basic Concepts

Two ideas have driven theories of the cortex for decades: the "column" and the
"canonical circuit." Although these concepts certainly have a grain of truth to
them, it is now clear that they are oversimplifications in need of improvement.

### The Cortical Column

The "cortical column" is an anatomical term that connotes a discrete set of
cells operating together to perform a computational function. As originally understood, columns are discrete entities, but there can be connections between
columns. In this classical view, it must be possible to define boundaries between columns by some means; however, this is almost never possible. Even
in the barrel cortex, where "columns" could be easily understood as being defined by the anatomically distinct regions that respond only to single whiskers,
boundaries can only really be defined in layer 4; in the other cortical layers,
there are no clear cytoarchitectural borders corresponding to the barrels, nor do
the cells show sudden transitions in their whisker preference across the cortical surface. In a classical cortical column, features are represented across the
surface of the cortex, and in vertical penetrations, the majority of the neurons
share the same feature selectivity.

Vernon Mountcastle was one of the first cortical neurophysiologists to emphasize the vertical organization of the cortical layers (Mountcastle 1957). His
view of modularity was that a given cortical structure is composed of modules
with neighbors of similar functionality. Functionality is determined more by
the input a module receives than in the interconnectivity of the cells compos-
ing the module. The idea here is that this neighborhood similarity is preserving
some type of topology. The basic element of a module is that of a minicolumn,
which is about 30 microns in diameter and contains about 100 cells. Larger
columns (1 mm diameter) may be composed of hundreds of minicolumns.

Multiple behavioral variables are mapped to the same cortical module,
which suggests that these modules can participate in different systems and that
this participation is probably dynamic. Mountcastle believed that small clusters of neurons corresponding to columns were fundamental components of
cortical function. For each of these cortical processing units, output was produced from structured input. He defined a microcircuit as "the small number

of neurons synaptically linked in a processing chain that leads from some particular input...to some particular output" (Mountcastle 1998). These modules were defined almost entirely using criteria such as cellular morphology, layer-specific input-output patterns, and afferent-efferent projection anatomy. He emphasized the importance of discovering the input-output or cortical "operation" of this circuit, considered to be the essence of cortical function, and offered a number of candidates as examples (e.g., differentiation, pattern recognition and generation, coincidence detection, and encoding-decoding of output and input).

To a good approximation, different regions of cortex consist of similar cell types that occur in similar layers, with similar patterns of physiology and gene expression, and are connected using similar rules of connectivity and plasticity. Indeed, recent transcriptomic analysis (Tasic et al. 2018) suggests that while excitatory neurons may be distinct between cortical regions, inhibitory cells are extremely well preserved. Further work to date suggests that the connectivity, physiology, and *in vivo* functions of these cell types are largely preserved between areas (Douglas and Martin 2007; Harris and Shepherd 2015).

It has been over fifty years since the concept of a cortical column was formulated, and we suggest that it needs to be modified to fit current experimental evidence. Rather than a discrete column, the fundamental unit of cortical computation could be described as a "laterally iterated processing unit" (LIPU). Here, the idea is that the synaptic connections of every cell are set up by rules of activity-dependent and chemically hardwired plasticity that are largely independent of the cell's position on the cortical surface. This does not imply that cortical connectivity has to be spatially isotropic. For example, in the visual cortex of carnivorans, connections have a "patchy" appearance that links regions representing similar orientations (Das and Gilbert 1995). Nevertheless, these non-isotropic connections can arise from isotropic rules—a phenomenon familiar in physics (known as symmetry breaking) that allows, for example, a spatially non-isotropic crystal to form even out of spherically symmetrical atomic components.

In the not too distant future, it should be possible to reconstruct connectivity in a cubic millimeter of cortex from electron microscope cross sections (Kornfeld and Denk 2018). This will provide evidence for the patterns of connections between different cell types and the degree to which they are repeated across the cortex. Differences in the spatial scale of repeat distances may be different in different areas of cortex.

As we have defined it, the LIPU is still an example of modularity. What are the boundaries that define this unit? Are they physical (anatomical), computational (e.g., a field of integration encompassed by a "convolutional kernel"), or merely conceptual (a device that makes a complex system easier for scientists to understand)? While the answer is of course still unknown, insights can be gained from computational models of artificial networks, to which we turn to next.

**What Does the LIPU Do?**

If there is a canonical circuit embedded in the LIPU, presumably it applies a similar processing strategy to diverse types of inputs, performing information processing functions that are useful to the rest of the brain. Of course, we still do not know what this function is but several candidate theories have been put forth. Below, we begin with four of the leading theories.

*LIPU Theory 1: Unsupervised Learning*

Perhaps the oldest hypothesis for cortical function, the roots of unsupervised learning can be traced to Barlow (1972), Marr (1970), and Konorski (1967). The hypothesis is that cortical excitatory neurons apply unsupervised learning rules to extract features from the data: the input patterns are of distinct statistical structure, which means they will be likely to correspond to features in the natural world of behavioral significance.

At the level of computational models, many unsupervised learning rules have been described. Neural instantiations of standard statistical procedures (including principal component analysis, cluster analysis, and independent component analysis) have all been formalized (Hertz et al. 1991; Dayan and Abbott 2001). The multiple excitatory cell types in the cortex might use different learning rules to instantiate different types of unsupervised learning, and perhaps this is the cause of their characteristically different tuning properties (Harris and Shepherd 2015). Furthermore, cortical cells could represent unsupervised rules that we have yet to imagine.

A simple modification of the theory allows for cortex to implement simple supervised learning rules to form, for example, more detailed representations of stimuli that are present at times of high behavioral salience, as signaled by the activity of neuromodulatory systems (at least in sensory cortex). Substantial evidence suggests not only that cortical plasticity is enhanced by neuromodulators, but that *in vivo* representations of stimuli are stronger and more persistent when neuromodulatory systems are active (Froemke et al. 2007).

This theory is viewed by many neuroscientists as a default. However, it does not account for many experimental facts; for instance, the diversity of inhibitory cells found in cortex and their diverse modulation by nonsensory factors (McGinley et al. 2015b). Indeed, while some unsupervised rules require inhibitory cells for their activity, there are none that need such extreme diversity. Another is the existence of recurrent and feedback excitatory connections, which are not required by such networks.

*LIPU Theory 2: Excitatory Recurrence Allows Bayesian Inference*

The second theory, which dates at least to computational models in the 1980s (Ackley et al. 1985), is based on the ubiquitous presence of excitatory recurrent connections in cortex. These connections come at a cost: misfunction in recurrent excitatory circuits is at the root of epilepsy. Subcortical structures, while often having recurrent inhibitory connections, do not have such extensive recurrent excitation, nor have recurrent excitatory connections been described in any species other than amniotes. Presumably, therefore, recurrent excitation plays an essential role in cortical computation, and perhaps one that allowed some amniotes to develop impressive cognitive capabilities.

At the heart of this theory is the idea that recurrent excitatory connections encode "priors" or "expectations" concerning relationships between stimuli that the animal is likely to experience. For example, visual scenes often contain extended contours. Recurrent connections between excitatory visual cortical neurons connect neurons whose receptive field centers are elongated parallel to the orientation of the receptive field (Iacaruso et al. 2017). These connections should therefore be able to "fill in the gaps" in continuous contours. Generalizing from this simple sensory example, one might expect recurrent connections to allow associations between higher-order cognitive concepts in multiple cortical regions.

These ideas have been formalized in a computational neural network architecture called the "Boltzmann machine" (Ackley et al. 1985). In a Boltzmann machine, Hebbian plasticity strengthens connections between coactive neurons; if an assembly of neurons is usually driven together by sensory stimuli, connections between them will later enable "filling in" of the activity of neurons whose activity is missing. Mathematically, it was possible to prove that the Boltzmann machine performs a formal process of Bayesian inference: given the available sensory evidence, it estimates possible causes that are compatible with them, sampling an activity pattern from a posterior probability distribution of possible states of the outside world, that are compatible with the available (incomplete) sensory data.

The Boltzmann machine is an attractive model for a function of the LIPU. Furthermore, some behaviors of this network bear an uncanny resemblance to brain activity (e.g., it produces spontaneous activity that mimics the structure of expected sensory inputs). Nevertheless, many features of actual cortical circuits are not required by Boltzmann machines. At least in its initial formulation, there are no inhibitory neurons in a Boltzmann machine, let alone a myriad of cell types. In addition, there is no structured connectivity of different cell types and cortical layers, and there are no spikes. Furthermore, while the original Boltzmann machine was able, in principle, to perform any inference given sufficient time, it learned too slowly in practice to be of use in real-world information processing. More complex versions of the Boltzmann machine architecture, which involve hierarchically repeated populations analogous to a hierarchy of cortical regions, are much more computationally powerful however (Hinton and Salakhutdinov 2006). Thus, a version of the Boltzmann machine that incorporates more complex features could be even more computationally powerful.

*LIPU Theory 3: The Liquid State Machine*

The third theory, unlike the first two, does not need synaptic plasticity to adapt network connectivity. It is based on the framework of "reservoir" computing (Maass et al. 2002) that uses a randomly structured recurrent neural network to nonlinearly transform a time-varying input signal into a spatial high-dimensional representation. At each time step, the network combines the incoming stimuli with a volley of recurrent signals containing a memory trace of recent inputs. For a network with $N$ neurons, the resulting activation vector at a discrete time $t$ can be regarded as a point in an $N$-dimensional space. Over time, these points form a pathway through the state space, also referred to as a neural trajectory. This computation serves a feature expansion (i.e., a projection based on many nonlinear basis functions) as well as an explicit implementation of fading memory of past states. While this feature expansion is not specific to a task, task-specific computation is implemented based on learned and weighted task-specific linear or nonlinear mappings of neuronal activity (linear combinations are sufficient).

Even though computational properties of reservoirs can, in principal, have universal computational properties—that is, they can implement any Turing computable function (Buonomano and Maass 2009)—the performance of a reservoir depends on the connectivity, properties of the dynamical elements (i.e., neurons), and the state of the dynamical system—that is, whether the system is behaving regularly, critically, or chaotically (Legenstein and Maass 2007). Moreover, the performance is often much smaller compared to recurrent systems that are optimized (e.g., LSTMs and backpropagation through time) (Hochreiter and Schmidhuber 1997). This gave rise to modified versions of the reservoir computing theory that use neuronal plasticity to self-organize connectivity and the dynamical state of the system (Lazar et al. 2009), implement unsupervised learning to implement noise-robust efficient computations (Toutounji and Pipa 2014), and reward-modulated optimization of reservoirs (Bellec et al. 2019).

There are several features that make this theory of reservoir computing special. The initial reservoir computing idea is useful because it shows that even fully random recurrent networks can compute. This is important from a developmental point of view, since an initially random network can be used to bootstrap the problem and optimize computations over time. This is especially relevant, since the feature expansion into a higher-dimensional representation, carried by a large number of cortical cells, enables random connections to re-code information into the "sparse" format helpful to render synaptic plasticity more efficient (Barth and Poulet 2012). Furthermore, because recurrent networks integrate and mix together activity across a range of times, they are able to transform patterns only distinguishable as temporal sequences into spatial patterns, such that a given neuron only fires in response to a specific temporal sequence. This framework has been given a memorable name, describing recurrent random networks as a "liquid state machine."

What the liquid state machine does not do, however, is produce behavior: it just reformats a code into a form that can then be used by downstream structures to learn appropriate behavioral responses. The hard work of learning appropriate responses to stimuli is thus left to downstream structures; the corticostriatal synapse, whose plasticity is well characterized and controlled by dopamine, may be one possible locus for this.

The liquid state machine does not predict the diversity and specific connectivity of different cell types and layers but it is not inconsistent with them. Indeed, computational experiments show that random recurrent circuits with structured connectivity perform better than networks with completely random connectivity (Lazar et al. 2009; see also Singer, this volume). It is therefore conceivable that the complex structure of connectivity found in the cortex evolved to help this function of pattern separation.

### LIPU Theory 4: Subtractive Predictive Coding

This hypothesis is in some ways an opposite of the second. A Boltzmann machine amplifies responses to expected stimuli: when an input arrives that matches the types of inputs seen before, the response is stronger, with missing neurons' activity filled in, and more vigorous than it would be to a completely novel type of input. The concept of subtractive predictive coding is the opposite: expected inputs are discarded, while responses to unexpected stimuli are amplified and passed on to downstream structures (Keller and Mrsic-Flogel 2018).

The best example of this processing scheme comes not from cortex, but from the lateral line lobe of weakly electric fish (Bell et al. 1997). These fish sense the surrounding environment by producing electric fields and sensing the disturbances in these fields caused by nearby objects or other organisms. However, most of the electric field impinging on their sensors does not reflect external objects but simply comes directly from their field generation organs, which the fish must subtract out to find the behaviorally relevant external signals. By generating artificial signals as filtered versions of the field which the fish generates, Bell et al. (1997) were able to show that the lateral line lobe performs this subtraction and does so in an adaptive way that also subtracts the signal presented by the experimenters.

The subtractive predictive coding hypothesis posits that the cortex performs a similar function, but it says more: Not only does the cortex subtract simple consequences of one's own actions (such as subtracting the sound of your own voice to hear other people talking over you). It is able to make more complex predictions, for example, computing an expected pattern of visual input based on high-level cognitive expectations, and subtracting it from the actual input pattern to detect subtle features that do not match expectation.

Some very widely observed phenomena can be seen as examples of subtractive predictive coding. For example, presentation of a steady, sustained tone will not cause sustained activity in auditory cortex; it will cause strong activity at its onset and again at its offset ("accommodation"). Given a model where silences and sounds are expected to be sustained, this can be interpreted as producing activity when the times in which a violation of this expectation occurred. Nevertheless, by this standard, accommodation is not a specific function of cortex: it happens in the sensory receptors themselves and again at many levels of the processing hierarchy. It may be that the cortex specializes in subtracting predictions of advanced statistical models of the outside world, but the experimental evidence for this is mixed. For example, Keller et al. (2012) reported that ~10% of neurons in mouse visual cortex respond to mismatches between self-motion and visual motion signals, whereas Saleem et al. (2013) reported that visual cortical neurons responded instead to a match between these two signals, which would be more consistent with hypothesis 2 than hypothesis 4.

## Spontaneous Activity

Another feature of cortical physiology which we refer to is *spontaneous cortical activity*. Clearly, it should come as no surprise to find spontaneous activity in the nervous system. If there was no spontaneous activity in the circuits controlling respiration, we would have a problem. However, the presence of spontaneous activity in the sensory systems is more surprising. Spontaneous activity in sensory systems, and the related phenomenon of variable responses to sensory stimuli, seem fairly specific to cortex: much lower levels of variability are seen in subcortical mammalian structures. As yet, there is no consensus on the function of structured spontaneous cortical activity, but it is possible to list some hypotheses, again non-exclusive.

### *Spontaneous Activity Theory 1: Nothing, or Worse*

The first possibility, which cannot be excluded based on current data, is that spontaneous cortical activity serves no function at all. The cortex is spontaneously active under anesthesia, and as far as we know performs no information processing in this state. Although spontaneous cortical activity costs some energy, it may be that this is so minor, in evolutionary terms, that an animal suffers little disadvantage, even if there is no need for it to occur at all.

An even more extreme view holds that spontaneous activity is worse than useless: it is a form of noise that actually impairs processing of sensory inputs by interfering with neuronal representations. In this view, neurons are noisy devices, and worse, this noise becomes correlated through the cortex's highly recurrent connectivity. One result that could be taken as evidence for this perspective is that correlated fluctuations in primate visual cortex get smaller when the subject is attending to a sensory stimulus (Harris and Thiele 2011).

An alternative interpretation of this result is discussed below. A related concept from motor neurophysiology is that neurons in the motor cortex can merge together such that their combined activity is a "null space" that is occupied specifically when muscle activity is absent (Kaufman et al. 2014).

### Spontaneous Activity Theory 2: Imagery, Memory Recall, and Consolidation

Spontaneous activity shares many features with sensory-evoked activity. For example, Kenet et al. (2003) have reported similarities between sensory-evoked and spontaneous activity patterns in anesthetized cats and, recently, consistent observations were made in awake ferrets, with spontaneous activity being more exuberant in the awake than in the (lightly) anesthetized state (Smith et al. 2018). Thus, one might hypothesize that spontaneous activity in sensory systems correlates with processes such as imagery and memory recall.

In this view, the brain spontaneously produces patterns of neural activity that mimic actual sensory responses and have similar consequences on downstream structures. These consequences might involve production of actions: for instance, when remembering the nature and location of an object currently hidden from view, the brain might reproduce activity patterns similar to those the object would itself produce, thus allowing motor actions to be performed similar to those the object would itself produce.

Even if a spontaneous activity event does not directly produce action, it can have other consequences, such as changes in synaptic strengths. For example, recapitulation of activity patterns that occurred in previous behavior could cause further consolidation of the synaptic changes that encoded this memory; consistent with this view, interruption of spontaneous events in hippocampus after behavioral experience disrupts formation of long-term memories of that experience (Girardeau et al. 2009; Jadhav et al. 2012). More complex possibilities exist: spontaneous activation of neuronal assemblies containing overlapping cell populations could cause changes in synaptic strengths linking these neurons, thereby forming associations between previously unrelated concepts. This process could be a basis for the process that humans subjectively describe as "thinking."

### Spontaneous Activity Theory 3: Nonsensory Context

While spontaneous activity shares some structural properties with sensory responses, they are far from identical (Scholvinck et al. 2015). Perhaps, then, a major function of spontaneous cortical activity in sensory systems has no direct connection to sensory processing, but instead encodes nonsensory variables, which are integrated with, and can modulate the detection of, sensory stimuli (McGinley et al. 2015a).

An important clue to this comes from the mouse visual cortex. Activity in visual cortex changes when mice run, even in complete darkness (Niell and Stryker 2010). This activity presumably has nothing to do with expected sensory stimuli. Furthermore, neurons in sensory cortex respond to rewards (Shuler and Bear 2006), and imaging of axons arriving in sensory cortex from elsewhere shows they convey very complex nonsensory information. It may be that spontaneous cortical activity is in fact a high-dimensional representation of an animal's current cognitive and behavioral state, which the cortex integrates with sensory information (Stringer et al. 2018). The optimal behavior to produce in any circumstance depends on a combination of sensory input and internal context; by integrating these two classes of information, the cortex may provide information allowing an animal to perform behaviors that integrate sensory and nonsensory data.

Cortical traveling waves, which have been observed in both sleep states and awake state, are another source of spontaneous activity (Muller et al. 2018). They modulate the membrane potentials of neurons in a spatially organized way and vary in frequency from theta (4–8 Hz; Lubenov and Siapas 2009) to gamma (30–80 Hz; Gabriel and Eckhorn 2003).

### Spontaneous Activity Theory 4: Housekeeping / Homeostasis

Our final theory suggests that spontaneous activity is not a reflection of information processing per se, but rather that it functions to maintain the biophysical and biochemical state of the network. Spontaneous electrical activity is prominent in the development of the nervous system from the earliest stages (Spitzer 2006). The function of this early spontaneous activity presumably has nothing to do with processing of sensory information, memory recall, or motor variables. Instead, it seems to function to specify neural circuits, for example, to determine the differentiation, migration, and wiring of developing neurons. Cortical spontaneous activity shows very sudden changes with development: adult patterns show a substantially different structure to those present earlier in development (Luhmann and Khazipov 2018). Nevertheless, it remains possible that cortical spontaneous activity in adults plays at least a partial role, similar to its role in early development, enabling and guiding low-level maintenance of cellular and circuit properties. Several studies have reported that axonal conduction delays in the cerebellum are tuned to allow complex spikes from the inferior olive to arrive in a precisely timed manner, despite differing physical lengths of these axons (Sugihara et al. 1993; Baker and Edgley 2006). If this is the case, some homeostatic mechanism must enforce these constant delays; spontaneous activity, perhaps during sleep, could be a key part of the process. Spontaneous activity during sleep has also been proposed to enable "downscaling" of firing rates and synaptic strengths built up during waking (Tononi and Cirelli 2014) or other metabolic functions (Vyazovskiy and Harris 2013). Enabling these low-level metabolic and circuit functions might be a key function of spontaneous activity in both waking and sleep, parallel to the information-processing roles described above.

## Considerations from Evolution

The brain of any species cannot be understood in isolation but is best considered in an evolutionary context. Two critical concepts related to evolution, in general, and the brain, in particular, are *inheritance* (i.e., features that have been continuously present in a given phylogenetic lineage) and *convergence* (i.e., features that arose independently multiple times but which accomplish similar functions).

For an example of inheritance, consider the molecular building blocks of nervous systems (e.g., ion channels), which can be found in a highly similar form in bacteria. The synaptic transmission machinery operates with the same molecular components and principles across all animals, as far as we know. In fact, many of the molecular elements and their functional interactions were worked out in yeast. Sponges have some cells, called flask cells, that contain many of the molecular components of the postsynaptic compartment (ionotropic receptors, e.g., are missing but metabotropic ones are present). Flask cells, however, are not neurons, and sponges have no nervous system or synapses (Sakarya et al. 2007). So, either synapses evolved by borrowing and adapting already existing components, or present-day sponges lost a nervous system that existed in one of their ancestors. Sponges diverged from us and other animals some 600 million years ago.

Short-term synaptic plasticity mechanisms, such as facilitation and depression, are found in simplest nervous systems. Spike timing-dependent plasticity exists in insect nervous systems but whether these use glutamate and NMDA receptors is not currently known. Synchronization has been discovered in mollusks, insects, etc. Spatiotemporal representations are found in invertebrate sensory systems (e.g., in locust olfaction; Wehr and Laurent 1996; Mazor and Laurent 2005), or leech motor and premotor systems (Briggman et al. 2005).

Examples of convergence include looming sensitivity in single neurons in insects and birds. It consists algorithmically as a division of angular velocity by an exponential of angular size (Gabbiani et al. 2002). This algorithmic description also applies to cells in thalamic nucleus rotundus in diving birds (Sun and Frost 1998). It is very unlikely that the same computation (or need) existed in their common ancestor, which was some sort of worm. Another example is Jayaraman's result (Seelig and Jayaraman 2015) on head direction-like cells in the central complex of insects, which is similar to models of head direction cells in mammalian hippocampus. The olfactory system is also interesting: despite the fact that the molecular nature of the olfactory receptor genes is different in invertebrates and mammals, the organization (convergence to glomeruli, divergence and random-like distributed projections to second structures—piriform cortex or mushroom bodies) is similar, probably through convergence.

Evaluating homology across species from very different lineages is critical for cross species comparison but is a challenging task. For example, a dorsal telencephalon or pallium is part of the vertebrate brain *bauplan*. Thus it can be found in fish, amphibians, reptiles, birds and mammals. To trace back the evolution of the mammalian cortex, one has to look first at the outgroup of mammals; that is, reptiles.

Unlike fish and amphibians, a large portion of reptilian pallium has a three-layered organization, indicating that a layered cerebral cortex emerged about 320 million years ago in the ancestor of mammals and reptiles (the amniote ancestor). In addition, reptiles and birds harbor a nonlaminated pallial region, called dorsal ventricular ridge (DVR), where neocortical-like circuits have been identified.

The structural and functional differences of reptilian and mammalian pallial regions have fueled controversies on the evolutionary origin of the mammalian neocortex. How can we compare reptilian and mammalian pallial regions, cell types, and circuits? Do similarities result from homology or convergent evolution? And how can this discussion inform us on the evolution of cortical function?

Homology hypotheses can be tested by comparing early development, gene expression, and connectivity. The existence of thalamo-recipient neurons in the anterior DVR led to the "equivalent circuits" hypothesis, stating the homology of anterior DVR and neocortical L4 neurons. The analysis of a small set of molecular markers supported this idea. However, anterior DVR and neocortex develop from two distinct regions of the embryonic pallium, and the conservation of developmental fields would predict the homology of anterior DVR with mammalian claustrum and parts of the pallial amygdala, derived from the ventrolateral pallium.

To test these hypotheses further in an unbiased manner, Molnár and colleagues compared gene networks in micro-dissected chick and mouse pallial regions (Belgard et al. 2013). Their results show that only five genes are shared between a L4 gene module and an anterior DVR module. Micro-dissected brain regions, however, may contain cells of different types in different proportions, and this might confound the analysis and hide similarities. To overcome this limitation, Tosches et al. (2018) applied single-cell RNA sequencing to the turtle and lizard pallium.

The single-cell approach allows the analysis of small and sparse cell populations such as cortical interneurons. The comparison of turtle and mouse data shows that the same classes of GABAergic interneurons exist in the two species: interneurons derived from medial (MGE) and caudal ganglionic eminences (CGE), including somatostatin, parvalbumin-like and vasoactive intestinal polypeptide-like types. This suggests that developmental and/or functional constraints led to the conservation of these interneuron types for over 320 million years.

High-level clusters of glutamatergic neurons map to distinct regions of the reptilian pallium: the hippocampus, dorsal cortex, olfactory cortex, the so-called "pallial thickening," and the DVR. These regions express different combinations of transcription factors, reflecting their distinct developmental and evolutionary histories. The comparison of regional transcription factor codes in reptiles and mammals supports the hypothesis that the anterior DVR is homologous to the mammalian lateral amygdala, as also indicated by the fact that these regions develop from homologous developmental fields and establish similar connections with the rest of the brain. Nevertheless, many effector genes (e.g., ion channels, cell adhesion molecules) are shared between the reptilian anterior DVR and the mammalian neocortex, indicating that the expression of the same gene sets in these two pallial regions is regulated by different transcription factors. In conclusion, different pallial regions expanded independently in the reptilian and mammalian lineages—ventral pallium (anterior DVR) versus dorsal pallium (neocortex)—resulting in the convergent evolution of gene expression and circuits.

## Representations and Neural Codes

### What Is a Code, Anyway?

*Encoding and Decoding Models: Definitions and Scientific Goals*

The terms *code*, *representation*, *encode*, and *decode* have become highly overloaded in neuroscience: different people use the same phrase to mean very different things, so that investigators often talk past each other rather than coming together to synthesize and integrate ideas. Grounding of these terms requires discussion of the goals and the assumptions in the models used to achieve those goals. A subset of our group had a lively discussion on these points, and here we attempt to explain those sometimes divergent viewpoints.

It is widely assumed that neural spikes are, for most problems of interest, the carriers of information to support moment-to-moment behavior. (Here "behavior" is broadly construed to include sensation, cognition, and action, and could be studied in an ethological or a laboratory context.) Under that assumption, three main types of data are typically measured and/or experimentally controlled:

1. energy patterns that impinge on sensory epithelia,
2. spike patterns in populations of neurons in one or more locations in the brain, and
3. the positions of one or more parts of the body (e.g., arms, eyes, vocal apparatus).

The goal of much of systems neuroscience is to use such data to "understand" how the internal parts of the system operate together to execute complex sensorimotor loops (i.e., "cognition," "complex behavior," "intelligence," etc.). A more modest goal may be to describe the information content contained in a population of neurons, without assumptions of the explicit role these neurons may play in the behavior generation. This more relaxed approach may mitigate many of the arguments between differing viewpoints that arise from invalid assumptions. Nonetheless, experiments define concepts derived from such measurements (e.g., "motivation," "reward expectation"), and it is important to keep in mind that such definitions are not direct measurements; they are only inferences, as they assume one or more underlying models of what the brain is doing. Indeed, all such assumed internal latent variables are inferred from the same three basic measurements above: stimuli, neural activity, and behavioral measurements. And, if judged at all, each model is judged on the accuracy of predictions it makes for other observed variables (typically neural spikes and/ or behavior).

The form of the understanding we seek is not usually explicitly stated. We argue, however, that it should ideally be in the form of inferred, neural mechanistic causal models that describe the linkages between those three types of measurements. A *neural mechanistic model* is a model that minimally contains approximations of neurons and their connections. A *causal neural mechanistic model* is one in which external perturbations can be injected or model parts removed, so that the resulting effects on the other parts of the model will be accurately predicted.

As a point of departure, we may begin to understand some aspects of the transformations taking place as raw sensory signals propagate through the nervous system. These transformations are usually considered mechanistically; that is, how an output of some entity (e.g., a neuron deep in the visual system) fires relative to an input (e.g., a pattern of light energy on the retina). As an example of a neural mechanistic causal model, consider a transfer function that is implemented as a set of modeled neural elements and their connections, which aim to describe and predict this transformation accurately. This model can (a) explain how the stimulus is responsible for the output, (b) make predictions of what other internal neural responses should be found along the way, and (c) predict how direct perturbations of those internal elements will lead to perturbations in the output. While such predictions may turn out to be incorrect, the model can drive a principled selection of future experiments, which would aim to reject this model in favor of one or more alternatives. We refer to this model as an "encoding model" (see Figure 13.1). An encoding model has the advantage of providing a concise description of the relationship between neural activity and the variables being encoded, but it may turn out that cause-and-effect relationships might be very complex in biological systems. Still, until someone proposes another way to make scientific progress, important work continues utilizing this framework in the hope that such complexity can be overcome.

This conceptual discussion and the three types of measurements listed above (stimuli, neuronal spikes, and behavior) lead one to see that there are two primary types of neural mechanistic causal models:

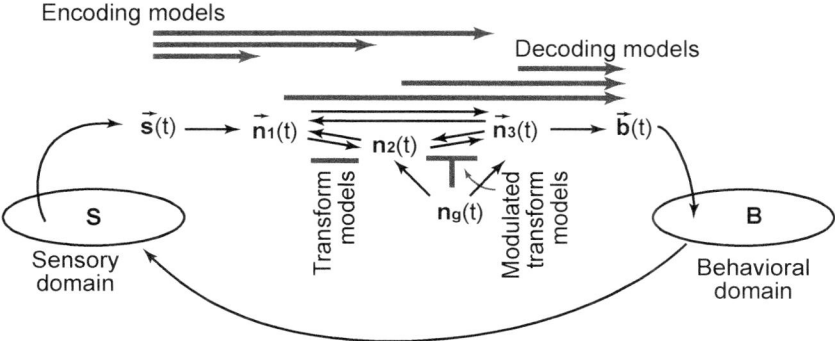

**Figure 13.1**   Nomenclature for different types of modeling paradigms of mechanisms underlying moment-to-moment sensory-cognitive-motor behavior. If we limit ourselves to spiking activity and moment-to-moment behavior, four modeling paradigms may encompass potentially all of the model-building activities in the field. **S** represents a defined sensory domain (e.g., the set of all possible visual movies of a specified size and resolution); $s(t)$ is a sample from such a set (e.g., a frame of one movie). Each $n(t)$ is a potentially time-varying population vector of neural activity in a brain region (e.g., the entire set of pyramidal neurons in layer 2/3 of primate visual area V1 that project to primate visual area V2), whereas $b(t)$ is a potentially time-varying vector that describes, e.g., the current positions of all body parts, which itself lives in a set of possible configurations of body parts (**B**). Of course, reduced description $b(t)$ is also possible and potentially useful (e.g., the subject's choice on each behavioral trial). Arrows show the neural populations that are assumed to have direct connections: behavior can influence the environment and thus the next impinging sensory sample (outer loop). The two most dominant forms of model building are (a) encoding models, which are causal neural mechanistic models that apply to **S** to generate predictions of the responses of neuron populations (**n**) and (b) decoding models, which are causal neural mechanistic models that apply to one or more specified **n**'s to generate predictions in a specified behavioral domain (**B**). Although the examples depicted above are for visual-cognitive-motor domains, similar examples can be readily defined for other sensory-cognitive-motor domains.

1.  Causal neural mechanistic models that link between the sensory epithelia and neural spiking activity at one or more places in the nervous system (which we refer to "encoding models").
2.  Causal neural mechanistic models that link between neural spiking activity at one or more places in the nervous system and the positions of the body parts (here termed "decoding models").

While much progress has been made on encoding models, disagreement surrounds decoding models: what they are, their utility, and how they should be interpreted. Here we emphasize that the goal of the decoding approach is to gather knowledge of the natural causal brain processes, not simply to control an externally attached device. We also emphasize that science must propose hypotheses that can be implemented (through models) so that they can be tested; that is, a downstream homunculus is *not* a scientific decoding model.

Broadly speaking, "neural decoding" is any implementable analysis that demonstrates the ability to predict some outcome by extracting information from neural recordings. The implementation is usually in the form of some kind of extraction algorithm (e.g., linear filter, population vector algorithm, Kalman filter, or deep learning neural network) and the success of the decoding analysis is measured by its accuracy of predicting the targeted outcome (e.g., arm trajectory, behavioral choice). *An above chance decoding performance shows that information about the predicted variable(s) is present in the recorded neural activity.* This has scientific value because information is exposed that may not be obviously present in individual neurons and because information identification constrains the hypothesis space of causal neural mechanistic models that may exist between the recorded neurons and the predicted behavior (i.e., "decoding models") (Majaj et al. 2015).

A common point of confusion in our field is how the modeler views the decoding analysis: whether the extraction algorithm itself is physically realized, in this case, within the nervous system (perhaps within a particular anatomical location or set of locations), or whether it is merely a tool used to identify the information content at a node defined by the neurons whose activity patterns were recorded as input to the extraction algorithm. Again, as defined above, decoding means extracting information from neural firing rates. This can be performed within the nervous system or by an outside observer. In either case, the code itself exists, independent of whether or not it is decoded.

Based on the principle of decoding, we hope to better understand the information contained in patterns of neural activity. Decoding analyses expose information that is transformed by the brain as it propagates through the system. This allows scientists to propose and construct models and to make predictions about the behavior (which the decoding test defined above has already at least partly achieved) as well as about neural activity that intervenes between the originally recorded neurons and the behavior. Similarly, scientists working on encoding models should propose alternative causal linkages from sensory epithelia to patterns of neural activity and conduct experiments to distinguish between those alternative encoding models. Of course, all models are wrong at some level, so that scientific work will continue indefinitely, at least until causal neural mechanistic models of sufficient accuracy and predictive power are obtained to support the application goals of the society that funds the scientific research (e.g., new ways to intervene causally in the system to ameliorate brain disorders). In what follows, we formalize these definitions a bit further in the hopes of grounding these key ideas.

We define a potentially time-varying, multidimensional stimulus in the external world as $s(t)$, where any specific $s(t)$ is an element from a domain $S$ (e.g., a set of natural images, movies, sound sequences). We define a potentially multivariate, potentially time-varying body state in (lagged) response to an element drawn from the domain $S$ as a behavior $b(t)$ that will reside in the range $B$ (e.g., the possible positions of the body parts that are chosen to be monitored).

In addition, we define the time-varying activity vector of a given subset of neurons (e.g., all layer 2/3 pyramidal cells in visual area V1 that project to layer 4 of V2) as **n**(t). For instance, we can imagine the following example population vectors: **n**[V1](t), **n**[IT](t), **n**[basal ganglia](t), etc. Given this framework, we propose the following possible model paradigms, which must be built and tested (Figure 13.1):

- encoding models, which map from **s**(t) to some **n**(t),
- decoding models, which map from some **n**(t) to **b**(t),
- neural population transform models, which map between **n1**(t) and **n2**(t), potentially recurrent, and
- modulated neural population transform models, which map between **n1**(t) and **n2**(t), subject to another neural population(s) **ng**(t).

We do not aim to be overly prescriptive, but to engender shared scientific terminology and associated goals.

Progress has resulted from, and some of us believe that it will continue to result from, making measurements and using those to estimate the parameters of different hypothesized model families in each one of these modeling paradigms. We note that this framework is highly general and inclusive. For example, it includes the notions of external time and the notion of recurrent processing (note the arrows in Figure 13.1). Our goal here is not to strongly limit the alternative model families that might be considered, but to define the expected inputs and the expected predictors of any actual model. Once a model paradigm is chosen (from the above list), and the parameters of a model are determined (i.e., the model is selected from a hypothesized model family), the accuracy of the model is evaluated by its ability to predict its output variable(s) from its input variable(s) on held-out input variable settings (i.e., values of the input variables that were not used to fix the model parameters). That is, the test is generalization within the domain (e.g., **S**) of the model, where generalization can be defined as tests of increasing distance from the inputs used to determine the parameters of the model (aka training data). In addition to its predictive accuracy (generalization), the model might also be judged by its elegance or simplicity (e.g., amount of training data needed to specify the model parameters), the degree to which the model components correspond to known neuroanatomy, and/or the minimum description length of the model.

For a given stimulus domain (**S**) and behavioral domain (**B**) and proposed intermediate (hidden) neural population vectors, nearly all models in systems neuroscience can be placed into one of these paradigms: *encoding, transform, modulated transform, decoding.* Note that this is a very general and inclusive framework for common ground, as it leaves open the questions of timescales, dimensionality, model families to be considered, etc. The definitions are also broad enough to include instantiated neural network models of all existing potential normative theories: predictive coding, generative models, Bayesian inference, etc. Each of these choices might be highly specific to an area of conceptual study, though it is hoped that groups working in similar domains (e.g., recurrent sensory systems) might build and test some common model families so that model families that naturally work well (highly accurate in predictions) in more than one sensory system (albeit with potentially different model parameters in each case) will be readily discovered.

To fit such a model, parameters do not need to be estimated entirely via empirical data (e.g., neural responses and/or behavior). Indeed, the currently most accurate (predictively accurate) models of the primate ventral visual stream were discovered by fitting the parameters of a global model that includes each of these model types: encoding, transform1, transform 2, transform 3,... transform 6, decoding. Notably, the parameters of each and every one of those component models were fit entirely from a large number of pairs (**s**, **b**) drawn from a large domain **S** (images in the central ten degrees) and **B** (category reports). The internal "neural" populations of these deep artificial neural network models turned out to be highly predictive (over new **s**'s taken from **S**) of the responses of the internal neural population vectors (**n**) at multiple stages of the ventral stream (**n**[IT], **n**[V4]) (Yamins et al. 2013, 2014) and **n**[v1] (Tolias et al. 2001). Critically, that success was enabled because the mapping from **S** to **B** was chosen in a way to make the tasks computationally challenging (invariant object categorization). This thus appears to be an example of convergent evolution—artificial neural networks for visual processing have "evolved" (under human organized optimization pressure) to have internal representations that look very much like the brain's internal representations. Stepping back, the organizing point is that models which map from **S** to **n** are all examples of *encoding models* (above), and the goodness of these models is not judged on just capturing data, but on predicting new data. Such models are still far from complete, even in the visual system.

We note that encoding models that take high-dimensional sensory data as input naturally contain notions of receptive fields (RF, i.e., fields can be measured by doing virtual electrophysiology on the model, or by tracing the connections in the model). However, these encoding models are much more accurate (at predicting their population vectors (n) than the basic RF encoding model (also see below on the limitations of tuning curves). Thus, while the concept of a RF is still a useful teaching concept that can predict some aspects of any neuron's response, it is not a very accurate encoding model.

In that regard, we note that current encoding models are still not able to explain all aspects of the neural responses, most notably, many current encoding models lack temporal dynamics. And modulatory transform models have not yet been incorporated in any serious way. Testing new encoding models in that expanded family of recurrent artificial neural networks is ongoing work in multiple groups (i.e., causal neural mechanistic models), and we hope that those model families will be informed by the discussions and the data presented at this Forum.

It is useful to consider encoding and decoding in terms of statistics as well as from a communication/information perspective. As an example, consider the classic behavioral paradigm of locomotion. Suppose we want to predict the phase of locomotion based on the firing rate of a single, isolated afferent fiber from a cutaneous pressure receptor in the skin of the foot (Werner and Mountcastle 1965). If we assume a quasi-steady state (e.g., walking at a constant speed on a treadmill), we should be able to make a step-phase prediction with some degree of accuracy. If the afferent fired a graded volley of action potentials that appeared as a single bump in an event-triggered histogram beginning 10 ms after foot-strike, one could try to match an instantaneous firing rate to a phase of the step cycle. However, even if the firing rate pattern was exactly the same for each step, that prediction would be uncertain, because the same firing rate occurs twice for each step (except at the top of the bump). The histogram can be thought of as a probability distribution (probability of firing at a time point during the step cycle). So, acting as an ideal observer under ideal conditions, you would have a 50% error rate. In more realistic situations, where the behavioral state is not limited to isolated walking on a treadmill, that skin stretch receptor is being activated continuously in many different behavioral contexts. Predictions of behavior based only on observing that afferent's firing rate would be very poor.

This thought experiment was based on a single primary afferent. In general, neurons are driven by many different sources. For example, neurons in the motor cortex have been shown to carry information about ten different components of arm and hand motion simultaneously (Wodlinger et al. 2015). These components include the velocity of the hand as it moves through 3D space, orientation of the wrist, and shaping of the fingers. Because of the high-dimensionality of this encoding, the same discharge rate is associated with many different weighted combinations of these parameters. Fortunately, better decoding (i.e., better ability to predict action) can be achieved by observing the firing rates of many neurons simultaneously. Even though parameter encoding by individual neurons is redundant, the bias of each neuron's firing rate to a particular combination of these parameters (i.e., its "tuning") is specific. This uniqueness of a neuron's tuning function makes it possible for extraction algorithms to decode parameters encoded simultaneously in single-unit activity. Furthermore, parameters that are weakly encoded by individual neurons, but have a consistent effect on the firing rate of many members of the population, can be extracted with these algorithms. This general principle is why populations of neurons are needed for more successful prediction of behavior from neural population.

While successful prediction of behavior (i.e., "decoding" by an observer) can add support to the inference that the recoded neurons are causally linked to the behavior in a neural mechanistic manner, such success does not guarantee that it is the correct inference. How might that inference be strengthened? We see two ways: First, requiring an ever more detailed prediction of the specified behavioral domain (B, Figure 13.1) should tend to show the limitations of incorrect causal inferences (i.e., incorrect decoding models as shown in Figure 13.1). This cannot prove the causal link, but it could still lead us to the correct neural mechanistic model that intervenes between the neural activity and the behavior. Second, direct perturbation of the recorded neural elements (e.g., silencing of multiple, targeted individual neurons) should produce behavioral effects that are precisely predicted by the decoding model. Ever more detailed perturbations can again be used to lead us to the correct neural mechanistic decoding model. Both approaches should ideally be applied to the brain subsystem under study. The larger point to keep in mind is that decoded information emerging from the processing of neural data, no matter how accurate it may be, does not guarantee that this information is used by the nervous system. It is only a starting point to a very large family of neural mechanistic causal models, and those causal relationships—and thus the most veridical models—are likely to be highly complex.

We can illustrate this type of problem with the walking experiment. Within this task are a series of behavioral events (e.g., a trigger cue, onset of movement, target acquisition, reward administration). A peri-event histogram of the skin afferent's firing rate, triggered on one of the behavioral events, will have a structure (e.g., a bump) representing the probability of the afferent having an action potential at a point in time relative to the event. In terms of communication theory, because there is a correlation between the event and the firing rate, there is information (reduction in uncertainty) being transmitted between the event and the neuron's firing rate. This relation can still be very noisy and may not mean that there is any kind of direct synaptic connectivity (direct causal relation) between the event and the change in firing rate. This is an important point, has led to a great deal of confusion, and is often contentious. Walking is a cyclic behavior in which the entire body oscillates with a period equal to the step-cycle length. Every part of the skeleton, every somatic pressure sensor, visual and auditory input, almost all muscles, and probably most neurons are going to be entrained by this periodic behavior. A histogram of almost any neuron's firing rate will show some kind of structure. That neuron is therefore transmitting information encoded as firing rate through the step cycle even though it is unlikely to be linked by any direct "circuitry" to the neural source driving locomotion. The foot receptor transduces pressure and, when activated by stretched skin, is "causing" the neuron to fire, but whether the signal in our afferent fiber is transmitted in a way that is decoded subsequently is unknown. Simplistically, we can record the firing rate from neurons in many other parts of the nervous system and those rates will be highly correlated to that of our afferent. This in no way means that our afferent is "causing" those other neurons to fire.

Given the definitions of encoding models and decoding models above, then a neural "code" is conceptually defined as a particular measure of information somewhere in the brain that is *both* a product of an encoding model and an input to a decoding model. In this formulation, the job of alternative decoding models is to specify *what* measure of information is causally critical (to a domain of behavior) and *how* it is causally critical (to that domain of behavior). And a key job of alternative encoding models is to explain and predict *that* information using neural mechanistic causal linkages from sensory epithelia to the specified measure of neural population activity. That is, a putative neural "code" is specified with respect to the knowledge of (or at least the hypothesis of) the downstream causal circuitry. In the next section, we provide another perspective on neural codes from the perspective of communication theory, which also relies inherently on the notion of a combined encoding and decoding model.

## *Coding from a Communication-Theoretic Perspective*

The concept of a "code" is one of the most commonly used terms in the neurosciences. Outside the neurosciences, we usually mean something very specific when we refer to a "code" or "coding"; these concepts have been formalized in communication and information theory. In the neurosciences, we have often a much more rudimentary notion of a "code": Does the communication theoretic notion of coding make any sense for the brain, and to what extent? What are the gaps in our knowledge about the "neural code"? Looking at various levels in the nervous system (retina, V1, M1) we see that in some cases the communication theoretical approach is sensible (retina), in others it is by approximation (V1), and in others becomes highly problematic (M1). Is it time to abandon the notion of a "code" because it erroneously carries with it all the communication theoretic baggage and the notion of "representation," or can we use this baggage?

Using concepts of information and communication theory, we can specify a number of conditions for coding in the classical communication theoretical sense and differentiate this from a more rudimentary concept of a code in the sense of merely containing information. We will distinguish two types of codes:

- Code Type 1: A code in the communication theoretical sense, which we call a *coding algorithm.*
- Code Type 2: A code that depends on the human observer, treating the link between the encoded variable and neural response as a virtual communication channel. A code in this sense means to contain information or, in other terminology, to have a tuning curve (see below).

There are five conditions for a Type 1 code in the communication theoretical sense:

- Condition 1: The sender receives some data.

- Condition 2: The sender encodes this data with a series of symbols (e.g., bit sequences), a "code," in a systematic manner (according to an algorithm).
- Condition 3: This achieves some transmission or storage goal.
- Condition 4: This code is constructed in a way that achieves data compression and allows error correcting at the receiver site (to solve the two practical problems that bandwidth is limited and communication channels are noisy).
- Condition 5: Finally, there needs to be some receiver (decoder) that can "understand" and do something meaningful with the transmitted symbols (e.g., decode the original signal or perform some action based on the message).

If such a model applies, then this permits us to use a powerful toolbox of techniques from communication theory to analyze the system. It also allows us to think about representations as a coding algorithm as well as to better understand other computational frameworks, like deep neural networks (Shwartz-Ziv and Tishby 2017). The definition of a Type 1 code makes clear that for the nervous system, we need both the notion of a "receptive field" and a "projective field" (Lehky and Sejnowski 1988); in our terminology used earlier, we need both an encoding and a decoding model. Below, we will further examine how the concepts of a Type 1 code apply to the nervous system (visual and motor system) and evaluate what limitations and gaps in our knowledge exist.

Usually, when neuroscientists talk about a "code" they take a stimulus (S), some neural activity (R), and compute a mutual information function $I(S;R)$ (or any other measure of dependence) to demonstrate that R "encodes" S. However, this is not equivalent to a demonstration of the existence of coding in the communication theoretical sense; it merely provides a generic measure of statistical dependence and shows that R contains information about S (Code Type 2). The fact that neural activity contains information does not mean that this information is being used or that the information can be easily decoded. Nevertheless, the exercise of quantifying whether neuronal populations contain information and which algorithms work best to decode generates very useful hypotheses about whether these neuronal populations may "encode" in the Code Type 1 sense and which algorithms the neural decoder may use. This is also important for constructing brain–computer interfaces (Schwartz et al. 2006).

## Coding in the Visual System

It has proven fruitful to model responses of photoreceptors or neurons in the retina as encoding the image that falls on the retina: The retina transforms the received image (Condition 1) in another signal, in a systematic way (electrical impulses/currents, organized topographically) (Condition 2), to achieve some goal (transmission of information about this image to the cortex over a channel with limited bandwidth and some noise) (Condition 3). The code is constructed according to some smart principles, achieving data compression (e.g., removing redundancies between pixels; Schwartz and Simoncelli 2001) (Condition 4), and there are receivers that understand the message and do something meaningful with it; namely the superior colliculus and the LGN, ultimately leading to behavior (Condition 5).

Whether and/or how the concept of coding applies to cortical areas like primary visual cortex or area IT remains far from clear. First, what does V1 encode (Condition 1)? Area V1 does not only encode the image on the retina but is sensitive to many other internal and external variables, such as arousal, movement, attention, and other sensory modalities (McAdams and Maunsell 1999; Niell and Stryker 2010; McGinley et al. 2015b; Stringer et al. 2018). In addition, V1 might not just encode the image on the retina but perform a type of Bayesian inference about the causes of the sensory data using priors and expectations (Rao and Ballard 1999; Friston and Kiebel 2009).

Second, how does V1 encode the data (Condition 2)? The population rate vectors are commonly assumed to form the coding substrate. However, it has not been demonstrated that they yield a "complete" representation of the image on relevant behavioral timescales (see, e.g., Van Rullen and Thorpe 2001; Resulaj et al. 2018), and there is evidence that there is additional stimulus information encoded in spike timing (discussed further below). Furthermore, we do not know which spikes are part of the code that is transmitted to other areas and which spikes are merely part of the coding process (e.g., spikes from interneurons), nor do we precisely understand the role that correlations play in coding. To make matters more complicated, there is an abundance of spontaneous ("dark") activity (discussed further below) that does not seem to encode any sensory information, and there exists tremendous state variability in sensory responses (Harris and Thiele 2011; McGinley et al. 2015a).

Third, what are the coding design principles in area V1 (Condition 4)? There is evidence that V1 receptive fields are optimized for sparse coding, and processes such as surround modulation have been interpreted from the perspective of efficient coding (Olshausen and Field 1996; Rao and Ballard 1999).

Fourth, who is the receiver of V1 information and what does this receiver do (Conditions 3 and 5)? There are many receivers of V1 information, including cortical areas (V2, V4, MT) and subcortical areas (e.g., cerebellum, striatum, superior colliculus). Cells may transmit different information depending on cortical/subcortical projection targets (Lur et al. 2016). Furthermore, it remains largely unclear which information in V1 responses is being used by which receiver and for what purpose. Finally, there are strong recurrent interactions between V1 and V2–V4, meaning that the (hierarchical) model of a unidirectional communication channel with a separable sender and receiver breaks down.

If we move forward to areas like IT (or a deep layer of a neural network), then one could say that the "neural code" (Type 1) becomes increasingly more "usable" higher up in the processing hierarchy, in the sense that it becomes easier to do something meaningful with it (e.g., a face-selective IT neuron or hippocampal place cell), although there's inevitable data loss compared to lower areas (data-processing inequality). This process can be thought of as a series of "unfolding" transformations that create increasingly linearly separable manifolds, corresponding to object categories, in high-dimensional spaces (e.g., Chung et al. 2015). The quantification of how "usable" and efficient a code is might be critical to interpret the significance of mutual information quantifications between stimuli and neural responses. For instance, we can, in principle, decode object categories more accurately from the retina than from area IT (data-processing inequality). This does not mean that the receiver of the retinal output uses this information to make decisions about object categories and act upon them. However, information about object categories, and the image in general, can be easily decoded from activity in area IT and is highly compressed.

### Coding in the Motor System

Turning toward the motor system, it becomes apparent that the notion of a communication theoretical code (Type 1) becomes problematic for many reasons. Information about many different movement parameters is present in the firing rates of motor cortical neurons. This information is encoded in the motor cortex. The problems begin with *how* the information was encoded: Did the encoding occur before arriving in the motor cortex, in local circuitry, or as input to the particular neuron, whose action potentials are being recorded? The time-varying values of different movement parameters tend to be correlated, reflecting the complex mechanics of movement where many degrees of freedom vary simultaneously. M1 neurons have high-dimensional tuning curves, so that the firing rates of individual neurons contain information about many of these parameters. This makes it difficult to parcel the encoded information into separate categories. Although the motor cortex is often considered to be composed of upper motoneurons projecting to the "final common pathway" (Sherrington 1906), in reality, it is one of many inputs to the subcortical neuronal substrate of muscle contraction. This presumed role of the motor cortex in muscle contraction has fostered historical controversies pertinent to the idea of whether decoding takes place at all in the projection targets of these neurons. As an extreme example, if motor cortical neurons function merely as upper motoneurons, then the information contained in their firing rates does not need to be decoded at all, since the putative role of these neurons is solely muscle activation. In contrast, if the encoded information is pertinent to more cognitive issues, such as the intended action of a hand on an object, then for this information to be realized as behavior output, "decoding" must take place as it is transformed by "downstream" structures to "cause" muscle contraction.

Further examination of implicit assumptions might help focus these issues. There is a general tendency in neuroscience to view the nervous system as discrete, separate components. This stems from historic anatomical descriptions of the system as well as the clinical observations underlying neurology and neuropsychology, which focus on finding localized lesions in the system. Furthermore, since the industrial age, we have become comfortable with the idea of machines composed of individual parts, each with a specific function. These factors come together to reinforce the general simplistic notion of cause and effect that underlie most functional descriptions of nervous system operation. Structure A projects to Structure B, contributing excitatory input to B's neurons, and these are diagrammed as sticks with plus and minus signs between boxes for each structure. Of course, these "circuit" diagrams rapidly increase in connections as more results are added, but the boundaries between the boxes remain fixed even as the number of sticks increases. Although it is obvious that many inputs interact to "cause" an output, such consideration is usually set aside to "simplify" neural functioning, keeping functional description within the bounds of simple causality.

This predilection toward simplistic causal circuitry is manifest in classic visual system neurophysiology. Here the concept of hierarchical organization prevails. Processing starts in the retina where coding begins, and rods and cones pixelate the visual scene. The pixel information is then transmitted to subcortical and cortical structures. As this information traverses successive brain structures, it is transformed successively. The concept here is that visual information is molded into a coherent image, one that is ultimately realized as a perceived, accurate description of the world. This concept originated with Hubel and Wiesel. They found that neurons in the cat thalamus and visual cortex had receptive fields of various complexity and hypothesized that increased complexity resulted from successive stages of processing. This concept prevailed in ensuing years during which researchers found that neurons in cortical areas anatomically farther from V1 seemed to have response properties that encompassed a wider set of visual filters. This was the motivation for attempts to organize the multitude of vision-related cortical structures into a coherent framework. Van Essen and colleagues developed a set of anatomical criteria to delineate different vision-related structures and to categorize the anatomical connections between them (Felleman and Van Essen 1991). Of particular relevance here was the idea that projections originating only from superficial cortical layers and terminating in layer 4 of the target area transmitted information in the *forward* direction, whereas those coming from both deep and superficial layers terminating outside layer 4 were receiving *feedback* information. In this case, *forward* means ascending the hierarchy with feedback in the opposite direction. Felleman and Van Essen (1991) considered the difficulty of resolving reciprocal and lateral connectivity into the scheme and suggested that hierarchical structure could exist even without stepwise serial processing. For this reason, they extended the basic anatomical criteria and added a third category of lateral connectivity to build the canonical Felleman–Van Essen diagram. This scheme consists of boxes, corresponding to specific structures, vertically arranged into hierarchical levels. The arrows connecting the boxes are based on anatomical tracing data.

Although this notion of hierarchy was inferred from neurophysiological experiments of function, in the Felleman–Van Essen diagram, only anatomical criteria were used. From a functional standpoint, reciprocal connectivity is not easily resolved into a flow of information. In terms of causation, the relative timing of discharge between interconnected sites might be indicative of transmission direction. As has become apparent from cross-correlation studies (Moore et al. 1966; Perkel et al. 1967; Gerstein and Perkel 1972), however, simple causal interaction between pairs of neurons is very rare. This issue is exacerbated with box-and-arrow diagrams, suggesting that information is processed in successive hierarchical levels with well-defined borders, implying that information enters as discrete input and leaves as transformed output with the complete operation taking place within the confines of the structures comprising that level. This logic is engrained in theories of sensory processing. In terms of encoding and decoding, this theme would suggest that input to a processing stage would need to be decoded and then encoded as output transmitted to the next stage.

This conceptual framework is difficult to apply to motor systems. Continuing the hierarchical logic, the general inference is that raw sensory input is processed successively to form a consciously perceived percept of the world. This takes place in well-defined anatomical structures, and according to the Felleman–Van Essen diagram, the hippocampus is the pinnacle where the percept crystallizes. From there, other cortical operations take place leading to a well-formed decision to achieve a particular goal. The goal is then transmitted to the motor system to produce the movement that achieves that goal. However, to find evidence for this scheme, it is necessary to identify the input to the system. Support for this type of post-decision signaling has proved elusive. Furthermore, many different anatomical structures project to the motor system and these projections do not follow a successive sequence of clearly defined serial processing steps.

Similar problems underlie the controversies of whether the primary motor cortex (M1) functions directly and primarily to generate muscle contraction or, instead, in the formulation of higher-level behavioral planning that gets transformed to muscle contraction as it "descends" a hierarchical structure to spinal motoneurons. Anatomical evidence shows that a small component of M1 output projects directly to spinal motoneurons and historic electrical stimulation of M1 results in somatotopic muscle contraction, which would support the idea that M1 functions to contract muscles. The counterargument is supported by recording experiments that extract movement information related to the velocity of the arm, wrist, and fingers during movement (Wodlinger et al. 2015). In the reverse hierarchical scheme, this would be "downstream" from muscle contraction in terms of execution (muscle contraction "causes" limb displacement), but "upstream" when considered as a plan (muscles are contracted to make the arm move according to a plan). Since this information appears in the motor cortex well before muscle contraction, this could support the argument that M1 functions in high-level movement planning. It should be noted that signals reflecting muscle EMG can also be extracted from M1 activity (Humphrey 1986; Townsend et al. 2006; Pohlmeyer et al. 2007) which further clouds this argument. At this point it is useful to reinforce the distinction between coding and code. The ability to extract muscle or movement information from M1 activity shows that this information exists and has been encoded (somewhere). What it might be used for (i.e., where it is being decoded) is a separate issue. As for the information content, there are at least two explanations for how muscle and movement information can be extracted from the same population of neurons. First, the disparate information may be encoded in a high-dimensional space, as seems to be the case for at least ten different kinematic parameters (Wodlinger et al. 2015). In this case, the muscle and kinematic parameters would simply occupy different dimensions. Second, the muscle and kinematic parameters are correlated (Todorov 2000; Reina et al. 2001; Scott 2003). This would suggest that both parameter sets share a single input source and that M1 activity is also related to that source. A subsequent "decoding" stage that separates these parameters may not even be needed, if the common muscle-kinematic signaling is formatted to contribute to muscle excitability.

The idea of hierarchy comes into play again in these issues. Area M1 may not be a singular node where information converges in an exclusive sense; this convergence may occur only in the executed movement. Instead, information about movement may be highly distributed throughout many interconnected structures of the motor system (and probably other parts), making it difficult (and perhaps improper) to designate a neural signal as an input or output. Since synaptic integration is a fundamental property of nervous systems, and in mammals there are typically thousands of converging dendritic inputs and as many diverging axonal terminals, the ability of any single or small group of synapses to "cause" a downstream event is small. This means that simple cause-and-effect arguments have limited utility in explaining function. It is important to consider the nervous system in its actual complexity and to realize that conventional concepts of discrete circuitry based on straightforward causal logic has placed severe limits on our understanding of the nervous system.

## Conclusion

In practice, for the vast majority of neuroscience studies, we are still at the stage of figuring out what information neuronal populations contain on longer timescales; the many unknowns stipulated (e.g., information on short timescales, goals of encoding, relevant receivers, design principles, which information is actually being used, distributed representations, assumptions about hierarchy) imply that by and large we do not know what the neural code (Type 1) is, and how useful the communication channel model will prove to be for different systems. Other models that have been successfully able to model neuronal responses, like deep neural networks, do not have any inherent notion of coding in the communication theoretical sense, although coding concepts have been used to improve our understanding of what these networks actually do (Shwartz-Ziv and Tishby 2017). If progress is to be made, future efforts will need to go beyond quantifying what information is contained, to quantifying what information is actually being transmitted to whom and for what use.

## Generalization to New Conditions and Failures Thereof

Encoding/decoding models, such as those described above, may fit the data they were trained on, but might not necessarily generalize; that is, they may not predict responses to new stimuli that have not been previously tested. We illustrate this point with an important *failure* of such generalization, using a situation which, according to the textbook understanding of vision as a feedforward representation of aspects of the retinal image, should not occur.

Since the pioneering work of Hartline, Kuffler, Hubel, Wiesel, and others (see Spillmann 2014), the notion that visual neurons have receptive fields anchored to particular positions on the retina has been a fundamental concept underpinning of visual neuroscience. Thus, measurement of the receptive field's position under one set of conditions might be expected to generalize accurately to the position of the receptive field tested under other conditions.

In structures such as V4, FEF, and parietal cortex, however, this generalization has not held. When the eyes move or maintain fixation at different orbital positions, the retinal location of the receptive field can shift to novel positions. The new receptive fields appear on varying timescales and may be either transiently present in conjunction with a change in eye position or exist stably for the duration of an epoch of fixation.

This finding has important implications for what needs to be included in models of encoding of visual information and suggests that the "label" on the line for such neurons is not an exact match to single particular retinal or eye-centered locations. Instead, eye position/movement is one aspect of the full "context vector" that needs to be incorporated into predicting how a neuron will respond under novel circumstances. Other factors in that context vector include attentional state, arousal, task context, recent stimulus history, stimuli from other modalities, and no doubt many as yet unexplored sensory and cognitive factors. We describe these variables here using human intuitive phrases, but ultimately they must be instantiated by aspects of neural architecture and neural firing, for which we do not yet have intuitive access (e.g., "$n_g$" in Figure 13.1).

Other examples come from earlier work in the visual system, in V1, where what is encoded depends critically on stimulus configuration. This challenges the concept of receptive fields: a simple cell, for example, will have unpredictable responses when challenged with complex scenes. Since the space of possible contextual modifications is close to infinite, there is no canonical definition of a receptive field. The same problem will hold for representations in general: they will change as a function of the content to be represented (encoded). This inability to establish 1:1 mapping will also pose problems for the analysis of the relation between a code and the respective behavioral consequences (as in decoding models, see Figure 13.1). We expect these challenges for models that link neural activity to behavior (i.e., decoding models) to be most severe at intermediate levels of processing, but to diminish as one builds decoding models that take as their input neural responses that are closer to the motor effectors (muscles) (see "b(t)" in Figure 13.1).

On the whole, we cannot at present assume that assessments of visual coding at the individual neuron level measured in one task will necessarily generalize to another.

### Reliability, Stability or Generalizations to Repetitions of the Same Conditions: Inferring the Stimulus from the Activity

Another widely recognized problem is that even repeating the same conditions does not produce the same activity pattern. This variability in neural firing is often referred to as noise, but it is increasingly understood that what appears as noise to the experimenter is not necessarily noise to the brain but could reflect signals related to aspects of the environment or state of the organism that are not under experimental control.

Put another way, this variability means that one might not be able to reliably predict the firing pattern of the population from the stimulus. Another way of asking the question is whether the stimulus can nevertheless be inferred from the neural activity, despite this variability. Judging the type of information present in a neural population in this manner provides insight into what knowledge an organism has access to.

*Reliability*, in general, is defined as an invariance of a classification or identification of a state in the presence of some kind of perturbation. This perturbation can result in a change of the code or representation, as a consequence of noise or unknown states of the system. Reliability of a representation characterizes the ability to identify the encoded information from noise-perturbed observation.

In contrast to reliability, *stability* refers to a change of the representation over time. Often a stable code is understood as a constant encoding model. Additionally, stability has been defined as an error correction property that reduces the noise of a perturbed system. This definition has often been used for dynamical systems that show dynamics governed by attractors of some kind.

## What Signals Constitute the Code or Are Relevant for Information Transmission?

That spikes are central elements of information transmission supporting moment-to-moment behavior seems beyond dispute. In recent decades, it has also become clear that there are temporal aspects to neural response patterns, and these temporal aspects can have powerful implications for information transmission between ensembles of neurons or between brain regions. The functional importance of spike timing has been explored in the following contexts:

- The transmission of information between neuronal populations.
- The encoding of information through relative timing among neurons, or to some external event.
- The formation of memories through spike timing-dependent plasticity (STDP) (see also Singer, this volume).

Synchronization can modulate information transmission through enhanced integration of EPSPs, as well as through dendritic nonlinearities (Salinas and Sejnowski 2001). Furthermore, synchronous volleys of excitatory inputs may effectively escape from feedforward inhibition (Fries 2015). Coherence between sending and receiving neuronal populations may bias information transmission by aligning the arrival of input spikes with windows of opportunity in the receiver (Fries 2005). One view is that information is encoded through population rate coding, but that the transmission of information is modulated by synchrony and coherence among neuronal populations, according to cognitive demands (Fries 2005). Support for selective information transmission according to cognitive demands comes from the finding that attention selectively and strongly modulates the inter-areal coherence in the gamma-frequency range, between areas V1 and V4 (Bosman et al. 2012). There is, however, ample evidence for encoding of information through spike timing. For instance, hippocampal CA1 place fields carry place information both through rate changes as well as through the spike phase relative to ongoing theta oscillations (Huxter et al. 2008). A similar phenomenon, in the gamma-frequency range, has been demonstrated in visual and frontal cortex (König et al. 1995; Siegel et al. 2009; Vinck et al. 2010; Havenith et al. 2011). Finally, the existence of STDP mechanisms shows that the timing of pre- and postsynaptic spikes critically governs synaptic plasticity formation (Markram et al. 1997; Sejnowski and Paulsen 2006). In hippocampus, neural activity shows extremely synchronous behavior during sharp-wave ripple complex, with sequential activation patterns mimicking the sequential activation of neurons during spatial navigation. These patterns are thought to be important for the consolidation of episodic memories, and interruption of hippocampal activity during sharp-wave ripples impairs spatial memory formation (Girardeau et al. 2009) as well as place field stability (Roux et al. 2017).

Relevant temporal dynamics can also be non-oscillatory. Examples of complex but not necessarily oscillatory dynamics include insect olfactory system (Wehr and Laurent 1996), leech motor decision making (Briggman et al. 2005), rodent hippocampal system/replay (O'Keefe and Recce 1993), birdsong motor system (Hahnloser et al. 2002), and primate motor cortex (Churchland et al. 2012; Mante et al. 2013; Suway et al. 2018). The concept of an "oscillation" in general suggests a static and clock-like behavior that does little justice to the nonstationary nature of neural activity (Burns et al. 2011; Xing et al. 2012) as well as to the spatiotemporal dynamics from which these "oscillations" are often the result. What appears to be an oscillation in recordings on single electrodes is often a traveling wave on arrays of electrodes (Gelperin and Tank 1990; Kleinfeld et al. 1994; Tank et al. 1994; Lubenov and Siapas 2009; Muller et al. 2016). However, as discussed by Singer (this volume), both empirical studies and simulation experiments indicate that the nonstationary and transient features of oscillations are actually advantageous for information processing and dynamic routing of neuronal activity. The underlying spiking activity is sparse, in contrast to the dense traveling waves in epileptiform activity (Muller et al. 2018).

## Codes, Constancies, and Control of Behavior

The behavioral responses evoked by a sensory stimulus may be relatively rapid, simple, and stimulus-locked, such as an eye movement bringing the fovea to bear on a visual stimulus of interest. Alternatively, they can be slower and the consequence of an extended period of internal and covert processing involving a multitude of factors, such as a real-world decision to attend one university over another.

Even for comparatively simple behaviors that lend themselves to laboratory study, there is likely redundancy in the code and degeneracy in the relationship between activity patterns and behavioral outcomes. For instance, there are many different ways to achieve the same action on the environment. Consider an arm that has 4 degrees of freedom (DOF) from the shoulder to the wrist: to reach in 3D space, there are more DOFs than movement dimensions. This means there is an infinite combination of DOFs that will achieve the same endpoint movement. However, psychophysics shows us that we tend to use the same combinations (approximately) in a reliable way.

Certain DOFs tend to be linked or correlated during movement. Why this happens is not always due to mechanical or anatomical constraints as some can be violated volitionally. These invariants reflect a "choice" made by the system. For motor control, these choices seem to reflect control efficiency, minimizing the amount of information that needs to be transmitted to accomplish a goal. This general concept is usually attributed to Nikolai Bernstein (1967), who studied the structure of movement using an early form of video motion tracking. Bernstein articulated the concept of "motor equivalence" in which the same movement could be produced in many different ways. He used drawing movements as a prime example and emphasized the difference between metrics and topology. His examination of repeated drawings showed that the metrics of the drawn object varied between repetitions, but that the shape of the object (topology) as drawn by the same individual was constant. Furthermore, if that object was drawn on a table top or blackboard, the personal topological features remained consistent. This was also true if the object was drawn with the dominant or nondominant hand. Although the set of effectors (muscles and joints) varied greatly, the behavioral outcome (the drawn object) was the same. Bernstein then used these findings to discuss locationism in the nervous system. Since topology was invariant, he argued that this was the dominant organizing principle of motor function. Thus, topological features of the movement would be expected to have an anatomical constancy instead of muscles. He proposed a thought experiment in which neural activity could be observed in the brain. If muscles were localized in the brain, then there would be complicated zig-zag patterns of activity across the cortex because muscular activity is highly variable between movements. If that was the case, he asked, what would be the advantage of having neurons spatially segregated according to the muscle they activate?

Indeed, experimental results show that extracting movement trajectories of the arm and hand from motor cortical activity during reaching and drawing is straightforward and robust (Georgopoulos et al. 1986; Schwartz 1994). Population decoding of these movements is the basis for neural prosthetics. In contrast, extraction of the muscle activity taking place as the arm moves freely through space has proven to be much more difficult. Such generalized motor representations bear a resemblance to constancies that are familiar in sensory processing, such as our ability to assess color as a comparatively stable object property despite variation in the wavelengths that reach our eyes under different illumination conditions, or the perception that the world is not moving despite massive shifts in the retinal image with every eye movement.

### The Single-Neuron Tuning Curve: A Motivating Idea Whose Time Has Passed?

Single-unit neurophysiology has, over the past four decades, focused a great deal of effort on describing the responses of each recorded neuron to a set (typically ~20) of experimental conditions chosen at evenly (typically) spaced intervals along a single, predetermined physical dimension (typically inspired by pilot studies or by prior work). Classic examples include:

- Responses of V1 to the orientation of a visually presented light bar (e.g., Hubel and Wiesel 1962)

- Responses of M1 cells during the performance of in-plane center-out arm movements (e.g., Georgopoulos et al. 1982)
- Responses of a visual area MT cell to the in-plane direction of visual motion
- Responses of "face neurons" in monkey inferior temporal cortex

In each case, a "tuning curve" (or tuning function) is determined by fitting (e.g., least squares error fit) the neural responses (dependent variable) with a smooth, low-parameter mathematical function of the value of prespecified experimental axis (independent variable). The mathematical function chosen by the experimenter for fitting typically has a single peak over the domain of the independent variable, which is taken to be the value of that variable that is predicted to give the maximum response for that unit (the so-called, preferred orientation). In the cases of discreet experimental conditions (e.g., "face neurons"), the tuning curve is implicitly assumed to be a step function (e.g., on the X axis, the tested images can be plotted from left to right, where all images containing a face are to the right of the step up). Other parameters of the tuning curve are also typically computed and reported (e.g., the standard deviation of a Gaussian function can be taken as the orientation tuning width). In such studies, the values of these tuning curve fits are typically summarized over the entire sampled set of single neurons.

These single tuning curves have been very useful for at least three reasons: First, they demonstrate that individual neurons have response sensitivity over the measured variable (e.g., response sensitivity to the orientation of a drifting, full field visual grating). Second, because of the smoothness prior implicitly contained in the chosen mathematical tuning functions (e.g., Gaussian, cosine), they predict how individual neurons will respond to similar conditions (e.g., orientations that were not tested; images containing faces that were not tested). Third, in some cases, the tuning curve can be used as a functional marker to ask if one is recording from a particular area (e.g., strong motion direction tuning as a functional signature of area MT).

It can be argued, however, that the tuning curve has outlived its scientific usefulness, although our group did not unanimously agree on this point. We note at least three key weaknesses: First, in all sensory systems, single neurons are clearly sensitive to experimental changes along many possible axes besides the one chosen by the experimenter. This is well known and attempts are often made to compensate for this by either relegating some of this "nuisance" sensitivity to the methods (e.g., we first found the receptive field of the neuron, which is itself a tuning function over two dimensions) or handled by trying to test one or two other stimulus axes (e.g., orientation bandwidth, hue). While such attempts can be valiant, they always underestimate the complexity of the neural responses because the experimental condition space is very large (e.g., the dimensions of image space). Even more problematic, the ability of the experimenter to guess at the "best" experimental axis rapidly diminishes after even just one nonlinearity in the neural processing (e.g., V2 in vision), and appears almost completely arbitrary once one reaches very deep levels of the processing (e.g., inferior temporal cortex in vision).

This constraint is closely related to the second serious limitation: tuning curves have very limited ability to predict the responses of individual neurons beyond interpolations of the specific conditions already tested (i.e., limited ability to generalize). Thus, by definition, tuning curves do not contain generalized knowledge of the neuron's processing function (i.e., the image-computable encoding function in vision). Again, this problem gets dramatically worse the deeper one goes into the system (more nonlinearities).

A third major limitation of tuning curves is that they promulgate the idea that the goal of the field is to discover the "optimal" stimulus of each single unit, as if the single neuron is a homunculus that can offer direct insight into questions of complex human behavior. This is clearly misguided in the contemporary context of population coding, and we believe that even our contemporary ideas of population coding will look naive in another twenty years.

All three limitations are the result of the understandable desire of the experimenter (indeed, of the field) to impose a human-interpretable prior on the responses of the neuron to help organize one's thinking before reporting those responses to the world. Simply put, we would prefer it to be the case that neural responses can be reduced to a few dimensions of the domain of interest (e.g., the domain of all images) so that we can more readily communicate our findings—our "story" and our discovered "principles"—to other members of our species. As cognitive scientists, we deeply appreciate the social primate value of storytelling. But that is not the same as science that gauges its progress through accurate prediction of the phenomena of interest (e.g., neural spikes, behavior). We see no reason to assume that evolution has left us with an adult brain whose complex internals are readily communicated to other humans in such simple forms as tuning curves.

Fortunately, when we set the tuning curve aside, we do not need to go back to square one. We now have better methods of estimating much more accurately (i.e., generalize to new images) the encoding functions of individual neurons using systems identification methods combined with modern artificial neural networks that provide much more appropriate (highly nonlinear) encoding bases (e.g., Yamins et al. 2013). These methods have rapidly spread in the visual system, and somewhat in the auditory system. They have not yet been applied to all sensory systems or to motor systems, but much active work is ongoing and we expect these advances to continue apace. We also note that even these currently cutting-edge approaches will still be incomplete without incorporating models of how internal states (e.g., ongoing neural activity within the local population) predict neural responses during presentation of sensory stimuli (Dechery and MacLean 2018). Indeed, the most current encoding models for visual processing are still not able to capture the temporal dynamics of visual system neurons.

Zooming out, we also believe that the human desire for interpretability should not be forgotten but that it should be redirected. While far from guaranteed, human interpretable "principles" might still be found, but modern artificial neural network methods and experimental progress both argue that human interpretable principles might best be found at the levels of cortical architecture, learning, development, and perhaps even evolution. For example, the principles could take the form not of a set of connections or activity patterns that allow the brain to perform a computation, but of the rules of activity-dependent plasticity that enable these connections to be set up (see, Singer, this volume).

Even though the tuning curve has limitations as a research tool, it has not outlived its usefulness as a pedagogical tool. Indeed, it provides an elementary introduction to the idea of encoding functions in sensory systems (and projection functions in motor systems), which then motivates the idea of a high-dimensional, predictive response function. In this vein, tuning curves helped to promote an important conceptual advance: the idea of population coding. Specifically, tuning curves considered from a population of neurons (as outlined above) naturally suggest that, when viewed as a group, the value of the currently presented stimulus can be "reported" to downstream brain regions (population code), and they have motivated ideas and testing of how alternative population codes might estimate that value to guide behavior (e.g., examples in motion discrimination, motor control).

In sum, the idea of a tuning curve has helped carry the field toward the contemporary goal of discovering accurately predictive neural response functions (e.g., image computable encoding functions in vision) as well as toward defining contemporary concepts of population coding. However, we now know that as soon as we step beyond the very earliest stages of a sensory system, the tuning curve becomes overly simplistic as to only maintain introductory pedagogical value. Fortunately, the contemporary approaches outlined above are ready to carry the research forward.

## Units of Analysis in Brain Tissue

### Are Circuits Well Defined and Amenable to Study?

Considerable interest in neuroscience in recent years has focused on the concept of circuits. The general idea of a circuit comes from electronics. In that system, circuits are composed of discrete components and the connections between them are concrete. In the brain, the physical connectivity can also be fairly concrete. In some cases, different structures can also be well defined.

In the functional domain, however, this is not clear. Do defined single anatomical structures have singular functions that are different from other structures that remain constant over time? This functional idea, as expressed in typical box-and-arrow diagrams (with plus and minus signs) to describe the functional pathways of information, is wrong because it implies a high degree of discreteness that is hardwired: In a brain containing billions of neurons, we cannot define nodes this cleanly unless every neuron has its own box. More importantly, given the high degree of recurrence in brain circuits, we cannot define input and output this way.

Historically, the cortical column was viewed as a key computational element of a cortical (micro)circuit. A cortical column was initially defined (in the visual system) as a small cross section of cortex, in which neurons in the different layers of cortex share some kind of common property, beyond similarities in their receptive field position. For example, in V1, there exists a retinotopic map in two dimensions along the cortical sheet. Across layers, however, there is a similarity in the ocular dominance and orientation selectivity of neurons at a given location on that cortical sheet. There can be discontinuities in the sensitivity to orientation and ocular dominance for adjacent locations along the cortical sheet, which can be thought of as the borders of columns.

In updating the concept of a column or canonical circuit in cortex, a key observation is that there are massive numbers of excitatory recurrent connections. They are mostly local, and there is some degree of stereotypy both within and across layers (potentially also including the thalamus). It should be noted that the probability of a connection falls off exponentially as a function of distance, calling into question the idea that there are regularly repeating boundaries between circuit elements. We should thus probably think of the cortical sheet as changing in a continuous fashion, with motifs of local connection patterns repeating smoothly.

Do such motifs perform basic sets of operations that are stereotypic across cortical areas, applied to whatever the inputs of that area may be? One such operation might be a convolution using a local kernel, followed by a static nonlinearity and normalization, as employed in artificial convolutional neural networks. This analogy, in fact, viably demonstrates how powerful such a concept, in principle, is when applied to real-world pattern recognition tasks (e.g., object recognition), or when transferring it from one specific set of inputs to another (e.g., images vs. sounds).

Whether this analogy is deep or valid only on a superficial level is currently under debate. What is clear, however, is that in cortex such an operation would perform a much more complex and flexible nonlinear operation involving a number of different cell types, recurrent excitatory and inhibitory feedback (within and across different layers) and potentially employing a whole range of different temporal delays to boost computation (see below); it would also be adjustable, for example, by neuromodulation. While the instantiation of this "convolutional" operator in cortex (e.g., precise wiring diagram) might vary from site to site, the plasticity/developmental rules by which such an operator could arise may be the same across cortex (i.e., translational invariant; see also discussion of LIPU, above).

Such a concept is comparatively familiar in the visual domain, but the extension to other domains is a more complicated question. For instance, in auditory processing, widely separated frequencies that are integer multiples of a common fundamental are likely to be grouped and processed similarly but may be processed by quite distinct neural populations at early stages of the pathway. In structures like prefrontal cortex, neurons are responsive to sensory stimuli, but there is no known similarity of tuning to stimulus features in nearby neurons. It could be the case that there is some other dimension of the input space that is well ordered, but this has yet to be established.

An open question, then, is how can we leverage modern experimental and computational tools to establish the existence of such an operator and to characterize its computational capabilities? If approached anatomically, the region in cortex that we would have to analyze would likely cover several millimeters. A functional characterization would require a tight control of both the various inputs and outputs to cortex. Inputs could be assessed by calcium imaging of axon terminals that provide input from other cortical areas or thalamus. Assessing the output would require identifying the anatomical projection patterns of putative output units. Developing adequate perturbations will certainly be crucial. An interesting first step in this direction has been conducted by Constantinople and Bruno (2013), who show that silencing pharmacologically layer 4 in barrel cortex affects response properties of layer 5/6 neurons (assessed with intracellular recordings) very little, suggesting that it might be possible to study some components of the operator independently from others.

## Recurrent Connections and Ongoing Activity

A key issue, which is arguably not a central element of many views of cortical coding, is the importance of recurrent connections and the elaboration of signaling in time that such recurrence necessarily involves (see Singer, this volume). Whether this recurrence is excitatory, inhibitory, or both has important implications for its impact on neural coding and function. A number of roles and effects of recurrence have been identified or hypothesized, including the preservation of information over various rather short timescales (millisecond to second range), the generation of "spontaneous" activity and activity fluctuations, as well as "handshaking" to reflect acknowledgment of signals passed from one ensemble to another, as in asynchronous computing.

As noted, not only are local neocortical and hippocampal circuitry distinguished by extensive recurrent excitatory connections, but there are also extensive long recurrent loops between cortex and thalamus, cortex and cerebellum, and cortex and basal ganglia, not to mention projections to and from attentional centers such as parietal cortex and frontal eye fields. Clearly, recurrent connections are a defining feature of neocortex. The prevalence of local recurrent connectivity has the downside of apparently making these structures predisposed to the pathophysiology of epilepsy, thus suggesting that recurrence must also have benefits that justify this cost.

## Dark Activity

Recurrent connections give rise, at least in part, to spontaneous, ongoing activity, or activity changes that are not locked to the stimulus presentation (Figure 13.2). Such aspects of neural signaling are called "dark activity" to reflect the fact that we have very little understanding of their functional role. Early studies of reduced slice preparations demonstrated that isolated circuitry in acute slices of neocortex have a capacity for spontaneous activity. Notably

### Stimulus-evoked spiking and trial-averaged LFP

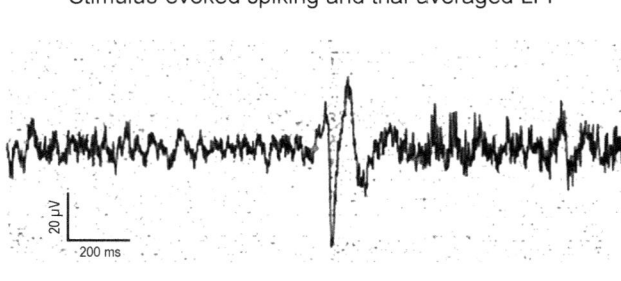

### On single trials, prominent low frequency LFP fluctations

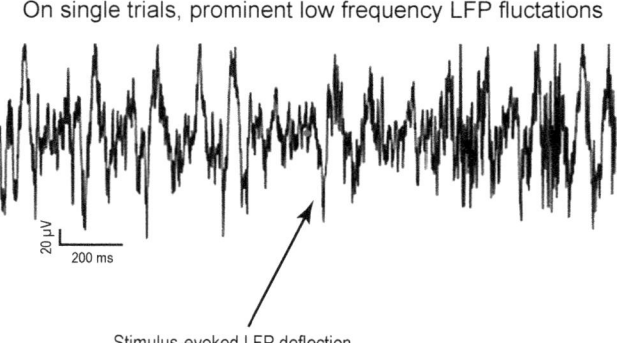

Stimulus-evoked LFP deflection

**Figure 13.2**  Dark activity is a prominent component of cortical activity. Upper panel: Black line shows local field potential (LFP) recorded on a single channel in marmoset middle temporal area, averaged over multiple representations of a drifting grating during passive fixation. Dots indicate spiking activity of a single middle temporal unit. Red dot indicates time of stimulus onset. The average LFP shows modest fluctuations prior to stimulus onset. Lower panel: LFP recorded on a single trial drawn from the trials that were averaged in the upper panel. LFP fluctuations are pronounced and similar in amplitude to the stimulus evoked response (arrow). Unpublished data from Z. Davis, L. Muller, T. J. Sejnowski, and J. H. Reynolds.

it remains unclear how much this predisposition for spontaneous activity, as a consequence of local recurrent connectivity, is engaged in the ongoing or spontaneous activity *in vivo*.

Dark activity is manifested, in part, in fluctuations in the activity of individual neurons or pairwise correlations that occurs even in the absence of a stimulus. Neural fluctuations have often been treated as a form of neural noise (Zohary et al. 1994; Shadlen and Newsome 1998; Bair et al. 2001; Kohn and Smith 2005). Consistent with this interpretation, attention can reduce neuronal variability (Cohen and Maunsell 2009; Mitchell et al. 2009), possibly by regulation of lateral inhibitory circuitry (Schmitz and Duncan 2018), consistent with the notion that attention may quench noise so as to improve sensory encoding. Such fluctuations have also recently been theorized to permit preservation of multiple items in neural ensembles as a form of time division multiplexing (Caruso et al. 2018).

These forms of variability may, however, reflect other computational modes that are computed by the same neural population. It has been proposed that synchronous neural responses may act to regulate input gain (Swadlow et al. 1998; Fries et al. 2001a; Salinas and Sejnowski 2001; Sohal et al. 2009) or aid information transfer between these populations (Fries 2015). Multichannel recording approaches have revealed traveling waves of neural activity in multiple cortical areas, from sensory to motor (Muller et al. 2018). These waves can be evoked by external stimuli and can also occur spontaneously. They are likely mediated by recurrent circuits, transiently modulating neural excitability as they pass.

Moreover, ongoing activity has implications for any consideration of cortical population encoding of stimuli. It is difficult, for instance, to predict single-trial activity of any individual neuron simply knowing its tuning properties (~15% variance explained). In contrast, local population activity, including both tuned and untuned neurons, can be used to predict individual neuron activity on a moment-to-moment basis very well (up to 85%) in awake ambulating mice (Dechery and MacLean 2018). This argues strongly for a multineuronal-based coding scheme that takes into account the state of the local population, presumably dictated in large part by local recurrent connections, rather than the stimulus alone. Cortical population responses can be seen as generative rather than passive.

At a larger spatial scale, inter-areal recurrent connections strongly regulate activity. For example, within the domain of vision, recurrent circuits from the frontoparietal attentional control network, including parts of the oculomotor system (Reynolds and Chelazzi 2004), impinge on visual cortical areas. In addition to modulating ongoing activity, as noted above, these feedback signals strongly modulate stimulus-evoked responses, increasing the strength of responses to attended neurons and, through local recurrent inhibitory circuits, suppressing responses to unattended stimuli (Moran and Desimone 1985; Reynolds and Heeger 2009; Ni and Maunsell 2017).

At a more general level, recurrence of excitatory connections has the capacity to promote both inference (lateral and hierarchical) and learning. Recurrence also plays an integral part in multiple recurrent neural network models, such as Boltzmann machines, liquid state machines, and echo networks.

## Activity Evolves in Time

In the past, delays were ignored in most models (especially in models that tried to identify computational principles), despite the fact that we all know that delays are omnipresent and range from sub-milliseconds to seconds. In addition, delays may result from several biological mechanisms: from conduction velocities, axonal delays, delays that result from the activation of neurons, or produced by modulatory processes in the system.

Recent models in machine learning and artificial intelligence, for instance, make use of recurrent networks to process time series, such as speech. For learning the optimal connectivity for a certain task, recurrence is unfolded over time to map time to space. For this, a constant delay between neurons and layers is assumed. Delays have also been used to implement auxiliary functions in dynamical systems. Among these is the use of delays to stabilize oscillatory dynamics and zero time lag synchronization. None of these approaches, however, addresses the effect of distribution of delays, which seem to be the most appropriate description of delays in recurrent networks.

Therefore, understanding the impact of delay distributions on computations remains a challenge. Recently a conceptual framework, *delay-coupled reservoir computing*, was introduced (Lagorce and Benosman 2015). It extends the computational principles from reservoir computing and explicitly uses single delays as well as delay distributions for the implementation of universal computation (Appeltant et al. 2011). The framework uses the concept of delay-coupled differential equations, which is a differential equation that receives a delayed and maybe transformed input back into the system.

In contrast to ordinary differential equations, this difference brings the system into a whole new category of dynamical systems that map functions onto functions (i.e., infinite dimensional mapping).

While the mathematical concept may be hard to understand in detail, there is a beautiful and simple analogical interpretation. The additional delay coupling of a single dynamical system to itself (i.e., neuron) can be interpreted as a network of virtual neurons that are recurrently connected with a constrained connectivity matrix. Thus, the combination of delay-coupled single elements into a network generates a larger recurrent network composed of the real neurons with real connectivity, and virtual neurons with virtual connectivity that are contributed by the delayed self-coupling. In other words, the effect of the delay coupling is a virtual increase in the number of neurons and an effective increase in the coding space.

In sum, the framework of delay-coupled reservoir computing helps us understand the effect of delays and distribution of delays on computation as a simple extension of the classical reservoir computing with an increased number of neurons.

## Context and Network State

Another aspect of the neural code that recurrent connections may contribute to is the dependence of response properties on context, both in space and time. What is happening around the neuron at a given moment and what has happened to it previously strongly influence the neuronal response. Cortical pyramidal cells are anatomically interconnected with thousands of other excitatory and inhibitory neurons; these connections likely mediate this spatiotemporal contextual influence. On the intracellular level, these contextual influences can be observed as synaptic barrages that change the likelihood of action-potential generation, for example, by changing the membrane potential.

Considering the entire cerebral cortex at the same moment, the map of membrane potentials of all of the cortical neurons could be visualized as an excitability map. The probability of activity flowing in a particular path through the cortical network will be an interaction between this excitability map and incoming activity, which subsequently changes the excitability map. Thus, the excitability map shapes interaction networks of cortical neurons on a moment-to-moment basis, allowing a great deal of flexibility to be incorporated into cortical networks. To obtain stable perceptions and behavior, however, these highly context- and history-dependent network states are expected to exhibit stable states of activity that correspond to the stable perception or action. We propose that an important feature of the cerebral cortex is the ability to generate both stable and highly flexible patterns of activity in space and time that allow behavior to occur in both a stereotyped and flexible manner.

## Cortex Cannot be Understood in Isolation

The cerebral cortex evolved in mammals, joining other more ancient structures in early vertebrates that previously supported autonomous behavior. Presumably, the cortex enhanced survival in ways that we would like to understand. Considering other parts of the brain with which the cortex interacts may help expand our understanding.

The cerebral cortex is tightly coupled to several important brain structures. The thalamus is the gateway to the cortex but it also receives cortical feedback. Interestingly, the feedback connections are more numerous but weaker than the more robust feedforward projections, with a wide range of time delays in the 10–100 ms range (Crick and Koch 1998). The basal ganglia are another partner with the cortex. The cortex projects to the striatum, which through a sequence of subcortical projections returns to the cortex through the thalamus, a loop that takes around 100 ms. A third loop between the cortex and the cerebellum, including the prefrontal cortex, is reciprocally connected with the lateral cerebellum. The contributions of these loops are essential for understanding what the neocortex contributes to brain function.

All of these structures are interconnected with brainstem and sensory periphery, where signals that originate in the brain can even result in changes to the sensory input. For example, pupil dilation is influenced by auditory stimuli, arousal, and likely other factors. By influencing pupil dilation, such factors can influence the light reaching the retina with consequences for subsequent visual processing. Pupil diameter is highly correlated with state of central neuromodulatory systems and the membrane potentials of cortical neurons (Reimer et al. 2016). In the auditory system, it has recently been shown that eye movements are accompanied by an eardrum oscillation (Gruters et al. 2018), again suggesting information exchange between sensory pathways can be implemented via the control of the mechanisms of transduction.

## Conclusions

At the time of the Dahlem Workshop (Rakic and Singer 1988), only rudimentary knowledge was available on how cortical circuits are organized, and this information was based on the concept of a cortical column. Today we have a better idea of how the different types of neurons are connected and how they influence each other, especially the many different types of inhibitory neurons. Thirty years ago, electrodes were placed in the cortex blindly and cortical neurons were recorded whose inputs and outputs were largely unknown. Although these recordings revealed diverse response properties, the observations were correlational, and it was difficult to determine how they influenced behavior. Today, optical recording techniques have made it possible to image activity in thousands of neurons simultaneously in dense clusters, and to influence their activity with optogenetics. The emphasis has shifted from the properties of single neurons to the dynamical trajectories of neural populations in state space. Although these recordings are no longer "blind," we are still far from having a functional account of cortical processing.

We have also gone from a paucity to a plethora of computational hypotheses for how information in cortical circuits is organized. There are many ways that the features of the world encoded by cortical neurons could be combined and used to produce complex behaviors. Conceptual frameworks from information theory, Bayesian probability theory, and dynamical systems theory might all give us useful insights and predictions for experiments. Machine learning algorithms are being used to analyze the big data being generated in physiological and anatomical experiments. The convergence of deep learning architectures in artificial intelligence with cortical architectures is generating insights into how cortical hierarchies could enable object recognition in images and recognition of speech.

Over the next ten years, we anticipate that progress should accelerate rapidly, both because of improved techniques for probing and manipulating neurons, and because of more sophisticated computational hypotheses for how to interpret neural recordings. Thus, in another thirty or so years time, the participants of a future Cortex Forum should have a much better understanding of how the cerebral cortex transforms dynamic patterns of input activity, how memories are organized, and how, in concert with other brain areas, the cortex gives rise to our cognitive abilities.

# Complexity and Computation
# in Human Cognition

# 14

# Complexity and Computation in the Brain

## The Knowns and the Known Unknowns

Karl J. Friston

## Abstract

This chapter sets the scene for the treatment of complexity and computation in human cognition and discusses how this treatment is informed by the neurobiological and functional properties of the cerebral cortex. Its agenda is to establish some guiding principles that may help identify hypotheses and computational architectures that go beyond mere descriptions of how the cortex underwrites the repertoire of functions we enjoy, such as action, perception, cognition, affect, and consciousness. In short, it explores the computational imperatives that form the basis for human experience. Complexity and computation are considered, as is how they organize our approach to neuronal dynamics. Criteria are identified that any tenable theoretical framework must respect. In addition, it discusses computational theories that can be entertained, and the degree to which they account for empirical data from anatomy and neurophysiology. Finally, some of the deeper issues that face sentient artifacts are considered that, ultimately, possess a sense of self, purpose, and agency.

## Introduction

The purpose of this chapter is to review the fundaments of complexity and computation in the brain and provide some pointers that frame other contributions in this volume. It may seem an almost impossible task to survey all the issues that attend action, perception, cognition, and consciousness in the human brain; however, there are some relatively straightforward principles that make our job much easier. We will pursue the basic theme of complexity and computation, considering carefully what these notions entail. This paves the way for a broad ontology of theories that can be separated into normative theories of *what* the brain is doing and process theories of *how* the brain implements normative imperatives. This separation is useful because it divides the conceptual (*what*) from the empirical (*how*) labor, and allows us to specify clearly the pressing questions that need to be answered.

This chapter comprises four sections. In the first, we will consider the nature of complexity, from the point of view of dynamical systems and self-organization, as well as from the perspective of inference and statistics. This section leaves us with an outstanding issue: How does dynamical complexity relate to structural complexity and vice versa? The second section turns to the notion of computation and the principles that could shed light on computation in the brain. In brief, we will consider computation from the point of view of inference and how this can be grounded to give a physics of computation that can be meaningfully applied to neuronal systems. The third section looks at prevalent normative theories of brain function with a special focus on currently dominant paradigms, such as predictive coding, the Bayesian brain, and active inference. We review these approaches in the light of preceding discussions on complexity and computation. Having addressed the normative side of the challenge, we then consider the more challenging issues of identifying process theories that are consistent with the principles of computation and endorsed by our growing knowledge of cortical and subcortical networks in the brain. This discussion is organized around two scales: large-scale connectomes and hierarchical architectures in the brain, which contextualize smaller-scale processing (e.g., the canonical cortical microcircuit). In the final section, outstanding issues are raised that largely turn on the remarkable capacity for human retrospection and epistemic planning. This, in turn, presents some key questions about the timing of representations and the representations of time. It is at this point that some of the known unknowns start to rear their heads. In other words, this chapter ceases to be a review of what we know and becomes a prospectus for future discussion and work.

## Complexity in the Brain

The origin of the word complexity derives from the Latin word *com* (meaning together) and *plex* (meaning woven). A complex system is therefore characterized by its dependencies and interactions, where characteristic, complex behavior *emerges*. This emergence is sometimes taken to mean that there are no high-order instructions or principles that prescribe the interactions—interactions that are generally considered to be "greater than the sum of their parts." However, as we will see later, this is probably not true. Complexity is itself a complex issue, famously reflected in the fact that there is no single definition of complexity. Having said this, in the physical sciences, there are several formal measures of complexity, depending upon the field of application.

Some common examples include computational complexity, usually cast in terms of minimum description lengths that allow people to classify computational problems by complexity class (e.g., P, NP). This is closely related to Kolmogorov complexity and minimum message length in algorithmic information theory (Hinton and Zemel 1993; MacKay 1995; Wallace and Dowe 1999). These are important measures that relate closely to statistical complexity which will play a key role later (Hinton and van Camp 1993). In statistical mechanics and probability theory, complexity is sometimes associated with the notion of entropy; however, this is a slightly naive assumption and misses the point that complexity is really about relationships (i.e., the dependency of entropy over different partitions), such as hierarchical entropy measures over time or the complexity measures that underpin integrated information theory (Tononi et al. 1994). This sort of complexity speaks to the complicated statistical dependencies among the states of the system in question. We will refer to this as *structural complexity* to distinguish it from the *dynamical complexity* that emerges from a system's dynamics.

In dynamical systems there are many forms of complicated (Latin: *com* meaning together and *plicare* meaning to fold) behaviors that rest upon attractor manifolds that are literally *folded* into some mathematical phase or state space. Three common sorts of complexity in dynamical systems theory are reviewed in Table 14.1. In brief, dynamical complexity usually entails an unpredictable space-filling trajectory that, paradoxically, has an attractor of low measure or volume. To unpack these technical terms, what we are saying here is that if one measures a complex dynamical system and plots its states, as in state space over time, the resulting trajectory traces out a path on an attractor or manifold. The peculiar thing about complex systems is that this manifold or attracting set reaches many corners of state space and yet has a very small volume. This is what is meant by "space-filling with low measure." Essentially, this means that complex systems have attracting sets of states which they visit time and time again; however, their paths through state space are convoluted and unpredictable (in the sense of deterministic chaos or other forms of itinerancy). Furthermore, the attracting states ensure that the system will only be found in a very small number of states, compared with the possible states in which it could be found. In many senses, nearly every system we encounter in daily life is an example of a complex system, ranging from the weather through the behavior of our children to nearly every aspect of our exchanges with the world (e.g., our eye movements). A key feature of complex dynamics is their wandering or *itinerant* aspect.

The importance of itinerancy for brain function has been articulated many times (Nara 2003), particularly from the perspective of computation and autonomy (van Leeuwen 2008). Itinerancy provides a link between exploration and foraging in ethology (Ishii et al. 2002) as well as dynamical systems theory approaches to the brain (Freeman 1994). These approaches variously emphasize the importance of chaotic itinerancy (Tsuda 2001) and self-organized criticality (Bak et al. 1988; Kitzbichler et al. 2009; Deco and Jirsa 2012). Itinerant dynamics also arise from metastability (Jirsa et al. 1994) and underlie important phenomena, like winnerless competition (Rabinovich et al. 2008). For a description of these phenomena, see Table 14.1.

**Table 14.1**   Phenomena that underlie dynamical complexity.

| Phenomenon | Description |
| --- | --- |
| Chaotic itinerancy | Chaotic itinerancy refers to the behavior of complicated (usually coupled nonlinear) systems that possess weakly attracting sets, *Milnor attractors*, with basins of attraction that are very close to each other. Their proximity destabilizes the Milnor attractors to create *attractor ruins*, which allow the system to leave one attractor and explore another, even in the absence of noise. A Milnor attractor is a chaotic attractor—onto which the system settles from a set of initial conditions—with positive measure (volume). However, another set of initial conditions (also with positive measure) that belongs to the basin of another attractor can be infinitely close; this is called *attractor riddling*. Itinerant orbits typically arise from unstable periodic orbits that reside in (are dense within) the attractor, where the heteroclines of unstable orbits typically connect to another attractor, or they wander out into state space and then back onto the attractor, giving rise to *bubbling*. In other words, unstable manifolds from saddles (i.e., fixed points attracting in one direction and repelling in another) densely embedded in the attractors become stable manifolds and connect different attractors. This is a classic scenario for *intermittency* in which the dynamics are characterized by long laminar (ordered) periods as the system approaches a Milnor attractor and brief turbulent phases, when it gets close to an unstable manifold. If the number of periodic orbits is large, then this can happen indefinitely. The term *ergodic* is used to describe a dynamical system that has the same behavior averaged over time as averaged over its states. The celebrated ergodic theorem (Birkhoff 1931) addresses the behavior of systems that have been evolving for a long time: intuitively, an ergodic system forgets its initial states, such that the probability that a system is found in any state becomes—for almost every state—the proportion of time that this state is occupied. |
| Heteroclinic cycling | In heteroclinic cycling there are no attractors, not even Milnor ones (or at least there is a large open set in state space with no attractors); there are only saddles connected one to the other by heteroclinic orbits. A saddle is a point (invariant set) that has both attracting (stable) and repelling (unstable) manifolds. A heteroclinic cycle is a topological circle of saddles connected by heteroclinic orbits. If a heteroclinic cycle is asymptotically stable, the system spends longer and longer in a neighborhood of successive saddles, producing a peripatetic wandering through state space. The resulting heteroclinic cycles have been proposed as a metaphor for neuronal dynamics that underlie cognitive processing (Rabinovich et al. 2012) and exhibit important behaviors such as winnerless competition, of the sort seen in central pattern generators in the motor system. |

**Table 14.1 (continued)**

| Phenomenon | Description |
|---|---|
| Multistability and switching | In multistability, there are typically a number of classical attractors which are stronger than Milnor attractors in the sense that their basins of attraction not only have positive measure but are also open sets. These attractors are not connected; they are separated by a basin boundary. However, they are weak in the sense that the basins are shallow (and topologically simple). System noise is then required to drive the system from one attractor to another; this is called *switching*. Noise plays an obligate role in switching but is not a prerequisite for heteroclinic cycling; noise acts to settle the excursion time around the cycle onto some characteristic timescale. Without noise, the system will gradually slow as it gets closer and closer (but never onto) the cycle. In chaotic itinerancy, the role of noise is determined by the geometry of the instabilities. Multistability underlies much of the work on attractor network models of perceptual decisions and categorization; for example, in binocular rivalry (Theodoni et al. 2011). |

We have focused on dynamical complexity using concepts that are usually applied to autonomous dynamical systems; that is, systems with dynamics that do not depend upon any independent (i.e., control) variable from outside the system (including time itself). Clearly, to drill down on any particular neuronal system, especially the cortex, we need to acknowledge that it will respond sensitively to outside influences (and may well show time-dependent effects, such as adaptation). In light of this, it might be important to consider nonautonomous dynamical systems, an emerging branch of applied mathematics (Kloeden and Rasmussen 2011), its application in the field of recurrent neuronal networks (Ørstavik and Stark 1998), as well as the analysis of interactive nonautonomous dynamical systems (Schumacher et al. 2012) and causality (Schumacher et al. 2015). The influence of coupled (and therefore nonautonomous) dynamical systems on each other, via the emergence of things like generalized synchrony may have a fundamental role in coordinating neuronal dynamics (Hunt et al. 1997; Schumacher et al. 2012; Friston and Frith 2015), as we will see below.

**Measures of Complexity**

The title of this section is actually quite loaded. Thus far we have not yet defined complexity; we have just described ways in which it is manifest or can be measured. This is an important distinction because the measurable characteristics of a phenomenon do not, necessarily afford teleological insight. We know that the brain is complex at many levels. Neuronal dynamics are itinerant, show self-organized criticality and metastability (Deco and Jirsa 2012; Cocchi et al. 2017). Furthermore, the dynamic coordination implied by a universe of biological and neuronal rhythms lends the brain a repertoire of complex dynamics and attracting sets that are possibly unparalleled in the universe (Singer and Gray 1995; von der Malsburg et al. 2010). But does this help us understand how the brain works? To harness complexity in a functionalist or teleological sense, it is useful to consider another form of complexity that is closely related to the algorithmic and computational complexity described above.

## Statistical Complexity

In statistics and probability theory, complexity has a very particular meaning. It essentially measures the degrees of freedom of a (statistical or mathematical) description of some phenomena or data. These degrees of freedom are technically measured with something like the Kullback-Leibler (KL) divergence between the posterior and prior probability distributions over the causes of data. To unpack this, we must first assume that the world in which we operate is a world of probability distributions and beliefs. Generally, these are distributions over the causes of data or sensory samples; namely, states of the world "out there." Once we describe things in terms of beliefs, we can then evaluate the change in beliefs induced by a measurement or sensory sample. This change is scored by the KL divergence between the posterior belief—after seeing the data or making a measurement—relative to the prior belief—before seeing the data.

In this setting, beliefs are just shorthand for probability distributions of the sort found in Bayesian statistics, or indeed quantum mechanics. This sort of complexity is an attribute of a belief, model, or hypothesis about the causes of outcomes or measures. This may seem a rather colloquial and restrictive sort of complexity; however, it has a much broader scope of application than one might initially guess. This follows from the fact that nearly all interesting physics (and daily life) reduces to some form of inference or measurement. In fact, from the point of view of quantum mechanics right through to general relativity, everything can be reduced to metrology or measurement (Cook 1994). In relation to algorithmic complexity, this means the imperative for efficient communication, decoding, modeling, or hypothesis testing is to *minimize complexity*; in other words, to account for the causes of our sensory interactions with the universe in terms of short messages of minimum complexity (Wallace and Dowe 1999; Schmidhuber 2010). This is nothing more than Ockham's principle.

## Putting the Complexities Together

This brief consideration of complexity poses a rather obvious dialectic. If all the principles of algorithmic and information complexity require complexity to be minimized—for example, the principle of maximum efficiency, the principle of minimum redundancy, and so on (Barlow 1961; Optican and Richmond 1987; Linsker 1990)—why is the world so replete with systems that have clear dynamical complexity? We will leave this as a question to be resolved. In fact, any worthy theory of brain function should be able to resolve this dialectic. Before turning to candidate theories, let us now consider the ground rules for computation.

## Computation in Humans and Other Animals

Originally, computers were people who made calculations during the Industrial Revolution (Latin: *com* meaning together and *putare* meaning thinking, or reckoning). This is important because computing was, and always has been, a competence of humans, even if the modern perspective on computation focuses on artificial (*in silico*) computation. Nowadays, computation is nearly synonymous with computer science; namely, any type of calculation that follows a well-defined model that can be articulated as an algorithm or scheme. But does this definition really help us?

Let us take a step back and think about what it means to compute or infer. On this view, one can think of computing in terms of deduction, induction, and abduction.[1] In relation to the *model* that underlies a well-defined computation, this translates into inferring the causes or meaning of some measurements or data through deductive, inductive, or abductive algorithms or reasoning. Most computer science treatments of computation would fall under the class of deductive or inductive (Turing style) computations. From the perspective of complexity and computation in the brain, this sort of computation is relatively uninteresting (because it is predicated on propositional logic, as opposed to dynamics and probability theory). We will therefore assume that the sort of computation that characterizes complex self-organizing systems is abductive in nature (e.g., the implicit algorithms we see playing out in meteorology, natural selection, and human perception). So what is abduction?

Loosely speaking, abduction can be thought of as inference to the best explanation. It is characteristically ampliative, in the sense that it often goes beyond the evidence or measurements at hand. In other words, it describes algorithms that appear to bring more to the table than is intrinsic to the computer's inputs. From a mathematical or statistical perspective, the closest algorithm we have to describe this form of computation is Bayesian belief updating, which calls on prior beliefs to contextualize the likelihood of sensory observations. This enables posterior beliefs to be formed that are an optimal assimilation of

---

[1]   Generally defined, *deduction* is "the deriving of a conclusion by reasoning," whereas *induction* refers specifically to "inference of a generalized conclusion from particular instances." *Abduction* is defined as "a syllogism in which the major premise is evident but the minor premise and therefore the conclusion only probable." The crucial distinction is that unlike deduction and induction, abduction is inherently probabilistic.

current evidence (computational inputs) into past experience (prior beliefs). This may seem to be another colloquial formulation of computation; however, it is difficult to think of any complex self-organizing system that cannot be cast in terms of probabilistic or Bayesian updating. Figure 14.1 provides an example of self-assembly and morphogenesis based purely on system dynamics that implicitly perform an elementary form of inference. This inclusive view of computation can be applied to evolution, motor control, and, perhaps, human experience itself (Ao 2009; Harper 2011; Frank 2012).

Figure 14.1 illustrates self-organization through (subpersonal) computation, based on minimizing variational free energy (an information theoretic quantity that measures the surprise or implausibility of some sensed data, given a model of how those data were generated). This simulation shows how simulated (colored) cells can self-organize into a particular form simply by computing and minimizing free energy: the target morphology is shown in the insert on the bottom right. Each time step in this simulation can be thought of as modeling the migration and differentiation of eight cells over several minutes. Upper panels show the time courses of system states encoding cell identity (left), the associated system states mediating migration and signal expression (middle), and the resulting trajectories, projected onto the first (vertical) direction and color-coded to show differentiation. These trajectories progressively minimize free energy (lower left panel), resulting in a differentiation of the ensemble (lower middle panel): the softmax function of the cells' internal states can be interpreted as the posterior beliefs; each cell (column) occupies a particular place in the ensemble (rows); white denotes a probability of one. The lower right panel shows the ensuing configuration: the trajectory is shown in small circles for each time step; the insert corresponds to the target configuration.

To the extent that one subscribes to this formulation of computation, it offers a useful and incontrovertible definition: computation can be defined as *any process that increases model evidence*. Model evidence will be a key concept in what follows and appears in many guises throughout the physical and life sciences. In brief, model evidence is the probability of some data or sensory state of a system, given the system in question. Conceptually, it is useful to treat a system and a model as synonymous. This allows one to think of any system as performing some computation on measurements of—or sensory exchanges with—its world. It is this model that lends computation its defining attribute. Examples of model evidence include the wave function in quantum mechanics, whose squared amplitude corresponds to the probability of a particular state given the quantum system or model in question (Ballentine 1970). In statistical mechanics, the negative log evidence becomes a thermodynamic (Gibbs) energy or potential (Ao 2008; Seifert 2012). In statistics per se, model evidence is also known as marginal likelihood (Beal 2003). In information theory, negative log evidence is known as self-information (Jones 1979); namely, the surprise (or surprisal) induced by an unlikely outcome. This definition is important because it means that the average of (negative log) model evidence

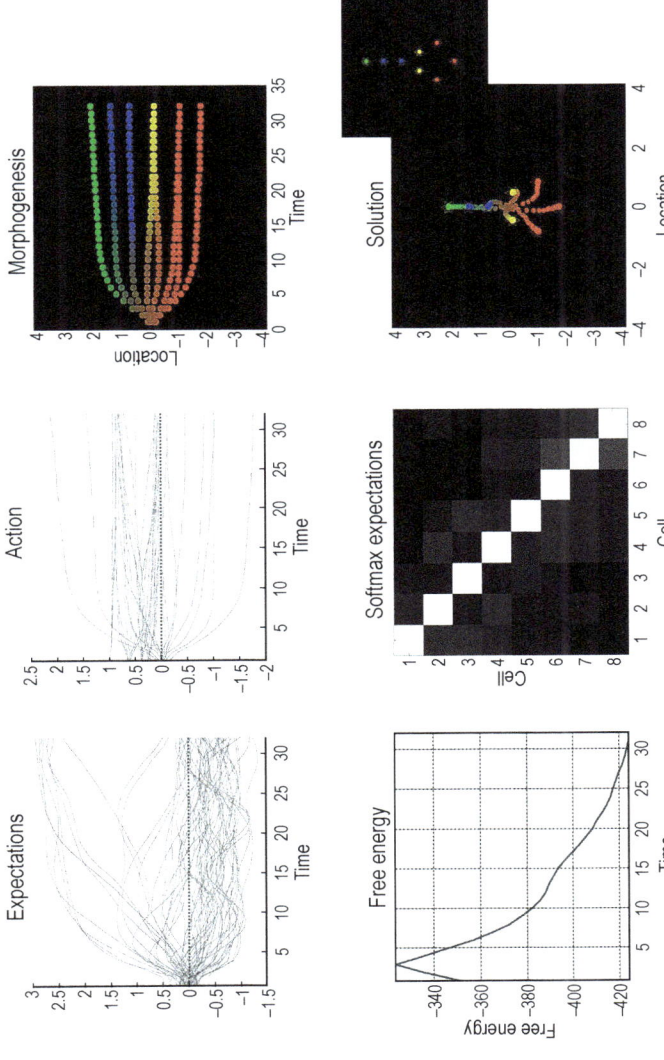

**Figure 14.1**  Self-assembly and morphogenesis.

is entropy. In turn, this means that any self-organizing system that resists the second law of thermodynamics is implicitly minimizing its informational and thermodynamic self-information on a moment-to-moment basis (Nicolis and Prigogine 1977; Friston 2013). Equivalently, interesting self-organizing systems with complex (and complicated) attracting sets must therefore maximize model evidence. So if we define computation as the maximization of model evidence, what does this tell us about the complexity of computation?

## Complexity and Computation

This is where the statistical and algorithmic definitions of complexity come into play. Put simply, model evidence can always be expressed as *accuracy minus complexity*. In other words, the evidence associated with some sensory input is just the difference between accuracy (the expected log probability of sensations, given a model of how they were caused) and complexity (the KL divergence between posterior and prior beliefs encoded by the model or system in question). On this view, we can regard self-organization in complex systems as internal responses to external perturbations; namely, sensory inputs. This means that one can associate the internal states of a system with belief states about what is causing or generating its sensory impressions on "the outside." This outside could be a heat bath in statistical thermodynamics (Seifert 2012) or the sensorium in human perception (Still et al. 2012). See Figure 14.2 for an illustration of how complexity minimization underlies action perception in the human brain.

Now let us take another look at the imperative that underlies computation. If computation necessarily increases model evidence, it must therefore entail a decrease in complexity. This is consistent with the minimum message length and algorithmic complexity reduction associated with maximum efficiency and Ockham's principle (Barlow 1974; Hinton and Zemel 1993; Wallace and Dowe 1999). Indeed, some people believe that all self-organization and adaptive behavior can be described in terms of minimizing complexity in one way or another (Schmidhuber 2006, 2010). This brief consideration of computation from the point of view of inference poses two fundamental challenges for any theory of brain function:

- How can we formulate neuronal computations that underlie action, perception, cognition, and consciousness to increase model evidence and the implicitly minimize complexity?
- How does the minimization of *algorithmic complexity* explain the emergence of *dynamical complexity* in sentient, self-organizing systems such as the brain?

To address these challenges, we now consider several global brain theories and see how they fare.

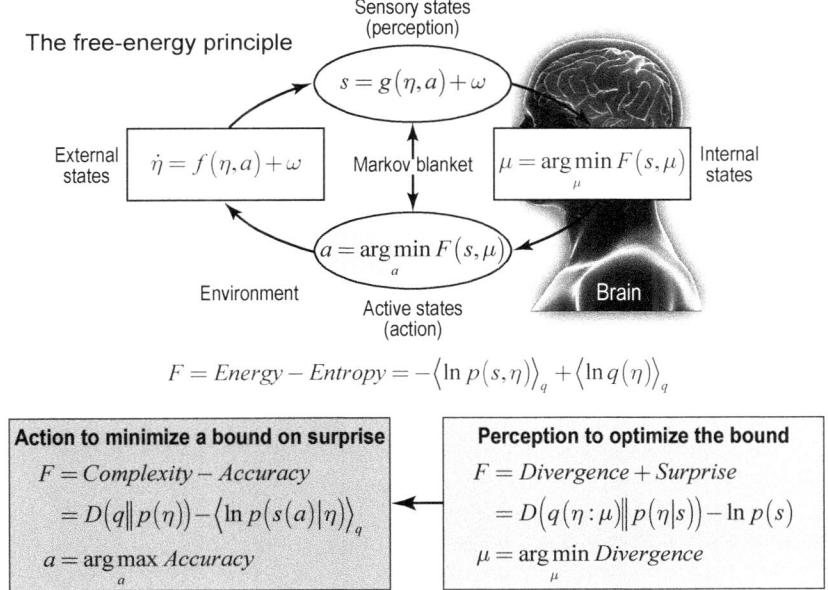

The free-energy principle

Sensory states
(perception)

$$s = g(\eta, a) + \omega$$

External states

$$\dot{\eta} = f(\eta, a) + \omega$$

Markov blanket

$$\mu = \arg\min_{\mu} F(s, \mu)$$

Internal states

$$a = \arg\min_{a} F(s, \mu)$$

Environment

Active states
(action)

Brain

$$F = Energy - Entropy = -\langle \ln p(s, \eta) \rangle_q + \langle \ln q(\eta) \rangle_q$$

**Action to minimize a bound on surprise**

$$F = Complexity - Accuracy$$

$$= D(q \| p(\eta)) - \langle \ln p(s(a) | \eta) \rangle_q$$

$$a = \arg\max_a Accuracy$$

**Perception to optimize the bound**

$$F = Divergence + Surprise$$

$$= D(q(\eta : \mu) \| p(\eta | s)) - \ln p(s)$$

$$\mu = \arg\min_{\mu} Divergence$$

**Figure 14.2** Bayesian computation in the brain. Upper panel: Schematic of the quantities that define a system and its coupling to the world. These quantities include the internal states of a system μ (e.g., a brain) and quantities describing exchange with the world; namely, sensory input $s = g(\eta, a) + \omega$ and action $a$ that changes the way the environment is sampled. The environment is described by equations of motion, $\dot{\eta} = f(\eta, a) + \omega$, which specify the dynamics of (hidden) states of the world, η. Here, ω denotes random fluctuations. Internal states and action both change to minimize free energy or self-information, which is a function of sensory input and a probabilistic belief $q(\eta : \mu)$ encoded by the internal states. Lower panel: Alternative expressions for free energy illustrating what its minimization entails. For action, free energy (i.e., self-information) can only be suppressed by increasing the accuracy of sensory data (i.e., selectively sampling data that are predicted). Conversely, optimizing internal states make the representation an approximate conditional density on the causes of sensory input (by minimizing KL divergence). This optimization makes the free energy bound to self-information tighter and enables action to avoid surprising sensations.

## Normative and Process Theories of Computation in the Brain

This section fleshes out the important distinction between *normative* and *process* theories that could be entertained in the neurosciences, with a special focus on the imperatives for complexity and computation described above. How might one approach theoretical frameworks for computation in the brain? Perhaps the easiest thing to do is to distinguish between theories or principles that describe *what* the brain does from process theories that describe *how* the brain does something. We will refer to these as normative (or state) and process theories, respectively.

The perspective in this section borrows heavily from the physical sciences, where it is almost self-evident that to understand any system—from the systems of quantum mechanics through to the canonical ensembles of statistical thermodynamics—it is necessary to specify the system's *Lyapunov function*. The notion of a single (Lyapunov) function that can describe the entire behavior of any system—indeed the universe—may seem fantastical; however, nearly all physics ultimately falls back on some form of Lyapunov function. Put simply, one can describe any (random) dynamical system—whether it is complex, self-organizing, or not—in terms of a set of (random) differential equations (Tomé 2006; Ao 2008; Seifert 2012). Furthermore, the flow or changes in the state of the system at any point in state space (i.e., for any given state) can be completely described by a (Lyapunov) function of those states.

Common examples here include the Lagrangian of gauge theories (e.g., general relativity) (Capozziello and De Laurentis 2011; Sengupta et al. 2016), the Hamiltonian of classical mechanics, the thermodynamic free energies of statistical mechanics, and the Schrödinger Hamiltonian of quantum mechanics (Ballentine 1970; Seifert 2012). All of these quantities are essentially the same thing and just score the improbability of occupying a particular state, such that the flow of the system will tend to vacate regimes of state space in which it is not typically found (i.e., which do not constitute parts of its attracting set). The physical analogies here are not terribly important. The important point is that a complete description of any system can be obtained if we understand what the system is doing in terms of the Lyapunov function it is continuously trying to minimize. The insight here is that we can describe any complex self-organizing system (including the brain) in terms of an apparent *optimization*. This is because—in virtue of decreasing its Lyapunov function—the brain will appear to be optimizing the Lyapunov function. So what is the Lyapunov function for the brain? The answer is exactly the same answer for any system: the self-information or negative log evidence.

There is a long and technical (and very interesting) back story to this assertion; however, we will simply accept this to be the case and recognize some common instances of the implicit self-evidencing implied by this formulation. First, self-evidencing, as the label suggests, implies that systems like the brain are in the game of maximizing the Bayesian model evidence for their models of the sensorium (Hohwy 2016). This is nothing more or less than the Bayesian brain hypothesis cast in terms of a state or normative theory (Ballard et al. 1983; Knill and Pouget 2004; Yuille and Kersten 2006). This has a long history dating back to the students of Plato, through Kant and Helmholtz (Helmholtz 1866/1962) to modern-day formulations in terms of perception as hypothesis testing and variational formulations, such as the free-energy principle and active inference (Gregory 1980; Hinton and Zemel 1993; Dayan et al. 1995; Friston 2010). Implicit optimization also subsumes dominant theories in psychology (e.g., reinforcement learning) and in economics (e.g., expected utility theory). In both reinforcement learning and expected utility theory, the underlying premise is that there is some reward, cost, value, or expected utility function that behavior is trying to realize. In active inference, this function is the expected model evidence over (prior) preferences about outcomes in the future (Friston et al. 2015b; Mirza et al. 2016). I will return to this later; for the moment, let us first see whether this account meets the challenges posed above.

## Algorithmic and Statistical Complexity: The Dialectic Resolved

If self-evidencing minimizes the complexity of our (generative) models of the world (Hohwy 2016), how does this explain the dynamical complexity of neuronal activity? It turns out that the answer is relatively straightforward. The explanation here has two parts: the first *dynamical* and the second *structural*. In terms of dynamical complexity, minimizing self-information or maximizing log evidence necessarily engenders *self-organized criticality* and the three mechanisms that underpin dynamical complexity (see Table 14.1). The reason is subtle but intuitive. If any system is trying to minimize its self-information, we can think of the system as tracing out a trajectory on a (self-information) function over its state space. For example, imagine a steel ball rolling over a curved surface, always searching for the lowest points. It can be seen immediately that the states which the system visits will repeatedly correspond to the (local) minima of the (self-information) surface, thereby creating an attracting set. Clearly, because the number of points that constitute a minimum is far less than the total number of points in state space, this attracting set will have low measure or volume. Now, here comes the clever bit. Because the curvature of this surface determines the precision or confidence of posterior beliefs, its curvature determines complexity. In other words, if the ball represents some neuronal population firing rate that is trapped in a very narrow ravine of the (self-information) surface, the system can be very confident about where it is located. Statistically, this is reflected in things like Fisher information and information geometry (Amari 1998). The important thing here is that as the curvature of self-information increases, the difference between the posterior and prior increases, and the complexity increases. This means that, by definition, regions of state space with low self-information must be relatively flat (much like river estuaries are broader than the hanging valleys from which their tributaries emerge). The flat aspect of these attracting minima means that the ball can roll around the local minima in any unconstrained, slowly oscillating or meandering fashion. These critical slowing and long-range fluctuations are the hallmark of self-organized criticality (Bak et al. 1988; Shin and Kim 2006). Furthermore, because the sites of the self-information minima are relatively shallow, this affords the opportunity to jump from local minima to local minima, thereby affording a mathematical image of metastability and other forms of critical dynamics (see Table 14.1 and Figure 14.3). We will see later that precision is itself a quantity that is optimized by the brain, and this optimization may be what we call attention. This lends attention an interesting

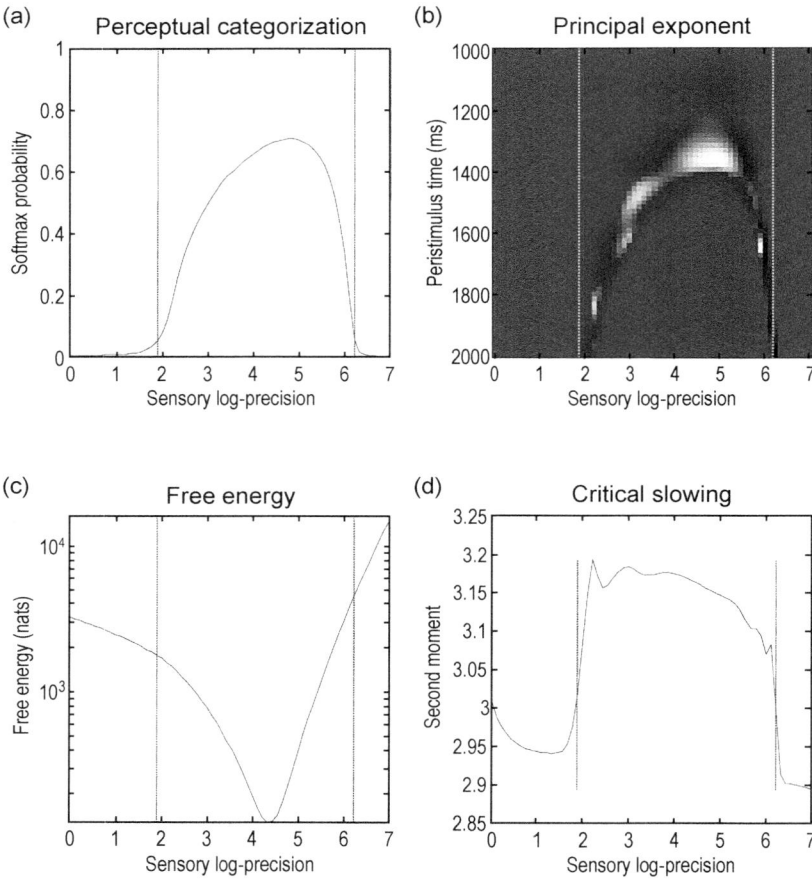

**Figure 14.3**    Self-organized criticality and computation: (a) The average probability, following stimulus onset, of correctly identifying a song over 64 values of precision on the motion of hidden attractor states. The two vertical lines correspond to the onset and offset of nontrivial categorization—a softmax probability of greater than 0.05. The variation in these average probabilities is due to the latency of the perceptual switch to the correct song. This can be seen in (b), which shows the principal conditional Lyapunov exponent (CLE) in image format as a function of peristimulus time (columns) and precision (rows). It can be seen that the principal CLE shows fluctuations in, and only in, the regime of veridical categorization. Crucially, these fluctuations appear earlier when the categorization probabilities were higher, indicating the prevalence of short latency perceptual switches. Time-averaged free energy is shown in (c) as a function of precision. As one might anticipate, this exhibits a clear minimum around the level of precision that produces the best perceptual categorization. In (d), a very clear critical slowing is shown in, and only in, the regime of correct categorization. In short, these results are consistent with the conjecture that free-energy minimization can induce instability and thereby provide a more responsive representation of hidden states in the world. Adapted from Friston et al. (2012), to which the reader is referred for further details.

interpretation; namely, it might be the psychological homologue of self-organized criticality that allows us to engage selectively with the sensory world in
both space and time (Coull and Nobre 1998; Feldman and Friston 2010).

In short, the very mathematical structure of computation in a Bayesian or
abductive sense necessarily entails self-organized criticality and fluctuations
of the sort that characterizes dynamical complexity. But what about structural
complexity and the form of neuronal architectures (e.g., connectivity)?

## Generative Models and Structural Complexity

In the search for accurate and minimally complex models of the sensorium,
the best solution is generally to recapitulate the causal or statistical structure of
the world "out there," within the system (e.g., the brain). In short, the best path
to self-evidence is to have a veridical and parsimonious model of the world
in which you are navigating. This is an old insight first articulated formally in
synergetics in terms of the good regulator theorem (Conant and Ashby 1970;
Seth and Friston 2015); namely, any system that can regulate its environment
must possess a model of that environment. This tells us something very interesting. It means that minimizing complexity—while maintaining an accurate
explanation or prediction of sensory inputs—will cause statistical regularities
and causal structure in the world to be transcribed into the system's internal
architecture. If one pursues this argument, then we have a natural explanation
for the finely crafted and interwoven connectivity in our brains that has all the
hallmarks of complexity (Friston and Buzsáki 2016). This is simply a restoration of the sparse, deep, or hierarchical structure of the world "out there," generating sensory impressions. In short, if we live in a complex and complicated
world, the minimization of complexity—in the algorithmic sense—not only
enforces self-organized criticality and dynamical complexity, it also mandates
a structural complexity that mirrors the world. For some, these may be pleasing
accounts of complexity and computation in the brain. So is this the end of the
story? Clearly not, it is only the beginning. Having a state or normative theory
is certainly very useful; however, it says nothing about the process theories that
actually explain neuronal computations.

## Process Theories

Clearly, there are many theories about neuronal processing that appeal to a
greater or lesser extent to normative theories. Happily, the dominant theory—
predictive coding—subsumes most available process theories. Predictive coding is not a normative theory; it is a particular algorithm or process theory
that has attracted a lot of attention over the past decades in explaining many
aspects of neuronal anatomy, physiology, psychophysics, and, more recently,
motor control (Srinivasan et al. 1982; Rao and Ballard 1999; Friston 2011b).
In brief, predictive coding was originally formulated for compressing large

files, based on the minimum description length notions above (Elias 1955). In the neurosciences, it now represents the most developed and established process theory for hierarchical message passing in the brain (Mumford 1992; Bastos et al. 2012). Predictive coding is just an algorithm or scheme that minimizes self-information or maximizes model evidence by updating internal states (i.e., representations) in the light of sensory evidence. It is distinguished from other formulations by calling on auxiliary variables termed *prediction errors*. Prediction errors are simply the difference between sensory inputs (or intermediate representations at hierarchical levels in hierarchical predictive coding) and predictions of those inputs based on internal representations or expectations. In the brain, predictive coding is usually described as reciprocal message passing among the levels of the cortical and subcortical hierarchy (Friston 2010). The recurrent aspect of this message passing is important and fundamentally asymmetric. In other words, top-down or descending messages convey predictions of expectations in the level below (or sensory input per se), whereas ascending or bottom-up signals communicate newsworthy prediction errors that update expectations, thereby improving predictions and resolving prediction errors throughout the hierarchy.

In engineering, the Kalman filter (a special linear case of Bayesian filtering) is the formal homologue of predictive coding. Predictive coding in its generalized form also provides a nice metaphor for several other important schemes used for data assimilation and uncertainty quantification; for example, reservoir computing and deep learning (Schmidhuber 2006; Hinton 2007; Tenenbaum et al. 2011; Salakhutdinov et al. 2013; LeCun et al. 2015). To understand this, we have to distinguish between *inference* and *learning*. In this chapter, inference corresponds to the estimation of (time-varying) causes in the world that are generating sensations, whereas learning corresponds to accumulating experience in the service of updating (time-invariant) model parameters that underwrite inference. Happily, when we put prediction errors into the algorithmic mix, things like *backpropagation of error* can be implemented using Hebbian or associative plasticity. Furthermore, schemes like reservoir computing and liquid state machines (Maass et al. 2002; Buonomano and Maass 2009) can also be considered as variants of predictive coding; for a discussion of how reservoir computing can self-organize to improve predictive coding, see Toutounji and Pipa (2014). The twist here is that instead of optimizing the parameters that enable predictive coding schemes to make better predictions, parameters mapping from a reservoir of dynamics are optimized to select the best prediction of temporally fluctuating inputs, or some supervised output. This sort of scheme (based on recurrent neural networks) has found a particularly powerful application in neurorobotics, reproducing many lifelike behaviors (Tani 2003; Tani et al. 2004).

Generalized predictive coding schemes also provide a nice vehicle for many other issues that attend the dynamic coordination of message passing in the brain (von der Malsburg et al. 2010). A key example here is the encoding of uncertainty through the precision or gain afforded by prediction errors. Technically, this corresponds to the Kalman gain or precision in Bayesian filtering formulations of evidence accumulation or assimilation (Clark 2013). Physiologically, there are a host of important mechanisms that may coordinate the implicit gain control of prediction errors. These range from the control of classical modulatory neurotransmitter systems through to excitation–inhibition balance in the coupling between superficial pyramidal cells and inhibitory interneurons (Yu and Dayan 2005). This is a particularly fascinating area that has clear correlates (of selective evidence accumulation) in terms of fast synchronized neuronal oscillations, which may be a crucial aspect of gating and communication in perceptual synthesis (Singer and Gray 1995; Fries 2005; Womelsdorf et al. 2007; Giraud and Poeppel 2012).

Oscillatory dynamics may also be a key player in process theories of forward and backward message passing in hierarchical predictive coding. The implicit mathematical structure of this message passing suggests that faster fluctuations in prediction errors may be communicated by high frequencies, whereas lower frequencies may convey descending connections (Bastos et al. 2012, 2015). If true, this puts the nonlinear integration of units encoding expectations and prediction errors within the same cortical column center stage in cortical computations (Kopell et al. 2011; Lee et al. 2013a).

**The Functional Anatomy of Predictive Coding**

So how does predictive coding fare as a process theory in relation to anatomy and physiology? Its explanatory scope is impressive. For example, it provides a principled explanation for the functional asymmetries between ascending and descending (forward and backward) extrinsic (between-area) connections in cortical hierarchies (Mesulam 1998; Hilgetag et al. 2000). These functional asymmetries entail the spectral asymmetries in neuronal oscillations above and established dissociations between driving (forward) and modulatory (backward) synaptic effects (Sherman and Guillery 1998, 2011; Bastos et al. 2012). Variants on different proposals for the integration of hierarchical or centrifugal patterns of extrinsic connections have emerged over the past few decades, starting with the seminal work of David Munford (1992) on the computational architecture of the neocortex. This work has been refined and embellished over the years, leading to detailed descriptions of canonical microcircuits for predictive coding that identify computational roles for individual cell types (Bastos et al. 2012; Shipp 2016); see also Figure 14.4, where the equations in the left panel provide a mathematical form for predictive coding and emphasize the key role of precision (see above) in coordinating and contextualizing the impact of prediction errors on belief updating. As noted above, the way that precision enters into belief updating in these schemes suggests a close link between optimizing precision and attention. In other words, some of the complexity associated with neuronal dynamics rests upon self-organized coupling,

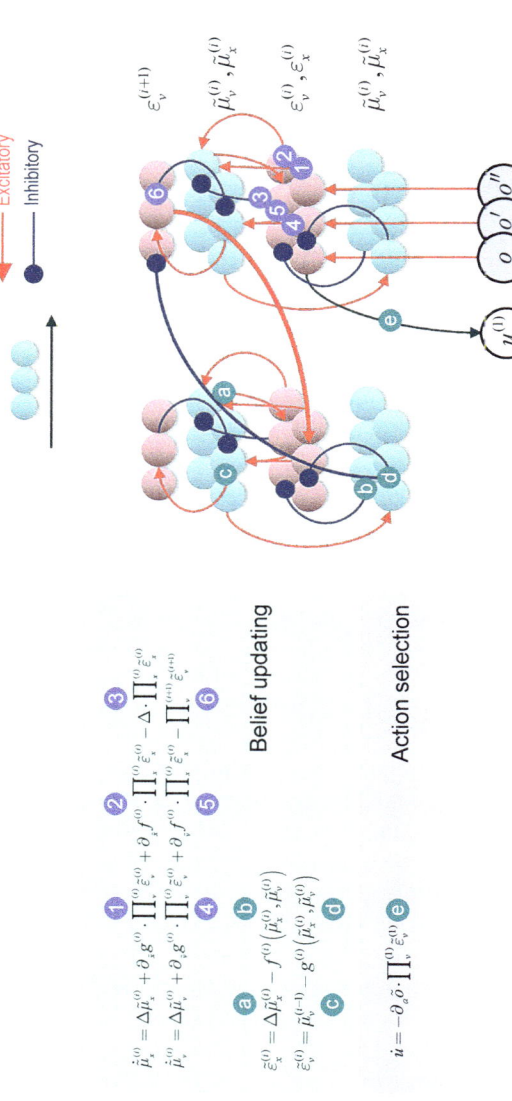

**Figure 14.4** Canonical microcircuits for predictive coding. This schematic illustrates how the mathematics of predictive coding schemes can be used to develop a detailed process theory for neuronal computation. This example is based on the anatomy of intrinsic and extrinsic connections described by Bastos et al. (2012). Left panels show the update dynamics for a hierarchical generalization of predictive coding, where the outputs of a higher level are the hidden causes of state dynamics in the level below. In this hierarchical setting, the prediction errors include prediction errors on both hidden causes and states. In the right panel, prediction errors have been assigned to granular layers that receive sensory afferents and ascending prediction errors from lower levels in the hierarchy, along with superficial pyramidal cells that broadcast ascending prediction errors. Prediction errors are denoted by $\varepsilon$, expectations by $\mu$, and precision by $\Pi$. The functions $f$ and $g$ embody a deep (i.e., hierarchical) generative model of sensory observations $o$ that are solicited by action $u$; $\Delta$ is a differential operator. Adapted from Friston et al. (2017a), to which the reader is referred for further details.

which may entail synchronous gain (Singer and Gray 1995; Fries et al. 2001b, 2008; Fries 2005; Bauer et al. 2006, 2014; Womelsdorf et al. 2007; Bendixen et al. 2012; Auksztulewicz and Friston 2015; Wildegger et al. 2017).

So, if predictive coding appears to capture so much of the brain's computational anatomy, do we have a complete picture (at least at a mesoscopic scale) of computation in the human brain? I would submit that we probably do not. In the final section, let us thus turn to some of the deeper challenges that constitute the focus of much current research.

## Beyond Predictive Coding

There are many reasons to suppose that predictive coding in and of itself is an incomplete process theory for neuronal computations. The most obvious shortcoming is its failure to account properly for action, planning, or intentions (Bernier et al. 2017). This problem can be finessed, in part, by appeal to active inference which, essentially, equips predictive coding schemes with classical reflexes (Friston et al. 2015b). This renders motor control a problem of predicting the proprioceptive consequences of action and speaks to purposeful behavior in terms of planning and inference (Attias 2003; Botvinick and Toussaint 2012; Mirza et al. 2016). This is an important extension of predictive processing and hierarchical Bayesian inference in the brain; however, it may raise more issues than it resolves. We will briefly consider a few of these key issues which are considered in other chapters in this volume. We start with some basic aspects of generative models that underlie purpose and selfhood and then consider some of their implications for neuronal dynamics.

### Temporal Thickness and Counterfactual Depth

Current trends in machine learning may be taken as a pointer for developments in computational neuroscience, exemplified by the success of deep convolution networks and deep learning (LeCun et al. 2015). This direction, however, is probably not fit for purpose for several reasons. First, deep *learning* and associated reinforcement learning paradigms do not address *inference*. In other words, although data is accumulated in the service of optimizing connection strengths, the problem of how hidden states of the world are inferred online (i.e., data assimilation) is, most often, not in the remit of machine learning. This speaks to the fact that there are usually no dynamics involved in the classification problems addressed in machine learning. In other words, the problem of recognizing static images of handwritten digits is very different to the problem of anticipating the intention and motor behavior of somebody writing digits by hand.

Second, current machine learning approaches using deep neural networks do not usually consider the encoding of uncertainty and, more importantly, epistemic value and intrinsic motivation (Ryan and Deci 1985; Oudeyer and Kaplan 2007; Schmidhuber 2010; Friston et al. 2015b). In other words, if the

brain is in the game of minimizing (expected) self-information, it is technically trying to minimize uncertainty. This follows because the time average
of self-information is entropy (Jones 1979). The minimization of uncertainty
through explorative behavior or epistemic foraging is probably one of the most
important imperatives for neuronal computation; thus, it has to be an integral
part of any normative theory. Unfortunately, most machine learning algorithms
do not accommodate this, leading to ad hoc ways of resolving the exploitation–exploration dilemma (Cohen et al. 2007). In contrast, to understand fully
how the brain recognizes and acts upon cues with epistemic affordance, we
need to have a clear understanding of how the brain models the future, even in
simple tasks like reading (Figure 14.5). Reading is particularly interesting because it speaks to the assimilation of sensory evidence at multiple hierarchical
or deep temporal scales (Poeppel et al. 2008), while at the same time calling
upon efficient foraging of the visual scene for salient information (Hassabis
and Maguire 2007; Mirza et al. 2016). This also brings us to the notion of
mnemonics and counterfactual depth which underlie purposeful and possibly
mindful behavior (Palmer et al. 2015; Seth 2015).

**Deep Temporal Models**

If we choose our behaviors based upon a model of the world, then that model
must entertain the consequences of action in the future. Furthermore, this future must be encoded at different temporal scales. This is literally a deep and
intriguing problem that has clear implications for the form of generative models embodied by neuronal connections and neurophysiology. Again, language
and reading provide excellent opportunities to understand how narratives are
synthesized in the brain and contextualize our active sampling of the sensorium (Barlow 1974; Beim Graben et al. 2008; Poeppel et al. 2008; Giraud and
Poeppel 2012; Dehaene et al. 2015; Konig and Buffalo 2016). There are some
fascinating issues when it comes to the details of the underlying process theories (O'Keefe and Recce 1993; Buzsáki et al. 2013; Murray et al. 2014; Friston
and Buzsáki 2016):

- How do we select models?
- How do we integrate the assimilation of sensory evidence over differ-
  ent timescales?
- Do we have a moving temporal frame of reference?
- Are there separate spatial and temporal representations (Dehaene et al.
  2015), or do we represent dynamic spatiotemporal trajectories?
- How do we contextualize our evidence accumulation and action
  selection?
- Do we have separate perceptual and motor representations or are these
  fundamentally integrated within generative models of the sensed world
  (Grafton and Hamilton 2007; Bernier et al. 2017; Cogan et al. 2017)?

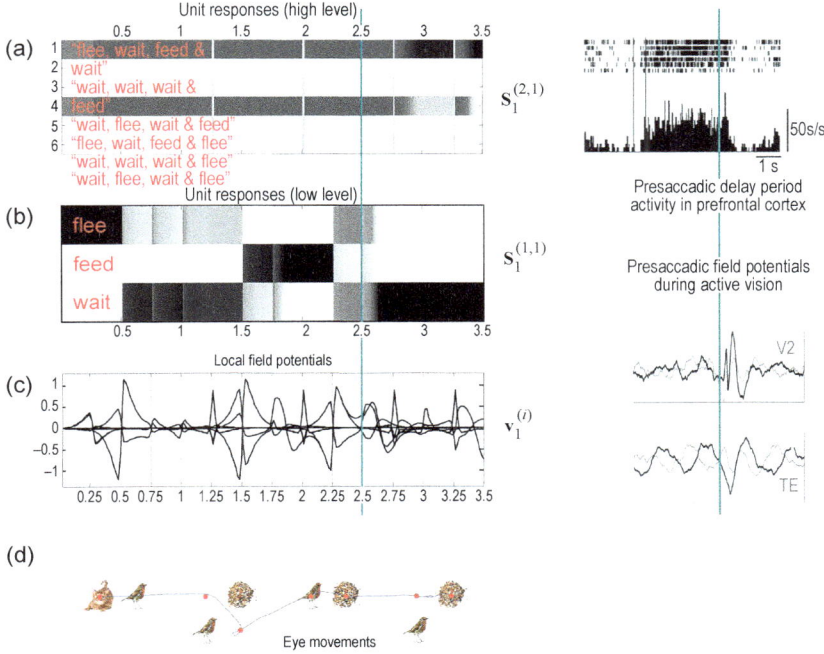

**Figure 14.5** Simulated electrophysiological responses during reading: this figure illustrates the sort of Bayesian belief updating that underlies reading. Neuronal dynamics are shown in terms of expectations about hidden states of a deep model (i.e., a model with hierarchical depth) generating words from sentences and pictograms from words. The upper panels show expectations are shown at the higher (a) and lower (b) hierarchical levels in raster format, where an expectation of one corresponds to black (i.e., the firing rate activity corresponds to image intensity). The horizontal axis is time over a reading trial, where each iteration corresponds roughly to 16 ms. The vertical axis corresponds to six sentences at the higher level and three words at the lower level. The resulting patterns of firing rate over time show a marked resemblance to delay period activity in the prefrontal cortex prior to saccades. Saccade onsets are shown by the vertical (cyan) lines. The inset on the upper right is based upon the empirical results reported in Funahashi (2014). The transients in (c) are the simulated firing rates in the upper panels filtered between 4 Hz and 32 Hz, and can be regarded as (band pass filtered) fluctuations in depolarization. These simulated local field potentials are again remarkably similar to empirical responses. The examples shown in the inset are based on the study of perisaccadic electrophysiological responses during activation reported in Purpura et al. (2003). The upper traces come from early visual cortex (V2), while the lower traces come from inferotemporal cortex (TE). Eye movement trajectories produced in this simulation of active inference are shown in (d). Adapted from Friston et al. (2017b), to which the reader is referred for further details.

- How does coevolution make the job of predicting the sensed world easier? Indeed, how do we take turns in predicting each other (see Ghazanfar and Takahashi 2014)?

- To what extent does our ability to predict rely upon neuromodulatory mechanisms, such as dopamine and noradrenaline, or indeed a level of consciousness (Strauss et al. 2015)?
- What is the role of brain oscillations (Mayer et al. 2016), such as theta-gamma coupling, between the prefrontal cortex and hippocampus?

These questions are particularly prescient in the context of predictive coding and language processing (see Arnal and Giraud 2012; Giraud and Poeppel 2012).

A clear utility of deep temporal models is the ability to understand neuronal dynamics in terms of message passing and belief updating. In other words, this formulation provides an explicit account of computations that are "about something." This top-down approach, starting from the generative model and then unpacking the requisite dynamics, can be contrasted with bottom-up approaches, such as deep learning, and the use of recurrent or unfolded networks that aspire to the same kind of functionality. However, these "black box" approaches do not necessarily admit the same level of algorithmic or functional interpretability. For a taxonomy of computational architectures that speaks to the recurrent neural networks used in machine learning and predictive coding-like schemes, based upon generative models, see Harris et al. (this volume).

Finally, it should be noted that the metaphor offered by predictive coding is only appropriate when dealing with continuous state spaces in continuous time. While this is perfectly fine for luminance contrast and perhaps visual motion as well as the detailed kinematics of muscle movements, it may be the wrong sort of parameterization for the lived world. In other words, our intentions, concepts, and sense of self usually come along as (sequences of) discrete or categorical states (see Dehaene et al. 2015; Wilson et al. 2017). Technically, these questions call for a completely different set of neuronal computations that inherit some similarities from predictive coding but are quintessentially different in their form. This is not a problem; indeed, it speaks to the known unknowns that can guide empirical research. For example, there are very particular predictions based on the belief propagation under discrete models (e.g., Markov decision processes) in comparison to continuous models (e.g., the state space models of predictive coding). There is clearly no right or wrong answer in terms of process theories, which means that empirical data will, ultimately, be in a position to adjudicate among the different hypotheses. To do this, however, one must be able to specify the dynamics and implicit coordination implied by various process theories. This is the challenge.

## Conclusion

In summary, we have taken the basic nature of complexity and computation to consider the constraints on state or normative theories of brain computation, with a special focus on self-evidencing and inference as the most promising formulation. From this, a number of different process theories—illustrated here with predictive coding—inform, or are informed by, empirical study and highlight the known unknowns in neuroscience. Further questions remain to be addressed regarding subconscious and conscious inference, minimal selfhood, interoceptive inference, emotions, and related areas in philosophy. Many ideas are emerging under the notion of the self-organizing and self-evidencing brain that promise to enrich this enquiry.

## Acknowledgments

K. J. F. is funded by the Wellcome Trust (Ref: 088130/Z/09/Z).

# 15

# Human Singularity and Symbolic Tree Structures

## The Demodularization Hypothesis

Stanislas Dehaene

### Abstract

Relative to other primates, humans exhibit a great variety of singular cognitive abilities for language, mathematics, music, tool use, theory of mind, and self-consciousness. What has brought about this singularity? This chapter examines the hypothesis that the human brain is unique in being endowed with a mental representation of nested, tree-like symbolic structures. Such syntactic structures are essential in the modern description of human languages, including natural languages as well as the artificial ones used in music or mathematics. Nonhuman animals may possess abstract representation of temporal sequences, but evidence suggests that those representations do not include the sort of nested tree structures typical of human grammars. Brain imaging, magneto-encephalography and intracranial recordings have begun to reveal the neural correlates of the nested structure of linguistic constituents, which involve Broca's area and the superior temporal sulcus of the left hemisphere. Importantly, the mental manipulation of musical and mathematical structures, which also involves nested trees, is not confined to such classical language areas. Instead, high-level mathematics involves bilateral intraparietal areas involved in elementary number sense and simple arithmetic as well as bilateral inferotemporal areas involved in processing Arabic numerals. This chapter proposes that several distinct circuits of the human brain have become attuned to nested tree structures for different domains, such as language, mathematics, or music. According to the demodularization hypothesis, during human brain evolution, primitive tree structures may have emerged within specialized neural circuits (e.g., those involved in spatial or geometrical computations) and were later exapted toward a more general role in language processing and conscious verbal report.

## Introduction: Hypotheses about Human Singularity

In many cognitive domains, humans are special among other primates. They are, for example, the only species that

- produce and understand language, a highly combinatorial communication system,
- create and use complex tools,
- formulate and test complex scientific theories, expressed in formal mathematical notations,
- possess a complex representation of other minds,
- exhibit a sophisticated representation of their own selves, and
- educate each other based on a representation of the gap between self versus other knowledge.

This list is not exhaustive. Can cognitive neuroscience shed some light on the origins of these remarkable human singularities?

Darwin's view, as expressed in *The Descent of Man*, was that the "difference in mind between man and the higher animals, great as it is, certainly is one of degree and not of kind" (Darwin 1888). In all of these domains, he argued, we can find nonhuman precursors in other animals. Nevertheless, Darwin would probably not have denied the importance of a research program searching for the evolutionary changes that made the acquisition of language, mathematics, theory of mind, or education radically easier for humans than for other animals. Indeed, Darwin defended the view that language ability reflects "an instinctive tendency to acquire an art," a view close to Noam Chomsky's notion of an innate "language acquisition device" or Peter Marler's idea of "learning by instinct."

In this chapter, I consider the possibility that human singularity relies on a novel type of mental representation, recursively nested trees, which makes humans capable of fast learning whenever the structure of the learned domain conforms to such nested trees (Bolhuis et al. 2014). What makes human thought complex, according to the nested tree hypothesis, is that words or symbols are not just strung together into a sequence, but are mentally represented as hierarchical trees, thus offering a much greater combinatorial diversity, as in distinguishing the concepts $3x + 1$ from $3(x + 1)$, or *un-lockable* from *unlockable*. I speculate that the human brain is singular because, during evolution, it acquired the ability to represent complex internal tree structures, evaluate their adequacy as hypotheses for a given domain, and even impose them onto simple incoming sequences of stimuli.

## Five Types of Sequence Representations in the Human Brain

As early as the 1950s, the problem of serial order in behavior was identified by Karl Lashley (1951) as one of the pressing questions that behavioral and neural sciences should address. The problem can be stated succinctly: How does the brain encode temporal sequences of actions, such that this knowledge can be used to retrieve a sequence from memory, recognize it, anticipate forthcoming items, and generalize this knowledge to novel sequences with a similar structure?

Lashley (1951) showed that a simple "associative chain," based on the hypothesis that each element in a sequence served as a cue for the next, was insufficient to account for the complexity of motor sequences that were produced by various animal species. In a recent review of sequence learning research (Dehaene et al. 2015), my colleagues and I have proposed a typology of five different levels of sequence representations, each representing sequences at increasingly high levels of abstraction (Figure 15.1):

1. *Transition and timing knowledge*: knowledge of the specific transitions from one item to the next; that is, the identity and approximate timing of the next item relative to the preceding ones.
2. *Chunking*: the grouping of several contiguous items into a single "chunk" which can be manipulated as a whole at the next hierarchical level.
3. *Ordinal knowledge*: knowledge of which item comes first, which comes second, and so on, independently of their timing.
4. *Algebraic patterns*: abstract schemas that capture the sequential regularities underlying a sequence of items; for instance, noticing that the word "beriberi" comprises repeated syllables that conform to the ABAB pattern.
5. *Nested tree structures generated by symbolic rules*: at this level, characteristic of human languages, a sequence can be "parsed" according to abstract grammatical rules into a set of groupings, possibly embedded within each other, forming a nested structure of arbitrary depth, and involving the recursive use of the same types of element at multiple levels; an example is the parsing of the mathematical equation a + b sin ωt as a nested set of parentheses (a + (b(sin(ωt)))) or, equivalently, a tree structure:

A variety of experiments have demonstrated that the first four levels in this typology are present in various nonhuman animals, particularly in primates. To give just a few examples:

1. Transition and timing knowledge is evident in a variety of experiments which show that monkeys develop precise temporal expectations of auditory or visual stimuli, and exhibit a neural "mismatch response" to violations of these expectations (Meyer and Olson 2011; Uhrig et al. 2014; Wilson et al. 2017).
2. Chunking knowledge was demonstrated, for instance, in tamarin monkeys using the same procedure as in human infants (Hauser et al. 2001). Both species can detect recurrent "words," that is, chunks

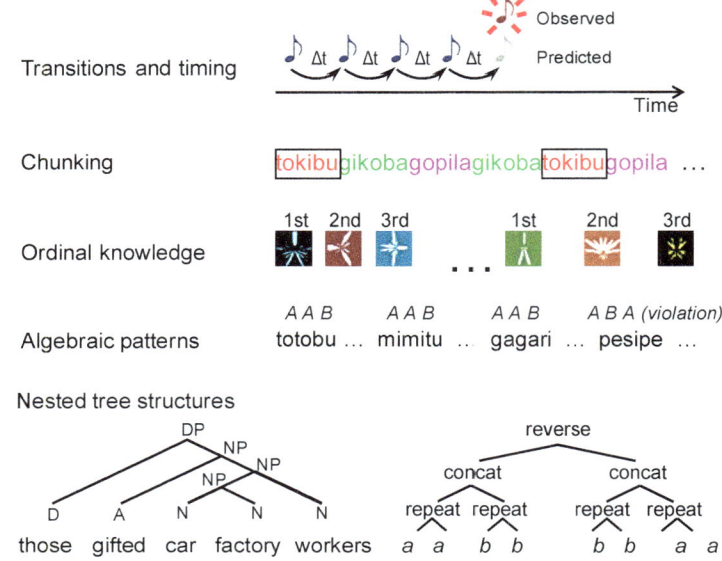

**Figure 15.1**    A typology of sequence representations. Five types of mental representations of sequences are postulated (for further details, see Dehaene et al. 2015). The first four are present in nonhuman animals, but a capacity to quickly acquire and manipulate nested tree structures may be unique to the human brain. From Dehaene et al. (2015), used with permission.

of syllables in a continuous stream such as *tokibu*gikobagopilatipolu-*tokibu*gopilatipolu*tokibu*gikobagopila.

3.  Ordinal knowledge is part of a general endowment for number processing in various animal species (Nieder and Dehaene 2009). Macaques, for instance, know which image in a list came first, second, third, or last (Chen et al. 1997), and they can string together the first item of a list with the second of another, and so on.

4.  Algebraic pattern knowledge has been observed in rats, macaques, and chimpanzees. All of these species can recognize when sequences of sounds or images obey a fixed pattern, such as *ABA* or xxxY (Murphy et al. 2008; Sonnweber et al. 2015; Wang et al. 2015).

Whether the fifth level (nested tree structures generated by symbolic rules) is present in nonhuman animals is a matter of great contention. The bulk of the evidence suggests that animals, even macaques or apes, do not easily learn natural or artificial languages whose structure is based on nested trees, or equivalently a context-free grammar (Penn et al. 2008; Fitch and Friederici 2012). Many primate species use vocal communication systems in the wild, but these do not seem to make use of a sophisticated syntax based on nested trees; at most, they involve the sequential concatenation of vocal sounds and possibly the use of a single modifier or "suffix" (Ouattara et al. 2009). For apes who have been taught artificial languages, comprising tens or even hundreds of token words in visual or sign language, it is noteworthy that they were unable to combine them systematically according to a complex grammar (Terrace et al. 1979; Yang 2013). Recently, it has been claimed that several species of birds have mastered a simple center-embedded context-free grammar called $A^nB^n$, yet this claim has been heavily criticized and, in my opinion, fully refuted (Beckers et al. 2016). Parrots may exhibit superior performance (Pepperberg 2013), including the comprehension of questions comprising several words, such as "how many green keys." However, such experiments are limited to a very small number of specific animals, with many years of training, and may still involve only a concatenation or intersection of abstract concepts, without genuine syntax.

## Evidence for Nested Tree Structures in Human Language

Comparatively, although still contested by a few linguists (Frank et al. 2012), there is considerable evidence that human language is "special" and requires covert, internal representations of abstract, symbolic syntactic structures (Haegeman 2005). Evidence for this arises from at least three distinct areas of research: linguistics, psycholinguistics, and cognitive neuroscience.

### Linguistic Descriptions of Language

As early as the 1950s, Noam Chomsky convincingly refuted the Skinnerian view of language as a chain of conditioned responses. The current view is that language syntax involves a nested tree of phrases or "constituents," which can be represented by parentheses: (the(big animal)(with (two horns)))). It is important to understand that this is a shorthand notation, as linguists today postulate that syntactic trees are not made of the raw input words, but of abstract covert objects, such as focus, topic, tense, complementizer, and trace. The case for nested constituent structures in linguistics rests on many observations:

- Cases of syntactic ambiguity, such as "looking at a man with binoculars," demonstrate that the same sequence of words can have two distinct internal representations, depending solely on tree attachment. This makes it impossible for those representations to rely solely on temporal order. Such tree-based ambiguity exists even within a single word whose morphemes can be ambiguously attached, as in un-(lock-able) versus (un-lock)-able.
- Ellipsis or substitution, whereby some strings of words—specifically those forming a subtree of the entire sentence structure (also called a

"phrase" or "constituent")—can be replaced by a single word: he (went (to (the store))) → he went to it, he went there, he did.

- Syntactic movement, whereby the same constituents can be moved to a distinct sentential location in order to form questions and relatives or to emphasize a specific topic: "*to the store*, that's where he went" or "*where* did he go").
- Long-distance dependencies, whereby the properties of two constituents (or rather, their top-level nodes or "heads") must be matched. One example is agreement in number (singular or plural) between a subject and a verb: he *was* tall, but they *were* tall. Such agreement relationships require skipping over intermediate subconstituents (e.g., "the car which passed the trucks *is* red") and therefore cannot be captured by linear structure alone. Many other long-distance phenomena, such as pronoun binding, also require considering a sentence's tree structure.
- Evidence that languages can be distinguished based on minimal differences in the language-specific rules or "parameters" that govern the formation of such tree structures. The syntactic differences between English and Japanese, for instance, can be largely accounted for by a single parametric difference in the ordering of words when tree structures are linearized during speech production: the head-first versus head-last parameter (for a very accessible account, see Baker 2001).

**Behavioral Studies**

Psycholinguistic studies (i.e., behavioral studies of language comprehension and production) have regularly observed that human verbal behavior reflects the underlying syntactic tree structures. To give but a few examples: During the comprehension of sentences with syntactic movements, human adults reactivate the antecedent of the trace of the displaced tree at precisely the moment when linguists postulate that such a trace should occur (e.g., Friedmann et al. 2008). During sentence production, they lengthen the duration of words and the intervals between sentences in direct, linear relation to the depth of the corresponding syntactic tree structure (Breen 2018). Finally, children's acquisition of language involves a systematic and nontrivial generalization over the tree structures governing, for instance, the placements of auxiliaries and verbs relative to negation and adverbs (Déprez and Pierce 1993).

**The Search for Neural Correlates of Syntactic Trees**

If the syntactic tree hypothesis is correct, there must be a set of specific brain circuits and neural codes, possibly unique to the human brain, that is engaged whenever humans process the syntactic structures of language. Functional MRI has provided strong converging evidence that a dedicated left hemispheric network is systematically associated with the formation and manipulation of nested syntactic and semantic structures (see Table 15.1 and Figure 15.2). These regions are part of an amodal network, spread all along the left superior temporal sulcus and inferior frontal gyrus, which is activated whenever humans process spoken or written sentences. Indeed, they are also active when deaf people process their native sign language.

Amidst this network, a core set of regions formed by the left inferior frontal gyrus and the left posterior superior temporal sulcus, often associated with a node in the left basal ganglia (Moreno et al. 2018), appears to be specifically involved in syntax. These regions exhibit a level of activation that is monotonically related to the number of nested constituents in the stimuli: they do not activate strongly to lists of words but show increasing levels of activity as the words are combined into syntactically correct constituents of 2, 3, 4, or 6 words, all the way to a full sentence of 12 words. Furthermore, they continue to respond in this manner even when the stimuli are delexicalized and comprised only of the function words and grammatical morphemes needed to parse them syntactically (Table 15.1, Jabberwocky condition; Pallier et al. 2011).

Many functional MRI studies indicate that these core regions are active whenever a subject represents or processes syntactic structures (even the simplest ones comprising a few words). Activation in those areas is proportional to syntactic complexity and to the amount of syntactic movement. Furthermore, they exhibit a clear dissociation from other more generic regions involved in cognitive effort and working memory (Fedorenko et al. 2011). They are also engaged when tree structures need to be manipulated internally to recover

**Table 15.1**   During functional MRI, adult volunteers read 12-word sequences of different length; corresponding brain activity is shown in Figure 15.2.

| Constituents | Examples (normal prose) |
|---|---|
| 12 words (c12) | I believe that you should accept the proposal of your new associate |
| 6 words (c06) | the mouse that eats our cheese two clients examine this nice couch |
| 4 words (c04) | mayor of the city he hates this color they read their names |
| 3 words (c03) | solving a problem repair the ceiling he keeps reading will buy some |
| 2 words (c02) | looking ahead who dies important task his dog few holes they write |
| 1 word (c01) | thing very tree where of watching copy tensed they states heart plus |

| Constituents | Examples (Jabberwocky) |
|---|---|
| 12 words (c12) | I tosieve that you should begept the tropufal of your tew viroflate |
| 6 words (c06) | the couse that rits our treeve fow plients afomine this kice bloch |
| 4 words (c04) | tuyor of the roty he futes this dator they gead their wames |
| 3 words (c03) | relging a grathem regair the fraping he meeps bouding will doy some |
| 2 words (c02) | troking ahead who mies omirpant fran his gog few biles they grite |
| 1 word (c01) | thang very gree where of wurthing napy gunsed they flotes blart trus |

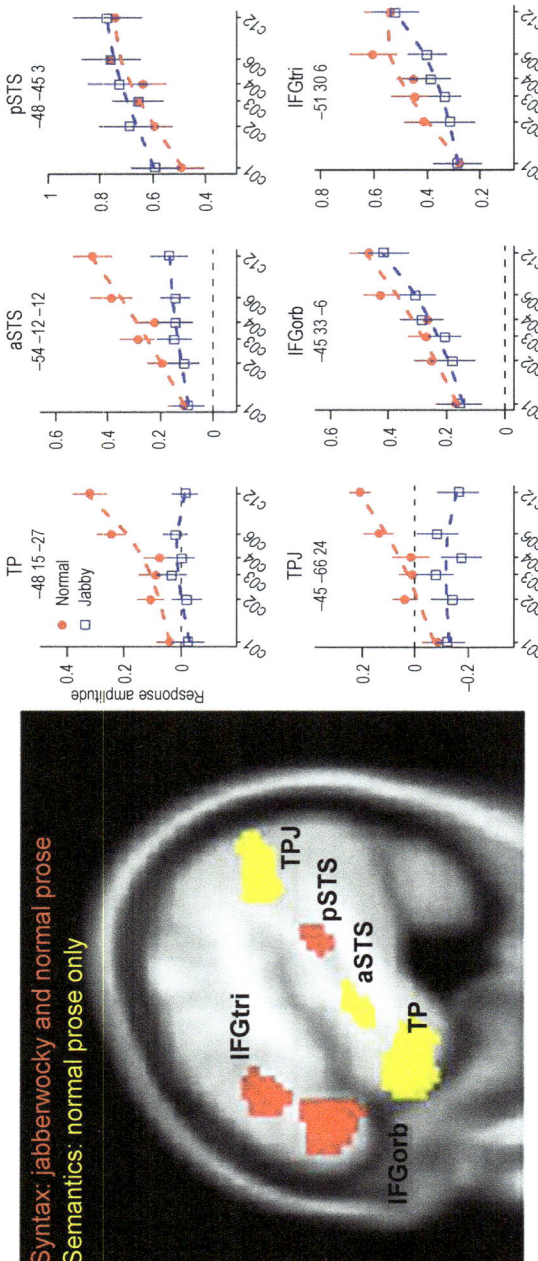

**Figure 15.2** Evidence for a core network for syntax in the human brain (from Pallier et al. 2011). The relevant language areas can be easily identified by having adult volunteers read 12-word sequences of varying length (Table 15.1) during functional MRI. Along the superior temporal sulcus and in the left inferior frontal region (Broca's area), activity increases monotonically with the complexity of the phrasal constituents. Red regions are more specifically involved in the syntax, because they continue to activate even when the phrases are formed of pseudo-words (a condition called "Jabberwocky," in honor of the eponymous poem by Lewis Carroll).

"who did what to whom" (Pattamadilok et al. 2015), or when syntactic ambiguities must be lifted. Finally, they appear to play a central role in agrammatism: patients with lesions to this core set of regions or the associated fiber tracts linking them are much more likely to develop agrammatic aphasia than other patients; even if these regions are not lesioned, they are strongly hypoactive in agrammatic patients (Tyler et al. 2011).

These studies pinpoint a set of highly specialized cortical and subcortical regions for sentence parsing and structural representation. Furthermore, they also largely validate some of the most important theoretical constructs postulated by linguists (e.g., the existence of nested constituents, syntactic movement, non-accusative verbs), since their presence or absence appears to be a key determinant of brain activity level in those regions. However, functional MRI has not, until now, revealed how these constructs are encoded by populations of neurons. Efforts are underway to clarify this point using patients with epilepsy whose brain signals can be recorded directly using epidural or intracortical electrodes.

In a recent study, we tracked the word-by-word changes in high gamma activity as patients processed each successive word in written sentences of controlled syntactic complexity. We observed that a subset of electrodes, largely confined to known cortical language areas, exhibited a systematic pattern reflecting the constituent structures of the stimulus sentences: their activity rose whenever a new word appeared, but it also decreased whenever several consecutive words or constituents could be merged into a large constituent structure. To take a simplified example: upon reading "two...sad...girls... often...cried," the activation progressively increased, but it collapsed after "girls" because those three words could be combined into a single noun phrase, the subject of the subsequent verb phrase. Importantly, the activity at any given time was proportional to the number of open nodes; that is, the items (words or multi-word constituents) that had not yet been merged together. Remarkably, in a regression that accounted for high gamma activity, similar weights were given to individual words (e.g., sad) as well as to temporary constituents (e.g., "two sad girls," "often cried"). The results were compatible with a bottom-up parsing system which applied rules of grammar in order to group words into nested constituent structures. Alternative models, for instance based solely on transition probabilities between individual words or between their grammatical categories, could be formally rejected, as they did not provide an equally good fit to the observed neurophysiological responses.

These findings provide strong evidence that constituents are the relevant units for syntactic structures. They also support the existence of the Merge operation, the most fundamental hypothesis of the recent "minimalist" approach to language. Merge is the basic tree-building operation hypothesized to take two words or constituents as input and form the binary tree whose leaves are those two constituents. Because this operation is recursive, it can represent an entire sentence as a tree with a nested set of embedded sub-trees (Figure 15.3).

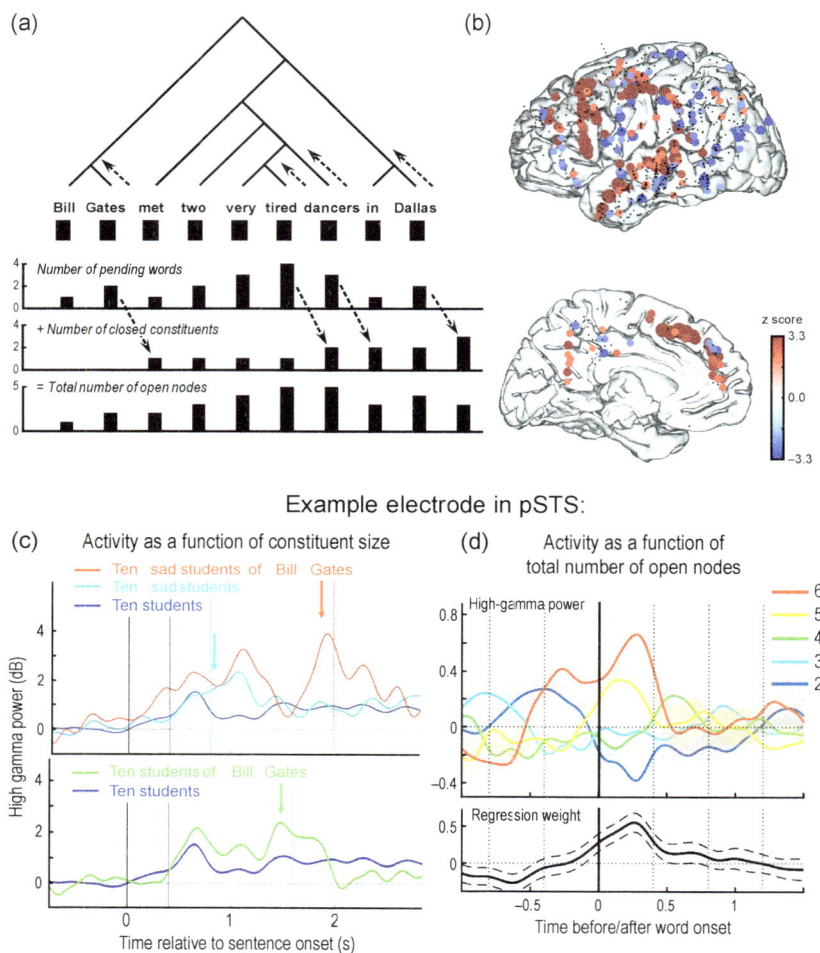

**Figure 15.3** Neurophysiological evidence for the online formation of constituent structures during sentence reading (from Nelson et al. 2017): (a) The proposed model hypothesizes that brain activity builds up in language areas for each successive word in a sentence, but collapses at phrase boundaries, when several words can be compressed into a closed constituent. (b) This prediction was upheld in the intracranial electrodes shown in red. (c) In an example electrode in left posterior superior temporal sulcus (pSTS), high gamma activity rises and falls according to constituent boundaries and (d) strongly correlates with the number of remaining open nodes shortly after the presentation of each word.

This is analogous to the multiscale compression operation which is performed, for instance, by the JPEG image compression algorithm. The human brain possesses a capacity to compress a sentence by identifying nested groups of words that act as constituents or phrases and operate together as a unit.

Indeed, this compression hypothesis, grounded on the observed rise and fall of intracranial brain signals during constituent structure building (Nelson et al. 2017), can account for two basic observations about language processing:

1. fMRI activity increases sublinearly, indeed close to logarithmically, with the number of words in a sentence (Pallier et al. 2011).
2. Memory for a sentence can be much better than a list of words of the same length and far exceed the typical working memory limit of ~7 items.

We do not yet know how syntactic trees are represented at the neural level, but it seems that the postulates of modern linguistics are largely vindicated.

## The Language of Mathematics

As noted earlier, outside of the human species, there is simply no behavioral or neural evidence so far to suggest that nonhuman animals are capable of a similar tree-based representation or compression operation. Indeed, the inferior frontal and superior/middle temporal regions involved in language processing are enormously expanded in the human brain, even relative to our closest cousins, the great apes (Smaers et al. 2017). Obviously, these regions were deeply transformed during evolution, and I speculate, as do others (Hauser et al. 2002; Fitch and Friederici 2012), that these regions may have acquired a novel representational tool: recursive tree structures. Is this property, however, unique to language? Did it evolve just once, in a single language-related brain circuit, thus placing language at the heart of human singularity? Is the emergence of language the sole factor responsible for all of our other talents for tool use, science, music, mathematics, or theory of mind?

To investigate this issue, my colleagues and I have performed a series of investigations of mathematical abilities and their brain mechanisms (Dehaene 2011). The results are very clear: the bulk of mathematics resides in a network of brain regions quite different from language areas. These regions encode nonlinguistic concepts of space, time, and number and are preempted or "recycled" during education to higher-level mathematics. Recently, for instance, we used functional MRI to investigate the brain networks for language and mathematics in professional mathematicians (Amalric and Dehaene 2016). During a brief period of reflection on mathematical statements, such as "the sine function is periodical: true or false?" mathematicians activated a bilateral dorsal network of parietal and frontal regions, as well as the bilateral lateral inferior temporal gyrus—regions that show no overlap with any language regions, as determined from a distinct localized area for written or spoken sentence processing (Figure 15.4). Indeed, the math-responsive network was also entirely different from the set of semantic regions involved in resolving similar general-knowledge verbal statements, such "London buses are red: true or false?" (Figure 15.5).

304

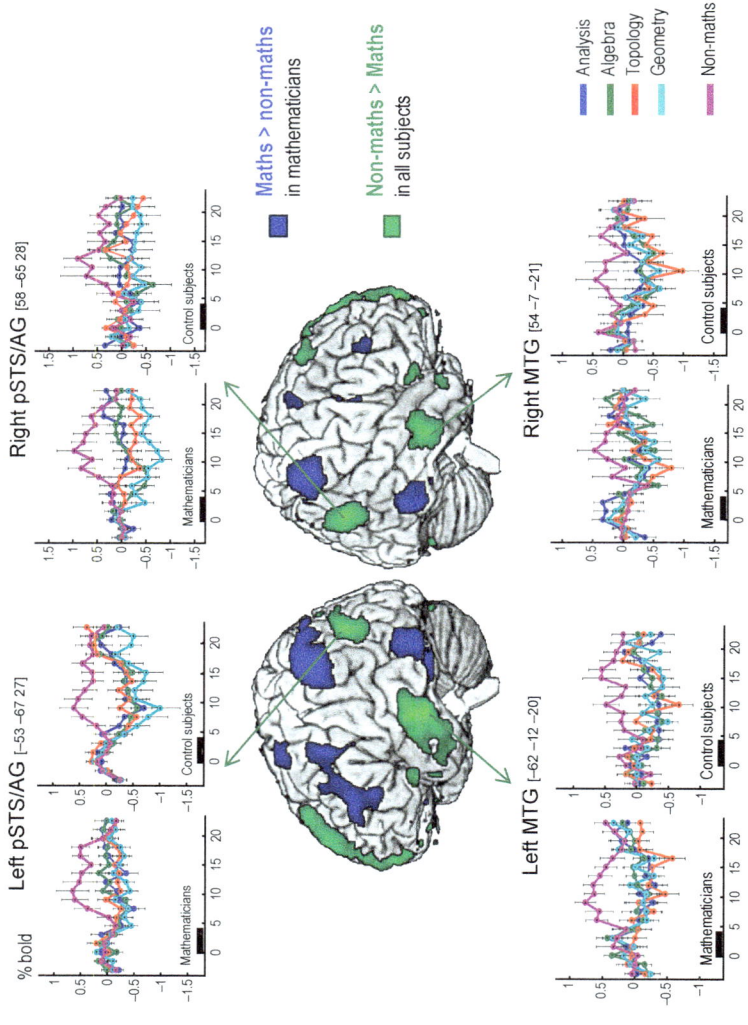

**Figure 15.4** Mathematical reflection recruits a network of brain areas distinct from language areas (from Amalric and Dehaene 2016). When mathematicians listen to mathematical statements and evaluate their veracity, the activated network (blue) differs from the one recruited when evaluating non-mathematical propositions (green).

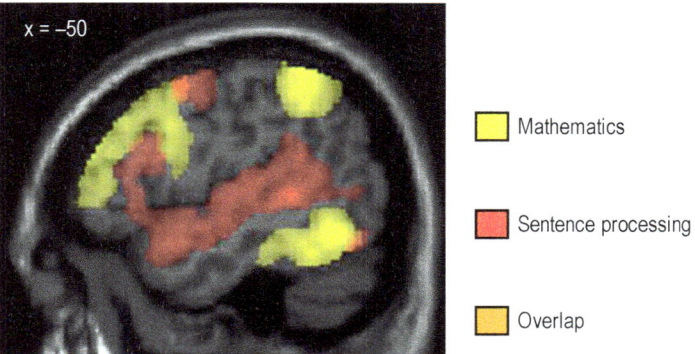

**Figure 15.5** Functional MRI demonstrates that the network active during mathematical operations (yellow) bypasses classical language areas (red).

Further experiments indicated that high-level mathematics activated the very same voxels involved in basic number recognition and arithmetic calculation, which are present in mathematicians and nonmathematicians alike, and can be traced to nonverbal representations of approximate number in infants. Thus, the results indicate a nonverbal origin for higher mathematical abilities and support the cortical recycling hypothesis, which postulates that they are grounded in prior circuits for processing number and space.

Our fMRI experiments with professional mathematicians did not explicitly study the format of representation for mathematical facts. However, there is general agreement that mathematic knowledge is organized as a language with nested tree structures. This point was already made by Galileo, who noted that "the book [of the universe] is written in the mathematical language, whose symbols are triangles, circles and other geometrical figures, without whose help it is impossible to comprehend a single word of it." Several behavioral and brain-imaging experiments have confirmed that to understand how subjects manipulate even the most basic structures of mathematics, an appeal to language-like syntax is indispensable. For instance, when processing two-digit numbers, subjects quickly and automatically assign each of the two digits to a syntactic position as decade versus unit, and weigh their quantity accordingly. When processing algebraic equations such as $3x^2 + y^3$, they automatically parse it into a nested set of constituents, thus failing to recognize that "$x^2 + y$" is, in fact, a subset of consecutive symbols in this string. In addition, when scanned with fMRI while manipulating such nested structures, they show a modulation of brain activity according to the number of nested constituents, not in language areas, but in bilateral intraparietal regions also involved in number sense (Maruyama et al. 2012).

In recent work, we investigated whether a "language of thought" comprising nested constituent structure is involved even in extremely basic tasks of spatial working memory (Amalric et al. 2017). We asked subjects to view a sequence of spatial locations on an octagon, and to remember the order in which the eight locations were presented. This is a variant of the "Corsi blocks" task for spatial working memory, except that we systematically vary the amount of geometrical structure in the stimuli. At one extreme of simplicity, the successive locations could simply run around the octagon in a serial manner, clockwise or counterclockwise. However, we also presented more complex but still regular patterns: a zigzag, two nested squares, two rectangles…all the way to a completely irregular sequence without any noticeable regularity.

Our results showed that even in this simple spatial memory setting, subjects already deployed a mental "language of thought" for geometry. Indeed, several aspects of their behavior were predicted by the amount of regularity in the sequence: with regular sequences, their memory was better; they could anticipate items that had not yet even been presented; and their eyes automatically moved to the next location, in a manner that directly reflected the underlying tree structure. Similar results were obtained with preschoolers and uneducated adults from the Amazon (the Mundurucu); for related experiments on the development of the perception of fractals, see Martins et al. (2014).

We formalized the results by proposing that during the spatial memory task, subjects search for a minimal formula for the spatial sequence within a mental "language of geometry." The idea is that subjects encode the sequence in memory according to its minimal description length (MDL), the shortest program capable of generating it. They use geometrical regularities such as repetitions and symmetries to compress the incoming sequence into a tree-like representation with a recursive structure (comprising, for instance, three nested levels of repetitions with variations). Indeed, MDL (also known as Kolmogorov complexity) was an excellent predictor of memory for spatial sequences.

We then used fMRI to identify which brain areas contribute to this "language of geometry." While subjects simply moved their eyes to the next target, sequence complexity (MDL) predicted activation in a broad set of nonlinguistic brain areas, primarily in dorsal parietal and prefrontal cortex, extending into the dorsal part of Brodmann's area 44. The latter area was the only region to survive once controls were introduced for eye movement length and for working memory per se. Thus, we believe that this dorsal part of the inferior frontal gyrus, distinct from other sectors of Broca's area involved in natural language processing, was involved in representing a spatial sequence as a rule-based tree structure.

Overall, the data suggest that Darwin was wrong when he noted that "a complex train of thought can be no more carried out without the use of words, whether spoken or silent, than a long calculation without the use of figures or algebra" (Darwin 1888). Mathematical reasoning requires nested structures of symbols, but those symbols need not be the words of natural language. Tree structures are present outside of the language domain, as in mathematics, but they recruit distinct brain networks.

I therefore suggest that the ability to represent and to manipulate recursive tree structures is not the property of a single brain circuit, but of multiple parallel systems. A similar conclusion has been reached by Peter Hagoort and his colleagues even *within* the language system (Hagoort 2013): distinct levels of phonology, syntax, and semantics may involve parallel temporofrontal circuits. In each of these domains, different sectors of the left inferior frontal gyrus may play a similar role of "unifying" the linguistic elements provided by distinct posterior brain, yet at different timescales and with different units of computation (phonemes, words, meanings).

## Conclusion: A Demodularization Hypothesis for the Emergence of Universal Tree Structures

The hypothesis that multiple circuits of the human brain exhibit a capacity for recursion raises the issue of how those circuits, during evolution, acquired this novel property. Two possibilities may be envisaged. First, it is possible that a single genetic event affected multiple brain regions. A mutation in the genes that control radial neuronal migration or cortical layer formation, for instance, may have jointly affected a large set of brain areas, possibly conferring novel computational properties to many parallel circuits at once. According to this possibility, preexisting circuits for spatial memory, auditory memory, and so forth would all have acquired, at some point in the human lineage, the ability to represent tree structures.

An alternative possibility is that tree structures evolved first within a specific brain circuit, and that this property was later extended to other circuits. The dissociation between cortical circuits for mathematical knowledge and natural language processing makes it possible for those circuits to have evolved at a different time in hominization. Is there any archeological evidence for this speculation? A sophisticated form of language is usually thought to have emerged only with *Homo sapiens*, accounting for this species' sudden cultural and geographical expansion. This recent origin for language (though speculative) contrasts sharply with the existence of very ancient proto-mathematical human-made artifacts. Tools with remarkable symmetry and geometrical regularity were already crafted as early as 2 million years ago and must therefore have been made not by *H. sapiens*, but by *H. ergaster* or archaic *H. erectus* (see Figure 15.6). Bifaces, for instance, are carefully crafted stone tools that present two orthogonal planes of symmetry and often a geometrically regular contour (e.g., an egg shape with a highly regular variation in curvature; or a pointed shape made of two lines and a circle). Around the same time, early humans also created polyhedral and spherical stone artifacts, sometimes close to a perfect sphere, suggesting that they could already conceive of regular mathematical objects before carefully sculpting them out of stone. While no other animal species creates such tools, it is still unclear whether these objects necessarily

**Figure 15.6**  Early in evolution, humans produced many "proto-mathematical" objects with regular shapes based on symmetry and geometry, as evidenced by bifacial and spheroid artifacts (top panels) dating back ca. 1.6 million yr BCE. Complex combinations of these forms became evident as early as 70,000 years ago, as illustrated by the engraved patterns of an ochre artifact (bottom), reported by Henshilwood et al. (2002).

imply a "language of geometry" capable of combining multiple concepts according to nested tree structures. A combinatorial system of geometry is only clearly attested in symbolic drawings with parallel lines and equilateral triangles on ochres from the Blombos cave in South Africa dating back at least 70,000 years (Henshilwood et al. 2009). The rudiments of the "language of geometry" must have been present at that time and, depending as to whether one accepts bifaces as evidence, perhaps as much as 1.5 million years before the emergence of spoken language.

Do these objects imply the existence of a spoken language? Can their emergence in the human species be explained by a broader capacity for mental representation based on a "language of thought?" I suggest that the capacity to form complex representation using nested tree structures first evolved in the domain of mathematical/scientific thought, using combinations of numbers and shapes, perhaps as early as 2 million years ago, and was later exapted as a broader, universal, nonmodular, generative ability for language.

There are several other possible origins for a nonverbal representation of nested tree structures prior to the emergence of language. Understanding human and animal bodies may require representing them as flexible tree structures; that is, as nested sets of "parts within parts within parts" (e.g., fingers within hands within limbs), in a recursive manner. Parsing visual scenes involves understanding the relative locations of objects, which may require nesting (e.g., a bird on the bush left of the rock). Similarly, navigating in space may require representing spaces as a nested structure of embedded places of increasing size (e.g., a spot in a room in a cave). Encoding of action plans may involve a nested tree structure of goals and subgoals. Finally, understanding social groups may involve representing the nested trees of family and dominance relationships.

For all these reasons, I find it plausible to propose a "demodularization hypothesis" for the emergence of universal tree structures in humans. The proposal is that a capacity for recursive tree-based representations emerged early on during primate evolution or hominization, within a specific module, before being extended to linguistic communication. Primitive tree structures would have emerged within specialized neural circuits (e.g., those involved in spatial or geometrical computations). Only much later would they have been exapted toward a more general role in language processing.

The peculiarity of language, indeed, is that its semantic structures span over and bring together a vast array of mental representations. Within a sentence, we can combine together any object, any action, any person, any idea, any logical connector or quantifier in countless manners. It seems that any conscious object of thought can be integrated in the language system. Indeed, this is why the criterion of verbal reportability, in humans at least, is considered by many as the primary evidence that some information is conscious (Weiskrantz 1997). The highest level of language processing is therefore nonmodular. As noted by Fodor (1983), it is not organized in a modular manner but as a "horizontal" system capable of interconnecting many, indeed virtually any, mental processors, and therefore participating in a nonmodular conscious "global neuronal workspace" (Dehaene 2014). This property, however, need not have been present in protohumans and may constitute an exaptation of a simpler, modular system.

What is exciting about this possibility is that, if true, we may reasonably hope to find precursors of the tree structures of language in nonhuman species. The neural code for linguistic structures lies beyond our reach, because obvious ethical reasons currently restrict our access to the massively parallel recordings of human neurons that would be needed to characterize it. However, by studying the representation of space or body parts in nonhuman animals, we may perhaps identify a simpler but similarly organized tree-based neural system.

# 16

# Phylogeny and Ontogeny in Human Neuroscience

Asif A. Ghazanfar

## Abstract

Currently we do not have a really good idea about what is special about the human brain and how this has led to uniquely human behaviors. To progress forward, we first need to ignore appeals to authority (e.g., Darwin) and accept that mammalian brains are not simply differently sized versions of the same thing. This does not mean that there are not commonalities between the brains of mammalians and other taxonomic groups, but that the only way to identify meaningful similarities and differences is through a comparative approach that looks at a number of different species. This chapter argues that two other lines of investigation are important in comparative neuroscience. First, investigating development will help to solve how evolution finds the same or different solutions. Homologous or convergent developmental trajectories reveal the constraints (or the lack of constraints) on how the brain reaches an adaptive solution. Second, investigating the body and its biomechanics will reveal how the structure of the body generates both constraints and advantages for the nervous system. Understanding the evolution of the human brain requires a comparative understanding of how it develops and operates in concert with the body.

## What Is Special about the Human Brain?

We know a lot about the human brain, but oddly enough, we do not know a lot about what makes it special (Preuss 2000b). As a result, there is no well-developed theory of how our brain generates behavior, unique or otherwise; that is, we lack clear ideas regarding the distinctive characteristics of the human brain that would illuminate how and why human behaviors may be similar or different from the behavior of a monkey, a rodent, or any other animal. One potential explanation for this is that we have wrong ideas about the nature of brain evolution, and perhaps this is Darwin's fault. Darwin argued that the neural differences between humans and other animals were not of kind but of degrees; that is, he argued in effect that all brains were the same, just differently sized (Darwin 1888). Thus, our ideas about brain evolution may be faulty because most neuroscience research has focused on the human brain, along with a few animal species (mice, rats, and macaque monkeys) whose brains are treated as uniformly structured, miniaturized versions of the human brain (Preuss 2000b).

To understand what it means to be human, it is essential to understand how our brains and behaviors evolved. The origins of the human brain and how it shapes (and is shaped by) our behavior are largely mysterious for two reasons. First, there are few ways to reconstruct ancestral human behaviors as no records of behavior (beyond stone tools, perhaps) have been preserved. Thus, for instance, we have no idea what Neanderthal vocalizations sounded like or whether word-like utterances were strung together syntactically by them. Second, soft tissues such as the brain and muscles do not fossilize, and the hominid fossil records of skulls and skeletons that we do have are woefully incomplete and representative of only a very few individuals. As such, an understanding of the origins of the human brain and behavior must resort to the comparative method: Understanding what it means to be human requires comparing our brains and behaviors to those of other extant species. With any behavioral and/or neuronal phenotype that two closely related species share, it can be inferred that their last common ancestor also exhibited that phenotype—the phenotype is *homologous*. Conversely, any behavioral and/or neuronal phenotype that two distantly related species share (i.e., a phenotype that is unlikely to be shared by a common ancestor) would be an instance of *convergent evolution*. In this manner, the comparative method reveals the behavioral and biological traits of extinct ancestors and helps us identify behavioral and neural homologies, species-unique capacities, and/or products of convergent evolution.

## The Few Things We Know about the Human Brain Are Special

Using the comparative approach, we have a handful of findings regarding what makes the human brain different from the brains of other animals. Our brains are much bigger than expected for our body size (Jerison 2012). Along with this overall increased size, evidence suggests that, over the course of human evolution, association areas in the neocortex grew disproportionately relative to primary sensory areas (Sherwood et al. 2008). For example, our prefrontal cortices are larger than expected when compared to other primates, although, as noted by Preuss (2000b), this finding is complicated by the different criteria that have been used to define "prefrontal" or "frontal" cortex across studies (Brodmann 1912; Blinkov and Glezer 1968; Semendeferi et al. 1997). Within cortical areas, we also see human specializations. For example, human primary visual cortex has modified magnocellular pathway components in layer 4—modifications that are absent in apes and monkeys—suggesting changes in how visual information was processed over the course of human evolution (Preuss et al. 1999).

Oddly enough, humans do not seem to possess any unique cortical areas relative to other Old World primates, at least when brain areas are defined using cytoarchitectural criteria (Zilles et al. 1995; Preuss 2000b; Petrides and Pandya 2002). For example, though humans alone possess linguistic abilities, macaques and apes share with humans homologous cortical areas located in what is defined as the language-related Broca (Deacon 1992; Petrides et al. 2005; Schenker et al. 2010) and Wernicke regions (Galaburda and Pandya 1982; Spocter et al. 2010). Similarities notwithstanding, connectivity between ventral motor and inferior parietal regions (a pathway known in humans as the arcuate fasciculus) is nevertheless quite different between humans, chimpanzees, and macaques (Rilling et al. 2008). In humans, the terminations of the arcuate fasciculus connect the superior, middle, and inferior gyri of the temporal lobe with the following frontal regions: ventral premotor cortex, pars opercularis, pars triangularis, and middle frontal gyrus. Examination of the same pathway in the chimpanzee revealed extensive frontal terminations similar to humans, but the terminations in the middle and frontal temporal gyri were much less numerous. In macaques, these temporal lobe terminations were entirely absent.

While there is evidence of human brain specializations such as those described above, we have no idea how such neural differences relate to behavior. It is typically assumed, for example, that the overall larger size of the human brain must (somehow) confer our uniquely human behaviors, such as language. Such facile ideas are easily dismissed. For instance, some microcephalic patients—whose brains are about the size of a chimpanzee's brain (i.e., one third the size of a normal human brain)—are able to produce language, within limits (Dobyns 2002; Allen 2009). How is this possible if brain size is of primary importance to language function? We should be careful not to be lulled into adopting the notion of the "cerebral rubicon" (Allen 2009)—the idea that once the size of our ancestral human brains reached some threshold, our cognitive abilities increased disproportionately giving us our uniquely human behaviors.

When comparing brains and behaviors of different species with humans, it is crucial to note that while comparative studies may reveal that rats, mice, and/or macaques share some phenotypes with humans, this does not mean that these species *are* the last common ancestors. Yes, of course, that should be obvious. What I mean is that we sometimes forget that a macaque monkey, for instance, does not by itself represent what a primate ancestral to humans would be like. This is because each extant species followed its own evolutionary trajectory for millions of years, yielding its own species-specific behavioral and neural specializations. Roughly speaking, the split between the evolutionary lineages that led to rodents versus primates occurred 90 million years ago. So, rats and mice have been evolving their own specializations in parallel with humans for that many years. Similarly, the lineages leading to marmoset monkeys (a New

World primate) and macaque monkeys versus humans separated 40 million and 25 million years ago, respectively. As a result, each primate species has a number of shared neural phenotypes with humans and other primates as well as a number of differences (Preuss 2009).

## Taking Development Seriously

We must always keep in mind that human behavior and neurobiology are products of phylogenetic as well as ontogenetic processes (Gould 1977). Evolution acts on developmental processes to produce adult phenotypes. Changing developmental trajectories is the only way to evolve phenotypic changes. By comparing developmental processes, we can thus compare more deeply the similarities or differences across species (Schneirla 1949; Deacon 1990; Finlay et al. 2001). A similar behavioral phenotype across two species may arise through different or identical developmental processes (i.e., exhibit multiple realizability), and different developmental trajectories *could* suggest entirely different neural mechanisms, even though the behaviors seem the same.

### Growing a Big Brain

Since brain size seems to be the obvious feature that distinguishes our brain from other species, let us start there. In terms of mass and neuronal numbers, the human brain appears to be a scaled-up version of a primate brain (Herculano-Houzel 2009). Although it is not exceptional in terms of its cellular composition, the human brain contains as many neuronal and nonneuronal cells as would be expected of a primate brain this size. In terms of absolute numbers of neurons, however, the human brain contains more neurons relative to other primates. It is through developmental processes that it achieves this difference. For example, the genesis of cortical neurons during primate development (including humans) is distinguished from mouse development by the appearance of a novel zone of cell proliferation known as the outer subventricular zone; this zone contains an additional population of neuronal stem cells that contribute to the increased size of primate neocortices relative to other mammals (Lui et al. 2011; Dehay et al. 2015). Another link between development and evolution, as it relates to brain size, is evident in the analysis of genes related to the nervous system. A number of studies have revealed that there are genes in the nervous system which show unique patterns of evolution in the primate lineage leading to humans (e.g., Kouprina et al. 2004; for a review, see Gilbert et al. 2005). Mutations in some of these genes lead to congenital microcephaly (Mochida and Walsh 2001; Dobyns 2002). This suggests that they may have a role in defining the most distinguishing characteristic of the human brain: its size.

## Development Timing

Often in contemporary comparative studies, the behavioral capacities of *adult* nonhuman animals are compared with those observed in human infants, with the guiding assumption being that similar behaviors would indicate the same underlying neural processes. Indeed, whole research programs devote much effort to comparing the behavioral capacities of adult monkeys or apes to those of developing humans (e.g., Egan et al. 2007; Herrmann et al. 2007). Unwittingly, I have also participated in this line of thinking. For example, my colleagues and I showed that adult Old World monkeys are able to match species-specific faces to voices (Ghazanfar and Logothetis 2003; Jordan et al. 2005); the implicit assumption in our work was that this is homologous to the ability of human infants to do so (Kuhl and Meltzoff 1982; Patterson and Werker 2003; Jordan and Brannon 2006). One possible source for thinking this way may have come from the elegant, pervasive, tenacious but ultimately incorrect idea that "ontogeny recapitulates phylogeny" (Gould 1977). Under this scenario, the human infant goes through stages of development that reflect all human ancestors (Haeckel 1866). In other words, the human infant brain must go through a stage that represents a "primitive" adult monkey brain; any uniquely human behavioral and neural capacities are "added on" after that stage (in evolutionary biology, this is referred to as "terminal addition"). We are left with the "triune" brain theory, which holds that new neural circuits get added on to a "reptilian brain" to generate mammalian behaviors (MacLean 1990). The validity of "ontogeny recapitulates phylogeny" and its attendant ideas (e.g., terminal addition, the triune brain theory) have been debunked many times as it relates to behavioral development (Medicus 1992). Consider this vivid debunking example: The coqui frog, found and heard all over Puerto Rico, skips the tadpole stage: it bypasses the "fish" stage in the "ontogeny recapitulates phylogeny" scenario during development and emerges from the egg as a fully formed but diminutive frog (Callery and Elinson 2000). Ontogeny does *not* recapitulate phylogeny.

The fundamental problem with making claims about homologous behaviors is the possibility that similar behavioral capacities may be mediated by different developmental processes. This alternative scenario is possible because brain development follows different trajectories in animals relative to humans, particularly with regard to timing. Old World monkey infants, for instance, are neurologically precocial relative to human infants. At birth, the rhesus monkey brain is heavily myelinated whereas the human brain is only moderately myelinated (Gibson 1991). Likewise, in the rhesus monkey, sensorimotor tracts are heavily myelinated by 2–3 postnatal months, whereas in humans they are not myelinated until 8–12 months of age. These facts suggest that postnatal myelination in the rhesus monkey brain is about three to four times faster than in the human brain (Gibson 1991; Malkova et al. 2006). Although the rate is different, the spatiotemporal sequence of myelination (and other indices of brain growth) along different neural pathways is the same between monkeys and humans (Clancy et al. 2000; Kingsbury and Finlay 2001) and generally coincides with the emergence and development of species-specific motor, socio-emotional, and cognitive behaviors (Antinucci 1989; Konner 1991). Finally, in terms of overall brain size at birth, Old World monkeys are among the most precocial of all mammals (Sacher and Staffeldt 1974): ~65% of their brain size is present at birth compared to only ~25% in human infants (Sacher and Staffeldt 1974; Malkova et al. 2006). Thus, human infants are born altricial relative to most other primates (Portmann 1990), due most likely to a maternal inability to provide necessary energy requirements to the developing fetus (Dunsworth et al. 2012).

This altriciality means that the human infant brain is shaped by postnatal experience to a much greater degree than other primates. Indeed, evidence suggests that even the adult human brain retains some of the plasticity that is typically exhibited in other species only in the developing brain; that is, humans evolved the capacity to maintain elevated levels of neural plasticity over a long lifetime (Bufill et al. 2011). For example, serotonergic innervation differs between adult humans, chimpanzees, and macaque monkeys, and increases in serotonin have been related to increases in plasticity. Humans and chimpanzees have a greater serotonergic innervation of the frontal cortex than macaques, suggesting selection for increased plasticity among hominoids (Raghanti et al. 2008). Moreover, genes related to synaptic plasticity increased their expression by sixfold in humans relative to chimpanzees and macaques, presumably leading to the greater synaptic density, higher synaptic turnover, and increased rates of dendritic growth found in human brains (Cáceres et al. 2003, 2006). In fact, the adult human brain has levels of gene expression that correspond to that of juvenile chimpanzees, suggesting that it evolved to retain higher levels of neural plasticity; in other words, it is "neotenic" (Gould 1977; Somel et al. 2009).

Developmental timing and species differences in the capacity to remain "plastic" may be the key to understanding the origin of at least some of our uniquely human behavioral capacities. Overall, this suggests that our brains—far more than any other species—can be molded by postnatal experience to a greater degree, not only in early life as a result of altriciality but also for longer periods even into adulthood. This is consistent with theories that link uniquely human neural structure–function relationships to the interactions between experience and neural development over the course of a lifetime (Dehaene and Cohen 2007; Anderson and Finlay 2014; Karmiloff-Smith 2015). The basic idea behind these theories is that while there are a number of evolved constraints as to how the human brain can organize itself over the course of development, experience (including cultural acquisitions such as reading and arithmetic) can exploit the human brain's plasticity to organize it in particular ways that are simply impossible in other species. The theories seem to differ in the extent to which they emphasize greater (Dehaene and Cohen 2007) or lesser (Anderson and Finlay 2014; Karmiloff-Smith 2015) degrees of evolutionarily related constraints. Dehaene and Cohen (2007) argue that the influence of experience is strongly constrained by an infant's domain-specific brain organization, whereas Karmiloff-Smith (2015) and Anderson and Finlay (2014) suggest that the organization of the infant's brain is a product of experience-dependent processes from the outset of its development, and thus is not constrained by preexisting organizational biases.

## How Species-Typical Bodies Shape Species-Typical Brains

Typically we think of the brain's (or, more accurately, the neocortex's) job as planning future actions based on its information processing of sensory signals from the environment, followed by the generation of commands for movements based on those plans. What we forget is that the body and its species-typical structure also play an important role in this process. Different parts of the body act as filters for both incoming and outgoing signals (Chiel and Beer 1997; Tytell et al. 2011).

Every part of primate anatomy—from the head to the feet, literally—exhibits species-specific specializations (Fleagle 2013). An obvious example is the outer ear. It is extremely variable in size, shape, and mobility, even among primates, and these factors determine how each species hears. In nocturnal primates (e.g., galagos), which rely primarily on hearing to catch prey, the ears are very large (relative to head size) and mobile, with mobility conferred through a special set of muscles. In humans, the ear is small and has only limited movement. How one hears is determined by the size and shape of the ears: the ridges and valleys of the outer ear filter sounds—making some parts of the sound louder and others softer—before they hit the eardrum (Batteau 1967). Critically, which parts of a given sound get louder or softer also depends on whether the sound is hitting the outer ear from above or below; thus we learn to associate acoustic differences with the vertical location of the sound source.

The importance of our bodies' physical conformation to behavior and experience is reflected in how it changes and guides the nervous system during development. Continuing with the ear example, we localize sounds well as of a very young age but since our ears are still growing and changing shape, the developing brain must recalibrate itself to account for these bodily changes (King and Moore 1991). In fact, the neural circuits of the auditory system are so dependent upon the shape of the ears to guide its function that it has to wait for the body to catch up to it. Neurophysiological recordings of auditory cortical neurons in very young ferrets listening to sounds revealed that these neurons encode spatial location poorly (Mrsic-Flogel et al. 2003). The natural assumption is that the neurons are poorly tuned because they are still developing (e.g., perhaps lacking inhibitory circuits that would sharpen tuning in the auditory cortex). In actuality, however, it is because the shape of the ears (the body) is still developing and has not yet achieved its adult-like form. Experimentally providing the same young ferrets the ears of an adult (via virtual acoustics: delivering sounds directly in the animal's ear canals after they have been filtered by a simulated adult ear) can drive quite suddenly those auditory cortical neurons to encode sound location accurately (Mrsic-Flogel et al. 2003). Thus, in this case, the developing body is guiding the sensory functions of the nervous system, not the other way around.

The developing body also shapes motor output. Human newborns are able to make well-coordinated stepping movements when held upright, but these movements disappear by ~2 months of age (Thelen et al. 1984). While it was assumed by many that the change in stepping behavior was due solely to the developing nervous system (e.g., McGraw 1945), Thelen and colleagues hypothesized that the loss of stepping behavior was due to body growth: infants' legs typically fatten up postnatally and they do not yet have the strength to move heavier legs. To test this hypothesis, Thelen et al. (1984) submerged the infants' legs in water, effectively decreasing their mass. This resulted in the reappearance of stepping behavior and thus falsified the alternative hypothesis that neural change was necessary: change in behavior was due to changes in the body. Along similar lines, it would typically (and reasonably) be presumed that changes in vocal production over the course of development are the results of learning and, thus, changes in the nervous system. In marmoset monkeys, however, computational modeling of sensory feedback from the lungs onto central pattern generators showed that the decline in the production of context-inappropriate vocalizations could simply be the result of lung growth (a change in body morphology) without any concomitant changes in central nervous system structure (Zhang and Ghazanfar 2018). The model's predictions were tested by placing the marmoset infants in a helium-oxygen environment to effectively decrease the load on the lungs, similar to submerging human infants' legs in water to decrease their effective mass (Thelen et al. 1984). This simulated a reversal in lung growth and, as predicted, resulted in a reversion back to immature vocal behavior (Zhang and Ghazanfar 2018). The developing body can create distinct behavioral changes without the need for concomitant changes in the nervous system.

Conversely, understanding the biomechanics of the body can also be enormously useful in identifying that neural changes *were* required to generate new behaviors. For example, a precision grip is a grasping behavior that is common among Old World monkeys and apes (including humans). Such a grip allows an object to be grasped with two fingers without the use of the palm (Napier and Napier 1985). Among New World monkeys, only the cebus monkey is known to use a precision grip—an example of convergent evolution (Costello and Fragaszy 1988). The question is: What is it about the cebus monkey that allows it to produce a precision grip like an Old World primate yet unlike closely related species, such as the squirrel monkey? The answer could be due to the natural selection of the necessary hand biomechanics, neural circuitry changes, or both. It turns out that the hand structure of cebus and squirrel monkeys is very similar: both have thumbs that cannot rotate around a joint in the manner that an Old World primate's thumb can. It was thus assumed that neither species could perform a precision grip (Napier and Napier 1985). The fact that the cebus monkey can indeed use a precision grip suggests that its difference with other New World monkeys (or at least the squirrel monkey) is strictly brain related.

In this particular case, the neural differences may be both general organizational differences and specific ones. For instance, cortical areas 2 and 5, associated with motor planning and coordination, are very well developed in macaques, an Old World monkey, as well as in cebus monkeys (Padberg et al. 2005). In other New World primates, however, areas 2 and 5 are either absent or poorly developed. The emergence of identical cortical areas, in this case areas 2 and 5, across species (cebus and Old World monkeys) separated by a common ancestor 40 million years ago suggests that there are rather strict developmental constraints on neocortical organization (Krubitzer and Kaas 2005; Finlay and Uchiyama 2015). A specific neural difference related to the precision grip is the organization of connections from the motor cortex to the spinal cord. Cebus monkeys have extensive corticomotoneuronal terminations in the ventral horn of the spinal cord; such connections are largely absent in squirrel monkeys (Bortoff and Strick 1993). Thus, there are important differences in corticospinal projections across species that may reflect features of the sensorimotor behavior that are characteristic of that species (Lemon and Griffiths 2005).

Similarly, the production of human speech sounds has long been thought to be due to the unique anatomy and configuration of the human vocal tract. This hypothesis states that the broad phonetic range used in modern human speech required key changes in peripheral vocal anatomy during recent human evolution. For example, no nonhuman primates have ever been trained to produce speech sounds, even in chimpanzees that have been raised from birth in human homes (Kellogg 1968). This biomechanical hypothesis was widely accepted, primarily due to a seminal study which used a computer program to explore the phonetic capability of a macaque cadaver and, by extension, other nonhuman primates (Lieberman et al. 1969). New data based on X-ray images from living macaque monkeys have challenged this hypothesis (Fitch et al. 2016). This study revealed that the basic primate vocal production apparatus is easily capable of producing five clearly distinguishable vowels (e.g., those in the English words "bit," "bet," "bat," "but," and "bought") and that the phonetic range inherent in a macaque vocal tract, based on actual observed vocal tract configurations, would not impede linguistic communication *if macaques possessed human-like neural control systems* (Fitch et al. 2016). Consistent with this idea, a recent study of baboon vocalizations shows that their acoustic range is much more similar to human vowel sounds than previously thought, despite having a different vocal biomechanical configuration (Boë et al. 2017). The inability of nonhuman primates to speak does not reflect biomechanical limitations but rather the lack of neural circuitry to enable sophisticated vocal control.

## Conclusions

At present we do not have a really good idea about what is special about the human brain and how this leads to uniquely human behaviors. To make progress in this area, we need to ignore appeals to authority (e.g., Darwin) and accept that mammalian brains are not simply differently sized versions of the same thing. These differences are not just of degree but of kind as well. Even the generic scaling laws of biological organisms would suggest that how a tiny mouse brain operates and is organized should be very different from that of a human brain (West 2017). This does not mean that there are not commonalities between them or with other species, but that the only way to identify the meaningful similarities and differences is through a comparative approach that looks at a number of different species (Preuss 2000a, b, 2009; Krubitzer and Kaas 2005), and not just in one part of the phylogenetic tree (Katz 2016b). For instance, the octopus' vertical lobe—a structure important for learning and memory—is organized in a fashion that is strikingly similar to the mammalian hippocampus (Shomrat et al. 2015). This is clearly a case of convergent evolution and suggests that there may be constraints on how a "learning and memory" structure can be assembled.

Another key to this comparative neuroscience approach is to incorporate developmental processes. We cannot assume, even in closely related species, that similarities in behavior translate to similarities in neural circuitry (Katz 2016a). Conversely, we cannot assume that distantly related species with similar behaviors exhibit those behaviors through wholly different neural mechanisms. Investigating development will help to solve how evolution finds the same or different solutions. Homologous or convergent developmental trajectories reveal the constraints (or the lack of constraints) on how the brain reaches an adaptive solution.

Finally, the body and its biomechanics are players in the evolution of behavior that are as important as the brain. The structure of the body generates constraints as well as advantages for the nervous system, and the evolution of any behavior must account for both as there is continuous feedback between the nervous system, body, and environment in any adaptive behavior (Chiel and Beer 1997). Developmental changes in the body alone can lead to radical changes in behavior. Moreover, understanding biomechanical constraints also illuminates what behavioral changes are strictly related to neurobiology differences.

# 17

# Computation and Its Neural Implementation in Human Cognition

Lucia Melloni, Elizabeth A. Buffalo,
Stanislas Dehaene, Karl J. Friston, Asif A. Ghazanfar,
Anne-Lise Giraud, Scott T. Grafton, Saskia Haegens,
Bijan Pesaran, Christopher I. Petkov,
and David Poeppel

## Abstract

How do the computations of the cerebral cortex and subcortical structures account for human perception, cognition, and affect? Answering this question requires understanding how the neurobiological and functional properties of the human brain give rise to the repertoire of human faculties and behavior, and hence, an understanding of the *neural mechanisms* that implement these functions. While research over the past decades has made substantial progress toward this end, significant challenges still lie ahead, and new opportunities open up daily as neuroscience and related fields develop and implement new theories and technologies. To (begin to) address these challenges, this chapter explores conceptual and methodological aspects inherent to the study of the neurobiology of the human mind that are at the core of the current "central paradigm" (Kuhn 1962) in neuroscience, but are often taken for granted and undergo little scrutiny. In particular, it discusses what defines or constitutes "uniquely human" mental capacities, the promises and pitfalls of using animal models to understand the human brain, whether neural solutions and computations are shared across species or repurposed for potentially uniquely human capacities, and what inspiration and information can be drawn from recent developments in artificial intelligence. Attention is given to laying out desiderata for future investigations into the human mind.

**Group photos (top left to bottom right)** Lucia Melloni, David Poeppel, Bijan Pesaran, Karl Friston, Elizabeth Buffalo, Scott Grafton, Asif Ghazanfar, Stan Dehaene, Anne-Lise Giraud, Lucia Melloni, Christopher Petkov, Saskia Haegens, Bijan Pesaran, Scott Grafton, Elizabeth Buffalo, Christopher Petkov, Stan Dehaene, Anne-Lise Giraud, Asif Ghazanfar, David Poeppel, Karl Friston

## Singular or Not? The Human Animal

To understand the structure and function of *human* brains, one must address the question of potential uniqueness or singularity, ever cognizant that the word "uniqueness" raises its own set of provocative questions. The human cortex has structural and physiological properties that underwrite neuronal activity which in turn underpins the implementation of computations that may or may not be specific to the function of the human brain. Further definition, however, is required when we ask whether the potential distinctions between human brain, computational inventory, or behavioral repertoire, compared to other species, are a matter of degree, as argued for by Darwin (1888), or systemic discontinuities (Fitch 2012; Parravicini and Pievani 2018; Ghazanfar, this volume). Discontinuity of evolution is clearly not an idea that is widely endorsed. Rather than rehearse potential uniqueness features, we might instead pursue the argument of commonalities. We contend, however, that this is just as difficult as identifying properties that are apparently found only in human cortex.

For the sake of argument, we take the position that some distinctions between humans and other species can be readily identified, and we take it as our task to understand how to account for such species-typical features. One example that points to human-specific organization concerns the suite of operations that comprise *combinatorics*. These come to light in language, mathematics, music, theory of mind, and potentially in other domains not yet understood as well in terms of formalization.

In the language domain, it has long been argued that only humans have the capacity to produce the kinds of representations characteristic of syntax (e.g., Merge). To date, there is no clear case of a nonhuman primate that has learned to combine words systematically according to a complex grammar (Terrace et al. 1979; Yang 2013). In the few cases in which nonhuman primates have been able to produce sequences adhering to a supra-regular grammar, this was only accomplished after extensive training (over 10,000 trials), whereas preschool children master this behavior in less than five trials (Wang et al. 2018). In numerical cognition (i.e., the sense of number and capacity for mathematical thinking), it is well established that monkeys and humans start life with a similar approximate number system (Dehaene 2011). The acquisition of verbal counting and a system of Arabic numerals allows human children to move from an approximate, compressive representation of numerical quantities to an exact, linear system of number (Siegler and Opfer 2003; Dehaene 2011). In the absence of formal education (e.g., in indigenous Amazon populations), the approximate system remains largely unchanged in human adults (Pica et al. 2004). Monkeys can be taught some number symbols, and this leads them to become somewhat more precise in a number comparison task (Livingstone et al. 2010). They may even begin to understand the rudiments of addition and subtraction (Livingstone et al. 2014). Yet, they continually make errors, even under highly motivating reinforcement schedules, and never perform at the level of precision and exactness attained even by young human children. Arguably, sharp distinctions between precise consecutive numbers, as those between truth and falsehood, may be unique to humans: the formation of a complex combinatorial system of arithmetic certainly is.

Accepting that there are domains of behavior that may be singular to humans leads us to ask in which ways those operations are different: Are they rooted in simpler forms of behavior that can be useful to study when trying to understand how uniquely human behaviors emerged? In which way are the structure of the cerebral cortex and its neural codes different, relative to the brains and codes that implement these more basic behaviors? Have neural codes been exapted or created *de novo* for new functions?

Addressing these questions is inextricably linked with exploring neural solutions in other species to establish convergence and divergence. The usefulness of a comparative approach to understand the human brain and its dysfunctions is clear, yet there are a number of outstanding challenges that complicate such an enterprise. These challenges must be factored into any discussion if progress is to be made in understanding which neural solutions and computations might be shared across species or repurposed for potentially uniquely human capacities.

## The Challenges of Understanding the Neurobiology of Human Cognition through Animal Models

The most straightforward approach to understanding the complexities of the human brain is to study the human brain itself, and with the emergence and refinement of a range of neuroimaging technologies, progress has been achieved over the last decades. For ethical and technological reasons, however, *direct* access to neural activity on a spatial scale, deemed necessary to unravel the neural computations that give rise to cognition and perception (i.e., for populations of individually resolvable neurons), is extremely limited in humans. Such recordings are currently only possible in patients who undergo brain surgery for tumor resection (e.g., Desmurget et al. 2009), implantation of deep-brain stimulation electrodes (e.g., Wahl et al. 2008; Cavanagh et al. 2011), or invasive epilepsy monitoring (e.g., Ding et al. 2016; Schwiedrzik et al. 2018); that is, only in brains that are affected by disease. These recordings are serendipitous in nature because they rely on recording sites that are selected for monitoring based solely on clinical considerations. Therefore, we must rely on animal models for the most part to leverage neuroscientific toolsets available for the study of the brain, including system perturbation and circuit manipulations. This necessitates making assumptions and compromises about the aspects of human cognition that can be realistically modeled by the species serving as a model.

One common implicit and often untested assumption in the field pertains to *homology*. Specifically, it is commonly assumed that differences among species are a matter of degree, such that mechanisms are conserved across phylogeny (perhaps with a scaling function). Thus, studies in other animals are thought to advance our understanding of human cognition and the human brain through a relatively straightforward translation of findings from one brain to another. Taking nonhuman animals as a sufficiently faithful model of human brain function or dysfunction (e.g., for depression or schizophrenia) is clearly useful and has provided important insight into the neurobiology of cognition. For example, although their evolutionary lineages split some 25 million years ago (Kumar and Hedges 1998), Old World nonhuman primates and humans both have a system to represent numerical quantities with striking similarities (Nieder and Dehaene 2009). At the same time, since we do not fully understand mammalian and cross-species homologies, we are often surprised at how challenging it is to translate pathophysiological mechanisms from murine animal models to primate models to humans for clinical purposes (Sena et al. 2010; van der Worp et al. 2010). This leads to questions on how readily insights in basic neuroscience obtained in another species are translatable to humans without a more complete understanding of homologies and specializations across the relevant species. Furthermore, as our understanding of human and nonhuman brains advances, more differences become apparent: the organizational principles of inter-areal connections, for example, seem to differ fundamentally between rodents and primates (Horvat et al. 2016; Gamanut et al. 2018), and there are multiple, potentially nontrivial differences in the structure of the visual system between macaque monkeys and humans (Preuss 2004). Still, commonalities also become more evident: even parts of prefrontal cortex thought to have specialized in humans, such as Broca's area, show remarkably conserved cyto- and receptor-architectonic patterns between monkeys, apes, and humans (Zilles and Amunts 2018).

The mere issue of establishing homology is complicated in and of itself (Rendall and Di Fiore 2007; Hall 2013). In the past, homology has predominantly been addressed on the level of morphological features (structural homology). Nowadays, the concept of homology has been expanded to other aspects, including behavior (phenotypical homology). However, as yet there is no consensus as to the level (neural, computational, algorithmic) needed for homologies to be useful or the criteria (e.g., genetic, developmental) required to constitute evidence for homology to distinguish it, for example, from analogy. In addition, phenotypical features can be superficially similar but of separate evolutionary origin because they have both experienced similar selective pressure. Thus, to test rigorously for homologies or identify the form of specialization, it seems pertinent to gather evidence from behavior, genetics, development, *and* neurobiological mechanisms. In addition, neurobiological evidence should be gathered across several levels (e.g., architectonic, morphological, neurophysiological).

As pointed out by Ghanzanfar (this volume), differences and commonalities between species should not only consider the brain but also the body, *and* sensorium. For example, it was widely accepted that nonhuman primates do not speak due to vocal limitations in the anatomy and the configuration of their vocal tract. Detailed X-ray studies of the vocal tract in nonhuman primates, however, led to a rejection of this hypothesis: the vocal production apparatus in primates is capable of producing five clearly distinguishable vowels (Fitch et al. 2016; Boë et al. 2017). Differences must therefore lie elsewhere and remain to be identified. This example clearly demonstrates the need to consider differences at several levels to draw firm conclusions about whether a set of behaviors is similar or different across species.

We note that similar to the problems inherent to a consideration, for example, of just the brain, there are limitations in relying on observed behavior alone to infer species uniqueness or, more importantly, non-uniqueness. One problem concerns multiple realizability. As mentioned above, superficial similarity does not guarantee shared evolutionary origins: the same behavior in two species can be due to profoundly different, underlying cognitive operations and neural mechanisms. Even if we focus solely on human behavior, there are many classic cases of multiple realizations in sequence processing (Grafton et al. 1995; Schendan et al. 2003), visual category, and procedural learning (Clower and Boussaoud 2000). For example, in visual category learning, a subject learns through feedback whether stimuli are members of one category or another. Critically, a subject can draw on at least two learning mechanisms to develop the skill (Ashby and O'Brien 2005). Depending on the literature, one mechanism is referred to as reinforcement learning, procedural learning, implicit learning, model-free learning, or information integration. The other mechanism is referred to as rule-based, explicit, or model-based learning. Both mechanisms draw on different neural circuits (roughly, dopamine/striatal mediated and cortical) during training, and final performance is dependent on different neural systems. Whether or not a given species will draw on each of these learning mechanisms is highly dependent on brain design and task complexity (Smith et al. 2012a). These distinctions cannot be formed by observing behavioral performance in isolation. The ambiguity of multiple realizations necessitates additional evidence via task decomposition, ontological approaches, or neurophysiological methods as well as approaches from artificial intelligence (AI). This leads to a reframing of the question to one that examines species-specific *means* for accomplishing a given behavior rather than one of uniqueness in any given species (Smith et al. 2012b).

Another challenge to consider is whether our experimental assays preclude us from seeing similarities between species (i.e., a Type 2 problem). Failure to detect relevant similarities may result from the fact that we are forcing experimental animals to execute tasks that are not part of their natural repertoire; if so, we would expect behavior to be optimized for their own species-typical learning apparatus (Krakauer et al. 2017). An alternative approach would be to use evolutionarily more remote species for specific traits that they may or may not share with us, rather than using monkeys (the closest available model system to the human brain) as a proxy for human behavior and cognitive functioning. In the language domain, we have tried to address certain aspects of language in apes and monkeys with limited success, specifically when it comes to higher-order combinatorics. For instance, chimpanzees using signs could not combine more than two symbols for communication (Terrace et al. 1979; Yang 2013). Importantly, there has been no systematic coevolution between monkeys and humans; monkeys have developed their own communication system, which does not possess key features of human language. There are species, however, that have coevolved with humans, that are under heavy pressure to understand human language, and which have developed speech perception skills (Andics et al. 2016). Dogs, in particular, are exposed to human language from birth and yet never acquire the ability to produce speech. Dogs do understand human orders made by specific word sequences (Bloom 2004; Kaminski et al. 2004; Pilley and Reid 2011). These types of animal models can serve to address questions about language processing in the human brain; for example, how much the speech production system contributes to speech perception and whether combinatorial properties are specifically human. Other remote species which have not coevolved with humans, yet show similar levels of encephalization as apes and humans, and have had specific pressures (unlike apes) to communicate by the auditory modality (e.g., cetaceans), may also be useful to study. Despite the absence of coevolution with humans, dolphins are able to understand word sequences (Herman and Morrel-Samuels 1995), which might mean that they also use temporally structured sequences of abstract symbols in their own cognitive functioning. These highly adapted mammals rely entirely on oral communication to maintain contact with their offspring and hence represent yet another alternative model of complex oral communication.

Another alternative is to explore repertoires of behaviors that animals exhibit in the wild as these may offer structural similarities to the computation under scrutiny. For instance, to understand whether recursivity is a feature exhibited in other animals, a potentially fruitful approximation of how animals establish hierarchies, even an atypical one (e.g., center-embedded dependencies), would be worth exploring. Work by Cheney and Seyfarth's group (e.g., Bergman et al. 2003) shows that baboons use their knowledge of social dominance hierarchies to evaluate vocal exchanges between animals with different social rank. Other work in the visual domain suggests that human infants evaluate object shape and color hierarchically (Werchan et al. 2015). Paradigms such as these provide a glimpse into combinatorial operations that respect certain hierarchical dependencies. In nonhuman animals, hierarchical dependencies may not capture the full complexity of the problem (e.g., manipulations of word classes in relation to syntactic knowledge) but they might permit us to get at the core neurobiological processes that support various aspects of these operations. The main advantage of this approach is that it builds on sets of behaviors and operations for which animals have evolved, that they naturally exhibit, thus taking us away from artificial paradigms which may constrain, and possibly misguide, the results that we get. One might argue that the approach somehow preempts the answer, as the implicit assumption is that the natural behavior is already a good approximation to the computation that we are trying to test in humans. This is not necessarily the case if one couples this approach with further tests that constrain the problem. For instance, one could test whether the neural instantiation of the specific operation under scrutiny is the same between humans and animals through neuroimaging and/or electrophysiology (Tsao et al. 2008; Yovel and Freiwald 2013; Schwiedrzik et al. 2015; Wilson et al. 2015; Kikuchi et al. 2017; Sliwa and Freiwald 2017).

Taken together, it seems worthwhile to reconsider how "uniqueness" is defined and how to determine whether discontinuities exist in evolution. Potential pitfalls in the current research program include negligence of non-brain aspects in the assessment of similarities and differences between animal models and humans as well as an overreliance on behavior alone. Finally, a potentially fruitful avenue is to expand the range of available model systems, specifically targeting animals that have evolved circumscribed capabilities that may help us understand aspects of functions, such as language, that we consider uniquely human, as well as tapping onto natural behaviors that animals exhibit in the wild as opposed to employing artificial tasks as is currently done.

Notwithstanding the challenges, countless examples have already demonstrated the usefulness of animal models in illuminating the human brain and its dysfunction: the discovery of spatial codes in rodent medial temporal structures by John O'Keefe, May-Britt Moser, and Edvard I. Moser has direct relevance on our understanding of human cognition (O'Keefe 1976; Fyhn et al. 2004); studies by Benabid and DeLong in monkeys paved the pathway for deep brain stimulation treatment in Parkinson patients (Benabid et al. 1991; Bergman et al. 1994); and the interdisciplinary work by Peter Dayan, Ray Dolan, and Wolfram Schultz in human and nonhuman primates identified the neural computations for reward-related learning with implications for addiction, gambling, and clinically impaired decision making (e.g., Schultz 2015). Below, we explore how insights gathered from animal models *can* help us understand the human brain, and its potential unique set of cognitive operations.

## Repurposing the Old: From Sequences to Combinatorics

The issue of uniqueness also arises at the level of basic neural computations. Is there a set of common neural computations across species? If so, can this set explain aspects of human cognition that are putatively unique? Or are there neural codes that are themselves uniquely human?

To begin to address these questions, we take the case of language, as we think it offers fruitful starting points for discussion in relation to computations that might be shared across species versus computations which might be an attribute of human cortex. Some of the best-supported evidence for neural codes comes from sensory domains. However, insight derived from sensory modalities (while relevant) does not currently offer an explanation (or even a satisfying clarification) of the problem of linguistic representation and computation. Specific neural computation "for language" must be examined at a level of abstraction that goes beyond sensory and motor coding, because linguistic representation and computation can stem from auditory (speech), visual (sign language, text), and somatosensory information (Braille). In all these cases, the sensory modalities provide interface information to linguistic computations, which have specific, abstract properties. Figure 17.1 schematizes the nature of the problem: sequential information is processed by the sensorimotor interfaces (the input and output strings), but the system must be able to traffic in structured representations that permit computations over representations that go well beyond linear strings.

To be sure, there are other aspects of perception and cognition that may capitalize on some of the operations on hierarchies that we discuss here, notably action planning and movement, spatial navigation, and visual scene analysis. However, we focus on language because cross-species work is particularly complicated. Aspects of language that merit explanation include the property of discrete infinity, (nested) hierarchy, structure dependence, constituency, and the organization of the mental lexicon. To operationalize these key concepts

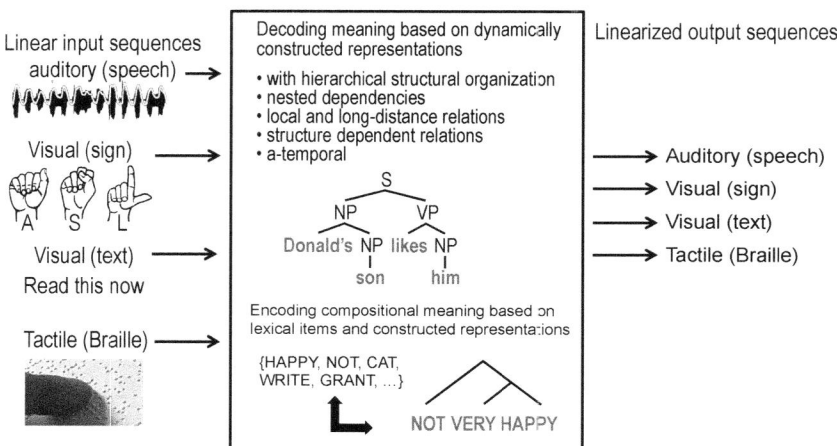

**Figure 17.1** Neural computation "for language" requires abstract coding schemes. Sequential and linear information processed by the sensorimotor interfaces, the input and output strings, must be transformed into structured, hierarchical representations, allowing for computations of representations that go well beyond linear strings (e.g., recursivity).

and provide examples to work through the challenges, we characterize the issues as follows:

1.  There are terminal (basic) elements, roughly words (e.g., the word "lock"). Even within words, there is compositional structure: strings (morphemes) can be concatenated to create different words, which, in turn, depending on the specific form of concatenation, result in different meanings due to structural ambiguity (e.g., "un-lockable" versus "unlock-able").

2.  Words can be combined (e.g., "red boat" or "bad example") such that the resultant item inherits the properties of just one of the elements (a subroutine sometimes called labeling). That is, a "bad example" is a constituent, and this constituent bears the label "type of example" but not "type of bad."

3.  The concatenation of words is based on structural (syntactic) and meaning-based (semantic) constraints: "new plans give hope" is parsed into the constituents [new plans] [give hope].

4.  Recursivity: In the phrase "fast red boat," "red" modifies "boat" and "fast" modifies "red boat." This is an example of the recursive application of a rule, in which first the modifier A ("red") is applied to object B ("boat") to yield a new object, B′: $[B' \wedge A B]$.

We advance the hypothesis that understanding such a generative system requires breaking the problem into formal operations (computations) that comprise the system. Those formal operations might map onto specific neural responses that may be amenable to neural coding research. As a starting point, we take the taxonomy of sequential operations illustrated by Stanislas Dehaene (see Figure 15.1, this volume): transitions and timing knowledge, chunking, ordinal knowledge, algebraic patterns, and nested tree structures.

Evidence shows that the first four levels represent sequence construction operations and sequence representations that are shared with other animals; in contrast, current evidence points to the conjecture that combinatorics which yield hierarchical nested tree structures might be a human singularity (for further discussion, see Dehaene, this volume). According to this hypothesis, what makes human thought complex (and, perhaps, of a certain kind) is that symbols (in language, mathematics, and perhaps music, action planning/motor control, and visual scene understanding) are not just strung together into a sequence (string-of-beads hypothesis); they are mentally represented as hierarchically structured trees (Calder-mobile hypothesis), thus offering combinatorial and interpretive diversity. This raises the question whether existing neural codes[1] might be used to represent such sequences and, if so, whether they are

---

[1] We explicitly do not concentrate on the question of what the neural code might mean: whether neurons really represent an external property or not, and what the neural code could be (e.g., rate code, state-space trajectories).

implemented similarly across different species. If the processing of tree structures that impose structure *dependence* is *not* shared across animals, it is still pertinent to determine whether neural codes have been exapted in evolution or created *de novo* for such purpose.

To get traction on the problem, we suggest reducing it to the establishment and representation of *relations*. The argumentation strategy we pursue here is to turn to the implementational level of description, in the sense of Marr (1982), looking to neurobiological properties that may motivate research on relations as they might be described at the algorithmic and computational levels. The desideratum for the language case is that the formal relation can express hierarchical nesting. The need to represent relations is also present for many other domains, including visual scene analysis, action planning, and motor control.

In vision, in particular for scene perception (and, more challengingly, scene understanding), relations need to be established: from exploring with our eyes to forming a scene representation to using it, say, to grasp an object. This includes the representation of spatial and topological relations: a pair of glasses *on* top of the table to the *left* of the cup. In even the simplest kinds of motor action, such as picking up the glasses on the table, a relation is formed through the intimate timing between the velocity of the arm as it reaches toward the object and the opening of the fingers which achieve a maximum aperture at a highly reproducible moment before enclosing the glasses (Paulignan et al. 1990). At first glance, this hierarchical relation is dominated by the kind of grasp the object requires, which regulates the arm speed. If, however, the cup is in the way of the glasses, then the hierarchical relation flips, with the limb trajectory dominating the timing of subsequent events. As simple and intuitive as these examples might seem, we do not fully understand how this is accomplished.

To further illustrate how considering the representation of relations in other domains can inform research on language, we explore here the prima facie similarity with spatial navigation in more detail. In spatial navigation, relations need to be established from moving around to forming a map to using it to navigate. Navigation requires chaining operations, or a series of sequences to find a path from A to B. Considering that superficial similarity, we turn to neural codes observed in the hippocampus. As rats run through a maze or forage in an open field, place cells in the hippocampus create a representation of the animal's environment (O'Keefe 1976), and ensembles of place cells fire in ordinal sequences that reflect the rat's ongoing experience (Dragoi and Buzsáki 2006). The hippocampal local field potential exhibits a prominent theta band (6–10 Hz) oscillation as the rat explores and actively processes incoming information. Importantly, within a theta cycle, the temporal offset between sequentially firing neurons is tightly correlated with the distance between each neuron's place field (Geisler et al. 2007), and these "theta sequences" incrementally advance across progressive theta cycles (Dragoi and Buzsáki 2006). These features of sequences within theta cycles allow the population of hippocampal neurons to link temporally disparate events with a sequentially active ensemble. That is, within each theta cycle, the sequential firing of place cells provides a representation of the rat's previous, current, and future location, thus providing a way to tile the gaps between experienced events (Buzsáki and Llinas 2017).

It is conceivable that these aspects of hippocampal activity map on to at least the first three stages of the sequences described by Dehaene et al. (2015), with the sequential firing of place cells reflecting the transition and timing of sensory experiences as the rat runs through a maze (Stage 1), and the repetition of sequences within a theta cycle reflecting both chunking (Stage 2) and ordinal knowledge (Stage 3). Interestingly, hippocampal sequences can be replayed forward as well as backward, and there is some evidence that the forward sweeps may reflect prospective coding whereas backward sweeps reflect retrospective coding (Diba and Buzsáki 2007), both of which are thought to support mechanisms of episodic memory. Theta sequences reflect ordinal knowledge (in conjunction with specific item features) because the timescale of the sequential replay of activity within a theta cycle is independent of the timescale of experienced events; the sequence, for instance, maintains only the relative temporal order of experienced places.

The codes described in the hippocampus could serve as pointers for further research in cortex. We note that the role of the hippocampus in language, music, or mathematics is poorly understood. Recent studies, however, have suggested that the hippocampus might play a role in language (Piai et al. 2016) as well as in statistical learning (Schapiro et al. 2014, 2017). The latter is thought to be a mechanism guiding the discoveries of words in continuous speech (Saffran and Kirkham 2018). Here, the question is how babies discover, parse, segment, and string units for further processing in the continuous acoustic stream. Neural responses observed in the hippocampus (e.g., theta phase precession, replay, pattern completion, and/or pattern separation) may provide starting points to understand how primitive sequential operations are implemented in the human brain. Still, although these neural processes have advanced our understanding of sequence representation, they only represent sequences as temporal successions of events (i.e., the string-of-beads hypothesis described above) and not as fully abstract, a-temporal, hierarchical representations, such as those observed in language and mathematics (i.e., the Calder-mobile hypothesis). Hence, further refinement of our understanding of those codes is needed to explain hierarchical relations.

Other potential mechanisms for the implementation of sequence processing can be considered. Below, we briefly outline a number of possible candidates for the implementation of different sequential operations.

## Chunking Operations

Here, two mechanisms are germane. The first is implemented through anatomical convergence. Feature-sensitive nodes A and B converge on conjunction-specific node C, which after appropriate adjustment of synaptic gain and thresholds will respond selectively to conjunctions of feature A and B. This strategy is commonly used in hierarchically structured feedforward networks, including deep convolutional networks. The wiring can either be genetically determined (e.g., motion detectors) or specified by experience, using an associative Hebbian mechanism of synaptic plasticity. This results in the implementation of a conjunction-specific node that reflects frequently occurring statistical contingencies of features. While this mechanism is extremely robust to establish learned and stable patterns of relations, an open question is whether and how such a mechanism could be extended to *online* sequence construction; that is, constructing a chunk or "type" based on sparse data. Another question is whether a convergence site "C" is even needed. It could be that the convergence sites merely hold a combinatorial code linking information available in representations A and B, wherever they may be held (Damasio 1989).

The second mechanism is the formation of Hebbian assemblies, via recurrent activity, consisting of reciprocally coupled nodes that respond to different features. Again, through Hebbian learning, coupling connections among the nodes of the future assembly will be strengthened such that those nodes will be coupled preferentially to represent frequently co-occurring sets of features. As a result, if the corresponding set of features is present, the assembly representing the conjunction will be ignited. This strategy requires recurrence, a typical property for cortex, and because recurrency generates additional dynamics it can also be used to associate (chunk) more complex features, such as particular sequences. Thereby, hippocampal recurrent activity could help to bind items A and B and make their sensory cortical neural representations sparser and more similar (Messinger et al. 2001).

**Establishment of Sequence Order**

Recurrent networks reverberate and are self-active as well as generative. They have fading memory (stimuli leave long-lasting traces in reverberating activity) and can therefore integrate (chunk) responses evoked by sequentially presented stimuli. If a node in such a network is activated, it produces "songs," (i.e., sequences of successively activated nodes), whereby the sequence depends on the functional architecture of the reciprocal coupling connections. As their weight distribution reflects statistical contingencies of previous input (experience), the "songs" correspond to the encoding of learned sequences. These can then be conjoined through the merging of different assemblies, using the same mechanisms of ensemble formation. Sequence-order judgments appear to depend on prefrontal cortex (Petrides 1995). Sequence-order neurons are seen in medial premotor cortex (Merchant et al. 2013) and the hippocampus during spatial exploration and memory tasks (Kraus et al. 2013; Aronov et al. 2017).

## Constructing Brackets

Brackets can be formed by different mechanisms. One possible mechanism is *attention* that selects one (out of many possible chunks) subset and then, through competition and winner-takes-all mechanisms, selects particular conjunctions over others. If these conjunctions contain a sequence, the network would automatically expect (produce) the sequence with the highest transition probabilities between successive states. This could, in part, address representing ordinal sequencing. However, this hypothesis is only pertinent when transition probability is a critical attribute of a sequence, and this is not always true, as in language.

An alternative solution is to distribute the bracketing task over different areas. We call this *anatomical factorization*, whereby the mapping rules for convergence determine the grain of the representations. Through convergence, the bracket around a chunk would now be a whole object. In this case, to get back to the relation of the components within the bracket, one would need to read out the nodes within the bracket, for instance through (top-down) feedback. To select the correct nodes at the lower level, some mechanism needs to be implemented that relates them to the big chunk in the bracket. This could be done by synchronizing ensembles across levels (as could be necessary, e.g., for mental imagery or silent speech).

Another possibility to form brackets is to use *time as coding space* and establish cross-frequency coupling. One could conceive slow rhythms as the bracket around a big chunk and the components to be represented by ensembles oscillating at higher frequencies. If consistent phase relations are assured between the slow and fast oscillations, it would be possible to decode which components belong to which of the bigger chunks within the bracket. Since recurrent networks can cope with the representation of sequences, the problem of ordinal coding can, in principle, be solved. Likewise, by having coupling across several different frequencies, nested relations can be specified. Such approaches to cross-frequency coupling (Hyafil et al. 2015), exemplified by the Lisman model (Lisman and Idiart 1995; Lisman 2005), are undoubtedly interesting, and perhaps even relevant, for some aspects of perception and cognition (e.g., Giraud and Poeppel 2012; Heusser et al. 2016). For this proposal, evidence is, however, scarce and contested.

## Summary

The problem of language, along with the set of sequential operations that it entails, illustrates how investigating the way in which (sequential) relations are encoded in cortex and other domains may inform us about a uniquely human cognitive function. Successfully employing this strategy involves asking to what extent underlying operations are shared across domains (e.g., with spatial navigation), and delineating which cognitive operations and neural codes are indeed unique to humans and which ones might be shared across species.

## The Formal Basis of Generative Models for Language

In the previous section, we established some key aspects of the way in which language is encoded, and the unique compositional architecture these codes must possess. Here, we consider the form of *computations*, with a special focus on the cognitive operations entailed by language processing. Again, in the spirit of Marr (1982), it seems imperative to consider this level of description to guide our search for the neural implementation of cognitive functions. We will entertain the concept of a normative perspective (i.e., framing all computation under the overarching imperative to optimize, in some sense, the encoding of beliefs) as an alternative framework to help understand not only the computations subserving language but also any form of computation. This optimization is defined in terms of an objective function that has various interpretations in terms of information theory (i.e., self-information), self-organization, and self-evidencing.

Crucially, this optimization can be cast in terms of inference (namely, optimizing beliefs that are parameterized or encoded by neuronal quantities) and brings about the concept of a generative model, which is necessary to define the objective function. Still, the question remains: What sorts of generative models might be used by the brain to parse, synthesize, and generate language? We will focus on the distinction between generative models of continuous and discrete states and how they lead to different forms of optimization and message passing. This is an important distinction, as the type of generative models apt for language processing rests upon discrete states of the world, equipped with symbolic or semantic labels. Finally, we turn to the implications for cortical computation in terms of the message passing required for the ordinal and nested structures above. The structure of these models will turn out to be a key attribute that defines the challenges for understanding—and modeling—language processing in the brain. Aspects of this structure include the difficult problem of structure learning, the accommodation of structural dependency in linguistics, and the way we carve nature at the joints—via a nested factorization of the latent causes of language (and, in more general terms, any narrative that underlies our active engagement with the world "out there").

### Encoding, Decoding, and the Neuronal Code

Modern versions of encoding (i.e., the mapping of a given stimulus onto a neural response) are associated with the notion of unconscious inference. On this view, there is a distinction between states of the world "out there" and the

sensory consequences of those states that are registered by sensory epithelia. Neuronal activity is taken to parameterize probabilistic beliefs over states of the world that are inferred through sensory impressions. If the neuronal code encodes beliefs about hidden or observed states of the world, what is computation?

One can develop a formal definition of neuronal computation in terms of an objective function that represents a lower bound on the evidence for a model and ensures the neuronal code is optimized with respect to states of affairs in the world. In Figure 14.2 (this volume), Karl Friston shows that this model is a probability distribution over the hidden states (i.e., causes) that generate sensory samples (i.e., consequences). Based on this model, one can compute the bound and specify neuronal dynamics in terms of a gradient descent of the ensuing objective function. This has a number of fundamental implications. First, it means that it should be possible, in principle, to specify neuronal dynamics in terms of self-evidencing (active inference) under some generative model. This implies that phenotyping a particular brain or creature boils down to specifying the generative model being used to navigate in their world. Under this view, emphasis is placed on understanding the form and nature of the generative model. Everything else should ideally follow from this model.

The second key observation is that any (probabilistic) generative model can be expressed as a Bayesian graph with nodes and edges. This is a simple construct that associates all hidden states (and sensations and actions) with nodes of a network, where the connections or edges denote conditional dependencies. The key point is that the form of the generative model defines, unambiguously, the requisite message passing among the nodes that constitute the gradient descent or neuronal dynamics. In other words, knowing the form of the generative mode means that we immediately know the computational or functional architecture of the brain, under the assumption that it is optimizing its beliefs about its world.

Thus, from basic principles we can arrive at a formal (if abstract) description of a brain that must be describable in terms of message passing among the nodes of a graph or network. Furthermore, in virtue of the causal structure in the world, the edges or connections will have a particular sparsity form (e.g., hierarchical structure). This means that we would expect to see self-evidencing computations play out on a relatively sparse (e.g., hierarchical) neuronal network. This resonates with the known neuroanatomy and neurophysiology of brains (Felleman and Van Essen 1991), which have this peculiar graphical structure, and could be contrasted with other organs such as the liver or blood. So what attributes might the generative model have?

At this point, we may consider the distinction between generative models of *discrete* and *continuous* states. This is a simple yet critical distinction based on the event space or support of the probability distributions (or densities). We can have states of the world that are categorical. In other words, we can be in one room or another room, but not both rooms at once. Our beliefs then would be a categorical *distribution* over a finite set of states. Alternatively, states can be continuous. The analogous probability *density* over some continuous state (e.g., the luminance contrast at a particular point in the visual field) yields variables ranging from 0 to infinity. The corresponding probability density may have some (e.g., lognormal) distribution depending on the uncertainty about the actual level of luminance contrast. In terms of the neuronal code, this means that if we adopt discrete state space models, one might associate neuronal (population) activity with the probability of being in a particular state at any particular time. Conversely, in continuous state space models, neuronal activity may encode the expectation or average of the probability distribution and scale with the intensity or level of the continuous hidden state (e.g., luminance contrast). In both cases, the gradient descent to understand neuronal dynamics applies. However, the nature of these dynamics depends sensitively on whether our generative model is over discrete or continuous states.

*Continuous state space models* would call upon some form of Bayesian filtering to implement gradient descent when sensory input fluctuates over time. Common examples of these belief updating schemes include predictive coding, Kalman-Bucy filtering, particle filtering, unscented filtering, and their hierarchical (and nonlinear) variants. Analogous schemes for *discrete state space models* include belief propagation and variational message passing. Variational message passing corresponds to the solutions to the neuronal dynamics that explicitly optimize variational free energy. Crucially, these are not filtering or predictive coding schemes; although they share many computational aspects.

Perhaps the most important aspect is that all belief updating schemes entail reciprocal message passing over the edges of the Bayesian graph. This has a fundamental implication for cortical computation. It means that reciprocal neuronal connectivity must (either directly or indirectly) be in play, if the brain engages in Bayesian belief updating. This is a strong constraint on neuronal dynamics, which mandates recurrent connectivity and reciprocal message passing between any neurons or neuronal populations that constitute sufficient statistics of conditional or posterior beliefs. A popular example here can be found in predictive coding: in this particular message passing scheme, prediction errors are passed forward (e.g., in cortical hierarchies), while descending predictions are reciprocated in the other direction. Exactly the same reciprocal or recurrent exchange is found in belief propagation and variational message passing.

Under this view the fundamentals of computational architectures in the brain rest upon the following:

- Adopt a *constructivist* perspective on neuronal computations so that neuronal activity encodes beliefs about something; namely, states of the world that generate sensations.

- Specify the *structure of a generative model*, which specifies the graphical form; that is, network architecture of reciprocal message passing or Bayesian belief updating.
- Formulate this *message passing* in terms of differential equations to specify the precise architecture and form of neuronal dynamics.

How would this recipe for understanding cortical computation, in terms of belief updating, play out in the context of higher cognitive functions such as language?

## Deep Generative Models for Language

We have tried to reduce the problem of understanding cortical computation to understanding the structure of generative models that explain how sensations are caused. We have posited that particular structures of generative models are necessary for language and detailed five structural aspects implicit to the generation of language, ranging from the ability to generate transitions among discrete states to the hierarchal nesting or parsing of tree structures. Furthermore, language has to deal with structural dependency, which could involve ordinal transposition and a particular form of parsing best understood in terms of hierarchical trees and their attendant decompositions.

From this arise two key implications for generative models that underlie neuronal dynamics in language processing. First, we are dealing with discrete state space models (e.g., hidden Markov models, Markov decision processes, hierarchal Dirichlet process models and their extensions), which immediately tells us that representations (i.e., expectations) about states of the world in the future (and past) are needed to support sequential transitions. Second, we need a hierarchical structure that allows for chunking and chaining within a particular (ordinal) temporal frame of reference. The requisite of deep temporal models brings with it some interesting functionality, along with some deep problems.

The capacity to represent sequences over time means that variational message passing builds, in effect, beliefs about the future (and the past). For example, reading the first word in a sentence already sets up a hypothesis space over all subsequent words, in virtue of message passing forward in time (and back again). This reciprocal message passing has, in part, a forward and backward aspect, in the sense that there is an explicit representation of the future. From a computational or cognitive perspective, it means that we have the capacity to hold in mind possible outcomes that are plausible given the sequential evidence sampled so far. Perhaps more interestingly, it also means that we can update our beliefs about initial experiences in the past. This provides an important opportunity to test hypotheses generated under these sorts of generative models, using prospective and retrospective inference, and to respond to unexpected evidence (i.e., violations at different levels of abstraction).

Simulating the variational message passing (under deep temporal models of this kind) exposes many issues related to the neurophysiological correlates of language processing. Perhaps the most interesting is the synchronous and asynchronous updating implied by discrete models. This necessarily involves the separation of timescales: a *fast* timescale for the optimization process itself and a *slower* timescale for sampling each new discrete sensory sample (e.g., through saccadic eye movements while reading or articulation of phonemes while talking). This discrete sampling of the world may progress at a theta frequency, while fast updating probably occurs with time constants associated with faster, for example, gamma frequencies (Melloni et al. 2009; Wang 2010; Giraud and Poeppel 2012). Furthermore, the hierarchal structure of these models necessarily entails a separation of temporal scales at different levels (e.g., delay period activity in the prefrontal cortex, in relation to fast dynamics lower in the auditory system). Empirically, this suggests a nesting of faster frequencies in slower frequencies, when belief updating is observed electrophysiologically with, necessarily, cross-frequency coupling and nested oscillations.

What are the special problems that accompany this sort of deep temporal model? These relate to the very structure or carving of this (linguistic) nature at its joints. The relational aspect of linguistic constructs (i.e., hidden causes or states) introduces a special problem that is probably best conceived of as a combinatorial explosion (e.g., discrete infinity). So what does this mean for the structure of the generative model?

One can finesse the complexity cost implicit in a combinatorial explosion by factorizing the generative model into conditionally independent causes and then binding these causes together, through convergent connectivity, to explain the particular pattern of sensory inputs at hand. A detour to vision may serve to clarify this point: an efficient way to address combinations of features (e.g., what and where) of an object in the visual field (i.e., low complexity encoding) would be to represent the nature (*what*) and location (*where*) attributes separately, then use the interaction or conjunction of these posterior expectations to predict the sensory input that would be sampled at any particular location in the visual field (Friston and Buzsáki 2016). This interaction between (roughly) orthogonal representational factors (i.e., a computational binding) entails second-order or multiplicative interactions between messages from the *what* and *where* parts of the generative model (e.g., the *what* and *where* pathways in the brain). In turn, this necessitates some form of modulatory or nonlinear optimization of synaptic efficacy of the sort associated with attentional selection mediated through dynamical mechanisms, as in communication through coherence (Fries 2005) or other neuromodulatory mechanisms. Thus, one important constraint of this view is that factorization implies multiplicative interactions when factorized features are combined.

What sort of factorization is in play in language? As indicated above, this factorization may be extremely complicated and must be hierarchically nested.

It might appear that certain syntactical structures are separated, by virtue of being associated with hidden factors from the actual semantic or phonological content. Furthermore, one has to consider the ordinal structure-dependent aspects of language. This speaks to the interesting possibility that we represent order or ordinal attributes in the same way that we represent locations in space. Put simply, there may be dedicated streams for encoding *when* that are combined with other factors encoding *what* at each level of hierarchal construction (Auksztulewicz et al. 2018). This is not unrelated to the notion of ordinal pointers involving convergent interactions between cortical language areas and the hippocampus (Friston and Buzsáki 2016).

At this point, one could start to speculate about the nested hierarchal and factorial form of generative models that would be fit for purpose in generating language. Perhaps, the most difficult problem in understanding the cortical computations that underlie language processing might not be in the details of the message passing or the biophysical implementation of the algorithms, but in understanding the very structure of the generative model and how this is acquired by a brain. This is known as structure learning or Bayesian model selection. These considerations emphasize the basic structure of generative models that possess the right sort of symmetry (i.e., invariances in conditional independencies) implicit in the right sort of carving or factorization. At present, simple symmetries have proven very effective in machine learning. Perhaps the most celebrated example of this is the weight sharing implicit in deep convolutional neuronal networks. This employs a simple factorization or invariance assumption that the weights of lateral connections at each level of the deep network are conditionally independent of their translational position. For the above arguments, we may be pressed to look for much more sophisticated symmetries that underlie our ability to parse and decompose invariance, when generating narratives in a world populated by creatures like us (who talk a lot). Let us now take a closer look at this issue from the perspective of neuroscience and AI.

## Every Happy Marriage Has Its Ups and Downs: Neuroscience and AI

Apart from studying the brains of humans and other animals, a complementary inroad into understanding the computations that underpin human cognition may lie *in silico*. Comparatively recent advances in computer algorithms and hardware have led to a massive increase in the capacity of computers to fulfill tasks at human or even superhuman performance levels in domains such as the recognition of images, letters, or speech (LeCun et al. 2015) to playing computer games (Mnih et al. 2015) or even Go (Silver et al. 2016). At the forefront of these advances are deep neural networks (DNN); that is, hierarchical stacks of convolutional neural networks. Because their performance in certain domains is so close to or even better than that of humans and can be built at will, DNNs seem to be promising tools to understand something more about human capacities. Especially in the domain of visual object recognition, DNNs now readily perform at the same level as humans or monkeys; interestingly, the properties of units in the higher layers of these networks show similar properties as neurons at the highest stages of object processing in monkey inferotemporal cortex (Yamins et al. 2014). This example suggests that there could be a fruitful, bidirectional exchange between AI and neurobiology to improve algorithms and to advance our understanding of the brain.

However, there are inherent differences between DNNs and human brains/behavior that are worth considering. For example, in terms of behavior: (a) deep learning algorithms need massive amounts of labeled training data (and regularization, etc.), whereas humans learn quickly and often in an unsupervised fashion; (b) DNNs typically learn specific tasks (e.g., recognizing cats) and generalize poorly to other, even similar tasks; (c) they do not have "common sense" and the domain of transfer learning is only emerging (Davis and Marcus 2015). In terms of biology, DNNs have many, sometimes hundreds of layers—more than the brain (e.g., the visual system is thought to consist of about 30 areas)—making it seem impossible to fit this number of layers into a skull. Furthermore, DNNs often rely on processing in massive data centers; running them on a small, autonomous device such as a phone immediately drains the battery. This serves to illustrate that brains and algorithms have evolved under different environmental pressures. Other long-standing arguments are that the backpropagation algorithms used to train DNNs are deemed biologically implausible (Crick 1989), although backpropagating action potentials and backward spread of plasticity have since been discovered (Fitzsimonds et al. 1997; Tao et al. 2000; Du et al. 2009). In addition, DNNs usually do not involve recurrent and long-range connections, which are characteristic of the cortex. Overall, this suggests that much remains to be accomplished before we can build machines that think and learn like humans (Lake et al. 2017). Still, comparing commonalities and differences between *in vivo* and *in silico* approaches to intelligence may be similarly fruitful as comparisons between species. What have new AI tools contributed to theories about human brain function? What do we learn about the brain by applying machine learning to neural data?

Much of AI today builds on neural networks models that were developed in the 1980s to understand features of human cognition based on aspects of what was known at the time about neural computations (e.g., multilayer structure, proximal connectivity). Hence, it may not be too surprising to find similarities between well-studied aspects of the neural processing and the way DNNs process data. One could thus argue that there is a fundamental circularity that we, neuroscientists, should remain aware of as we use these tools. These models were originally supposed to help us generate new testable hypotheses. Since these models were able to solve some (simple) computational operations, they have been reused for engineering purposes and refined to optimize machine performance, no longer considering neurophysiological plausibility. This pragmatic and laudable use of neuronal networks or other brain-like algorithms may become problematic when we start applying them to "model" or "analyze" the brain. The constraints these models impose and the hypotheses they imply become implicit, and ignoring them might profoundly mislead us. For example, the apparent solution of using multiple layers to address data complexity may not be the (only) solution the brain uses to solve the same problem. Alternatively, using multiple layers may, at a certain level of abstraction, simply be a different formulation of the same solution the brain is also believed to use (Liao and Poggio 2016).

Although they are increasingly being used as analysis tools in neuroscience, what do machine learning techniques actually tell us about the brain? Research on animals and in humans with brain lesions has taught us that we need to know what is necessary as well as what is sufficient (Bouton et al. 2018). In animals, we get this information by considering both loss and gain in function studies. As this cannot be done in humans, computational approaches may help to address this question, building networks/models and perturbing them to try to find out what is necessary and sufficient for people to do a task. As discussed, the specific case of language is particularly challenging. To test whether some of the predictions regarding how humans process language are plausible, therefore, we have to build a model and then show that the predictions it makes, regarding neural responses to novel stimuli, are accurate. If successful, the model will have captured some of what actually happens in the human brain. In recent work, Pereira et al. (2018) developed a decoding model based on a limited amount of training data and showed that it can infer the meaning of new words, phrases, or sentences from patterns of brain activation. To do this, they described a high-dimensional semantic space and used a representative sample from this semantic space. If this is indeed how the brain represents such relationships, then a decoder trained in this way should be able to generalize from a relatively small training set to new concepts/relationships (as these are all dimensions of the semantic space). Pereira et al. were able to show that their decoder, trained only on a limited set of individual word meanings, can use this strategy to decode meanings of sentences in this way. These representations allow the decoder to distinguish between semantically similar sentences as well as to capture the similarity structure of inter-sentence semantic relationships. Thus, it may be a method by which the brain itself carries out these computations. This illustrates how such a method can be used to generate new hypotheses for neuroscientists to test in the actual brain, much along the lines for which these models were originally developed.

Another example of how machine learning paradigms could be used to tell us something about the brain lies in their ability to rescue function, again something that previously has mainly been shown in animals. For instance, Ezzyat et al. (2018) have shown that one can use a closed-loop system to decode intracranially recorded neural activity from humans while they were learning lists of words, and then to implant artificial memories into lateral temporal cortex based on the patterns the machine learning algorithm extracted. Specifically, the system learned patterns associated with both successful encoding/recall and unsuccessful encoding/forgetting. Once the system had learned, the algorithm could then test whether or not this information was sufficient to induce memory: when it detected a pattern associated with forgetting, it stimulated the patients' brain to induce memory. Measured in terms of behavioral outcomes (i.e., words remembered), Ezzyat et al. showed that lateral temporal cortex—the site of stimulation—is sufficient to induce recall in humans.

Taken together, AI seems to offer both promises and pitfalls for neuroscience. Trying to understand what a DNN does comes with its own caveats. Clearly, neuroscientists should not naively apply DNNs for model building or analysis without considering the design principles of these networks. Nevertheless, reverse engineering neural networks that can solve tasks at human-level performance may provide a unique opportunity to grasp algorithmic and computational aspects of human intelligent behavior. As such, artificial neural networks should perhaps be treated like another species and not like a one-to-one model of the human brain. Finally, neuroscience should continue to build models that are solely made of a biological plausible set of submodels/routines, agnostic to neuroengineering tools, and provide biologically plausible options for engineering new algorithms.

## Desiderata for the Future

We end our discussion by considering desiderata for future studies with a focus on pressing opportunities for further discoveries:

1. Understanding the coding of relations: The coding of relations between objects is a common theme across domains: vision, motor control, spatial navigation, cognitive/semantic maps, language, etc. How relations are coded on the fly for flexible and purposeful behavior remains, however, one of the next frontiers of knowledge. At the same time, whether similar or different mechanisms are repurposed for the encoding of relations across domains is unknown. Future studies will hopefully be able to close this gap in our understanding of a fundamental brain operation.

2. Development of mesoscopic measurements: Human neuroscience studies rest upon noninvasive, macroscopic measurements (e.g., fMRI, MEG, EEG) that are detached from the detailed microscopic measurements found in animal models. The wonderful assortment of (molecular) tools used in rodents and increasingly in nonhuman primates to understand mechanistically cortical circuitry and operations (and,

where possible, their causal relevance for behavior) cannot be used in humans, preempting a mechanistic understanding of the same processes and principles at the same level directly in the human brain. The development of measurement technologies at the mesoscopic scale that are safe and minimally invasive (e.g., multicontact recording arrays) may help bridge some gaps between human and animal studies and are needed more than ever. Progress in understanding the human brain may be fundamentally impeded without the development of such tools.

3.  Ecological validity of behaviors in the laboratory versus in the wild: We question whether the experimental designs/tasks currently being used in much of neuroscience inappropriately constrain the type of answers that one might get. Specifically, how can we be certain that animals do or do not exhibit a specific behavior? Reductionist experimental paradigms, the reward schedules used to motivate animals to perform, as well as other variables may provide us with misguided answers simply because they do not tap into behaviors that an animal is equipped to produce. A possible alternative would be to access behaviors that intrinsically motivate animals to perform and exhibit specific behaviors. For example, we ask whether nonhuman primates would exhibit primitive forms of combinatorics when encouraged to teach conspecifics. Another venue for exploration would be to study behaviors in the wild, as those relate to the specific needs of the animals for survival. Inherently related to this question is whether training animals to perform human-like behaviors is informative or misleading our efforts to understand whether behavior across animals is similar or different.

4.  Targeted and explicit interspecies comparisons: Darwin's idea that differences between species are a matter of gradation permeates most of the scientific practice. It is implicitly assumed that mechanisms will translate across species once we understand the evolutionary changes that have occurred. In practice, parallel strains of studies on rodents, nonhuman primates, and humans are often conducted without an adequate exchange or engagement between groups to permit explicit interspecies comparisons. Of course, such comparisons come with challenges. As discussed, homologies and analogies need to be carefully delineated using multimodal evidence; a focus on only one aspect of the organism (e.g., only brain structure) without considering other relevant factors (e.g., mechanics of the body, genetics, development) can be misleading. Especially when insights from preclinical animal studies (e.g., on psychiatric or neurological diseases) are to be translated to humans, evolutionary factors need to be more explicitly considered; this could rescue a human treatment doomed for failure because of a key evolutionary change that occurred after the split from a common ancestor to murine species. For many neuroscience questions, technologies that allow explicit interspecies comparisons are

already available (e.g., comparative fMRI and intracranial recordings) and should be increasingly used for that purpose.

5.   Illumination of canonical principles in minds and machines through AI: Perhaps we did not recognize how hard the problem of visual recognition was until we tried to build a machine that could do it. Thus, an attempt at building a machine (AI) that could exhibit comparable behaviors to those of the human brain may be a fruitful approach to understand basic principles of human cognition. Such a research program entails using computational models tested on increasingly exquisite sets of behaviors to decide, among the family of models, which model best approximates human behavior. Principles extracted from those models could be used as hypotheses for further cognitive experiments, to help guide additional insight into the computations performed. In parallel, efforts should be made to relate properties of the computational models explicitly to neural architecture, and the other way around (e.g., Nayebi et al. 2018). This is certainly not an easy task, and whether efforts will be successful remains to be determined.

These are exciting times in neuroscience. Over thirty years have passed since the seminal Dahlem Workshop on the neurobiology of neocortex (Rakic and Singer 1988). Although we are far from a full understanding of brain function and how it enables cognition, we are optimistic that the next thirty years will bring important insights. The right ingredients are there: a rapid pace in the development of neurotechnologies for studying the brain, a flourishing field in AI, the capacity to build algorithms that match human behavior, and a scientific community that is willing to rethink how cross-species comparisons are used to understand what the cerebral cortex does, how it evolved to do so, and how it can afford high-level cognition. Together, this holds promise in helping us understand how the cerebral cortex and its rich set of connections operates, and how this makes us human.

# Bibliography

Note: Numbers in square brackets denote the chapter in which an entry is cited.

Abbott, L. F. 2008. Theoretical Neuroscience Rising. *Neuron* **60**:489–495. [7]

Abeles, M. 1991. Corticonics. Cambridge: Cambridge Univ. Press. [10]

Abeles, M., H. Bergman, I. Gat, et al. 1995. Cortical Activity Flips among Quasi Stationary States. *PNAS* **92**:8616–8620. [12]

Abeles, M., and G. L. Gerstein. 1988. Detecting Spatiotemporal Firing Patterns among Simultaneously Recorded Single Neurons. *J. Neurophysiol.* **60**:909–924. [12]

Ables, J. L., J. J. Breunig, A. J. Eisch, and P. Rakic. 2011. Not(Ch) Just Development: Notch Signalling in the Adult Brain. *Nat. Rev. Neurosci.* **12**:269–283. [2]

Aboitiz, F., and J. Montiel. 2003. One Hundred Million Years of Interhemispheric Communication: The History of the Corpus Callosum. *Braz. J. Med. Biol. Research* **36**:409–420. [3]

Abu-Khalil, A., L. Fu, E. A. Grove, N. Zecevic, and D. H. Geschwind. 2004. Wnt Genes Define Distinct Boundaries in the Developing Human Brain: Implications for Human Forebrain Patterning. *J. Comp. Neur.* **474**:276–288. [2]

Achard, S., R. Salvador, B. Whitcher, J. Suckling, and E. Bullmore. 2006. A Resilient, Low-Frequency, Small-World Human Brain Functional Network with Highly Connected Association Cortical Hubs. *J. Neurosci.* **26**:63–72. [7]

Ackley, D. H., G. E. Hinton, and T. J. Sejnowski. 1985. A Learning Algorithm for Boltzmann Machines. *Cogn. Neurosci.* **9**:147–169. [13]

Ainsworth, M., S. Lee, M. O. Cunningham, et al. 2012. Rates and Rhythms: A Synergistic View of Frequency and Temporal Coding in Neuronal Networks. *Neuron* **75**:572–583. [12]

Alberini, C. M. 2009. Transcription Factors in Long-Term Memory and Synaptic Plasticity. *Physiol. Rev.* **89**:121–145. [9]

Albert, E., and A.-L. Barabasi. 2002. Statistical Mechanics of Complex Networks. *Rev. Mod. Phys.* **74**:47–97. [7, 9]

Albert, M., N. Kalebic, M. Florio, et al. 2017. Epigenome Profiling and Editing of Neocortical Progenitor Cells during Development. *EMBO J.* **36**:2642–2658. [2]

Albert, R., H. Jeong, and A. L. Barabasi. 2000. Error and Attack Tolerance of Complex Networks. *Nature* **406**:378–382. [7]

Algan, O., and P. Rakic. 1997. Radiation-Induced, Lamina-Specific Deletion of Neurons in the Primate Visual Cortex. *J. Comp. Neur.* **381**:335–352. [5]

Alkuraya, F. S., X. Cai, C. Emery, et al. 2011. Human Mutations in NDE1 Cause Extreme Microcephaly with Lissencephaly. *The American Journal of Human Genetics* **88**:536–547. [4]

Allen, J. S. 2009. The Lives of the Brain: Human Evolution and the Organ of Mind. Cambridge, MA: Belknap Press. [16]

Almeder, R. 2007. Pragmatism and Philosophy of Science: A Critical Survey. *Int. Stud. Philos. Sci.* **21**:171–195. [7]

Alstott, J., M. Breakspear, P. Hagmann, L. Cammoun, and O. Sporns. 2009. Modeling the Impact of Lesions in the Human Brain. *PLoS Comput. Biol.* **5**:e1000408. [7]

Amalric, M., and S. Dehaene. 2016. Origins of the Brain Networks for Advanced Mathematics in Expert Mathematicians. *PNAS* **113**:4909–4917. [15]

Amalric, M., L. Wang, P. Pica, et al. 2017. The Language of Geometry: Fast Comprehension of Geometrical Primitives and Rules in Human Adults and Preschoolers. *PLoS Comput. Biol.* **13**:e1005273. [15]

Amari, S. 1998. Natural Gradient Works Efficiently in Learning. *Neural Comput.* **10**:251–276. [14]

Amir, R. E., I. B. Van den Veyver, M. Wan, et al. 1999. Rett Syndrome Is Caused by Mutations in X-Linked MECP2, Encoding Methyl-CpG-Binding Protein 2. *Nat. Genet.* **23**:185–188. [4]

Anderson, M. L., and B. L. Finlay. 2014. Allocating Structure to Function: The Strong Links between Neuroplasticity and Natural Selection. *Front. Hum. Neurosci.* **7**:918. [16]

Anderson, S. A., D. D. Eisenstat, L. Shi, and J. L. Rubenstein. 1997. Interneuron Migration from Basal Forebrain to Neocortex: Dependence on Dlx Genes. *Science* **278**:474–476. [5, 9]

Andics, A., A. Gabor, M. Gacsi, et al. 2016. Neural Mechanisms for Lexical Processing in Dogs. *Science* **353**:1030–1032. [17]

Andrews, T. J., S. D. Halpern, and D. Purves. 1997. Correlated Size Variations in Human Visual Cortex, Lateral Geniculate Nucleus, and Optic Tract. *J. Neurosci.* **17**:2859–2868. [5]

Andrews, W. D., M. Barber, and J. G. Parnavelas. 2007. Slit-Robo Interactions during Cortical Development. *J. Anat.* **211**:188–198. [5]

Ang, E. S., Jr., T. F. Haydar, V. Gluncic, and P. Rakic. 2003. Four-Dimensional Migratory Coordinates of GABAergic Interneurons in the Developing Mouse Cortex. *J. Neurosci.* **23**:5805–5815. [2]

Angevine, J. B., and R. L. Sidman. 1961. Autoradiographic Study of Cell Migration during Histogenesis of Cerebral Cortex in the Mouse. *Nature* **192**:766–768. [5]

Antinucci, F. 1989. Systematic Comparison of Early Sensorimotor Development. In: Cognitive Structure and Development in Nonhuman Primates, ed. F. Antinucci, pp. 67–85. Hillsdale, NJ: Lawrence Erlbaum Associates. [16]

Antonini, A., and C. J. Shatz. 1990. Relation between Putative Transmitter Phenotypes and Connectivity of Subplate Neurons During Cerebral Cortical Development. *Eur. J. Neurosci.* **2**:744–761. [5]

Ao, P. 2008. Emerging of Stochastic Dynamical Equalities and Steady State Thermodynamics. *Commun. Theor. Phys.* **49**:1073–1090. [14]

———. 2009. Global View of Bionetwork Dynamics: Adaptive Landscape. *J. Genet. Genomics* **36**:63–73. [14]

Appeltant, L., M. C. Soriano, G. V. d. Sande, et al. 2011. Information Processing Using a Single Dynamical Node as Complex System. *Nat. Commun.* **2**:468. [11, 13]

Arai, Y., J. N. Pulvers, C. Haffner, et al. 2011. Neural Stem and Progenitor Cells Shorten S-Phase on Commitment to Neuron Production. *Nat. Commun.* **2**:154. [5]

Arellano, J. I., B. Harding, and J.-L. Thomas. 2018 Adult Human Hippocampus: No New Neurons in Sight. *Cereb. Cortex* **28**:2479–2481. [5]

Arieli, A., D. Shoham, R. Hildesheim, and A. Grinvald. 1995. Coherent Spatiotemporal Patterns of Ongoing Activity Revealed by Real-Time Optical Imaging Coupled with Single-Unit Recording in the Cat Visual Cortex. *J. Neurophysiol.* **73**:2072–2093. [9, 12]

Arieli, A., A. Sterkin, A. Grinvald, and A. Aertsen. 1996. Dynamics of Ongoing Activity: Explanation of the Large Variability in Evoked Cortical Responses. *Science* **273**:1868–1871. [12]

Arnal, L. H., and A. L. Giraud. 2012. Cortical Oscillations and Sensory Predictions. *Trends Cogn. Sci.* **16**:390–398. [14]

Aronov, D., R. Nevers, and D. W. Tank. 2017. Mapping of a Non-Spatial Dimension by the Hippocampal-Entorhinal Circuit. *Nature* **543**:719–722. [17]

Aronson, D. G., G. B. Ermentrout, and N. Kopell. 1990. Amplitude Response of Coupled Oscillators. *Physica D* **41**:403–449. [10]

Artola, A., S. Bröcher, and W. Singer. 1990. Different Voltage-Dependent Thresholds for the Induction of Long-Term Depression and Long-Term Potentiation in Slices of the Rat Visual Cortex. *Nature* **347**:69–72. [10]

Artola, A., and W. Singer. 1987. Long-Term Potentiation and NMDA Receptors in Rat Visual Cortex. *Nature* **330**:649–652. [10]

Aru, J., V. Priesemann, M. Wibral, et al. 2015. Untangling Cross-Frequency Coupling in Neuroscience. *Curr. Opin. Neurobiol.* **31**:51–61. [10]

Asai, H., S. Ikezu, S. Tsunoda, et al. 2015. Depletion of Microglia and Inhibition of Exosome Synthesis Halt Tau Propagation. *Nat. Neurosci.* **18**:1584–1593. [4]

Ashby, F. G., and J. B. O'Brien. 2005. Category Learning and Multiple Memory Systems. *Trends Cogn. Sci.* **9**:83–89. [17]

Ashley, J., B. Cordy, D. Lucia, et al. 2018. Retrovirus-Like Gag Protein Arc1 Binds RNA and Traffics across Synaptic Boutons. *Cell* **172**:262–274. [4]

Assimacopoulos, S., E. A. Grove, and C. W. Ragsdale. 2003. Identification of a Pax6-Dependent Epidermal Growth Factor Family Signaling Source at the Lateral Edge of the Embryonic Cerebral Cortex. *J. Neurosci.* **23**:6399–6403. [2]

Assimacopoulos, S., T. Kao, N. P. Issa, and E. A. Grove. 2012. Fibroblast Growth Factor 8 Organizes the Neocortical Area Map and Regulates Sensory Map Topography. *J. Neurosci.* **32**:7191–7201. [5]

Aswolinskiy, W., and G. Pipa. 2015. RM-SORN: A Reward-Modulated Self-Organizing Recurrent Neural Network. *Front. Comput. Neurosci.* **9**:36. [11]

Atallah, B. V., and M. Scanziani. 2009. Instantaneous Modulation of Gamma Oscillation Frequency by Balancing Excitation with Inhibition. *Neuron* **62**:566–577. [10]

Attias, H. 2003. Planning by Probabilistic Inference. In: Proc. of the 9th Int. Workshop on Artificial Intelligence and Statistics. Key West: AISTATS. [14]

Auksztulewicz, R., and K. Friston. 2015. Attentional Enhancement of Auditory Mismatch Responses: A DCM/MEG Study. *Cereb. Cortex* **25**:4273–4283. [14]

Auksztulewicz, R., C. M. Schwiedrzik, T. Thesen, et al. 2018. Not All Predictions Are Equal: "What" and "When" Predictions Modulate Activity in Auditory Cortex through Different Mechanisms. *J. Neurosci.* **38**:8680–8693. [17]

Averbeck, B. B., P. E. Latham, and A. Pouget. 2006. Neural Correlations, Population Coding and Computation. *Nat. Rev. Neurosci.* **7**:358–366. [10]

Axmacher, N., D. P. Schmitz, T. Wagner, C. E. Elger, and J. Fell. 2008. Interactions between Medial Temporal Lobe, Prefrontal Cortex, and Inferior Temporal Regions during Visual Working Memory: A Combined Intracranial EEG and Functional Magnetic Resonance Imaging Study. *J. Neurosci.* **28**:7304–7312. [10]

Azevedo, F. A. C., L. R. B. Carvalho, L. T. Grinberg, et al. 2009. Equal Numbers of Neuronal and Nonneuronal Cells Make the Human Brain an Isometrically Scaled-up Primate Brain. *J. Comp. Neur.* **513**:532–541. [3]

Bae, B.-I., D. Jayaraman, and C. A. Walsh. 2015. Genetic Changes Shaping the Human Brain. *Dev. Cell* **32**:423–434. [4]

Bae, B.-I., I. Tietjen, K. D. Atabay, et al. 2014. Evolutionarily Dynamic Alternative Splicing of GPR56 Regulates Regional Cerebral Cortical Patterning. *Science* **343**:76476–76478. [3, 4]

Bailey, C. H., E. R. Kandel, and K. M. Harris. 2015. Structural Components of Synaptic Plasticity and Memory Consolidation. *Cold Spring Harb. Perspect. Biol.* 7:a021758. [9]

Bair, W., E. Zohary, and W. T. Newsome. 2001. Correlated Firing in Macaque Visual Area Mt: Time Scales and Relationship to Behavior. *J. Neurosci.* **21**:1676–1697. [13]

Bak, P., C. Tang, and K. Wiesenfeld. 1988. Self-Organized Criticality. *Phys. Rev. A* **38**:364–374. [14]

Baker, M. C. 2001. The Atoms of Language. New York: Basic Books. [15]

Baker, M. R., and S. A. Edgley. 2006. Non-Uniform Olivocerebellar Conduction Time in the Vermis of the Rat Cerebellum. *J. Physiol.* **570**:501–506. [13]

Baker, S. N., J. M. Kilner, E. M. Pinches, and R. Lemon. 1999. The Role of Synchrony and Oscillations in the Motor Output. *Exp. Brain Res.* **128**:109–117. [12]

Bakircioglu, M., O. P. Carvalho, M. Khurshid, et al. 2011. The Essential Role of Centrosomal NDE1 in Human Cerebral Cortex Neurogenesis. *The American Journal of Human Genetics* **88**:523–535. [4]

Bakken, T. E., J. A. Miller, S.-L. Ding, et al. 2016. A Comprehensive Transcriptional Map of Primate Brain Development. *Nature* **535**:367–375. [5]

Balasubramanian, K., V. Papadourakis, W. Liang, et al. 2019. Propagating Patterns of Activity across Motor Cortex Facilitate Movement Initiation. *bioRxiv*, preprint doi: https://doi.org/10.1101/549568. [12]

Balestra, F. R., P. Strnad, I. Flückiger, and P. Gönczy. 2013. Discovering Regulators of Centriole Biogenesis through Sirna-Based Functional Genomics in Human Cells. *Dev. Cell* **25**:555–571. [4]

Ballard, D. H., G. E. Hinton, and T. J. Sejnowski. 1983. Parallel Visual Computation. *Nature* **306**:21–26. [14]

Ballentine, L. E. 1970. The Statistical Interpretation of Quantum Mechanics. *Rev. Mod. Phys.* **42**:358–381. [14]

Balzeau, A., R. L. Holloway, and D. Grimaud-Hervé. 2012. Variations and Asymmetries in Regional Brain Surface in the Genus Homo. *J. Hum. Evol.* **62**:696–706. [3]

Banino, A., C. Barry, B. Uria, et al. 2018. Vector-Based Navigation Using Grid-Like Representations in Artificial Agents. *Nature* **557**:429–433. [10]

Banyai, M., A. Lazar, W. Singer, and G. Orbán. 2018. Cortical Response Statistics Is Shaped by Stimulus Context. *in preparation* [10]

Barber, M., and A. Pierani. 2016. Tangential Migration of Glutamatergic Neurons and Cortical Patterning during Development: Lessons from Cajal-Retzius Cells. *Dev. Neurobiol.* **76**:847–881. [5]

Barbrousse, A., and P. Ludwig. 2009. Fictions and Models. In: Fictions in Science: Philosophical Essays on Modelling and Idealization, ed. M. Suarez, pp. 56–75. London: Routledge. [7]

Bargmann, C. I., and E. Marder. 2013. From the Connectome to Brain Function. *Nat. Methods* **10**:483–490. [6]

Barlow, H. B. 1961. Possible Principles Underlying the Transformations of Sensory Messages. In: Sensory Communication, ed. W. Rosenblith, pp. 217–234. Cambridge, MA: MIT Press. [14]

———. 1972. Single Units and Sensation: A Neuron Doctrine for Perceptual Psychology? *Perception* **1**:371–394. [10, 13]

———. 1974. Inductive Inference, Coding, Perception, and Language. *Perception* **3**:123–134. [14]

Barth, A. L., and J. F. Poulet. 2012. Experimental Evidence for Sparse Firing in the Neocortex. *Trends Neurosci.* **35**:345–355. [13]

Bartolini, G., G. Ciceri, and O. Marín. 2013. Integration of GABAergic Interneurons into Cortical Cell Assemblies: Lessons from Embryos and Adults. *Neuron* **79**:849–864. [3]

Barton, R. A. 2012. Embodied Cognitive Evolution and the Cerebellum. *Phil. Trans. R. Soc. B* **367**:2097–2107. [5]

Barton, R. A., and C. Venditti. 2014. Rapid Evolution of the Cerebellum in Humans and Other Great Apes. *Curr. Biol.* **24**:2440–2444. [5]

Bassett, D. S., and E. T. Bullmore. 2017. Small-World Brain Networks Revisited. *Neuroscientist* **23**:499–516. [7, 9]

Bassett, D. S., D. L. Greenfield, A. Meyer-Lindenberg, et al. 2010. Efficient Physical Embedding of Topologically Complex Information Processing Networks in Brains and Computer Circuits. *PLoS Comput. Biol.* **6**:e1000748. [7]

Bassett, D. S., and O. Sporns. 2017. Network Neuroscience. *Nat. Neurosci.* **20**:353–364. [6, 7, 9]

Bassett, D. S., N. F. Wymbs, M. A. Porter, P. J. Mucha, and S. T. Grafton. 2014. Cross-Linked Structure of Network Evolution. *Chaos* **24**:013112. [7]

Bassett, D. S., P. Zurn, and J. I. Gold. 2018. On the Nature and Use of Models in Network Neuroscience. *Nat. Rev. Neurosci.* **19**:566–578. [7]

Bastos, A. M., W. M. Usrey, R. A. Adams, et al. 2012. Canonical Microcircuits for Predictive Coding. *Neuron* **76**:695–711. [9, 14]

Bastos, A. M., J. Vezoli, C. A. Bosman, et al. 2015. Visual Areas Exert Feedforward and Feedback Influences through Distinct Frequency Channels. *Neuron* **85**:390–401. [10, 14]

Bathellier, B., L. Ushakova, and S. Rumpel. 2012. Discrete Neocortical Dynamics Predict Behavioral Categorization of Sounds. *Neuron* **76**:435–449. [12]

Batista-Brito, R., and G. Fishell. 2009. The Developmental Integration of Cortical Interneurons into a Functional Network. *Curr. Top. Dev. Biol.* **87**:81–118. [5]

Batteau, D. W. 1967. The Role of the Pinna in Human Localization. *Proc. R. Soc. Lond. B* **168**:158–180. [16]

Bauer, M., R. Oostenveld, M. Peeters, and P. Fries. 2006. Tactile Spatial Attention Enhances Gamma-Band Activity in Somatosensory Cortex and Reduces Low-Frequency Activity in Parieto-Occipital Areas. *J. Neurosci.* **26**:490–501. [14]

Bauer, M., M. P. Stenner, K. Friston, and R. J. Dolan. 2014. Attentional Modulation of Alpha/Beta and Gamma Oscillations Reflect Functionally Distinct Processes. *J. Neurosci.* **34**:16117–16125. [14]

Bavelier, D., D. M. Levi, R. W. Li, Y. Dan, and T. K. Hensch. 2010. Removing Brakes on Adult Brain Plasticity: From Molecular to Behavioral Interventions. *J. Neurosci.* **30**:14964–14971. [9]

Beal, M. J. 2003. Variational Algorithms for Approximate Bayesian Inference thesis, PhD. Thesis, University College London. [14]

Bear, M. F., A. Kleinschmidt, Q. Gu, and W. Singer. 1990. Disruption of Experience-Dependent Synaptic Modifications in Striate Cortex by Infusion of an NMDA Receptor Antagonist. *J. Neurosci.* **10**:909–925. [10]

Becker, S. 1996. Mutual Information Maximization: Models of Cortical Self-Organization. *Network* **7**:7–31. [11]

Beckers, G. J. L., R. C. Berwick, K. Okanoya, and J. J. Bolhuis. 2016. What Do Animals Learn in Artificial Grammar Studies? *Neurosci. Biobehav. Rev.* **81**:238–246. [15]

Beggs, J. M., and D. Plenz. 2004. Neuronal Avalanches Are Diverse and Precise Activity Patterns That Are Stable for Many Hours in Cortical Slice Cultures. *J. Neurosci.* **24**:5216–5229. [12]

Beim Graben, P., S. Gerth, and S. Vasishth. 2008. Towards Dynamical System Models of Language-Related Brain Potentials. *Cogn. Neurodyn.* **2**:229–255. [14]

Belcher, A. M., C. C. Yen, H. Stepp, et al. 2013. Large-Scale Brain Networks in the Awake, Truly Resting Marmoset Monkey. *J. Neurosci.* **33**:16796–16804. [9]

Belgard, T. G., J. F. Montiel, W. Z. Wang, et al. 2013. Adult Pallium Transcriptomes Surprise in Not Reflecting Predicted Homologies across Diverse Chicken and Mouse Pallial Sectors. *PNAS* **110**:13150–13155. [13]

Bell, C. C., V. Z. Han, Y. Sugawara, and K. Grant. 1997. Synaptic Plasticity in a Cerebellum-Like Structure Depends on Temporal Order. *Nature* **387**:278–281. [13]

Bellec, G., F. Scherr, E. Hajek, et al. 2019. Biologically Inspired Alternatives to Backpropagation through Time for Learning in Recurrent Neural Nets. *arXiv* 1901.09049. [13]

Bellomo, N., A. Elaiw, A. M. Althiabi, and M. A. Alghamdi. 2015. On the Interplay between Mathematics and Biology: Hallmarks toward a New Systems Biology. *Phys. Life Rev.* **12**:44–64. [7]

Belluscio, M. A., K. Mizuseki, R. Schmidt, R. Kempter, and G. Buzsáki. 2012. Cross-Frequency Phase–Phase Coupling between Theta and Gamma Oscillations in the Hippocampus. *J. Neurosci.* **32**:423–435. [10]

Beltrán-Valero de Bernabé, D. 2004. Mutations in the *FKRP* Gene Can Cause Muscle-Eye-Brain Disease and Walker-Warburg Syndrome. *J. Med. Genet.* **41**:e61. [4]

Beltrán-Valero de Bernabé, D., S. Currier, A. Steinbrecher, et al. 2002. Mutations in the O-Mannosyltransferase Gene *POMT1* Give Rise to the Severe Neuronal Migration Disorder Walker-Warburg Syndrome. *The American Journal of Human Genetics* **71**:1033–1043. [4]

Belzung, C., and M. Lemoine. 2011. Criteria of Validity for Animal Models of Psychiatric Disorders: Focus on Anxiety Disorders and Depression. *Biol. Mood Anxiety Disord.* **1**:9. [7]

Benabid, A. L., P. Pollak, C. Gervason, et al. 1991. Long-Term Suppression of Tremor by Chronic Stimulation of the Ventral Intermediate Thalamic Nucleus. *Lancet* **337**:403–406. [17]

Benavides-Piccione, R., F. Hamzei-Sichani, I. Ballesteros-Yanez, J. DeFelipe, and R. Yuste. 2006. Dendritic Size of Pyramidal Neurons Differs among Mouse Cortical Regions. *Cereb. Cortex* **16**:990–1001. [8]

Bendixen, A., I. SanMiguel, and E. Schroger. 2012. Early Electrophysiological Indicators for Predictive Processing in Audition: A Review. *Int. J. Psychophysiol.* **83**:120–131. [14]

Bergman, H. G., T. Wichmann, and M. R. DeLong. 1994. The Primate Subthalamic Nucleus: II. Neural Activity in the Subthalamic Nucleus and Pallidum in the MPTP Model of Parkinsonism. *J. Neurophysiol.* **72**:507–652. [17]

Bergman, T. J., J. C. Beehner, D. L. Cheney, and R. M. Seyfarth. 2003. Hierarchical Classification by Rank and Kinship in Baboons. *Science* **302**:1234–1236. [17]

Berkes, P., G. Orbán, M. Lengyel, and J. Fiser. 2011. Spontaneous Cortical Activity Reveals Hallmarks of an Optimal Internal Model of the Environment. *Science* **331**:83–87. [10]

Bernardo, K. L., J. S. McCasland, T. A. Woolsey, and R. N. Strominger. 1990. Local Intra- and Interlaminar Connections in Mouse Barrel Cortex. *J. Comp. Neur.* **291**:231–255. [8]

Bernards, A., and J. F. Gusella. 1994. The Importance of Genetic Mosaicism in Human Disease. *New Engl. J. Med.* **331**:1447–1449. [4]

Bernier, P. M., K. Whittingstall, and S. T. Grafton. 2017. Differential Recruitment of Parietal Cortex during Spatial and Non-Spatial Reach Planning. *Front. Hum. Neurosci.* **11**:249. [14]

Bernstein, N. A. 1967. The Co-Ordination and Regulation of Movements. New York: Pergamon Press. [13]

Bertram, R., A. Sherman, and L. S. Satin. 2007. Metabolic and Electrical Oscillations: Partners in Controlling Pulsatile Insulin Secretion. *Am J Physiol Endocrinol Metab* **293**:E890–900. [6]

Best, M. D., A. J. Suminski, K. Takahashi, K. A. Brown, and N. G. Hatsopoulos. 2017. Spatio-Temporal Patterning in Primary Motor Cortex at Movement Onset. *Cereb. Cortex* **27**:1491–1500. [12]

Betizeau, M., V. Cortay, D. Patti, et al. 2013. Precursor Diversity and Complexity of Lineage Relationships in the Outer Subventricular Zone of the Primate. *Neuron* **80**:442–457. [2, 5]

Bettencourt, L. M., G. J. Stephens, M. I. Ham, and G. W. Gross. 2007. Functional Structure of Cortical Neuronal Networks Grown *in Vitro*. *Phys Rev E* **75**:021915. [7]

Betzel, R. F., A. Avena-Koenigsberger, J. Goni, et al. 2016. Generative Models of the Human Connectome. *Neuroimage* **124**:1054–1064. [7]

Betzel, R. F., and D. S. Bassett. 2017. Generative Models for Network Neuroscience: Prospects and Promise. *J. R. Soc. Interface* **14**:20170623. [7]

———. 2018. Specificity and Robustness of Long-Distance Connections in Weighted, Interareal Connectomes. *PNAS* **115**:E4880–E4889. [7, 9]

Betzel, R. F., J. D. Medaglia, and D. S. Bassett. 2018. Diversity of Meso-Scale Architecture in Human and Non-Human Connectomes. *Nat. Commun.* **9**:346. [7]

Beul, S. F., H. Barbas, and C. C. Hilgetag. 2017. A Predictive Structural Model of the Primate Connectome. *Sci. Rep.* **7**:43176. [7]

Beul, S. F., S. Grant, and C. C. Hilgetag. 2015. A Predictive Model of the Cat Cortical Connectome Based on Cytoarchitecture and Distance. *Brain Struct. Funct.* **220**:3167–3184. [7]

Bezgin, G., A. Solodkin, R. Bakker, P. Ritter, and A. R. McIntosh. 2017. Mapping Complementary Features of Cross-Species Structural Connectivity to Construct Realistic "Virtual Brains." *Hum. Brain Mapp.* **38**:2080–2093. [7]

Bhat, V., S. C. Girimaji, G. Mohan, et al. 2011. Mutations in *WDR62*, Encoding a Centrosomal and Nuclear Protein, in Indian Primary Microcephaly Families with Cortical Malformations. *Clin. Genet.* **80**:532–540. [4]

Bi, G. Q., and M. Poo. 1998. Synaptic Modifications in Cultured Hippocampal Neurons: Dependence on Spike Timing, Synaptic Strength, and Postsynaptic Cell Type. *J. Neurosci.* **18**:10464–10472. [10]

———. 2001. Synaptic Modification by Correlated Activity: Hebb's Postulate Revisited. *Annu. Rev. Neurosci.* **24**:139–166. [7]

Bianchi, S., C. D. Stimpson, A. L. Bauernfeind, et al. 2013. Dendritic Morphology of Pyramidal Neurons in the Chimpanzee Neocortex: Regional Specializations and Comparison to Humans. *Cereb. Cortex* **23**:2429–2436. [3]

Biederlack, J., M. Castelo-Branco, S. Neuenschwander, et al. 2006. Brightness Induction: Rate Enhancement and Neuronal Synchronization as Complementary Codes. *Neuron* **52**:1073–1083. [12]

Bielle, F., P. Marcos-Mondejar, M. Keita, et al. 2011. Slit2 Activity in the Migration of Guidepost Neurons Shapes Thalamic Projections during Development and Evolution. *Neuron* **69**:1085–1098. [5]

Bienenstock, E. L., L. N. Cooper, and P. W. Munro. 1982. Theory for the Development of Neuron Selectivity: Orientation Specificity and Binocular Interaction in Visual Cortex. *J. Neurosci.* **2**:32–48. [10]

Bilgüvar, K., A. K. Öztürk, A. Louvi, et al. 2010. Whole-Exome Sequencing Identifies Recessive WDR62 Mutations in Severe Brain Malformations. *Nature* **467**:207–210. [4]

Binzegger, T., R. J. Douglas, and K. A. Martin. 2004. A Quantitative Map of the Circuit of Cat Primary Visual Cortex. *J. Neurosci.* **24**:8441–8453. [8]

Bird, C. P., B. E. Stranger, M. Liu, et al. 2007. Fast-Evolving Noncoding Sequences in the Human Genome. *Genome Biol.* **8**:R118. [3]

Birkhoff, G. D. 1931. Proof of the Ergodic Theorem. *PNAS* **17**:656–660. [14]

Bishop, G. 1932. Cyclic Changes in Excitability of the Optic Pathway of the Rabbit. *Am. J. Phys.* **103**:213–224. [9]

Bishop, K. M., G. Goudreau, and D. D. O'Leary. 2000. Regulation of Area Identity in the Mammalian Neocortex by Emx2 and Pax6. *Science* **288**:344–349. [5]

Biswal, B., F. Z. Yetkin, V. M. Haughton, and J. S. Hyde. 1995. Functional Connectivity in the Motor Cortex of Resting Human Brain Using Echo-Planar MRI. *Magn. Reson. Med.* **34**:537–541. [6, 9]

Bittman, K. S., D. F. Owens, A. R. Kriegstein, and J. J. LoTurco. 1997. Cell Coupling and Uncoupling in the Ventricular Zone of Developing Neocortex. *J. Neurosci.* **17**:7037–7744. [2]

Bittman, K. S., and J. J. LoTurco. 1999. Differential Regulation of Connexin 26 and 43 in Murine Neocortical Precursors. *Cereb. Cortex* **9**:188–195. [2]

Bittner, K. C., A. D. Milstein, C. Grienberger, S. Romani, and J. C. Magee. 2017. Behavioral Time Scale Synaptic Plasticity Underlies CA1 Place Fields. *Science* **357**:1033–1036. [10]

Blanton, M. G., and A. R. Kriegstein. 1991. Morphological Differentiation of Distinct Neuronal Classes in Embryonic Turtle Cerebral Cortex. *J. Comp. Neur.* **310**:558–570. [5]

Blinkov, S. M., and I. I. Glezer. 1968. The Human Brain in Figures and Tables: A Quantitative Handbook. New York: Basic Books. [16]

Bliss, T. V. P., and T. Lomo. 1973. Long-Lasting Potentiation of Synaptic Transmission in the Dentate Area of the Anaesthetized Rabbit Following Stimulation of the Perforant Path. *J. Physiol.* **232**:331–356. [9, 10]

Bloom, P. 2004. Behavior: Can a Dog Learn a Word? *Science* **304**:1605–1606. [17]

Bock, D. D., W. C. Lee, A. M. Kerlin, et al. 2011. Network Anatomy and *in Vivo* Physiology of Visual Cortical Neurons. *Nature* **471**:177–182. [8]

Bock, O. 2013. Cajal, Golgi, Nansen, Schäfer and the Neuron Doctrine. *Endeavour* **37**:228–234. [9]

Boë, L.-J., F. Berthommier, T. Legou, et al. 2017. Evidence of a Vocalic Proto-System in the Baboon (*Papio papio*) Suggests Pre-Hominin Speech Precursors. *PLoS One* **12**:e0169321. [16, 17]

Bolhuis, J. J., I. Tattersall, N. Chomsky, and R. C. Berwick. 2014. How Could Language Have Evolved? *PLoS Biol.* **12**:e1001934. [15]

Bollobás, B. 1979. Graph Theory: An Introductory Course. New York, NY: Springer-Verlag. [7]

———. 1985. Random Graphs: Academic Press. [7]

Bond, J., E. Roberts, G. H. Mochida, et al. 2002. ASPM Is a Major Determinant of Cerebral Cortical Size. *Nat. Genet.* **32**:316–320. [4]

Bond, J., E. Roberts, K. Springell, et al. 2005. A Centrosomal Mechanism Involving CDK5RAP2 and CENPJ Controls Brain Size. *Nat. Genet.* **37**:353–355. [4]

Börgers, C., and N. J. Kopell. 2008. Gamma Oscillations and Stimulus Selection. *Neural Comput.* **20**:383–414. [10]

Borrell, V., and M. Götz. 2014. Role of Radial Glial Cells in Cerebral Cortex Folding. *Curr. Opin. Neurobiol.* **27**:39–46. [5]

Borrell, V., and I. Reillo. 2012. Emerging Roles of Neural Stem Cells in Cerebral Cortex Development and Evolution. *Dev. Neurobiol.* **72**:955–971. [5]

Borroto-Escuela, D. O., L. F. Agnati, K. Bechter, et al. 2015. The Role of Transmitter Diffusion and Flow versus Extracellular Vesicles in Volume Transmission in the Brain Neural-Glial Networks. *Phil. Trans. R. Soc. B* **370**:20140183. [7]

Bortoff, G. A., and P. L. Strick. 1993. Corticospinal Terminations in Two New-World Primates: Further Evidence That Corticomotoneuronal Connections Provide Part of the Neural Substrate for Manual Dexterity. *J. Neurosci.* **13**:5105–5118. [16]

Bosking, W. H., Y. Zhang, B. Schofield, and D. Fitzpatrick. 1997. Orientation Selectivity and the Arrangement of Horizontal Connections in Tree Shrew Striate Cortex. *J. Neurosci.* **17**:2112–2127. [10]

Bosman, C. A., J. M. Schoffelen, N. Brunet, et al. 2012. Attentional Stimulus Selection through Selective Synchronization between Monkey Visual Areas. *Neuron* **75**:875–888. [13]

Bosman, C. A., T. Womelsdorf, R. Desimone, and P. Fries. 2009. A Microsaccadic Rhythm Modulates Gamma-Band Synchronization and Behavior. *J. Neurosci.* **29**:9471–9480. [10]

Bostan, A. C., and P. L. Strick. 2018. The Basal Ganglia and Cerebellum: Nodes in an Integrated Network. *Nat. Rev. Neurosci.* **19**:338–350. [9]

Botvinick, M., and M. Toussaint. 2012. Planning as Inference. *Trends Cogn. Sci.* **16**:485–488. [14]

Bourgeois, J. P., P. S. Goldman-Rakic, and P. Rakic. 1994. Synaptogenesis in the Prefrontal Cortex of Rhesus Monkeys. *Cereb. Cortex* **4**:78–96. [5]

Bourgeois, J. P., P. J. Jastreboff, and P. Rakic. 1989. Synaptogenesis in Visual Cortex of Normal and Preterm Monkeys: Evidence for Intrinsic Regulation of Synaptic Overproduction. *PNAS* **86**:4297–4301. [5]

Bourne, J. N., and K. M. Harris. 2012. Nanoscale Analysis of Structural Synaptic Plasticity. *Curr. Opin. Neurobiol.* **22**:372–382. [9]

Bouton, S., V. Chambon, R. Tyrand, et al. 2018. Focal versus Distributed Temporal Cortex Activity for Speech Sound Category Assignment. *PNAS* **115**:E1299–E1308. [17]

Boyd, J. L., S. L. Skove, J. P. Rouanet, et al. 2015. Human-Chimpanzee Differences in a FZD8 Enhancer Alter Cell-Cycle Dynamics in the Developing Neocortex. *Curr. Biol.* **25**:772–779. [3, 5]

Braun, U., A. Schaefer, R. F. Betzel, et al. 2018. From Maps to Multi-Dimensional Network Mechanisms of Mental Disorders. *Neuron* **97**:14–31. [7]

Breakspear, M. 2017. Dynamic Models of Large-Scale Brain Activity. *Nat. Neurosci.* **20**:340–352. [7]

Breen, M. 2018. Effects of Metric Hierarchy and Rhyme Predictability on Word Duration in the Cat in the Hat. *Cognition* **174**:71–81. [15]

Brenner, S. 2003. Nature's Gift to Science (Nobel Lecture). *Chembiochem* **4**:683–687. [4]

Breuss, M., J. I.-T. Heng, K. Poirier, et al. 2012. Mutations in the ß-Tubulin Gene *TUBB5* Cause Microcephaly with Structural Brain Abnormalities. *Cell Rep.* **2**:1554–1562. [4]

Briggman, K. L., H. D. Abarbanel, and W. B. J. Kristan. 2005. Optical Imaging of Neuronal Populations during Decision-Making. *Science* **307**:896–901. [12, 13]

Britten, K. H., M. N. Shadlen, W. T. Newsome, and J. A. Movshon. 1993. Responses of Neurons in Macaque Mt to Stochastic Motion Signals. *Vis. Neurosci.* **10**:1157–1169. [12]

Bröcher, S., A. Artola, and W. Singer. 1992. Intracellular Injection of Ca2+ Chelators Blocks Induction of Long-Term Depression in Rat Visual Cortex. *PNAS* **89**:123–127. [10]

Brodmann, K. 1909. Vergleichende Lokalisationslehre Der Grosshirnrinde in Ihren Prinzipien Dargestellt Aufgrund Des Zellenbaues. Leipzig: Barth. [5, 8, 9]

———. 1912. Neue Ergebnisse Über Die Vergleichende Histologische Lokalisation Der Grosshirnrinde Mit Besonderer Berücksichtigung Des Stirnhirns. *Anat. Anz.* **41**:157–216. [16]

Brody, C. D. 1999a. Correlations without Synchrony. *Neural Comput.* **11**:1537–1551. [7]

———. 1999b. Disambiguating Different Covariation Types. *Neural Comput.* **11**:1527–1535. [7]

Bromer, C., T. M. Bartol, J. B. Bowden, et al. 2018. Long-Term Potentiation Expands Information Content of Hippocampal Dentate Gyrus Synapses. *PNAS* **115**:E2410–E2418. [9]

Brown, K. N., S. Chen, Z. Han, et al. 2011. Clonal Production and Organization of Inhibitory Interneurons in the Neocortex. *Science* **334**:480–486. [5]

Brox, A., L. Puelles, B. Ferreiro, and L. Medina. 2004. Expression of the Genes Emx1, Tbr1, and Eomes (Tbr2) in the Telencephalon of Xenopus Laevis Confirms the Existence of a Ventral Pallial Division in All Tetrapods. *J. Comp. Neur.* **474**:562–577. [5]

Bruinsma, T. J., V. V. Sarma, Y. Oh, et al. 2018. The Relationship between Dopamine Neurotransmitter Dynamics and the Blood-Oxygen-Level-Dependent (BOLD) Signal: A Review of Pharmacological Functional Magnetic Resonance Imaging. *Front. Neurosci.* **12**:238. [7]

Brunet, N., C. Bosman, M. Roberts, et al. 2015. Visual Cortical Gamma-Band Activity during Free Viewing of Natural Images. *Cereb. Cortex* **25**:918–926. [10]

Brunetti-Pierri, N., J. S. Berg, F. Scaglia, et al. 2008. Recurrent Reciprocal 1q21.1 Deletions and Duplications Associated with Microcephaly or Macrocephaly and Developmental and Behavioral Abnormalities. *Nat. Genet.* **40**:1466–1471. [3]

Bruno, R. M. 2011. Synchrony in Sensation. *Curr. Opin. Neurobiol.* **21**:701–708. [9]

Buckner, R. L., J. L. Roffman, and J. W. Smoller. 2014. Brain Genomics Superstruct Project (GSP) Dataverse. Harvard University. https://doi.org/10.7910/DVN/25833. [6]

Budd, J. M. 1998. Extrastriate Feedback to Primary Visual Cortex in Primates: A Quantitative Analysis of Connectivity. *Proc. Biol. Sci.* **265**:1037–1044. [12]

Budnik, V., C. Ruiz-Cañada, and F. Wendler. 2016. Extracellular Vesicles Round Off Communication in the Nervous System. *Nat. Rev. Neurosci.* **17**:160–172. [4]

Bufill, E., J. Agustí, and R. Blesa. 2011. Human Neoteny Revisited: The Case of Synaptic Plasticity. *Am. J. Human Biol.* **23**:729–739. [16]

Buhl, E. H., and W. Singer. 1989. The Callosal Projection in Cat Visual Cortex as Revealed by a Combination of Retrograde Tracing and Intracellular Injection. *Exp. Brain Res.* **75**:470–476. [10]

Bullmore, E., and O. Sporns. 2009. Complex Brain Networks: Graph Theoretical Analysis of Structural and Functional Systems. *Nat. Rev. Neurosci.* **10**:186–198. [7]

Buonomano, D. V., and W. Maass. 2009. State-Dependent Computations: Spatiotemporal Processing in Cortical Networks. *Nat. Rev. Neurosci.* **10**:113–125. [10, 11, 13, 14]

Buracas, G., A. Zador, M. Deweese, and T. Albright. 1998. Efficient Discrimination of Temporal Patterns by Motion-Sensitive Neurons in Primate Visual Cortex. *Neuron* **20**:959–969. [10]

Burns, S. P., S. Santaniello, R. B. Yaffe, et al. 2014. Network Dynamics of the Brain and Influence of the Epileptic Seizure Onset Zone. *PNAS* **111**:E5321–E5330. [7]

Burns, S. P., D. Xing, and R. M. Shapley. 2011. Is Gamma-Band Activity in the Local Field Potential of V1 Cortex a "Clock" or Filtered Noise? *J. Neurosci.* **31**:9658–9664. [10, 13]

Burns, S. P., D. Xing, M. J. Shelley, and R. M. Shapley. 2010. Searching for Autocoherence in the Cortical Network with a Time-Frequency Analysis of the Local Field Potential. *J. Neurosci.* **30**:4033–4047. [10]

Buschman, T. J., and E. K. Miller. 2007. Top-Down versus Bottom-up Control of Attention in the Prefrontal and Posterior Parietal Cortices. *Science* **315**:1860–1862. [10]

Bustamante, C. D., A. Fledel-Alon, S. Williamson, et al. 2005. Natural Selection on Protein-Coding Genes in the Human Genome. *Nature* **437**:1153–1157. [3]

Butler, A. B., and Z. Molnár. 2002. Development and Evolution of the Collopallium in Amniotes: A New Hypothesis of Field Homology. *Brain Res. Bull.* **57**:475–479. [5]

Butts, C. T. 2009. Revisiting the Foundations of Network Analysis. *Science* **325**:414–416. [7]

Buzsáki, G. 2006. Rhythms of the Brain. Oxford: Oxford Univ. Press. [10]

Buzsáki, G., and R. Llinas. 2017. Space and Time in the Brain. *Science* **358**:482–485. [17]

Buzsáki, G., N. Logothetis, and W. Singer. 2013. Scaling Brain Size, Keeping Timing: Evolutionary Preservation of Brain Rhythms. *Neuron* **80**:751–764. [10, 14]

Buzsáki, G., and X.-J. Wang. 2012. Mechanisms of Gamma Oscillations. *Annu. Rev. Neurosci* **35**:203–225. [10]

Bystron, I., C. Blakemore, and P. Rakic. 2008. Development of the Human Cerebral Cortex: Boulder Committee Revisited. *Nat. Rev. Neurosci.* **9**:110–122. [5]

Bystron, I., P. Rakic, Z. Molnár, and C. Blakemore. 2006. The First Neurons of the Human Cerebral Cortex. *Nat. Neurosci.* **9**:880–886. [5]

Cáceres, M., J. Lachuer, M. A. Zapala, et al. 2003. Elevated Gene Expression Levels Distinguish Human from Non-Human Primate Brains. *PNAS* **100**:13030–13035. [16]

Cáceres, M., C. Suwyn, M. Maddox, J. W. Thomas, and T. M. Preuss. 2006. Increased Cortical Expression of Two Synaptogenic Thrombospondins in Human Brain Evolution. *Cereb. Cortex* **17**:2312–2321. [16]

Calabrese, A., and S. M. Woolley. 2015. Coding Principles of the Canonical Cortical Microcircuit in the Avian Brain. *PNAS* **112**:3517–3522. [5]

Callery, E. M., and R. P. Elinson. 2000. Thyroid Hormone-Dependent Metamorphosis in a Direct Developing Frog. *PNAS* **97**:2615–2620. [16]

Calvin, W. H., and C. F. Stevens. 1968. Synaptic Noise and Other Sources of Randomness in Motoneuron Interspike Intervals. *J. Neurophysiol.* **31**:574–587. [11]

Camp, J. G., F. Badsha, M. Florio, et al. 2015. Human Cerebral Organoids Recapitulate Gene Expression Programs of Fetal Neocortex Development. *PNAS*201520760–201520766. [5]

Campbell, K., J. Vowinckel, M. Muelleder, et al. 2015. Self-Establishing Communities Enable Cooperative Metabolite Exchange in a Eukaryote. *eLife* **4**:e09943. [6]

Canolty, R. T., E. Edwards, S. S. Dalal, et al. 2006. High Gamma Power Is Phase-Locked to Theta Oscillations in Human Neocortex. *Science* **313**:1626–1628. [10]

Capecchi, M. R., and A. Pozner. 2015. ASPM Regulates Symmetric Stem Cell Division by Tuning Cyclin E Ubiquitination. *Nat. Commun.* **6**:8763. [4]

Capozziello, S., and M. De Laurentis. 2011. Extended Theories of Gravity. *Phys. Rep.* **509**:167–321. [14]

Capra, J. A., G. D. Erwin, G. McKinsey, J. L. R. Rubenstein, and K. S. Pollard. 2013. Many Human Accelerated Regions Are Developmental Enhancers. *Phil. Trans. R. Soc. B* **368**:20130025. [3–5]

Caputi, A., S. Melzer, M. Michael, and H. Monyer. 2013. The Long and Short of GABAergic Neurons. *Curr. Opin. Neurobiol.* **23**:179–186. [10]

Cardoso, M. M., Y. B. Sirotin, B. Lima, E. Glushenkova, and A. Das. 2012. The Neuroimaging Signal Is a Linear Sum of Neurally Distinct Stimulus- and Task-Related Components. *Nat. Neurosci.* **15**:1298–1306. [6]

Caronia-Brown, G., M. Yoshida, F. Gulden, S. Assimacopoulos, and E. A. Grove. 2014. The Cortical Hem Regulates the Size and Patterning of Neocortex. *Development* **141**:2855–2865. [5]

Caruso, V. C., J. T. Mohl, C. Glynn, et al. 2018. Evidence for Time Division Multiplexing: Single Neurons May Encode Simultaneous Stimuli by Switching between Activity Patterns. *bioRxiv*107185. [13]

Cavanagh, J. F., T. V. Wiecki, M. X. Cohen, et al. 2011. Subthalamic Nucleus Stimulation Reverses Mediofrontal Influence over Decision Threshold. *Nat. Neurosci.* **14**:1462–1467. [17]

Chakrabarti, L., Z. Galdzicki, and T. F. Haydar. 2007. Defects in Embryonic Neurogenesis and Initial Synapse Formation in the Forebrain of the Ts65Dn Mouse Model of Down Syndrome. *J. Neurosci.* **27**:11483–11495. [2]

Chambers, B., and J. N. MacLean. 2015. Multineuronal Activity Patterns Identify Selective Synaptic Connections under Realistic Experimental Constraints. *J. Neurophysiol.* **114**:1837–1849. [12]

———. 2016. Higher-Order Synaptic Interactions Coordinate Dynamics in Recurrent Networks. *PLoS Comput. Biol.* **12**:e1005078. [12]

Chan, Y. K., K. H. Sy, C. Y. Wong, et al. 2015. *In Vitro* Modeling of Emulsification of Silicone Oil as Intraocular Tamponade Using Microengineered Eye-on-a-Chip. *Invest. Ophthalmol. Vis. Sci.* **56**:3314–3319. [7]

Chance, S. A. 2014. The Cortical Microstructural Basis of Lateralized Cognition: A Review. *Front. Psychol.* **5**:820. [5]

Charrier, C., K. Joshi, J. Coutinho-Budd, et al. 2012. Inhibition of SRGAP2 Function by Its Human-Specific Paralogs Induces Neoteny during Spine Maturation. *Cell* **149**:923–935. [3, 5]

Chedotal, A., and L. J. Richards. 2010. Wiring the Brain: The Biology of Neuronal Guidance. *Cold Spring Harb. Perspect. Biol.* **2**:a001917. [3]

Chen, B., L. R. Schaevitz, and S. K. McConnell. 2005a. Fezl Regulates the Differentiation and Axon Targeting of Layer 5 Subcortical Projection Neurons in Cerebral Cortex. *PNAS* **102**:17184–17189. [2]

Chen, J. G., M. R. Rasin, K. Y. Kwan, and N. Sestan. 2005b. Zfp312 Is Required for Subcortical Axonal Projections and Dendritic Morphology of Deep-Layer Pyramidal Neurons of the Cerebral Cortex. *PNAS* **102**:17792–17797. [2]

Chen, J. L., S. Carta, J. Soldado-Magraner, B. L. Schneider, and F. Helmchen. 2013. Behaviour-Dependent Recruitment of Long-Range Projection Neurons in Somatosensory Cortex. *Nature* **499**:336–340. [8]

Chen, S., K. B. Swartz, and H. S. Terrace. 1997. Knowledge of the Ordinal Position of List Items in Rhesus Monkeys. *Psychol. Sci.* **8**:80–86. [15]

Chenn, A., and S. K. McConnell. 1995. Cleavage Orientation and the Asymmetric Inheritance of Notch1 Immunoreactivity in Mammalian Neurogenesis. *Cell* **82**:631–641. [2]

Chenn, A., and C. A. Walsh. 2002. Regulation of Cerebral Cortical Size by Control of Cell Cycle Exit in Neural Precursors. *Science* **297**:365–369. [5]

Chiel, H. J., and R. D. Beer. 1997. The Brain Has a Body: Adaptive Behavior Emerges from Interactions of Nervous System, Body and Environment. *Trends Neurosci.* **20**:553–557. [16]

Chklovskii, D. B. 2004. Synaptic Connectivity and Neuronal Morphology: Two Sides of the Same Coin. *Neuron* **43**:609–617. [9]

Choi, B. H., and L. W. Lapham. 1978. Radial Glia in the Human Fetal Cerebrum: A Combined Golgi, Immunofluorescent and Electron Microscopic Study. *Brain Res.* **148**:295–311. [5]

Cholfin, J. A., and J. L. Rubenstein. 2007. Patterning of Frontal Cortex Subdivisions by Fgf17. *PNAS* **104**:7652–7657. [5]

———. 2008. Frontal Cortex Subdivision Patterning Is Coordinately Regulated by Fgf8, Fgf17, and Emx2. *J. Comp. Neur.* **509**:144–155. [5]

Chou, S. J., Z. Babot, A. Leingartner, et al. 2013. Geniculocortical Input Drives Genetic Distinctions between Primary and Higher-Order Visual Areas. *Science* **340**:1239–1242. [5]

Chu, C. J., M. A. Kramer, J. Pathmanathan, et al. 2012. Emergence of Stable Functional Networks in Long-Term Human Electroencephalography. *J. Neurosci.* **32**:2703–2713. [7]

Chung, S. Y., D. D. Lee, and H. Sompolinsky. 2015. Linear Readout of Object Manifolds. *Phys. Rev. E* **93**:060301 [13]

Churchland, M. M., J. P. Cunningham, M. T. Kaufman, et al. 2012. Neural Population Dynamics during Reaching. *Nature* **487**:51–56. [12, 13]

Churchland, M. M., B. M. Yu, J. P. Cunningham, et al. 2010. Stimulus Onset Quenches Neural Variability: A Widespread Cortical Phenomenon. *Nat. Neurosci.* **13**:369–378. [10]

Churchland, M. M., B. M. Yu, M. Sahani, and K. V. Shenoy. 2007. Techniques for Extracting Single-Trial Activity Patterns from Large-Scale Neural Recordings. *Curr. Opin. Neurobiol.* **17**:609–618. [12]

Cisneros, L., J. Jimenez, M. G. Cosenza, and A. Parravano. 2002. Information Transfer and Nontrivial Collective Behavior in Chaotic Coupled Map Networks. *Phys. Rev. E* **65**:045204. [7]

Clancy, B., R. B. Darlington, and B. L. Finlay. 2000. The Course of Human Events: Predicting the Timing of Primate Neural Development. *Dev. Sci.* **3**:57–66. [16]

Clark, A. 2013. The Many Faces of Precision. *Front. Psychol.* **4**:270. [14]

Clower, D. M., and D. Boussaoud. 2000. Selective Use of Perceptual Recalibration versus Visuomotor Skill Acquisition. *J. Neurophysiol.* **84**:2703–2708. [17]

Clowry, G., Z. Molnár, and P. Rakic. 2010. Renewed Focus on the Developing Human Neocortex. *J. Anat.* **217**:276–288. [5]

Clowry, G. J., A. Alzu'bi, L. F. Harkin, et al. 2018. Charting the Protomap of the Human Telencephalon. *Semin. Cell Dev. Biol.* **76**:3–14. [2, 5]

Cobos, I., L. Puelles, and S. Martinez. 2001. The Avian Telencephalic Subpallium Originates Inhibitory Neurons That Invade Tangentially the Pallium (Dorsal Ventricular Ridge and Cortical Areas). *Dev. Biol.* **239**:30–45. [5]

Cocchi, L., L. L. Gollo, A. Zalesky, and M. Breakspear. 2017. Criticality in the Brain: A Synthesis of Neurobiology, Models and Cognition. *Prog. Neurobiol.* **158**:132–152. [14]

Cogan, G. B., A. Iyer, L. Melloni, et al. 2017. Manipulating Stored Phonological Input during Verbal Working Memory. *Nat. Neurosci.* **20**:279–286. [14]

Cohen, J. D., S. M. McClure, and A. J. Yu. 2007. Should I Stay or Should I Go? How the Human Brain Manages the Trade-Off between Exploitation and Exploration. *Phil. Trans. R. Soc. B* **362**:933–942. [14]

Cohen, M. R., and J. H. Maunsell. 2009. Attention Improves Performance Primarily by Reducing Interneuronal Correlations. *Nat. Neurosci.* **12**:1594–1600. [13]

Cohen, R., and S. Havlin. 2010. Complex Networks: Structure, Robustness and Function. New York: Cambridge University Press. [7]

Collingridge, G., and W. Singer. 1990. Excitatory Amino Acid Receptors and Synaptic Plasticity. *Trends Pharmacol. Sci.* **11**:290–296. [10]

Collins, C. E., E. C. Turner, E. K. Sawyer, et al. 2016. Cortical Cell and Neuron Density Estimates in One Chimpanzee Hemisphere. *PNAS* **113**:740–745. [3]

Conant, R. C., and W. R. Ashby. 1970. Every Good Regulator of a System Must Be a Model of That System. *Int. J. Syst. Sci.* **1**:89–97. [14]

Cong, L., F. A. Ran, D. Cox, et al. 2013. Multiplex Genome Engineering Using CRISPR/Cas Systems. *Science* **339**:819–823. [5]

Constantinople, C. M., and R. M. Bruno. 2013. Deep Cortical Layers Are Activated Directly by Thalamus. *Science* **340**:1591–1594. [9, 13]

Cook, A. 1994. The Observational Foundations of Physics. Cambridge: Cambridge Univ. Press. [14]

Costello, M. B., and D. M. Fragaszy. 1988. Prehension in Cebus and Saimiri: I. Grip Type and Hand Preference. *Am. J. Primatol.* **15**:235–245. [16]

Cotney, J., J. Leng, J. Yin, et al. 2013. The Evolution of Lineage-Specific Regulatory Activities in the Human Embryonic Limb. *Cell* **154**:185–196. [3]

Coull, J. T., and A. C. Nobre. 1998. Where and When to Pay Attention: The Neural Systems for Directing Attention to Spatial Locations and to Time Intervals as Revealed by Both PET and fMRI. *J. Neurosci.* **18**:7426–7435. [14]

Coulter, M. E., C. M. Dorobantu, G. A. Lodewijk, et al. 2018. The ESCRT-III Protein CHMP1A Mediates Secretion of Sonic Hedgehog on a Distinctive Subtype of Extracellular Vesicles. *Cell Rep.* **24**:973–986. [4]

Creutzfeldt, O. D. 1977. Generality of the Functional Structure of the Neocortex. *Naturwissenschaften* **64**:507–517. [5]

Crick, F. 1989. The Recent Excitement About Neural Networks. *Nature* **337**:129–132. [17]

Crick, F., and C. Koch. 1998. Constraints on Cortical and Thalamic Projections: The No-Strong-Loops Hypothesis. *Nature* **391**:245–250. [13]

Criminisi, A., J. Shotton, and E. Konukoglu. 2012. Decision Forests: A Unified Framework for Classification, Regression, Density Estimation, Manifold Learning and Semi-Supervised Learning. *Found. Trends Comput. Graph. Vis.* **7**:81–227. [11]

Crossley, P. H., S. Martinez, Y. Ohkubo, and J. L. Rubenstein. 2001. Coordinate Expression of Fgf8, Otx2, Bmp4, and Shh in the Rostral Prosencephalon during Development of the Telencephalic and Optic Vesicles. *Neuroscience* **108**:183–206. [5]

Cubelos, B., A. Sebastián-Serrano, L. Beccari, et al. 2010. Cux1 and Cux2 Regulate Dendritic Branching, Spine Morphology, and Synapses of the Upper Layer Neurons of the Cortex. *Neuron* **66**:523–535. [4]

Cuntz, H., F. Forstner, A. Borst, and M. Häusser. 2010. One Rule to Grow Them All: A General Theory of Neuronal Branching and Its Practical Application. *PLoS Comput. Biol.* **6**:e1000877. [8, 9]

Curran, E. J. 1909. A New Association Fiber Tract in the Cerebrum with Remarks on the Fiber Tract Dissection Method of Studying the Brain. *J. Comp. Neur.* **19**:645–656. [9]

Curto, C., V. Itskov, A. Veliz-Cuba, and N. Youngs. 2013. The Neural Ring: An Algebraic Tool for Analyzing the Intrinsic Structure of Neural Codes. *Bull. Math. Biol.* **75**:1571–1611. [7]

D'Huys, O., I. Fischer, J. Danckaert, and R. Vicente. 2012. Spectral and Correlation Properties of Rings of Delay-Coupled Elements: Comparing Linear and Nonlinear Systems. *Phys. Rev. E* **85**:056209. [10]

Damasio, A. R. 1989. The Brain Binds Entities and Events by Multiregional Activation from Convergence Zones. *Neural Comput.* **1**:123–132. [17]

Damoiseaux, J., S. Rombouts, and F. Barkhof. 2006. Consistent Resting-State Networks across Healthy Subjects. *PNAS* **103**:13848–13853. [9]

Darwin, C. 1888. The Descent of Man and Selection in Relation to Sex, vol. 1. London: John Murray. [15–17]

Das, A., and C. D. Gilbert. 1995. Long-Range Horizontal Connections and Their Role in Cortical Reorganization Revealed by Optical Recording of Cat Primary Visual Cortex. *Nature* **375**:780–784. [13]

Davis, E., and G. Marcus. 2015. Commonsense Reasoning and Commonsense Knowledge in Artificial Intelligence. *Comm. ACM* **58**:92–103. [17]

Dayan, P., and L. F. Abbott. 2001. Theoretical Neuroscience: Computational and Mathematical Modeling of Neural Systems. Cambridge, MA: MIT Press. [13]

Dayan, P., G. E. Hinton, and R. M. Neal. 1995. The Helmholtz Machine. *Neural Comput.* **7**:889–904. [14]

de Juan Romero, C., C. Bruder, U. Tomasello, J. M. Sanz-Anquela, and V. Borrell. 2015. Discrete Domains of Gene Expression in Germinal Layers Distinguish the Development of Gyrencephaly. *EMBO J.* **34**:1859–1874. [2]

de la Rocha, J., B. Doiron, E. Shea-Brown, K. Josic, and A. Reyes. 2007. Correlation between Neural Spike Trains Increases with Firing Rate. *Nature* **448**:802–806. [12]

de la Torre-Ubieta, L., J. L. Stein, H. Won, et al. 2018. The Dynamic Landscape of Open Chromatin during Human Cortical Neurogenesis. *Cell*1–35. [5]

De Ridder, D., S. Perera, and S. Vanneste. 2017. State of the Art: Novel Applications for Cortical Stimulation. *Neuromodulation* **20**:206–214. [7]

De Vico Fallani, F., L. Astolfi, F. Cincotti, et al. 2006. Brain Connectivity Structure in Spinal Cord Injured: Evaluation by Graph Analysis. *Conf. Proc. IEEE Eng. Med. Biol. Soc.* **1**:988–991. [7]

Deacon, T. W. 1990. Rethinking Mammalian Brain Evolution. *Integr. Comp. Biol.* **30**:629–705. [16]

———. 1992. Cortical Connections of the Inferior Arcuate Sulcus Cortex in the Macaque Brain. *Brain Res.* **573**:8–26. [16]

Dechery, J. B., and J. N. MacLean. 2017. Emergent Cortical Circuit Dynamics Contain Dense, Interwoven Ensembles of Spike Sequences. *J. Neurophysiol.* **118**:1914–1925. [12]

———. 2018. Functional Triplet Motifs Underlie Accurate Predictions of Single-Trial Responses in Populations of Tuned and Untuned V1 Neurons. *PLoS Comput. Biol.* **14**: e1006153. [13]

Deco, G., and V. K. Jirsa. 2012. Ongoing Cortical Activity at Rest: Criticality, Multistability, and Ghost Attractors. *J. Neurosci.* **32**:3366–3375. [14]

DeFelipe, J. 2011. The Evolution of the Brain, the Human Nature of Cortical Circuits, and Intellectual Creativity. *Front. Neuroanat.* **5**:1–17. [5]

DeFelipe, J., L. Alonso-Nanclares, and J. I. Arellano. 2002. Microstructure of the Neocortex: Comparative Aspects. *J. Neurocytol.* **31**:299–316. [3]

DeFelipe, J., and I. Fariñas. 1992. The Pyramidal Neuron of the Cerebral Cortex: Morphological and Chemical Characteristics of the Synaptic Inputs. *Prog. Neurobiol.* **39**:563–607. [3]

Dehaene, S. 2011. The Number Sense: How the Mind Creates Mathematics. Oxford: Oxford Univ. Press. [15, 17]

———. 2014. Consciousness and the Brain: Deciphering How the Brain Codes Our Thoughts (reprint edition). New York: Viking. [15]

Dehaene, S., and L. Cohen. 2007. Cultural Recycling of Cortical Maps. *Neuron* **56**:384–398. [16]

Dehaene, S., F. Meyniel, C. Wacongne, L. Wang, and C. Pallier. 2015. The Neural Representation of Sequences: From Transition Probabilities to Algebraic Patterns and Linguistic Trees. *Neuron* **88**:2–19. [14, 15, 17]

Dehay, C., P. Giroud, M. Berland, H. Killackey, and H. Kennedy. 1996. Contribution of Thalamic Input to the Specification of Cytoarchitectonic Cortical Fields in the Primate: Effects of Bilateral Enucleation in the Fetal Monkey on the Boundaries, Dimensions, and Gyrification of Striate and Extrastriate Cortex. *J. Comp. Neur.* **367**:70–89. [2]

Dehay, C., and H. Kennedy. 2007. Cell-Cycle Control and Cortical Development. *Nat. Rev. Neurosci.* **8**:438–450. [5]

Dehay, C., H. Kennedy, and K. S. Kosik. 2015. The Outer Subventricular Zone and Primate-Specific Cortical Complexification. *Neuron* **85**:683–694. [16]

Dehay, C., P. Savatier, and V. Cortay. 2001. Cell-Cycle Kinetics of Neocortical Precursors Are Influenced by Embryonic Thalamic Axons. *J. Neurosci.* **21**:201–214. [5]

Del Toro, D., T. Ruff, E. Cederfjäll, et al. 2017. Regulation of Cerebral Cortex Folding by Controlling Neuronal Migration via FLRT Adhesion Molecules. *Cell* **169**:621–624. [2, 5]

Delaunay, D., A. Kawaguchi, C. Dehay, and F. Matsuzaki. 2016. Division Modes and Physical Asymmetry in Cerebral Cortex Progenitors. *Curr. Opin. Neurobiol.* **42**:75–83. [5]

Deneve, S., P. E. Latham, and A. Pouget. 1999. Reading Population Codes: A Neural Implementation of Ideal Observers. *Nat. Neurosci.* **2**:740–745. [12]

Denker, M., S. Roux, H. Linden, et al. 2011. The Local Field Potential Reflects Surplus Spike Synchrony. *Cereb. Cortex* **21**:2681–2695. [12]

Dennis, M. Y., L. Harshman, B. J. Nelson, et al. 2017. The Evolution and Population Diversity of Human-Specific Segmental Duplications. *Nat. Ecol. Evol.* **1**:0069–0110. [5]

Dennis, M. Y., X. Nuttle, P. H. Sudmant, et al. 2012. Evolution of Human-Specific Neural SRGAP2 Genes by Incomplete Segmental Duplication. *Cell* **149**:912–922. [3]

Déprez, V., and A. Pierce. 1993. Negation and Functional Projections in Early Grammar. *Linguistic Inq.* **24**:25–67. [15]

des Portes, V., J. M. Pinard, P. Billuart, et al. 1998. A Novel CNS Gene Required for Neuronal Migration and Involved in X-Linked Subcortical Laminar Heterotopia and Lissencephaly Syndrome. *Cell* **92**:51–61. [4]

Desmurget, M., K. T. Reilly, N. Richard, et al. 2009. Movement Intention after Parietal Cortex Stimulation in Humans. *Science* **324**:811–813. [17]

de Zwart, J. A., A. C. Silva, P. van Gelderen, et al. 2005. Temporal Dynamics of the BOLD fMRI Impulse Response. *Neuroimage* **24**:667–677. [6]

Diba, K., and G. Buzsáki. 2007. Forward and Reverse Hippocampal Place-Cell Sequences during Ripples. *Nat. Neurosci.* **10**:1241–1242. [17]

DiCarlo, J. J., and D. D. Cox. 2007. Untangling Invariant Object Recognition. *Trends Cogn. Sci* **11**:333–341. [10]

Di Costanzo, S., A. Balasubramanian, H. L. Pond, et al. 2014. Pomk Mutations Disrupt Muscle Development Leading to a Spectrum of Neuromuscular Presentations. *Hum. Mol. Genet.* **23**:5781–5792. [4]

Diesmann, M., M.-O. Gewaltig, and A. Aertsen. 1999. Stable Propagation of Synchronous Spiking in Cortical Neural Networks. *Nature* **402**:529–533. [10]

Ding, N., L. Melloni, H. Zhang, X. Tian, and D. Poeppel. 2016. Cortical Tracking of Hierarchical Linguistic Structures in Connected Speech. *Nat. Neurosci.* **19**:158–164. [17]

Doan, R. N., B.-I. Bae, B. Cubelos, et al. 2016. Mutations in Human Accelerated Regions Disrupt Cognition and Social Behavior. *Cell* **167**:341–354. [3–5]

Dobyns, W. B. 2002. Primary Microcephaly: New Approaches for an Old Disorder. *Am. J. Med. Genet.* **112**:315–317. [16]

Doe, C. Q., and B. Bowerman. 2001. Asymmetric Cell Division: Fly Neuroblast Meets Worm Zygote. *Curr. Opin. Cell Biol.* **13**:68–75. [2]

Dong, G., J. Gao, R. Du, et al. 2013. Robustness of Network of Networks under Targeted Attack. *Phys. Rev. E* **87**:052804. [7]

Dotson, N. M., R. F. Salazar, and C. M. Gray. 2014. Frontoparietal Correlation Dynamics Reveal Interplay between Integration and Segregation during Visual Working Memory. *Journal of Neuroscience.* **34**:13600–13613. [10]

Douglas, R. J., and K. A. Martin. 2004. Neuronal Circuits of the Neocortex. *Annu. Rev. Neurosci.* **27**:419–451. [12]

———. 2007. Mapping the Matrix: The Ways of Neocortex. *Neuron* **56**:226–238. [13]

Dragoi, G., and G. Buzsáki. 2006. Temporal Encoding of Place Sequences by Hippocampal Cell Assemblies. *Neuron* **50**:145–157. [17]

Du, J. L., H. P. Wei, Z. R. Wang, S. T. Wong, and M. M. Poo. 2009. Long-Range Retrograde Spread of LTP and LTD from Optic Tectum to Retina. *PNAS* **106**:18890–18896. [17]

Dugas-Ford, J., J. J. Rowell, and C. W. Ragsdale. 2012. Cell-Type Homologies and the Origins of the Neocortex. *PNAS* **109**:16974–16979. [5]

Dum, R. P., D. J. Levinthal, and P. L. Strick. 2016. Motor, Cognitive, and Affective Areas of the Cerebral Cortex Influence the Adrenal Medulla. *PNAS* **113**:9922–9927. [9]

Dum, R. P., and P. L. Strick. 2013. Transneuronal Tracing with Neurotropic Viruses Reveals Network Macroarchitecture. *Curr. Opin. Neurobiol.* **23**:245–249. [9]

Dunsworth, H. M., A. G. Warrener, T. Deacon, P. T. Ellison, and H. Pontzer. 2012. Metabolic Hypothesis for Human Altriciality. *PNAS* **109**:15212–15216. [16]

Duque, A., Z. Krsnik, I. Kostovic, and P. Rakic. 2016. Secondary Expansion of the Transient Subplate Zone in the Developing Cerebrum of Human and Nonhuman Primates. *PNAS* **113**:9892–9897. [5]

Dwyer, N. D., B. Chen, S.-J. Chou, et al. 2016. Neural Stem Cells to Cerebral Cortex: Emerging Mechanisms Regulating Progenitor Behavior and Productivity. *J. Neurosci.* **36**:11394–11401. [3]

Egan, L. C., L. R. Santos, and P. Bloom. 2007. The Origins of Cognitive Dissonance: Evidence from Children and Monkeys. *Psychol. Sci.* **18**:978–983. [16]

Ehrens, D., D. Sritharan, and S. V. Sarma. 2015. Closed-Loop Control of a Fragile Network: Application to Seizure-Like Dynamics of an Epilepsy Model. *Front. Neurosci.* **9**:58. [7]

Eiraku, M., K. Watanabe, M. Matsuo-Takasaki, et al. 2008. Self-Organized Formation of Polarized Cortical Tissues from Escs and Its Active Manipulation by Extrinsic Signals. *Cell Stem Cell* **3**:519–532. [5]

Elgin, C. Z. 2010. Telling Instances. In: Beyond Mimesis and Convention: Representation in Art and Science, ed. R. Frigg and M. C. Hunter, pp. 1–17. Heidelberg: Springer. [7]

Elias, P. 1955. Predictive Coding. *IRE Tran. Info. Theory* **1**:16–24. [14]

Elsen, G. E., R. D. Hodge, F. Bedogni, et al. 2013. The Protomap Is Propagated to Cortical Plate Neurons through an Eomes-Dependent Intermediate Map. *PNAS* **110**:4081–4086. [2]

Elston, G. N. 2003. Cortex, Cognition and the Cell: New Insights into the Pyramidal Neuron and Prefrontal Function. *Cereb. Cortex* **13**:1124–1138. [8]

Elston, G. N., R. Benavides-Piccione, A. Elston, P. R. Manger, and J. Defelipe. 2011. Pyramidal Cells in Prefrontal Cortex of Primates: Marked Differences in Neuronal Structure among Species. *Front. Neuroanat.* **5**:2. [3, 5]

Elston, G. N., R. Tweedale, and M. G. Rosa. 1999. Cellular Heterogeneity in Cerebral Cortex: A Study of the Morphology of Pyramidal Neurones in Visual Areas of the Marmoset Monkey. *J. Comp. Neur.* **415**:33–51. [8]

Emera, D., J. Yin, S. K. Reilly, J. Gockley, and J. P. Noonan. 2016. Origin and Evolution of Developmental Enhancers in the Mammalian Neocortex. *PNAS* E2617–E2626. [5]

Enard, W., A. Fassbender, F. Model, et al. 2004. Differences in DNA Methylation Patterns between Humans and Chimpanzees. *Curr. Biol.* **14**:R148–149. [3]

Enard, W., S. Gehre, K. Hammerschmidt, et al. 2009. A Humanized Version of Foxp2 Affects Cortico-Basal Ganglia Circuits in Mice. *Cell* **137**:961–971. [3]

Enard, W., M. Przeworski, S. E. Fisher, et al. 2002. Molecular Evolution of FOXP2, a Gene Involved in Speech and Language. *Nature* **418**:869–872. [4]

ENCODE Project Consortium. 2012. An Integrated Encyclopedia of DNA Elements in the Human Genome. *Nature* **489**:57–74. [5]

Engel, A. K., P. König, A. K. Kreiter, and W. Singer. 1991. Interhemispheric Synchronization of Oscillatory Neuronal Responses in Cat Visual Cortex. *Science* **252**:1177–1179. [10]

Englund, C., A. Fink, C. Lau, et al. 2005. Pax6, Tbr2, and Tbr1 Are Expressed Sequentially by Radial Glia, Intermediate Progenitor Cells, and Postmitotic Neurons in Developing Neocortex. *J. Neurosci.* **25**:247–251. [2]

Ermentrout, G. B., and D. Kleinfeld. 2001. Traveling Electrical Waves in Cortex: Insights from Phase Dynamics and Speculation on a Computational Role. *Neuron* **29**:33–44. [10]

Espinosa, J. S., and M. P. Stryker. 2012. Development and Plasticity of the Primary Visual Cortex. *Neuron* **75**:230–249. [5]

Eugenides, J. 2002. Middlesex. New York: Farrar, Straus and Giroux. [7]

Evrony, G. D., E. Lee, B. K. Mehta, et al. 2015. Cell Lineage Analysis in Human Brain Using Endogenous Retroelements. *Neuron* **85**:49–59. [5]

Ezzyat, Y., P. A. Wanda, D. F. Levy, et al. 2018. Closed-Loop Stimulation of Temporal Cortex Rescues Functional Networks and Improves Memory. *Nat. Commun.* **9**:365. [17]

Falcon, M. I., V. Jirsa, and A. Solodkin. 2016. A New Neuroinformatics Approach to Personalized Medicine in Neurology: The Virtual Brain. *Curr Opin Neurol* **29**:429–436. [7]

Faraday, M. 1933. Faraday's Diary: Being the Various Philosophical Notes of Experimental Investigation Made by Michael Faraday, vol. IV (Nov. 12, 1839 - June 26, 1847). London: G. Bell and Sons, Ltd. [6]

Farcas, R., E. Schneider, K. Frauenknecht, et al. 2009. Differences in DNA Methylation Patterns and Expression of the CCRK Gene in Human and Nonhuman Primate Cortices. *Mol. Biol. Evol.* **26**:1379–1389. [3]

Fedorenko, E., M. K. Behr, and N. Kanwisher. 2011. Functional Specificity for High-Level Linguistic Processing in the Human Brain. *PNAS* **108**:16428–16433. [15]

Feldman, H., and K. Friston. 2010. Attention, Uncertainty, and Free-Energy. *Front. Hum. Neurosci.* **4**:215. [14]

Feldmeyer, D., G. Qi, V. Emmenegger, and J. F. Staiger. 2018. Inhibitory Interneurons and Their Circuit Motifs in the Many Layers of the Barrel Cortex. *Neuroscience* **368**:132–151. [8]

Feldt, S., P. Bonifazi, and R. Cossart. 2011. Dissecting Functional Connectivity of Neuronal Microcircuits: Experimental and Theoretical Insights. *Trends Neurosci.* **34**:225–236. [7]

Fell, J., E. Ludowig, B. P. Staresina, et al. 2011. Medial Temporal Theta/Alpha Power Enhancement Precedes Succesful Memory Encoding: Evidence Based on Intracranial EEG. *J. Neurosci.* **31**:5392–5397. [10]

Felleman, D. J., and D. C. Van Essen. 1991. Distributed Hierarchical Processing in the Primate Cerebral Cortex. *Cereb. Cortex* **1**:1–47. [7, 9, 13, 17]

Ferezou, I., F. Haiss, L. J. Gentet, et al. 2007. Spatiotemporal Dynamics of Cortical Sensorimotor Integration in Behaving Mice. *Neuron* **56**:907–923. [12]

Fernando, C., and S. Sojakka. 2003. Pattern Recognition in a Bucket. In: Advances in Artificial Life, Ecal 2003, ed. W. Banzhaf et al., vol. 2801, pp. 588–597. Lecture Notes in Computer Science, J. G. Carbonell and J. Siekmann, series ed. Berlin: Springer. [10]

Fertuzinhos, S., Z. Krsnik, Y. I. Kawasawa, et al. 2009. Selective Depletion of Molecularly Defined Cortical Interneurons in Human Holoprosencephaly with Severe Striatal Hypoplasia. *Cereb. Cortex* **19**:2196–2207. [3, 5]

Fietz, S. A., and W. B. Huttner. 2011. Cortical Progenitor Expansion, Self-Renewal and Neurogenesis: A Polarized Perspective. *Curr. Opin. Neurobiol.* **21**:23–35. [5]

Fietz, S. A., I. Kelava, J. Vogt, et al. 2010. OSVZ Progenitors of Human and Ferret Neocortex Are Epithelial-Like and Expand by Integrin Signaling. *Nat. Neurosci.* **13**:690–699. [2, 4, 5]

Fietz, S. A., R. Lachmann, H. Brandl, et al. 2012. Transcriptomes of Germinal Zones of Human and Mouse Fetal Neocortex Suggest a Role of Extracellular Matrix in Progenitor Self-Renewal. *PNAS* **109**:11836–11841. [5]

Finlay, B. L., R. B. Darlington, and N. Nicastro. 2001. Developmental Structure of Brain Evolution. *Behav. Brain Sci.* **24**:263–308. [16]

Finlay, B. L., and R. Uchiyama. 2015. Developmental Mechanisms Channeling Cortical Evolution. *Trends Neurosci.* **38**:69–76. [16]

Fitch, W. T. 2012. Evolutionary Developmental Biology and Human Language Evolution: Constraints on Adaptation. *Evol. Biol.* **39**:613–637. [17]

Fitch, W. T., B. de Boer, N. Mathur, and A. A. Ghazanfar. 2016. Monkey Vocal Tracts Are Speech-Ready. *Sci. Adv.* **2**:e1600723. [16, 17]

Fitch, W. T., and A. D. Friederici. 2012. Artificial Grammar Learning Meets Formal Language Theory: An Overview. *Phil. Trans. R. Soc. B* **367**:1933–1955. [15]

Fitzpatrick, D. C., R. Batra, T. R. Stanford, and S. Kuwada. 1997. A Neuronal Population Code for Sound Localization. *Nature* **388**:871–874. [12]

Fitzsimonds, R. M., H. J. Song, and M. M. Poo. 1997. Propagation of Activity-Dependent Synaptic Depression in Simple Neural Networks. *Nature* **388**:439–448. [17]

Flames, N., J. E. Long, A. N. Garratt, et al. 2004. Short- and Long-Range Attraction of Cortical GABAergic Interneurons by Neuregulin-1. *Neuron* **44**:251–261. [5]

Flames, N., R. Pla, D. M. Gelman, et al. 2007. Delineation of Multiple Subpallial Progenitor Domains by the Combinatorial Expression of Transcriptional Codes. *J. Neurosci.* **27**:9682–9695. [2]

Fleagle, J., and G. Ghazanfar. 2013. Primate Adaptation and Evolution. Cambridge, MA: Academic Press. [16]

Florio, M., M. Albert, E. Taverna, et al. 2015. Human-Specific Gene ARHGAP11B Promotes Basal Progenitor Amplification and Neocortex Expansion. *Science* **347**:1465–1470. [2, 3, 5]

Florio, M., V. Borrell, and W. B. Huttner. 2017. Human-Specific Genomic Signatures of Neocortical Expansion. *Curr. Opin. Neurobiol.* **42**:33–44. [5]

Florio, M., M. Heide, A. Pinson, et al. 2018. Evolution and Cell-Type Specificity of Human-Specific Genes Preferentially Expressed in Progenitors of Fetal Neocortex. *eLife* **7**:e32332. [5]

Florio, M., and W. B. Huttner. 2014. Neural Progenitors, Neurogenesis and the Evolution of the Neocortex. *Development* **141**:2182–2194. [5]

Florio, M., T. Namba, S. Pääbo, M. Hiller, and W. B. Huttner. 2016. A Single Splice Site Mutation in Human-Specific ARHGAP11B Causes Basal Progenitor Amplification. *Sci. Adv.* **2**:e1601941–e1601941. [5]

Fodor, J. A. 1983. The Modularity of Mind. Cambridge, MA: MIT Press. [15]

Fornito, A., E. T. Bullmore, and A. Zalesky. 2017. Opportunities and Challenges for Psychiatry in the Connectomic Era. *Biol. Psychiatry Cogn. Neurosci. Neuroimaging* **2**:9–19. [7, 9]

Fornito, A., A. Zalesky, and E. T. Bullmore. 2016. Fundamentals of Brain Network Analysis. Cambridge, MA: Academic Press. [7]

Fossati, M., R. Pizzarelli, E. R. Schmidt, et al. 2016. SRGAP2 and Its Human-Specific Paralog Co-Regulate the Development of Excitatory and Inhibitory Synapses. *Neuron* **91**:356–369. [3, 5]

Foster, J. A., L. Rinaman, and J. F. Cryan. 2017. Stress and the Gut-Brain Axis: Regulation by the Microbiome. *Neurobiol. Stress* **7**:124–136. [9]

Fox, M. D., and M. Raichle. 2007. Spontaneous Fluctuations in Brain Activity Observed with Functional Magnetic Resonance Imaging. *Nat. Rev. Neurosci.* **8**:700–711. [6]

Fox, M. D., A. Z. Snyder, J. L. Vincent, et al. 2005. The Human Brain Is Intrinsically Organized into Dynamic, Anticorrelated Functional Networks. *PNAS* **102**:9673–9678. [6]

Fox, P. T., and M. E. Raichle. 1986. Focal Physiological Uncoupling of Cerebral Blood Flow and Oxidative Metabolism during Somatosensory Stimulation in Human Subjects. *PNAS* **83**:1140–1144. [6]

Francis, M. M., J. E. Mellem, and A. V. Maricq. 2003. Bridging the Gap between Genes and Behavior: Recent Advances in the Electrophysiological Analysis of Neural Function in *Caenorhabditis elegans*. *Trends Neurosci.* **26**:90–99. [7]

Franco, S. J., C. Gil-Sanz, I. Martinez-Garay, et al. 2012. Fate-Restricted Neural Progenitors in the Mammalian Cerebral Cortex. *Science* **337**:746–749. [2, 5]

Frank, S. A. 2012. Natural Selection V: How to Read the Fundamental Equations of Evolutionary Change in Terms of Information Theory. *J. Evol. Biol.* **25**:2377–2396. [14]

Frank, S. L., R. Bod, and M. H. Christiansen. 2012. How Hierarchical Is Language Use? *Proceedings of the Royal Society B* **279**:4522–4531. [15]

Freeman, W. J. 1994. Characterization of State Transitions in Spatially Distributed, Chaotic, Nonlinear, Dynamical Systems in Cerebral Cortex. *Integr. Physiol. Behav. Sci.* **29**:294–306. [14]

Frenkel, M. Y., N. B. Sawtell, A. C. Diogo, et al. 2006. Instructive Effect of Visual Experience in Mouse Visual Cortex. *Neuron* **51**:339–349. [5]

Friedmann, N., G. Taranto, L. P. Shapiro, and D. Swinney. 2008. The Leaf Fell (the Leaf): The Online Processing of Unaccusatives. *Linguistic Inq.* **39**:355–377. [15]

Fries, P. 2005. A Mechanism for Cognitive Dynamics: Neuronal Communication through Neuronal Coherence. *Trends Cogn. Sci.* **9**:474–480. [10, 13, 14, 17]

———. 2009. Neuronal Gamma-Band Synchronization as a Fundamental Process in Cortical Computation. *Annu. Rev. Neurosci.* **32**:209–224. [10]

———. 2015. Rhythms for Cognition: Communication through Coherence. *Neuron* **88**:220–235. [13]

Fries, P., S. Neuenschwander, A. K. Engel, R. Goebel, and W. Singer. 2001a. Rapid Feature Selective Neuronal Synchronization through Correlated Latency Shifting. *Nat. Neurosci.* **4**:194–200. [10, 13]

Fries, P., D. Nikolić, and W. Singer. 2007. The Gamma Cycle. *Trends Neurosci.* **30**:309–316. [10]

Fries, P., J. H. Reynolds, A. E. Rorie, and R. Desimone. 2001b. Modulation of Oscillatory Neuronal Synchronization by Selective Visual Attention. *Science* **291**:1560–1563. [10, 14]

Fries, P., T. Womelsdorf, R. Oostenveld, and R. Desimone. 2008. The Effects of Visual Stimulation and Selective Visual Attention on Rhythmic Neuronal Synchronization in Macaque Area V4. *J. Neurosci.* **28**:4823–4835. [14]

Frigg, R. 2010a. Fiction and Scientific Representation. In: Beyond Mimesis and Convention: Representation in Art and Science, ed. R. Frigg and M. C. Hunter, pp. 97–138. Heidelberg: Springer. [7]

———. 2010b. Fiction in Science. In: Fictions and Models: New Essays, ed. J. Woods, pp. 247–287. Munich: Philosophia Verlag. [7]

———. 2010c. Models and Fiction. *Synthese* **172**:251–268. [7]

Frigg, R., and S. Hartmann. 2012. Models in Science. In: The Stanford Encyclopedia of Philosophy. E. N. Zalta, series ed. Stanford: Stanford Univ. [7]

Frigg, R., and M. C. Hunter. 2010. Beyond Mimesis and Convention: Representation in Art and Science. Netherlands: Springer. [7]

Frigg, R., and J. Nguyen. 2016. The Fiction View of Models Reloaded. *Monist* **99**:225–242. [7]

———. 2017. Scientific Representation Is Representation-As. In: Philosophy of Science in Practice, ed. H. K. Chao and J. Reiss, vol. 379, pp. 149–179. Studies in Epistemology, Logic, Methodology, and Philosophy of Science. Cham: Springer. [7]

Friston, K. 2010. The Free-Energy Principle: A Unified Brain Theory? *Nat. Rev. Neurosci.* **11**:127–138. [14]

———. 2011a. Functional and Effective Connectivity: A Review. *Brain Connect.* **1**:13–36. [7]

———. 2011b. What Is Optimal About Motor Control? *Neuron* **72**:488–498. [14]

———. 2013. Life as We Know It. *J. R. Soc. Interface* **10**:20130475. [14]

Friston, K., M. Breakspear, and G. Deco. 2012. Perception and Self-Organized Instability. *Front. Comput. Neurosci.* **6**:44. [14]

Friston, K., and G. Buzsáki. 2016. The Functional Anatomy of Time: What and When in the Brain. *Trends Cogn. Sci.* **20**:500–511. [14, 17]

Friston, K., and C. Frith. 2015. A Duet for One. *Conscious. Cogn.* **36**:390–405. [14]

Friston, K., and S. Kiebel. 2009. Predictive Coding under the Free-Energy Principle. *Phil. Trans. R. Soc. B* **364**:1211–1221. [13]

Friston, K., M. Levin, B. Sengupta, and G. Pezzulo. 2015a. Knowing One's Place: A Free-Energy Approach to Pattern Regulation. *J. R. Soc. Interface* **12**:pii: 20141383. [14]

Friston, K., T. Parr, and B. de Vries. 2017a. The Graphical Brain: Belief Propagation and Active Inference. *Netw. Neurosci.* **1**:381–414. [14]

Friston, K., F. Rigoli, D. Ognibene, et al. 2015b. Active Inference and Epistemic Value. *Cogn. Neurosci.* **6**:187–214. [14]

Friston, K., R. Rosch, T. Parr, C. Price, and H. Bowman. 2017b. Deep Temporal Models and Active Inference. *Neurosci. Biobehav. Rev.* **77**:388–402. [14]

Friston, K., G. Tononi, O. Sporns, and G. Edelman. 1995. Characterising the Complexity of Neuronal Interactions. *Hum. Brain Mapp.* **59**:229–243. [9]

Froemke, R. C., M. M. Merzenich, and C. E. Schreiner. 2007. A Synaptic Memory Trace for Cortical Receptive Field Plasticity. *Nature* **450**:425–429. [13]

Frühbeis, C., D. Fröhlich, W. P. Kuo, et al. 2013. Neurotransmitter-Triggered Transfer of Exosomes Mediates Oligodendrocyte–Neuron Communication. *PLoS Biol.* **11**:e1001604. [4]

Fu, J., I. M. Hagan, and D. M. Glover. 2015. The Centrosome and Its Duplication Cycle. *Cold Spring Harb. Perspect. Biol.* **7**:a015800. [4]

Fujimori, A., K. Itoh, S. Goto, et al. 2014. Disruption of *Aspm* Causes Microcephaly with Abnormal Neuronal Differentiation. *Brain Dev.* **36**:661–669. [4]

Fujita, E., Y. Tanabe, A. Shiota, et al. 2008. Ultrasonic Vocalization Impairment of Foxp2 (R552h) Knockin Mice Related to Speech-Language Disorder and Abnormality of Purkinje Cells. *PNAS* **105**:3117–3122. [4]

Fukuchi-Shimogori, T., and E. A. Grove. 2001. Neocortex Patterning by the Secreted Signaling Molecule FGF8. *Science* **294**:1071–1074. [5]

———. 2003. Emx2 Patterns the Neocortex by Regulating FGF Positional Signaling. *Nat. Neurosci.* **6**:825–831. [5]

Fukushima, M., R. C. Saunders, D. A. Leopold, M. Mishkin, and B. B. Averbeck. 2012. Spontaneous High-Gamma Band Activity Reflects Functional Organization of Auditory Cortex in the Awake Macaque. *Neuron* **74**:899–910. [9]

Funahashi, S. 2014. Saccade-Related Activity in the Prefrontal Cortex: Its Role in Eye Movement Control and Cognitive Functions. *Front Integr Neurosci* **8**:54. [14]

Fyhn, M., S. Molden, M. P. Witter, E. I. Moser, and M. B. Moser. 2004. Spatial Representation in the Entorhinal Cortex. *Science* **305**:1258-1264. [17]

Gabbiani, F., H. G. Krapp, C. Koch, and G. Laurent. 2002. Multiplicative Computation in a Visual Neuron Sensitive to Looming. *Nat. Biotechnol.* **420**::320–324. [13]

Gabi, M., K. Neves, C. Masseron, et al. 2016. No Relative Expansion of the Number of Prefrontal Neurons in Primate and Human Evolution. *PNAS* **113**:9617–9622. [5]

Gabriel, A., and R. Eckhorn. 2003. A Multi-Channel Correlation Method Detects Traveling Waves in Monkey Visual Cortex. *Journal of Neuroscience. Methods* **131**:171–184. [13]

Gal, J. S., Y. M. Morozov, A. E. Ayoub, et al. 2006. Molecular and Morphological Heterogeneity of Neural Precursors in the Mouse Neocortical Proliferative Zones. *J. Neurosci.* **26**:1045–1056. [2, 5]

Galaburda, A. M., and D. N. Pandya. 1982. Role of Architectonics and Connections in the Study of Primate Brain Evolution. In: Primate Brain Evolution, pp. 203–216. New York: Plenum Press. [16]

Gallego Romero, I., B. J. Pavlovic, I. Hernando-Herraez, et al. 2015. A Panel of Induced Pluripotent Stem Cells from Chimpanzees: A Resource for Comparative Functional Genomics. *eLife* **4**:e07103. [5]

Galuske, R. A. W., M. H. J. Munk, and W. Singer. in prep. Gamma Oscillations Gate Use-Dependent Changes of Orientation Maps in Cat Visual Cortex. *PNAS*, in press. [10]

Gamanut, R., H. Kennedy, Z. Toroczkai, et al. 2018. The Mouse Cortical Connectome, Characterized by an Ultra-Dense Cortical Graph, Maintains Specificity by Distinct Connectivity Profile. *Neuron* **97**:698–715. [17]

Ganmor, E., R. Segev, and E. Schneidman. 2011. Sparse Low-Order Interaction Network Underlies a Highly Correlated and Learnable Neural Population Code. *PNAS* **108**:9679–9684. [7]

Gansel, K. S., and W. Singer. 2012. Detecting Multineuronal Temporal Patterns in Parallel Spike Trains. *Front. Neuroinform.* **6**:18. [10, 12]

Gao, P., M. P. Postiglione, T. G. Krieger, et al. 2014. Deterministic Progenitor Behavior and Unitary Production of Neurons in the Neocortex. *Cell* **159**:775–788. [5]

Gao, P., K. T. Sultan, X. J. Zhang, and S. H. Shi. 2013. Lineage-Dependent Circuit Assembly in the Neocortex. *Development* **140**:2645–2655. [2]

Gao, P., E. Trautmann, M. Y. Byron, et al. 2017. A Theory of Multineuronal Dimensionality, Dynamics and Measurement. *bioRxiv* 214262. [12]

Garcia-Carpintero, M. 2010. Fictional Entities, Theoretical Models, and Figurative Truth. In: Beyond Mimesis and Convention: Representation in Art and Science, ed. R. Frigg and M. C. Hunter, pp. 139–168. Netherlands: Springer. [7]

Garcia-Moreno, F., E. Anderton, M. Jankowska, et al. 2018. Absence of Tangentially Migrating Glutamatergic Neurons in the Developing Avian Brain. *Cell Rep.* **22**:96–109. [5]

Garcia-Moreno, F., and Z. Molnár. 2015. Subset of Early Radial Glial Progenitors That Contribute to the Development of Callosal Neurons Is Absent from Avian Brain. *PNAS* **112**:E5058–5067. [5]

Garcia-Moreno, F., N. A. Vasistha, N. Trevia, J. A. Bourne, and Z. Molnár. 2012. Compartmentalization of Cerebral Cortical Germinal Zones in a Lissencephalic Primate and Gyrencephalic Rodent. *Cereb. Cortex* **22**:482–492. [2, 5]

Garel, S., K. J. Huffman, and J. L. Rubenstein. 2003. Molecular Regionalization of the Neocortex Is Disrupted in FGF8 Hypomorphic Mutants. *Development* **130**:1903–1914. [5]

Garel, S., and J. L. Rubenstein. 2004. Intermediate Targets in Formation of Topographic Projections: Inputs from the Thalamocortical System. *Trends Neurosci.* **27**:533–539. [5]

Ge, T., A. J. Holmes, R. L. Buckner, J. W. Smoller, and M. R. Sabuncu. 2017. Heritability Analysis with Repeat Measurements and Its Application to Resting-State Functional Connectivity. *PNAS* **114**:5521–5526. [6]

Geisler, C., D. Robbe, M. Zugaro, A. Sirota, and G. Buzsáki. 2007. Hippocampal Place Cell Assemblies Are Speed-Controlled Oscillators. *PNAS* **104**:8149–8154. [17]

Gelfert, A. 2016. How to Do Science with Models: A Philosophical Primer. Switzerland: Springer. [7]

Gelman, D. M., and O. Marin. 2010. Generation of Interneuron Diversity in the Mouse Cerebral Cortex. *Eur. J. Neurosci.* **31**:2136–2141. [5]

Gelperin, A., and D. W. Tank. 1990. Odour-Modulated Collective Network Oscillations of Olfactory Interneurons in a Terrestrial Mollusc. *Nature* **345**:437–440. [13]

Georgopoulos, A. P., J. F. Kalaska, R. Caminiti, and J. T. Massey. 1982. On the Relations between the Direction of Two-Dimensional Arm Movements and Cell Discharge in Primate Motor Cortex. *J. Neurosci.* **2**:1527–1537. [13]

Georgopoulos, A. P., A. B. Schwartz, and R. E. Kettner. 1986. Neuronal Population Coding of Movement Direction. *Science* **233**:1416–1419. [12, 13]

Gerhard, F., G. Pipa, B. Lima, S. Neuenschwander, and W. Gerstner. 2011. Extraction of Network Topology from Multi-Electrode Recordings: Is There a Small-World Effect? *Front. Comput. Neurosci.* **5**:4. [11]

Gerstein, G. L., and A. M. Aertsen. 1985. Representation of Cooperative Firing Activity among Simultaneously Recorded Neurons. *J. Neurophysiol.* **54**:1513–1528. [9]

Gerstein, G. L., and D. H. Perkel. 1972. Mutual Temporal Relationships among Neuronal Spike Trains: Statistical Techniques for Display and Analysis. *Biophys. J.* **12**:453–473. [13]

Geschwind, D. H., and P. Rakic. 2013. Cortical Evolution: Judge the Brain by Its Cover. *Neuron* **80**:633–647. [3–5]

Ghazanfar, A. A., and N. K. Logothetis. 2003. Facial Expressions Linked to Monkey Calls. *Nature* **423**:937–938. [16]

Ghazanfar, A. A., and D. Y. Takahashi. 2014. The Evolution of Speech: Vision, Rhythm, Cooperation. *Trends Cogn. Sci.* **18**:543–553. [14]

Giandomenico, S. L., and M. A. Lancaster. 2017. Probing Human Brain Evolution and Development in Organoids. *Curr. Opin. Cell Biol.* **44**:36–43. [5]

Gibson, K. R. 1991. Myelination and Behavioral Development: A Comparative Perspective on Questions of Neoteny, Altriciality and Intelligence. In: Brain Maturation and Cognitive Development: Comparative and Cross-Cultural Perspectives, ed. K. R. Gibson and A. C. Petersen, pp. 29–63. New York: Aldine de Gruyter. [16]

Gilbert, C. D., and T. N. Wiesel. 1989. Columnar Specificity of Intrinsic Horizontal and Corticocortical Connections in Cat Visual Cortex. *J. Neurosci.* **9**:2432–2442. [10]

———. 1990. The Influence of Contextual Stimuli on the Orientation Selectivity of Cells in Primary Visual Cortex of the Cat. *Vision Res.* **30**:1689–1701. [12]

Gilbert, S. L., W. B. Dobyns, and B. T. Lahn. 2005. Genetic Links between Brain Development and Brain Evolution. *Nat. Rev. Genet.* **6**:581–590. [16]

Gil-Sanz, C., A. Espinosa, S. P. Fregoso, et al. 2015. Lineage Tracing Using Cux2-Cre and Cux2-Creert2 Mice. *Neuron* **86**:1091–1099. [5]

Girardeau, G., K. Benchenane, S. I. Wiener, G. Buzsáki, and M. B. Zugaro. 2009. Selective Suppression of Hippocampal Ripples Impairs Spatial Memory. *Nat. Neurosci.* **12**:1222–1223. [13]

Giraud, A. L., and D. Poeppel. 2012. Cortical Oscillations and Speech Processing: Emerging Computational Principles and Operations. *Nat. Neurosci.* **15**:511–517. [14, 17]

Giusti, C., R. Ghrist, and D. S. Bassett. 2016. Two's Company, Three (or More) Is a Simplex: Algebraic-Topological Tools for Understanding Higher-Order Structure in Neural Data. *J. Comput. Neurosci.* **41**:1–14. [7, 9]

Glass, L., and J. Sun. 1994. Periodic Forcing of a Limit-Cycle Oscillator: Fixed Points, Arnold Tongues, and the Global Organization of Bifurcations. *Phys. Rev. E* **50**:5077–5084. [10]

Glasser, M. F., T. S. Coalson, E. C. Robinson, et al. 2016. A Multi-Modal Parcellation of Human Cerebral Cortex. *Nature* **536**:171–178. [3, 6]

Gleeson, J. G., K. M. Allen, J. W. Fox, et al. 1998. Doublecortin, a Brain-Specific Gene Mutated in Human X-Linked Lissencephaly and Double Cortex Syndrome, Encodes a Putative Signaling Protein. *Cell* **92**:63–72. [4]

Glickstein, S. B., H. Moore, B. Slowinska, et al. 2007. Selective Cortical Interneuron and GABA Deficits in Cyclin D2-Null Mice. *Development* **134**:4083–4093. [5]

Godfrey-Smith, P. 2009. Models and Fictions in Science. *Philos. Stud.* **143**:101–116. [7]

Goffinet, A., and P. Rakic, eds. 2000. Mouse Brain Development. Results and Problems in Cell Differentiation, vol. Berlin: Springer. [5]

Gokhman, D., E. Lavi, K. Prüfer, et al. 2014. Reconstructing the DNA Methylation Maps of the Neandertal and the Denisovan. *Science* **344**:523–527. [3]

Goldbeter, A. 1996. Biochemical Oscillations and Cellular Rhythms. Cambridge: Cambridge Univ. Press. [6, 9]

Goldman-Rakic. 1987. Motor control function of the prefrontal cortex. Ciba Found Symp. **132**:187–200. [5]

Gomez-Gardenes, J., Y. Moreno, and A. Arenas. 2007. Paths to Synchronization on Complex Networks. *Phys. Rev. Lett.* **98**:034101. [7]

Gould, S. J. 1977. Ontogeny and Phylogeny. Cambridge, MA: Harvard Univ. Press. [16]

Gourevitch, B., and J. J. Eggermont. 2010. Maximum Decoding Abilities of Temporal Patterns and Synchronized Firings: Application to Auditory Neurons Responding to Click Trains and Amplitude Modulated White Noise. *J. Comput. Neurosci.* **29**:253–277. [12]

Goyal, M. S., M. Hawrylycz, J. A. Miller, A. Z. Snyder, and M. E. Raichle. 2014. Aerobic Glycolysis in the Human Brain Is Associated with Development and Neotenous Gene Expression. *Cell Metab.* **19**:49–57. [9]

Grafton, S. T., and A. F. Hamilton. 2007. Evidence for a Distributed Hierarchy of Action Representation in the Brain. *Hum. Mov. Sci.* **26**:590–616. [14]

Grafton, S. T., E. Hazeltine, and R. Ivry. 1995. Functional Mapping of Sequence Learning in Normal Humans. *J. Cogn. Neurosci.* **7**:497–510. [17]

Gray, C. M., P. Konig, A. K. Engel, and W. Singer. 1989. Oscillatory Responses in Cat Visual Cortex Exhibit Inter-Columnar Synchronization Which Reflects Global Stimulus Properties. *Nature* **338**:334–337. [10–12]

Gray, C. M., and D. A. McCormick. 1996. Chattering Cells: Superficial Pyramidal Neurons Contributing to the Generation of Synchronous Oscillations in the Visual Cortex. *Science* **274**:109–113. [10]

Gray, C. M., and W. Singer. 1989. Stimulus-Specific Neuronal Oscillations in Orientation Columns of Cat Visual Cortex. *PNAS* **86**:1698–1702. [10]

Gregoriou, G. G., S. J. Gotts, H. Zhou, and R. Desimone. 2009. High-Frequency, Long-Range Coupling between Prefrontal and Visual Cortex during Attention. *Science* **324**:1207–1210. [10]

Gregory, R. L. 1980. Perceptions as Hypotheses. *Phil Trans R Soc Lond B* **290**:181–197. [14]

Gregory, T. R. 2005. Synergy between Sequence and Size in Large-Scale Genomics. *Nat. Rev. Genet.* **6**:699–708. [4]

Greig, L. C., M. B. Woodworth, M. J. Galazo, H. Padmanabhan, and J. D. Macklis. 2013. Molecular Logic of Neocortical Projection Neuron Specification, Development and Diversity. *Nat. Rev. Neurosci.* **14**:755–769. [3, 5]

Grienberger, C., X. Chen, and K. A. 2015. Dendritic Function *in Vivo*. *Trends Neurosci.* **38**:45–54. [10]

Grigoryeva, L., J. Henriques, L. Larger, and J.-P. Ortegad. 2014. Stochastic Nonlinear Time Series Forecasting Using Time-Delay Reservoir Computers: Performance and Universality. *Neural Netw.* **55**:59–71. [11]

Grillner, S. 2006. Biological Pattern Generation: The Cellular and Computational Logic of Networks in Motion. *Neuron* **52**:751–766. [10]

Grothe, I., S. D. Neitzel, S. Mandon, and A. K. Kreiter. 2012. Switching Neuronal Inputs by Differential Modulations of Gamma-Band Phase-Coherence. *Journal of Neuroscience.* **32**:16172–16180. [10]

Grove, E. A., and T. Fukuchi-Shimogori. 2003. Generating the Cerebral Cortical Area Map. *Annu. Rev. Neurosci.* **26**:355–380. [3]

Grove, E. A., and E. S. Monuki. 2013. Morphogens, Patterning Centers, and Their Mechanisms of Action. In: Patterning and Cell Type Specification in the Developing CNS and Pns, ed. J. L. R. Rubenstein and P. Rakic, pp. 25–44. San Diego: Elsevier. [5]

Grove, E. A., S. Tole, J. Limon, L. Yip, and C. W. Ragsdale. 1998. The Hem of the Embryonic Cerebral Cortex Is Defined by the Expression of Multiple Wnt Genes and Is Compromised in Gli3-Deficient Mice. *Development* **125**:2315–2325. [2]

Gruters, K. G., D. L. K. Murphy, C. D. Jenson, et al. 2018. The Eardrums Move When the Eyes Move: A Multisensory Effect on the Mechanics of Hearing. *PNAS* **115**:E1309–e1318. [13]

Gu, S., F. Pasqualetti, M. Cieslak, et al. 2015. Controllability of Structural Brain Networks. *Nat. Commun.* **6**:8414. [7, 9]

Guillamon-Vivancos, T., W. A. Tyler, M. Medalla, et al. 2018. Distinct Neocortical Progenitor Lineages Fine-Tune Neuronal Diversity in a Layer-Specific Manner. *Cereb. Cortex* **29**:1121–1138. [2]

Guillery, R. W., and H. J. Ralston. 1964. Fibers and Terminals: Electron Microscopy after Nauta Staining. *Science* **143**:1331–1332. [9]

Guo, C., M. J. Eckler, W. L. McKenna, et al. 2013. Fezf2 Expression Identifies a Multipotent Progenitor for Neocortical Projection Neurons, Astrocytes, and Oligodendrocytes. *Neuron* **80**:1167–1174. [2]

Guo, J., and E. S. Anton. 2014. Decision Making during Interneuron Migration in the Developing Cerebral Cortex. *Trends Cell Biol.* **24**:342–351. [3]

Guo, S., and J. Wu. 2013. Bifurcation Theory of Functional Differential Equations, vol. 184. Applied Mathematical Sciences. New York: Springer. [11]

Hacker, C. D., T. O. Laumann, N. P. Szrama, et al. 2013. Resting State Network Estimation in Individual Subjects. *Neuroimage* **82**:616–633. [6]

Hadj-Bouziane, F., N. Liu, A. H. Bell, et al. 2012. Amygdala Lesions Disrupt Modulation of Functional MRI Activity Evoked by Facial Expression in the Monkey Inferior Temporal Cortex. *PNAS* **109**:E3640–E3648. [9]

Haeckel, E. 1866. Generelle Morphologie Der Organismen: Allgemeine Grundzüge Der Organischen Formen-Wissenschaft, Mechanisch Begrundet Durch Die Von Charles Darwin Reformierte Descendenz-Theorie: II Allgemeine Entwickelungsgeschichte Der Organismen. Berlin: Georg Reimer. [16]

Haegeman, L. 2005. Thinking Syntactically: A Guide to Argumentation and Analysis. Blackwell Textbooks in Linguistics. Oxford: Wiley-Blackwell. [15]

Hagoort, P. 2013. Muc (Memory, Unification, Control) and Beyond. *Front. Psychol.* **4**:416. [15]

Hahnloser, R. H., A. A. Kozhevnikov, and M. S. Fee. 2002. An Ultra-Sparse Code Underlies the Generation of Neural Sequences in a Songbird. *Nature* **419**:65–70. [13]

Hall, B. K. 2013. Homology, Homoplasy, Novelty, and Behavior. *Dev. Psychobiol.* **55**:4–12. [17]

Hamasaki, T., A. Leingartner, T. Ringstedt, and D. D. O'Leary. 2004. EMX2 Regulates Sizes and Positioning of the Primary Sensory and Motor Areas in Neocortex by Direct Specification of Cortical Progenitors. *Neuron* **43**:359–372. [5]

Han, W., K. Y. Kwan, S. Shim, et al. 2011. TBR1 Directly Represses Fezf2 to Control the Laminar Origin and Development of the Corticospinal Tract. *PNAS* **108**:3041–3046. [2]

Han, W., and N. Sestan. 2013. Cortical Projection Neurons: Sprung from the Same Root. *Neuron* **80**:1103–1105. [3]

Han, Y., J. M. Kebschull, R. A. A. Campbell, et al. 2018. The Logic of Single-Cell Projections from Visual Cortex. *Nature* **556**:51–56. [9]

Hanashima, C., S. C. Li, L. Shen, E. Lai, and G. Fishell. 2004. Foxg1 Suppresses Early Cortical Cell Fate. *Science* **303**:56–59. [2]

Hanashima, C., L. Shen, S. C. Li, and E. Lai. 2002. Brain Factor-1 Controls the Proliferation and Differentiation of Neocortical Progenitor Cells through Independent Mechanisms. *J. Neurosci.* **22**:6526–6536. [2]

Hansel, C., A. Artola, and W. Singer. 1996. Different Threshold Levels of Postsynaptic [Ca2+]I Have to Be Reached to Induce LTP and LTD in Neocortical Pyramidal Cells. *J. Physiol. Paris* **90**:317–319. [10]

———. 1997. Relation between Dendritic Ca2+ Levels and the Polarity of Synaptic Long-Term Modifications in Rat Visual Cortex Neurons. *Eur. J. Neurosci.* **9**:2309–2322. [10]

Hansen, A. H., C. Duellberg, C. Mieck, M. Loose, and S. Hippenmeyer. 2017. Cell Polarity in Cerebral Cortex Development: Cellular Architecture Shaped by Biochemical Networks. *Front. Cell. Neurosci.* **11**:886–816. [5]

Hansen, D. V., J. H. Lui, P. Flandin, et al. 2013. Non-Epithelial Stem Cells and Cortical Interneuron Production in the Human Ganglionic Eminences. *Nat. Neurosci.* **16**:1576–1587. [3]

Hansen, D. V., J. H. Lui, P. R. L. Parker, and A. R. Kriegstein. 2010. Neurogenic Radial Glia in the Outer Subventricular Zone of Human Neocortex. *Nature* **464**:554–561. [2, 4, 5]

Harary, F. 1969. Graph Theory. Reading, MA: Addison-Wesley. [7]

Harper, M. 2011. Escort Evolutionary Game Theory. *Physica D* **240**:1411–1415. [14]

Harris, K. D., and G. M. Shepherd. 2015. The Neocortical Circuit: Themes and Variations. *Nat. Neurosci.* **18**:170–181. [8, 13]

Harris, K. D., and A. Thiele. 2011. Cortical State and Attention. *Nat. Rev. Neurosci.* **12**:509–523. [13]

Hartfuss, E., R. Galli, N. Heins, and M. Gotz. 2001. Characterization of CNS Precursor Subtypes and Radial Glia. *Dev. Biol.* **229**:15–30. [2]

Hartmann, C., A. Lazar, B. Nessler, and J. Triesch. 2015. Where's the Noise? Key Features of Spontaneous Activity and Neural Variability Arise through Learning in a Deterministic Network. *PLoS Comput. Biol.* **e1004640**:1–35. [10]

Hartmann, S., and R. Frigg. 2012. Scientific Models. In: The Philosophy of Science: An Encyclopedia, ed. S. Sarkar and J. Pfeifer, vol. 2, pp. 740–749. New York: Routledge. [7]

Harvey, C. D., F. Collman, D. A. Dombeck, and D. W. Tank. 2009. Intracellular Dynamics of Hippocampal Place Cells during Virtual Navigation. *Nature* **461**:941–946. [12]

Harwell, C. C., L. C. Fuentealba, A. Gonzalez-Cerrillo, et al. 2015. Wide Dispersion and Diversity of Clonally Related Inhibitory Interneurons. *Neuron* **87**:999–1007. [5]

Harwell, C. C., P. R. L. Parker, S. M. Gee, et al. 2012. Sonic Hedgehog Expression in Corticofugal Projection Neurons Directs Cortical Microcircuit Formation. *Neuron* **73**:1116–1126. [4]

Hassabis, D., and E. A. Maguire. 2007. Deconstructing Episodic Memory with Construction. *Trends Cogn. Sci.* **11**:299–306. [14]

Hassan, M. J., M. Khurshid, Z. Azeem, et al. 2007. Previously Described Sequence Variant in Cdk5rap2gene in a Pakistani Family with Autosomal Recessive Primary Microcephaly. *BMC Med. Genet.* **8**:58. [4]

Hatsopoulos, N. G., C. L. Ojakangas, L. Paniniski, and J. P. Donoghue. 1998. Information About Movement Direction Obtained from Synchronous Activity of Motor Cortical Neurons. *PNAS* **95**:15706–15711. [12]

Hatsopoulos, N. G., L. Paninski, and J. P. Donoghue. 2003. Sequential Movement Representations Based on Correlated Neuronal Activity. *Exp. Brain Res.* **149**:478–486. [12]

Hauser, M. D., N. Chomsky, and W. T. Fitch. 2002. The Faculty of Language: What Is It, Who Has It, and How Did It Evolve? *Science* **298**:1569–1579. [15]

Hauser, M. D., E. L. Newport, and R. N. Aslin. 2001. Segmentation of the Speech Stream in a Non-Human Primate: Statistical Learning in Cotton-Top Tamarins. *Cognition* **78**:B53–64. [15]

Havenith, M. N., S. Yu, J. Biederlack, et al. 2011. Synchrony Makes Neurons Fire in Sequence, and Stimulus Properties Determine Who Is Ahead. *J. Neurosci.* **31**:8570–8584. [10, 13]

Hay, E., S. Hill, F. Schurmann, H. Markram, and I. Segev. 2011. Models of Neocortical Layer 5b Pyramidal Cells Capturing a Wide Range of Dendritic and Perisomatic Active Properties. *PLoS Comput. Biol.* **7**:e1002107. [8]

Haygood, R., C. C. Babbitt, O. Fedrigo, and G. A. Wray. 2010. Contrasts between Adaptive Coding and Noncoding Changes during Human Evolution. *PNAS* **107**:7853–7857. [3, 5]

He, B. J., A. Z. Snyder, J. M. Zempel, M. D. Smyth, and M. E. Raichle. 2008. Electrophysiological Correlates of the Brain's Intrinsic Large-Scale Functional Architecture. *PNAS* **105**:16039–16044. [6]

He, S., Z. Li, S. Ge, Y. C. Yu, and S. H. Shi. 2015. Inside-out Radial Migration Facilitates Lineage-Dependent Neocortical Microcircuit Assembly. *Neuron* **86**:1159–1166. [2]

Hearst, M. A., S. T. Dumais, E. Osuna, J. Platt, and B. Scholkopf. 1998. Support Vector Machines. *IEEE Intell. Syst.* **13**:18–28. [11]

Hebb, D. O. 1949. Organization of Behavior. New York: John Wiley and Sons. [10, 12]

Heggelund, P., and K. Albus. 1978. Response Variability and Orientation Discrimination of Single Cells in Striate Cortex of Cat. *Exp. Brain Res.* **32**:197–211. [12]

Heide, M., W. B. Huttner, and F. Mora-Bermudez. 2018. Brain Organoids as Models to Study Human Neocortex Development and Evolution. *Curr. Opin. Cell Biol.* **55**:8–16. [5]

Helmholtz, H. 1866/1962. Concerning the Perceptions in General. In: Treatise on Physiological Optics, vol. III, pp. 1–37. New York: Dover. [14]

Hensch, T. K., and E. M. Quinlan. 2018. Critical Periods in Amblyopia. *Vis. Neurosci.* **35**:E014. [5]

Henshilwood, C. S., F. d'Errico, and I. Watts. 2009. Engraved Ochres from the Middle Stone Age Levels at Blombos Cave, South Africa. *J. Hum. Evol.* **57**:27–47. [15]

Henshilwood, C. S., F. d'Errico, R. Yates, et al. 2002. Emergence of Modern Human Behavior: Middle Stone Age Engravings from South Africa. *Science* **295**:1278–1280. [15]

Herculano-Houzel, S. 2009. The Human Brain in Numbers: A Linearly Scaled-up Primate Brain. *Front. Hum. Neurosci.* **3**:31. [9, 16]

———. 2012. Neuronal Scaling Rules for Primate Brains: The Primate Advantage. *Prog. Brain Res.* **195**:325–340. [9]

Herculano-Houzel, S., C. E. Collins, P. Wong, and J. H. Kaas. 2007. Cellular Scaling Rules for Primate Brains. *PNAS* **104**:3562–3567. [5]

Herculano-Houzel, S., B. Mota, P. Wong, and J. H. Kaas. 2010. Connectivity-Driven White Matter Scaling and Folding in Primate Cerebral Cortex. *PNAS* **107**:19008–19013. [5]

Herculano-Houzel, S., M. H. J. Munk, S. Neuenschwander, and W. Singer. 1999. Precisely Synchronized Oscillatory Firing Patterns Require Electroencephalographic Activation. *Journal of Neuroscience.* **19**:3992–4010. [10]

Herman, L. M., and P. Morrel-Samuels. 1995. Knowledge Acquisition and Asymmetry between Language Comprehension and Production: Dolphins and Apes as General Models for Animals. In: Readings in Animal Cognition, pp. 289–306. M. Bekoff and D. Jamieson, series ed. Cambridge, MA: MIT Press. [17]

Hernando-Herraez, I., R. Garcia-Perez, A. J. Sharp, and T. Marques-Bonet. 2015. DNA Methylation: Insights into Human Evolution. *PLoS Genet.* **11**:e1005661. [3]

Herrmann, E., J. Call, M. V. Hernández-Lloreda, B. Hare, and M. Tomasello. 2007. Humans Have Evolved Specialized Skills of Social Cognition: The Cultural Intelligence Hypothesis. *Science* **317**:1360–1366. [16]

Hertz, J., A. Krogh, and R. G. Palmer. 1991. Introduction to the Theory of Neural Computation: Westview Press. [13]

Heusser, A. C., D. Poeppel, Y. Ezzyat, and L. Davachi. 2016. Episodic Sequence Memory Is Supported by a Theta-Gamma Phase Code. *Nat. Neurosci.* **19**:1374–1380. [17]

Hevner, R. F. 2005. The Cerebral Cortex Malformation in Thanatophoric Dysplasia: Neuropathology and Pathogenesis. *Acta Neuropathol.* **110**:208–221. [5]

Hevner, R. F., and T. F. Haydar. 2012. The (Not Necessarily) Convoluted Role of Basal Radial Glia in Cortical Neurogenesis. *Cereb. Cortex* **22**:465–468. [3, 5]

Heyer, C. B., and H. D. Lux. 1976. Properties of a Facilitating Calcium Current in Pacemaker Neurons of the Snail Helix Pomatia. *J. Physiol.* **262**:319–348. [10]

Hilgetag, C. C., M. Medalla, S. F. Beul, and H. Barbas. 2016. The Primate Connectome in Context: Principles of Connections of the Cortical Visual System. *Neuroimage* **134**:685–702. [7]

Hilgetag, C. C., M. A. O'Neill, and M. P. Young. 2000. Hierarchical Organization of Macaque and Cat Cortical Sensory Systems Explored with a Novel Network Processor. *Phil. Trans. R. Soc. B* **355**:71–89. [7, 14]

Hill, R. S., and C. A. Walsh. 2005. Molecular Insights into Human Brain Evolution. *Nature* **437**:64–67. [5]

Hines, M. L., T. Morse, M. Migliore, N. T. Carnevale, and G. M. Shepherd. 2004. Modeldb: A Database to Support Computational Neuroscience. *J. Comput. Neurosci.* **17**:7–11. [9]

Hinton, G. E. 2007. Learning Multiple Layers of Representation. *Trends Cogn. Sci.* **11**:428–434. [14]

Hinton, G. E., and R. R. Salakhutdinov. 2006. Reducing the Dimensionality of Data with Neural Networks. *Science* **313**:504–507. [13]

Hinton, G. E., and D. van Camp. 1993. Keeping Neural Networks Simple by Minimizing the Description Length of Weights. In: Proc. of 6th Colt-93, pp. 5–13. New York: ACM. [14]

Hinton, G. E., and R. S. Zemel. 1993. Autoencoders, Minimum Description Length and Helmholtz Free Energy. In: Advances in Neural Information Processing Systems 6 (Nips 1993), pp. 3–10. Denver, CO: Morgan Kaufmann. [14]

Hochreiter, S., and J. Schmidhuber. 1997. Long Short-Term Memory. *Neural Comput.* **9**:1735–1780. [10, 13]

Hoerder-Suabedissen, A., and Z. Molnár. 2015. Development, Evolution and Pathology of Neocortical Subplate Neurons. *Nat. Rev. Neurosci.* **16**:133–146. [5]

Hoerder-Suabedissen, A., W. Z. Wang, S. Lee, et al. 2009. Novel Markers Reveal Subpopulations of Subplate Neurons in the Murine Cerebral Cortex. *Cereb. Cortex* **19**:1738–1750. [5]

Hohwy, J. 2016. The Self-Evidencing Brain. *Nous* **50**:259–285. [14]

Holland, M. A., K. E. Miller, and E. Kuhl. 2015. Emerging Brain Morphologies from Axonal Elongation. *Ann. Biomed. Eng.* **43**:1640–1653. [2]

Holme, P., and J. Saramaki. 2012. Temporal Networks. *Phys Rep* **519**:97–125. [7]

Holmgren, C., T. Harkany, B. Svennenfors, and Y. Zilberter. 2003. Pyramidal Cell Communication within Local Networks in Layer 2/3 of Rat Neocortex. *J. Physiol.* **551**:139–153. [12]

Holtmaat, A., and K. Svoboda. 2009. Experience-Dependent Structural Synaptic Plasticity in the Mammalian Brain. *Nat. Rev. Neurosci.* **10**:647–658. [9]

Homman-Ludiye, J., and J. A. Bourne. 2017. The Marmoset: An Emerging Model to Unravel the Evolution and Development of the Primate Neocortex. *Dev. Neurobiol.* **77**:263–272. [5]

Hopfield, J. J. 1982. Neural Networks and Physical Systems with Emergent Collective Computational Abilities. *PNAS* **79**:2554–2558. [11]

———. 1987. Learning Algorithms and Probability Distributions in Feed-Forward and Feed-Back Networks. *PNAS* **84**:8429–8433. [10]

———. 1988. Artificial Neural Networks. *IEEE Circuits Devices Mag.* **4**:3–10. [11]

Horvat, S., R. Gamanut, M. Ercsey-Ravasz, et al. 2016. Spatial Embedding and Wiring Cost Constrain the Functional Layout of the Cortical Network of Rodents and Primates. *PLoS Biol.* **14**:e1002512. [17]

Howard, B., Y. Chen, and N. Zecevic. 2006. Cortical Progenitor Cells in the Developing Human Telencephalon. *Glia* **53**:57–66. [5]

Hu, C. K., and H. E. Hoekstra. 2017. Peromyscus Burrowing: A Model System for Behavioral Evolution. *Semin. Cell Dev. Biol.* **61**:107–114. [5]

Hu, J. S., D. Vogt, M. Sandberg, and J. L. Rubenstein. 2017. Cortical Interneuron Development: A Tale of Time and Space. *Development* **144**:3867–3878. [3, 5]

Hu, W. F., M. H. Chahrour, and C. A. Walsh. 2014a. The Diverse Genetic Landscape of Neurodevelopmental Disorders. *Annual Review of Genomics and Human Genetics* **15**:195–213. [4]

Hu, W. F., O. Pomp, T. Ben-Omran, et al. 2014b. Katanin P80 Regulates Human Cortical Development by Limiting Centriole and Cilia Number. *Neuron* **84**:1240–1257. [4]

Huang, J., F. Liu, H. Tang, et al. 2017. Tranylcypromine Causes Neurotoxicity and Represses Bhc110/Lsd1 in Human-Induced Pluripotent Stem Cell-Derived Cerebral Organoids Model. *Front. Neurol.* **8**:626. [7]

Hubel, D. H., and T. N. Wiesel. 1962. Receptive Fields, Binocular Interaction and Functional Architecture in the Cat's Visual Cortex. *J. Physiol.* **160**:106–154. [13]

Humphrey, D. R. 1986. Representation of Movements and Muscles within the Primate Precentral Motor Cortex: Historical and Current Perspectives. *Fed. Proc.* **45**:2687–2699. [13]

Hunt, B., E. Ott, and J. Yorke. 1997. Differentiable Synchronisation of Chaos. *Phys. Rev. E* **55**:4029–4034. [14]

Huntley, G. W., and E. G. Jones. 1991. Relationship of Intrinsic Connections to Forelimb Movement Representations in Monkey Motor Cortex: A Correlative Anatomic and Physiological Study. *J. Neurophysiol.* **66**:390–413. [8]

Hutchison, R. M., and S. Everling. 2012. Monkey in the Middle: Why Non-Human Primates Are Needed to Bridge the Gap in Resting-State Investigations. *Front. Neuroanat.* **6**:29. [9]

Hutchison, R. M., T. Womelsdorf, E. A. Allen, et al. 2013. Dynamic Functional Connectivity: Promise, Issues, and Interpretations. *Neuroimage* **80**:360–378. [6]

Huttenlocher, P. R. 1979. Synaptic Density in Human Frontal Cortex: Developmental Changes and Effects of Aging. *Brain Res.* **163**:195–205. [3]

Huttenlocher, P. R., and C. de Courten. 1987. The Development of Synapses in Striate Cortex of Man. *Hum. Neurobiol.* **6**:1–9. [5]

Huttner, W. B., and M. Brand. 1997. Asymmetric Division and Polarity of Neuroepithelial Cells. *Curr. Opin. Neurobiol.* **7**:29–39. [2]

Huxter, J. R., T. J. Senior, K. Allen, and J. Csicsvari. 2008. Theta Phase-Specific Codes for Two-Dimensional Position, Trajectory and Heading in the Hippocampus. *Nat. Neurosci.* **11**:587–594. [13]

Hyafil, A., A. L. Giraud, L. Fontolan, and B. Gutkin. 2015. Neural Cross-Frequency Coupling: Connecting Architectures, Mechanisms, and Functions. *Trends Neurosci.* **38**:725–740. [17]

Hyvärinen, A., and E. Oja. 1998. Independent Component Analysis by General Nonlinear Hebbian-Like Learning Rules. *Signal Process.* **64**:301–313. [11]

Iacaruso, M. F., I. T. Gasler, and S. B. Hofer. 2017. Synaptic Organization of Visual Space in Primary Visual Cortex. *Nature* **547**:449–452. [10, 13]

Ishii, S., W. Yoshida, and J. Yoshimoto. 2002. Control of Exploitation-Exploration Meta-Parameter in Reinforcement Learning. *Neural Netw.* **15**:665–687. [14]

Ito, J., P. Maldonado, W. Singer, and S. Grün. 2011. Saccade-Related Modulations of Neuronal Excitability Support Synchrony of Visually Elicited Spikes. *Cereb. Cortex* **21**:2482–2497. [10]

Ito, M. 2008. Control of Mental Activities by Internal Models in the Cerebellum. *Nat. Rev. Neurosci.* **9**:304–313. [5]

Jabaudon, D. 2017. Fate and Freedom in Developing Neocortical Circuits. *Nat. Commun.* **8**:16042. [3]

Jackson, A., V. J. Gee, S. N. Baker, and R. N. Lemon. 2003. Synchrony between Neurons with Similar Muscle Fields in Monkey Motor Cortex. *Neuron* **38**:115–125. [12]

Jadhav, S. P., C. Kemere, P. W. German, and L. M. Frank. 2012. Awake Hippocampal Sharp-Wave Ripples Support Spatial Memory. *Science* **336**:1454–1458. [13]

Jaeger, H., and H. Haas. 2004. Harnessing Nonlinearity: Predicting Chaotic Systems and Saving Energy in Wireless Communication. *Science* **304**:78–80. [11]

Jamuar, S. S., and C. A. Walsh. 2015. Genomic Variants and Variations in Malformations of Cortical Development. *Pediatr. Clin. North Am.* **62**:571–585. [5]

Jayaraman, D., A. Kodani, D. M. Gonzalez, et al. 2016. Microcephaly Proteins Wdr62 and Aspm Define a Mother Centriole Complex Regulating Centriole Biogenesis, Apical Complex, and Cell Fate. *Neuron* **92**:813–828. [4]

Jeong, Y., F. C. Leskow, K. El-Jaick, et al. 2008. Regulation of a Remote Shh Forebrain Enhancer by the Six3 Homeoprotein. *Nat. Genet.* **40**:1348–1353. [4]

Jerison, H. 2012. Evolution of the Brain and Intelligence. London: Elsevier. [16]

Jia, H., N. L. Rochefort, X. Chen, and A. Konnerth. 2010. Dendritic Organization of Sensory Input to Cortical Neurons *in Vivo*. *Nature* **464**:1307–1312. [8]

Jia, X., S. Tanabe, and A. Kohn. 2013a. Gamma and the Coordination of Spiking Activity in Early Visual Cortex. *Neuron* **77**:762–774. [10]

Jia, X., D. Xing, and A. Kohn. 2013b. No Consistent Relationship between Gamma Power and Peak Frequency in Macaque Primary Visual Cortex. *J. Neurosci.* **33**:17–25. [10]

Jiang, X., S. Shen, C. R. Cadwell, et al. 2015. Principles of Connectivity among Morphologically Defined Cell Types in Adult Neocortex. *Science* **350**:aac9462. [9]

Jirsa, V. K., R. Friedrich, H. Haken, and J. A. Kelso. 1994. A Theoretical Model of Phase Transitions in the Human Brain. *Biol. Cybern.* **71**:27–35. [14]

Jobst, B. C., R. Kapur, G. L. Barkley, et al. 2017. Brain-Responsive Neurostimulation in Patients with Medically Intractable Seizures Arising from Eloquent and Other Neocortical Areas. *Epilepsia* **58**:1005–1014. [7]

Johnson, M. B., Y. I. Kawasawa, C. E. Mason, et al. 2009. Functional and Evolutionary Insights into Human Brain Development through Global Transcriptome Analysis. *Neuron* **62**:494–509. [3, 5]

Johnson, M. B., X. Sun, A. Kodani, et al. 2018. ASPM Knockout Ferret Reveals an Evolutionary Mechanism Governing Cerebral Cortical Size. *Nature* **556**:370–375. [4, 5]

Johnson, M. B., and C. A. Walsh. 2017. Cerebral Cortical Neuron Diversity and Development at Single-Cell Resolution. *Curr. Opin. Neurobiol.* **42**:9–16. [3, 5]

Johnson, M. B., P. P. Wang, K. D. Atabay, et al. 2015. Single-Cell Analysis Reveals Transcriptional Heterogeneity of Neural Progenitors in Human Cortex. *Nat. Neurosci.* **18**:637–646. [4, 5]

Jones, D. S. 1979. Elementary Information Theory. Oxford: Clarendon Press. [14]

Jordan, K. E., and E. M. Brannon. 2006. The Multisensory Representation of Number in Infancy. *PNAS* **103**:3486–3489. [16]

Jordan, K. E., E. M. Brannon, N. K. Logothetis, and A. A. Ghazanfar. 2005. Monkeys Match the Number of Voices They Hear with the Number of Faces They See. *Curr. Biol.* **15**:1034–1038. [16]

Kadmon Harpaz, N., D. Ungarish, N. G. Hatsopoulos, and T. Flash. 2018. Movement Decomposition in the Primary Motor Cortex. *Cereb. Cortex* **29**:1619–1633. [12]

Kailath, T. 1980. Linear Systems. Englewood Cliffs: Prentice Hall. [7]

Kaiser, M. 2017. Mechanisms of Connectome Development. *Trends Cogn. Sci.* **21**:703–717. [7]

Kaiser, M., and C. C. Hilgetag. 2006. Nonoptimal Component Placement, but Short Processing Paths, Due to Long-Distance Projections in Neural Systems. *PLoS Comput. Biol.* **2**:e95. [7]

Kaiser, M., C. C. Hilgetag, and A. van Ooyen. 2009. A Simple Rule for Axon Outgrowth and Synaptic Competition Generates Realistic Connection Lengths and Filling Fractions. *Cereb. Cortex* **19**:3001–3010. [7]

Kalebic, N., E. Taverna, S. Tavano, et al. 2016. CRISPR/Cas9-Induced Disruption of Gene Expression in Mouse Embryonic Brain and Single Neural Stem Cells *in Vivo*. *EMBO Rep.* **17**:338–348. 5]

Kalmbach, B. E., A. Buchin, B. Long, et al. 2018. h-Channels Contribute to Divergent Intrinsic Membrane Properties of Supragranular Pyramidal Neurons in Human versus Mouse Cerebral Cortex. *Neuron* **100**:1194–1208.e1195. [8]

Kaminski, J., J. Call, and J. Fischer. 2004. Word Learning in a Domestic Dog: Evidence for "Fast Mapping". *Science* **304**:1682–1683. [17]

Kamm, G. B., R. López-Leal, J. R. Lorenzo, and L. F. Franchini. 2013a. A Fast-Evolving Human NPAS3 Enhancer Gained Reporter Expression in the Developing Forebrain of Transgenic Mice. *Phil. Trans. R. Soc. B* **368**:20130019. [3]

Kamm, G. B., F. Pisciottano, R. Kliger, and L. F. Franchini. 2013b. The Developmental Brain Gene NPAS3 Contains the Largest Number of Accelerated Regulatory Sequences in the Human Genome. *Mol. Biol. Evol.* **30**:1088–1102. [3]

Kaneko, M., and M. P. Stryker. 2017. Homeostatic Plasticity Mechanisms in Mouse V1. *Phil. Trans. R. Soc. B* **372**:20160504. [5]

Kang, H. J., Y. I. Kawasawa, F. Cheng, et al. 2011. Spatio-Temporal Transcriptome of the Human Brain. *Nature* **478**:483–489. [3, 5]

Karmiloff-Smith, A. 2015. An Alternative to Domain-General or Domain-Specific Frameworks for Theorizing About Human Evolution and Ontogenesis. *AIMS Neurosci.* **2**:91–104. [16]

Karten, H. J. 1969. The Organization of the Avian Telencephalon and Some Speculations on the Phylogeny of the Amniote Telencephalon. *Ann. NY Acad. Sci.* **167**:164–179. [5]

Katz, P. S. 2016a. Evolution of Central Pattern Generators and Rhythmic Behaviours. *Phil. Trans. R. Soc. B* **371**:20150057. [16]

———. 2016b. Model Organisms in the Light of Evolution. *Curr. Biol.* **26**:R641–R666. [16]

Kaufman, M. T., M. M. Churchland, S. I. Ryu, and K. V. Shenoy. 2014. Cortical Activity in the Null Space: Permitting Preparation without Movement. *Nat. Neurosci.* **17**:440–448. [12, 13]

Kawaguchi, A., T. Ikawa, T. Kasukawa, et al. 2008. Single-Cell Gene Profiling Defines Differential Progenitor Subclasses in Mammalian Neurogenesis. *Development* **135**:3113–3124. [2]

Keays, D. A., G. Tian, K. Poirier, et al. 2007. Mutations in α-Tubulin Cause Abnormal Neuronal Migration in Mice and Lissencephaly in Humans. *Cell* **128**:45–57. [4]

Kebschull, J. M., P. Garcia da Silva, A. P. Reid, et al. 2016. High-Throughput Mapping of Single-Neuron Projections by Sequencing of Barcoded RNA. *Neuron* **91**:975–987. [5]

Keeney, J. G., J. M. Davis, J. Siegenthaler, et al. 2015. DUF1220 Protein Domains Drive Proliferation in Human Neural Stem Cells and Are Associated with Increased Cortical Volume in Anthropoid Primates. *Brain Struct. Funct.* **220**:3053–3060. [3]

Kelava, I., I. Reillo, A. Y. Murayama, et al. 2012. Abundant Occurrence of Basal Radial Glia in the Subventricular Zone of Embryonic Neocortex of a Lissencephalic Primate, the Common Marmoset *Callithrix jacchus*. *Cereb. Cortex* **22**:469–481. [2]

Keller, A. 1993. Intrinsic Synaptic Organization of the Motor Cortex. *Cereb. Cortex* **3**:430–441. [8]

Keller, G. B., T. Bonhoeffer, and M. Hubener. 2012. Sensorimotor Mismatch Signals in Primary Visual Cortex of the Behaving Mouse. *Neuron* **74**:809–815. [13]

Keller, G. B., and T. D. Mrsic-Flogel. 2018. Predictive Processing: A Canonical Cortical Computation. *Neuron* **100**:424–435. [13]

Kellis, M., B. Wold, M. P. Snyder, et al. 2014. Defining Functional DNA Elements in the Human Genome. *PNAS* **111**:6131–6138. [4]

Kellogg, W. N. 1968. Communication and Language in the Home-Raised Chimpanzee. *Science* **162**:423–427. [16]

Kemere, C., G. Santhanam, B. M. Yu, et al. 2008. Detecting Neural-State Transitions Using Hidden Markov Models for Motor Cortical Prostheses. *J. Neurophysiol.* **100**:2441–2452. [12]

Kenet, T., D. Bibitchkov, M. Tsodyks, A. Grinvald, and A. Arieli. 2003. Spontaneously Emerging Cortical Representations of Visual Attributes. *Nature* **425**:954–956. [9, 10, 13]

Kepecs, A., and G. Fishell. 2014. Interneuron Cell Types Are Fit to Function. *Nature* **505**:318–326. [9]

Khambhati, A. N., K. A. Davis, T. H. Lucas, B. Litt, and D. S. Bassett. 2016. Virtual Cortical Resection Reveals Push-Pull Network Control Preceding Seizure Evolution. *Neuron* **91**:1170–1182. [7]

Khambhati, A. N., A. E. Sizemore, R. F. Betzel, and D. S. Bassett. 2017. Modeling and Interpreting Mesoscale Network Dynamics. *Neuroimage* **180**:337–349. [7]

Kikuchi, Y., A. Attaheri, B. Wilson, et al. 2017. Sequence Learning Modulates Neural Responses and Oscillatory Coupling in Human and Monkey Auditory Cortex. *PLoS Biol.* **15**:e2000219. [17]

Kilavik, B. E., S. Roux, A. Ponce-Alvarez, et al. 2009. Long-Term Modifications in Motor Cortical Dynamics Induced by Intensive Practice. *J. Neurosci.* **29**:12653–12663. [12]

Kim, J. Z., J. M. Soffer, A. E. Kahn, et al. 2018. Role of Graph Architecture in Controlling Dynamical Networks with Applications to Neural Systems. *Nat. Phys.* **14**:91–98. [7, 9]

Kim, S., M. K. Lehtinen, A. Sessa, et al. 2010. The Apical Complex Couples Cell Fate and Cell Survival to Cerebral Cortical Development. *Neuron* **66**:69–84. [2]

Kim, S. Y., and W. Lim. 2015. Fast Sparsely Synchronized Brain Rhythms in a Scale-Free Neural Network. *Phys Rev E* **92**:022717. [7]

King, A. J., and D. R. Moore. 1991. Plasticity of Auditory Maps in the Brain. *Trends Neurosci.* **14**:31–37. [16]

King, M.-C., and A. C. Wilson. 1975. Evolution at Two Levels in Humans and Chimpanzees. *Science* **188**:107–116. [3, 5]

Kingsbury, M. A., and B. L. Finlay. 2001. The Cortex in Multidimensional Space: Where Do Cortical Areas Come From? *Dev. Sci.* **4**:125–157. [16]

Kitzbichler, M. G., M. L. Smith, S. R. Christensen, and E. Bullmore. 2009. Broadband Criticality of Human Brain Network Synchronization. *PLoS Comput. Biol.* **5**:e1000314. [14]

Kivel, M., A. Arenas, M. Barthelemy, et al. 2014. Multilayer Networks. *J. Complex Netw.* **2**:203–271. [7]

Kleinfeld, D., A. Bharioke, P. Blinder, et al. 2011. Large-Scale Automated Histology in the Pursuit of Connectomes. *J. Neurosci.* **31**:16125–16138. [8]

Kleinfeld, D., K. R. Delaney, M. S. Fee, et al. 1994. Dynamics of Propagating Waves in the Olfactory Network of a Terrestrial Mollusk: An Electrical and Optical Study. *J. Neurophysiol.* **72**:1402–1419. [13]

Kleinschmidt, A., M. F. Bear, and W. Singer. 1987. Blockade of "NMDA" Receptors Disrupts Experience-Dependent Plasticity of Kitten Striate Cortex. *Science* **238**:355–358. [10]

Klimm, F., D. S. Bassett, J. M. Carlson, and P. J. Mucha. 2014. Resolving Structural Variability in Network Models and the Brain. *PLoS Comput. Biol.* **10**:e1003491. [7]

Kloeden, P. E., and M. Rasmussen, eds. 2011. Nonautonomous Dynamical Systems. Mathematical Surveys and Monographs, vol. 176. R. H. Cohen et al., series ed. Providence: American Mathematical Society. [14]

Knill, D. C., and A. Pouget. 2004. The Bayesian Brain: The Role of Uncertainty in Neural Coding and Computation. *Trends Neurosci.* **27**:712–719. [14]

Knutsen, A. K., C. D. Kroenke, Y. V. Chang, L. A. Taber, and P. V. Bayly. 2013. Spatial and Temporal Variations of Cortical Growth during Gyrogenesis in the Developing Ferret Brain. *Cereb. Cortex* **23**:488–498. [2]

Ko, H., S. B. Hofer, B. Pichler, et al. 2011. Functional Specificity of Local Synaptic Connections in Neocortical Networks. *Nature* **473**:87–91. [12]

Kodani, A., T. W. Yu, J. R. Johnson, et al. 2015. Centriolar Satellites Assemble Centrosomal Microcephaly Proteins to Recruit Cdk2 and Promote Centriole Duplication. *eLife* **4**:e07519. [4]

Kohn, A., and M. A. Smith. 2005. Stimulus Dependence of Neuronal Correlation in Primary Visual Cortex of the Macaque. *Journal of Neuroscience.* **25**:3661–3673. [10, 13]

Kohwi, M., M. A. Petryniak, J. E. Long, et al. 2007. A Subpopulation of Olfactory Bulb GABAergic Interneurons Is Derived from Emx1- and Dlx5/6-Expressing Progenitors. *J. Neurosci.* **27**:6878–6891. [5]

Kole, M. H. 2011. First Node of Ranvier Facilitates High-Frequency Burst Encoding. *Neuron* **71**:671–682. [8]

Koles, K., J. Nunnari, C. Korkut, et al. 2012. Mechanism of Evenness Interrupted (Evi)-Exosome Release at Synaptic Boutons. *J. Biol. Chem.* **287**:16820–16834. [4]

König, P., A. K. Engel, P. R. Roelfsema, and W. Singer. 1995. How Precise Is Neuronal Synchronization? *Neural Comput.* **7**:469–485. [13]

König, P., and T. B. Schillen. 1993. Assembly Formation and Segregation by a Self-Organizing Neuronal Oscillator Model. In: Computation and Neural Systems, ed. F. H. Eeckman and J. M. Bower, pp. 509–514. Boston: Kluwer Academic Publishers. [10]

Konig, S. D., and E. A. Buffalo. 2016. Modeling Visual Exploration in Rhesus Macaques with Bottom-up Salience and Oculomotor Statistics. *Front Integr Neurosci* **10**:23. [14]

Konner, M. 1991. Universals of Behavioral Development in Relation to Brain Myelination. In: Brain Maturation and Cognitive Development: Comparative and Cross-Cultural Perspectives, ed. K. R. Gibson and A. C. Petersen, pp. 181–223. New York: Aldine de Gruyter. [16]

Konopka, G., T. Friedrich, J. Davis-Turak, et al. 2012. Human-Specific Transcriptional Networks in the Brain. *Neuron* **75**:601–617. [4, 5]

Konorski, J. 1967. Integrative Activity of the Brain: An Interdisciplinary Approach. Chicago: Univ. of Chicago Press. [13]

Kopan, R., and M. X. Ilagan. 2009. The Canonical Notch Signaling Pathway: Unfolding the Activation Mechanism. *Cell* **137**:216–233. [2]

Kopell, N. J., G. B. Ermentrout, M. A. Whittington, and R. D. Traub. 2000. Gamma Rhythms and Beta Rhythms Have Different Synchronization Properties. *PNAS* **97**:1867–1872. [10]

Kopell, N., M. A. Whittington, and M. A. Kramer. 2011. Neuronal Assembly Dynamics in the Beta1 Frequency Range Permits Short-Term Memory. *PNAS* **108**:3779–3784. [14]

Kopell, N. J., H. J. Gritton, M. A. Whittington, and M. A. Kramer. 2014. Beyond the Connectome: The Dynome. *Neuron* **83**:1319–1328. [7]

Korkut, C., B. Ataman, P. Ramachandran, et al. 2009. Trans-Synaptic Transmission of Vesicular Wnt Signals through Evi/Wntless. *Cell* **139**:393–404. [4]

Kornack, D. R., and P. Rakic. 1998. Changes in Cell-Cycle Kinetics during the Development and Evolution of Primate Neocortex. *PNAS* **95**:1242–1246. [5]

Korndörfer, C., E. Ullner, J. García-Ojalvo, and G. Pipa. 2017. Cortical Spike Synchrony as a Measure of Input Familiarity. *Neural Comput.* **29**:2491–2510. [10, 11]

Kornfeld, J., and W. Denk. 2018. Progress and Remaining Challenges in High-Throughput Volume Electron Microscopy. *Curr. Opin. Neurobiol.* **50**:261–267. [13]

Kostovic, I., and M. E. Molliver. 1974. A New Interpretation of Laminar Development of Cerebral-Cortex: Synaptogenesis in Different Layers of Neopallium in Human Fetus. *Anat. Rec.* **178**:395. [5]

Kostovic, I., and P. Rakic. 1990. Developmental History of the Transient Subplate Zone in the Visual and Somatosensory Cortex of the Macaque Monkey and Human Brain. *J. Comp. Neur.* **297**:441–470. [2, 5]

Kouprina, N., A. Pavlicek, G. H. Mochida, et al. 2004. Accelerated Evolution of the ASPM Gene Controlling Brain Size Begins Prior to Human Brain Expansion. *PLoS Biol.* **2**:e126. [4, 16]

Kowalczyk, T., A. Pontious, C. Englund, et al. 2009. Intermediate Neuronal Progenitors (Basal Progenitors) Produce Pyramidal-Projection Neurons for All Layers of Cerebral Cortex. *Cereb. Cortex* **19**:2439–2450. [2, 5]

Krakauer, J. W., A. A. Ghazanfar, A. Gomez-Marin, M. A. MacIver, and D. Poeppel. 2017. Neuroscience Needs Behavior: Correcting a Reductionist Bias. *Neuron* **93**:480–490. [16, 17]

Kraus, B. J., R. J. Robinson, 2nd, J. A. White, H. Eichenbaum, and M. E. Hasselmo. 2013. Hippocampal "Time Cells": Time versus Path Integration. *Neuron* **78**:1090–1101. [17]

Kroenke, C. D., and P. V. Bayly. 2018. How Forces Fold the Cerebral Cortex. *J. Neurosci.* **38**:767–775. [5]

Krubitzer, L., and J. Kaas. 2005. The Evolution of the Neocortex in Mammals: How Is Phenotypic Diversity Generated? *Curr. Opin. Neurobiol.* **15**:444–453. [3, 16]

Krubitzer, L., and D. S. Stolzenberg. 2014. The Evolutionary Masquerade: Genetic and Epigenetic Contributions to the Neocortex. *Curr. Opin. Neurobiol.* **24**:157–165. [5]

Kruskal, P. B., L. Li, and J. N. Maclean. 2013. Circuit Reactivation Dynamically Regulates Synaptic Plasticity in Neocortex. *Nat. Commun.* **4**:2574. [12]

Kuhl, P. K., and A. N. Meltzoff. 1982. The Bimodal Perception of Speech in Infancy. *Science* **218**:1138–1141. [16]

Kuhn, T. S. 1962. The Structure of Scientific Revolutions. Chicago: Univ. of Chicago Press. [17]

Kuida, K., T. F. Haydar, C. Y. Kuan, et al. 1998. Reduced Apoptosis and Cytochrome C-Mediated Caspase Activation in Mice Lacking Caspase 9. *Cell* **94**:325–337. [5]

Kumar, S., and S. B. Hedges. 1998. A Molecular Timescale for Vertebrate Evolution. *Nature* **392**:917–920. [17]

Kundu, B., D. W. Sutterer, S. M. Emrich, and B. R. Postle. 2013. Strengthened Effective Connectivity Underlies Transfer of Working Memory Training to Tests of Short-Term Memory and Attention. *Journal of Neuroscience.* **33**:8705–8715. [10]

Kuramoto, Y. 1990. Collective Synchronization of Pulse-Coupled Oscillators and Excitable Units. *Physica D* **50**:15–30. [10]

Kwan, K. Y., M. M. Lam, Z. Krsnik, et al. 2008. SOX5 Postmitotically Regulates Migration, Postmigratory Differentiation, and Projections of Subplate and Deep-Layer Neocortical Neurons. *PNAS* **105**:16021–16026. [2]

Kwan, K. Y., M. M. S. Lam, M. B. Johnson, et al. 2012a. Species-Dependent Posttranscriptional Regulation of Nos1 by FMRP in the Developing Cerebral Cortex. *Cell* **149**:899–911. [3]

Kwan, K. Y., N. Sestan, and E. S. Anton. 2012b. Transcriptional Co-Regulation of Neuronal Migration and Laminar Identity in the Neocortex. *Development* **139**:1535–1546. [5]

Lachenal, G., K. Pernet-Gallay, M. Chivet, et al. 2011. Release of Exosomes from Differentiated Neurons and Its Regulation by Synaptic Glutamatergic Activity. *Mol. Cell. Neurosci.* **46**:409–418. [4]

Laclef, C., and C. Metin. 2018. Conserved Rules in Embryonic Development of Cortical Interneurons. *Semin. Cell Dev. Biol.* **76**:86–100. [5]

Lagorce, X., and R. Benosman. 2015. Stick: Spike Time Interval Computational Kernel, a Framework for General Purpose Computation Using Neurons, Precise Timing, Delays, and Synchrony. *Neural Comput.* **27**:2261–2317. [13]

Lai, C. S. L., S. E. Fisher, J. A. Hurst, F. Vargha-Khadem, and A. P. Monaco. 2001. A Forkhead-Domain Gene Is Mutated in a Severe Speech and Language Disorder. *Nature* **413**:519–523. [4]

Lakatos, P., G. Karmos, A. D. Mehta, I. Ulbert, and C. E. Schroeder. 2008. Entrainment of Neuronal Oscillations as a Mechanism of Attentional Selection. *Science* **320**:110–113. [10]

Lakatos, P., G. Musacchia, M. N. O'Connel, et al. 2013. The Spectrotemporal Filter Mechanism of Auditory Selective Attention. *Neuron* **77**:750–761. [10]

Lake, B. B., R. Ai, G. E. Kaeser, et al. 2016. Neuronal Subtypes and Diversity Revealed by Single-Nucleus RNA Sequencing of the Human Brain. *Science* **352**:1586–1590. [3]

Lake, B. M., T. D. Ullman, J. B. Tenenbaum, and S. J. Gershman. 2017. Building Machines That Learn and Think Like People. *Behav. Brain Sci.* **40**:e253. [17]

Lako, M., S. Lindsay, P. Bullen, et al. 1998. A Novel Mammalian *wnt* Gene, WNT8B, Shows Brain-Restricted Expression in Early Development, with Sharply Delimited Expression Boundaries in the Developing Forebrain. *Hum. Mol. Genet.* **7**:813–822. [2]

LaMantia, A. S., and P. Rakic. 1990. Axon Overproduction and Elimination in the Corpus Callosum of the Developing Rhesus Monkey. *J. Neurosci.* **10**:2156–2175. [9]

Lancaster, M. A., M. Renner, C.-A. Martin, et al. 2013. Cerebral Organoids Model Human Brain Development and Microcephaly. *Nature* **501**:373–379. [5]

Landau, A. N., and P. Fries. 2012. Attention Samples Stimuli Rhythmically. *Curr. Biol.* **22**:1000–1004. [10]

Landau, I. D., R. Egger, V. J. Dercksen, M. Oberlaender, and H. Sompolinsky. 2016. The Impact of Structural Heterogeneity on Excitation-Inhibition Balance in Cortical Networks. *Neuron* **92**:1106–1121. [8]

Lander, E. S., L. M. Linton, B. Birren, et al. 2001. Initial Sequencing and Analysis of the Human Genome. *Nature* **409**:860–921. [3, 4]

Lange, C., W. B. Huttner, and F. Calegari. 2009. Cdk4/Cyclind1 Overexpression in Neural Stem Cells Shortens G1, Delays Neurogenesis, and Promotes the Generation and Expansion of Basal Progenitors. *Cell Stem Cell* **5**:320–331. [5]

Larger, L., M. C. Soriano, D. Brunner, et al. 2012. Photonic Information Processing Beyond Turing: An Optoelectronic Implementation of Reservoir Computing. *Opt. Express* **20**:3241–3249. [11]

Larkum, M. 2013. A Cellular Mechanism for Cortical Associations: An Organizing Principle for the Cerebral Cortex. *Trends Neurosci.* **36**:141–151. [9, 10]

Lashley, K. S. 1951. The Problem of Serial Order in Behavior. In: Cerebral Mechanisms in Behavior; the Hixon Symposium, ed. L. A. Jeffress, p. 311. New York: Wiley. [15]

Laumann, T. O., A. Z. Snyder, A. Mitra, et al. 2016. On the Stability of BOLD fMRI Correlations. *Cereb. Cortex* **27**:4719–4732. [6]

Lazar, A., G. Pipa, and J. Triesch. 2009. SORN: A Self-Organizing Recurrent Neural Network. *Front. Comput. Neurosci.* **3**:1–9. [10, 13]

LeCun, Y., Y. Bengio, and G. Hinton. 2015. Deep Learning. *Nature* **521**:436–444. [10, 14, 17]

Lee, J. H., M. A. Whittington, and N. J. Kopell. 2013a. Top-Down Beta Rhythms Support Selective Attention via Interlaminar Interaction: A Model. *PLoS Comput. Biol.* **9**:e1003164. [14]

Lee, S., I. Kruglikov, Z. J. Huang, G. Fishell, and B. Rudy. 2013b. A Disinhibitory Circuit Mediates Motor Integration in the Somatosensory Cortex. *Nat. Neurosci.* **16**:1662–1670. [9]

Lee, W.-C. A., V. Bonin, M. Reed, et al. 2016. Anatomy and Function of an Excitatory Network in the Visual Cortex. *Nature* **532**:370–374. [4]

Lefort, S., C. Tomm, J. C. Floyd Sarria, and C. C. Petersen. 2009. The Excitatory Neuronal Network of the C2 Barrel Column in Mouse Primary Somatosensory Cortex. *Neuron* **61**: 301–316. [9]

Legenstein, R., and W. Maass. 2007. Edge of Chaos and Prediction of Computational Performance for Neural Circuit Models. *Neural Netw.* **20**:323–334. [13]

Lehky, S. R., and T. J. Sejnowski. 1988. Network Model of Shape-from-Shading: Neural Function Arises from Both Receptive and Projective Fields. *Nature* **333**:452–454. [13]

Lehtinen, M. K., M. W. Zappaterra, X. Chen, et al. 2011. The Cerebrospinal Fluid Provides a Proliferative Niche for Neural Progenitor Cells. *Neuron* **69**:893–905. [2]

Lein, E. S., L. E. Borm, and S. Linnarsson. 2017. The Promise of Spatial Transcriptomics for Neuroscience in the Era of Molecular Cell Typing. *Science* **358**:64–69. [3]

Lemon, R. N., and J. Griffiths. 2005. Comparing the Function of the Corticospinal System in Different Species: Organizational Differences for Motor Specialization? *Muscle Nerve* **32**:261–279. [16]

Leone, D. P., K. Srinivasan, B. Chen, E. Alcamo, and S. K. McConnell. 2008. The Determination of Projection Neuron Identity in the Developing Cerebral Cortex. *Curr. Opin. Neurobiol.* **18**:28–35. [3, 5]

Leopold, D. A., Y. Murayama, and N. K. Logothetis. 2003. Very Slow Activity Fluctuations in Monkey Visual Cortex: Implications for Functional Brain Imaging. *Cereb. Cortex* **13**:422–433. [6]

Letzkus, J. J., S. B. Wolff, E. M. Meyer, et al. 2011. A Disinhibitory Microcircuit for Associative Fear Learning in the Auditory Cortex. *Nature* **480**:331–335. [9]

Leugering, J., and G. Pipa. 2018. A Unifying Framework of Synaptic and Intrinsic Plasticity in Neural Populations. *Neural Comput.* **30**:945–986. [11]

Levine, M. W., and J. M. Shefner. 1977. A Model for the Variability of Interspike Intervals during Sustained Firing of a Retinal Neuron. *Biophys. J.* **19**:241–252. [11]

Levinthal, D. J., and P. L. Strick. 2012. The Motor Cortex Communicates with the Kidney. *J. Neurosci.* **32**:6726–6731. [9]

Levitt, P., and P. Rakic. 1980. Immunoperoxidase Localization of Glial Fibrillary Acidic Protein in Radial Glial Cells and Astrocytes of the Developing Rhesus Monkey Brain. *J. Comp. Neur.* **193**:815–840. [5]

Lewis, C. M., A. Baldassarre, G. Committeri, G. L. Romani, and M. Corbetta. 2009. Learning Sculpts the Spontaneous Activity of the Resting Human Brain. *PNAS* **106**:17558–17563. [10]

Lewitus, E., I. Kelava, A. T. Kalinka, P. Tomancak, and W. B. Huttner. 2014. An Adaptive Threshold in Mammalian Neocortical Evolution. *PLoS Biol.* **12**:e1002000. [5]

Li, Q., S. Zheng, A. Han, et al. 2014. The Splicing Regulator Ptbp2 Controls a Program of Embryonic Splicing Required for Neuronal Maturation. *eLife* **3**:e01201. [4]

Liao, Q., and T. Poggio. 2016. Bridging the Gaps between Residual Learning, Recurrent Neural Networks and Visual Cortex. *arXiv* 1604.03640. [17]

Liao, X., A. V. Vasilakos, and Y. He. 2017. Small-World Human Brain Networks: Perspectives and Challenges. *Neurosci. Biobehav. Rev.* **77**:286–300. [7]

Licatalosi, D. D., M. Yano, J. J. Fak, et al. 2012. Ptbp2 Represses Adult-Specific Splicing to Regulate the Generation of Neuronal Precursors in the Embryonic Brain. *Genes Dev.* **26**:1626–1642. [4]

Lieberman, P. H., D. H. Klatt, and W. H. Wilson. 1969. Vocal Tract Limitations on the Vowel Repertoires of Rhesus Monkey and Other Nonhuman Primates. *Science* **164**:1185–1187. [16]

Liegeois, R., T. O. Laumann, A. Z. Snyder, J. Zhou, and B. T. T. Yeo. 2017. Interpreting Temporal Fluctuations in Resting-State Functional Connectivity MRI. *Neuroimage* **163**:437–455. [6]

Lima, B., W. Singer, N.-H. Chen, and S. Neuenschwander. 2010. Synchronization Dynamics in Response to Plaid Stimuli in Monkey V1. *Cereb. Cortex* **20**:1556–1573. [10]

Lima, B., W. Singer, and S. Neuenschwander. 2011. Gamma Responses Correlate with Temporal Expectation in Monkey Primary Visual Cortex. *Journal of Neuroscience.* **31**:15919–15931. [10]

Lin, I. C., M. Okun, M. Carandini, and K. D. Harris. 2015. The Nature of Shared Cortical Variability. *Neuron* **87**:644–656. [12]

Lindblad-Toh, K., M. Garber, O. Zuk, et al. 2011. A High-Resolution Map of Human Evolutionary Constraint Using 29 Mammals. *Nature* **478**:476–482. [5]

Linsker, R. 1990. Perceptual Neural Organization: Some Approaches Based on Network Models and Information Theory. *Annu. Rev. Neurosci.* **13**:257–281. [14]

Lisman, J. E. 2005. The Theta/Gamma Discrete Phase Code Occuring during the Hippocampal Phase Precession May Be a More General Brain Coding Scheme. *Hippocampus* **15**:913–922. [17]

Lisman, J. E., and M. A. Idiart. 1995. Storage of 7 +/- 2 Short-Term Memories in Oscillatory Subcycles. *Science* **267**:1512–1515. [17]

Liu, Y.-Y., J.-J. Slotine, and A.-L. Barabási. 2011. Controllability of Complex Networks. *Nature* **473**:167–173. [7]

Livingstone, M. S., W. W. Pettine, K. Srihasam, et al. 2014. Symbol Addition by Monkeys Provides Evidence for Normalized Quantity Coding. *PNAS* **111**:6822–6827. [17]

Livingstone, M. S., K. Srihasam, and I. A. Morocz. 2010. The Benefit of Symbols: Monkeys Show Linear, Human-Like, Accuracy When Using Symbols to Represent Scalar Value. *Anim. Cogn.* **13**:711–719. [17]

Lizarraga, S. B., S. P. Margossian, M. H. Harris, et al. 2010. CDK5RAP2 Regulates Centrosome Function and Chromosome Segregation in Neuronal Progenitors. *Development* **137**:1907–1917. [4]

Llinás, R. 2001. I of the Vortex. Cambridge, MA: MIT Press. [6]

Lo, C. Y., T. W. Su, C. C. Huang, et al. 2015. Randomization and Resilience of Brain Functional Networks as Systems-Level Endophenotypes of Schizophrenia. *PNAS* **112**:9123–9128. [7]

Lodato, S., and P. Arlotta. 2015. Generating Neuronal Diversity in the Mammalian Cerebral Cortex. *Annu. Rev. Cell. Dev. Biol.* **31**:699–720. [3]

Lodato, S., C. Rouaux, K. B. Quast, et al. 2011. Excitatory Projection Neuron Subtypes Control the Distribution of Local Inhibitory Interneurons in the Cerebral Cortex. *Neuron* **69**:763–779. [5]

Lodato, S., A. S. Shetty, and P. Arlotta. 2015. Cerebral Cortex Assembly: Generating and Reprogramming Projection Neuron Diversity. *Trends Neurosci.* **38**:117–125. [5]

Logothetis, N. K. 2008. What We Can Do and What We Cannot Do with fMRI. *Nature* **453**:869–878. [6, 9]

London, M., and M. Häusser. 2005. Dendritic Computation. *Annu. Rev. Neurosci.* **28**:503–532. [9, 11]

Long, K. R., B. Newland, M. Florio, et al. 2018. Extracellular Matrix Components HAPLN1, Lumican, and Collagen I Cause Hyaluronic Acid-Dependent Folding of the Developing Human Neocortex. *Neuron* **99**:702–719. [5]

Lo Nigro, C., S. S. Chong, A. C. M. Smith, et al. 1997. Point Mutations and an Intragenic Deletion in Lis1, the Lissencephaly Causative Gene in Isolated Lissencephaly Sequence and Miller-Dieker Syndrome. *Hum. Mol. Genet.* **6**:157–164. [4]

Lord, L. D., P. Expert, H. M. Fernandes, et al. 2016. Insights into Brain Architectures from the Homological Scaffolds of Functional Connectivity Networks. *Front. Syst. Neurosci.* **10**:85. [7]

Löwel, S., and W. Singer. 1992. Selection of Intrinsic Horizontal Connections in the Visual Cortex by Correlated Neuronal Activity. *Science* **255**:209–212. [10]

Lowet, E., M. J. Roberts, C. A. Bosman, P. Fries, and P. De Weerd. 2016. Areas V1 and V2 Show Microsaccade-Related 3–4-Hz Covariation in Gamma Power and Frequency. *Eur. J. Neurosci.* **43**:1286–1296. [10]

Lowet, E., M. J. Roberts, B. Gips, P. De Weerd, and A. Peter. 2017. A Quantitative Theory of Gamma Synchronization in Macaque V1. *eLife* **6**:e26642. [10]

Lu, H., Q. Zou, H. Gu, et al. 2012. Rat Brains Also Have a Default Mode Network. *PNAS* **109**:3979–3984. [6]

Lubenov, E. V., and A. G. Siapas. 2009. Hippocampal Theta Oscillations Are Travelling Waves. *Nature* **459**:534–539. [12, 13]

Luczak, A., P. Bartho, S. L. Marguet, G. Buzsáki, and K. D. Harris. 2007. Sequential Structure of Neocortical Spontaneous Activity *in Vivo*. *PNAS* **104**:347–352. [12]

Luczak, A., and J. N. Maclean. 2012. Default Activity Patterns at the Neocortical Microcircuit Level. *Front. Integr. Neurosci.* **6**:30. [12]

Luczak, A., B. L. McNaughton, and K. D. Harris. 2015. Packet-Based Communication in the Cortex. *Nat. Rev. Neurosci.* **16**:745–755. [12]

Luhmann, H. J., and R. Khazipov. 2018. Neuronal Activity Patterns in the Developing Barrel Cortex. *Neuroscience* **368**:256–267. [13]

Lui, J. H., D. V. Hansen, and A. R. Kriegstein. 2011. Development and Evolution of the Human Neocortex. *Cell* **146**:18–36. [3, 5, 16]

Lui, J. H., T. J. Nowakowski, A. A. Pollen, et al. 2014. Radial Glia Require PDGFD–PDGFR ℜ Signalling in Human but Not Mouse Neocortex. *Nature* **515**:264–268. [5]

Lukaszewicz, A., P. Savatier, V. Cortay, et al. 2005. G1 Phase Regulation, Area-Specific Cell Cycle Control, and Cytoarchitectonics in the Primate Cortex. *Neuron* **47**:353–364. [2, 5]

Lukoševičius, M., and H. Jaeger. 2009. Reservoir Computing Approaches to Recurrent Neural Network Training. *Comput. Sci. Rev.* **3**:127–149. [10]

Lundqvist, M., J. Rose, P. Herman, et al. 2016. Gamma and Beta Bursts Underlie Working Memory. *Neuron* **90**:152–164. [10]

Lur, G., M. A. Vinck, L. Tang, J. A. Cardin, and M. J. Higley. 2016. Projection-Specific Visual Feature Encoding by Layer 5 Cortical Subnetworks. *Cell Rep.* **14**:2538–2545. [8, 13]

Luskin, M. B., and C. J. Shatz. 1985. Studies of the Earliest Generated Cells of the Cat's Visual Cortex: Cogeneration of Subplate and Marginal Zones. *J. Neurosci.* **5**:1062–1075. [5]

Lynall, M. E., D. S. Bassett, R. Kerwin, et al. 2010. Functional Connectivity and Brain Networks in Schizophrenia. *J. Neurosci.* **30**:9477–9487. [7]

Ma, T., C. Wang, L. Wang, et al. 2013. Subcortical Origins of Human and Monkey Neocortical Interneurons. *Nat. Neurosci.* **16**:1588–1597. [3]

Maass, W. 2006. On the Computational Power of Winner-Take-All. *Neural Comput.* **12**:2519–2535. [11]

Maass, W., T. Natschläger, and H. Markram. 2002. Real-Time Computing without Stable States: A New Framework for Neural Computation Based on Perturbations. *Neural Comput.* **14**:2531–2560. [11, 13, 14]

MacKay, D. J. 1995. Free-Energy Minimisation Algorithm for Decoding and Cryptoanalysis. *Electron. Lett.* **31**:445–447. [14]

MacLean, J. N., B. O. Watson, G. B. Aaron, and R. Yuste. 2005. Internal Dynamics Determine the Cortical Response to Thalamic Stimulation. *Neuron* **48**:811–823. [12]

MacLean, P. D. 1990. The Triune Brain in Evolution: Role in Paleocerebral Functions. New York and London: Plenum Press. [16]

Macosko, E. Z., A. Basu, R. Satija, et al. 2015. Highly Parallel Genome-Wide Expression Profiling of Individual Cells Using Nanoliter Droplets. *Cell* **161**:1202–1214. [5]

Magistretti, P. J. 2000. Cellular Bases of Functional Brain Imaging: Insights from Neuron-Glia Metabolic Coupling. *Brain Res.* **886**:108–112. [9]

Mahadevan, A. R., N. E. Grandel, J. T. Robinson, and A. A. Qutub. 2017. Living Neural Networks: Dynamic Network Analysis of Developing Neural Progenitor Cells. *bioRxiv* [7]

Mainen, Z. F., and T. J. Sejnowski. 1995. Reliability of Spike Timing in Neocortical Neurons. *Science* **268**:1503–1506. [10]

Majaj, N. J., H. Hong, E. A. Solomon, and J. J. DiCarlo. 2015. Simple Learned Weighted Sums of Inferior Temporal Neuronal Firing Rates Accurately Predict Human Core Object Recognition Performance. *J. Neurosci.* **35**:13402–13418. [13]

Malatesta, P., E. Hartfuss, and M. Gotz. 2000. Isolation of Radial Glial Cells by Fluorescent-Activated Cell Sorting Reveals a Neuronal Lineage. *Development* **127**:5253–5263. [2, 5]

Maldonado, P., C. Babul, W. Singer, et al. 2008. Synchronization of Neuronal Responses in Primary Visual Cortex of Monkeys Viewing Natural Images. *J. Neurophysiol.* **100**:1523–1532. [10]

Malenka, R. C., and A. R. A. Nicoll. 1999. Long-Term Potentiation: A Decade of Progress? *Science* **285**:1870–1874. [9]

Malkova, L., E. Heuer, and R. C. Saunders. 2006. Longitudinal Magnetic Resonance Imaging Study of Rhesus Monkey Brain Development. *Eur. J. Neurosci.* **24**:3204–3212. [16]

Mante, V., D. Sussillo, K. V. Shenoy, and W. T. Newsome. 2013. Context-Dependent Computation by Recurrent Dynamics in Prefrontal Cortex. *Nat. Biotechnol.* **503**:78–84. [13]

Manzini, M. C., D. E. Tambunan, R. S. Hill, et al. 2012. Exome Sequencing and Functional Validation in Zebrafish Identify Gtdc2 Mutations as a Cause of Walker-Warburg Syndrome. *The American Journal of Human Genetics* **91**:541–547. [4]

Marchetto, M. C. N., I. Narvaiza, A. M. Denli, et al. 2013. Differential L1 Regulation in Pluripotent Stem Cells of Humans and Apes. *Nature* **503**:525–529. [5]

Marder, E., and D. Buchner. 2001. Central Pattern Generators and the Control of Rhythmic Movements. *Curr. Biol.* **11**:R986–R996. [10]

Marder, E., and J. M. Goaillard. 2006. Variability, Compensation and Homeostasis in Neuron and Network Function. *Nat. Rev. Neurosci.* **7**:563–574. [8]

Marder, E., and A. L. Taylor. 2011. Multiple Models to Capture the Variability in Biological Neurons and Networks. *Nat. Neurosci.* **14**:133–138. [8]

Mariani, J., G. Coppola, P. Zhang, et al. 2015. FOXG1-Dependent Dysregulation of GABA/Glutamate Neuron Differentiation in Autism Spectrum Disorders. *Cell* **162**:375–390. [5]

Mariani, J., M. V. Simonini, D. Palejev, et al. 2012. Modeling Human Cortical Development *in Vitro* Using Induced Pluripotent Stem Cells. *PNAS* **109**:12770–12775. [5]

Marin, O., A. Yaron, A. Bagri, M. Tessier-Lavigne, and J. L. Rubenstein. 2001. Sorting of Striatal and Cortical Interneurons Regulated by Semaphorin-Neuropilin Interactions. *Science* **293**:872–875. [5]

Marin-Padilla, M. 1971. Early Prenatal Ontogenesis of the Cerebral Cortex (Neocortex) of the Cat (Felis Domestica). A Golgi Study. I. The Primordial Neocortical Organization. *Z. Anat. Entwicklungsgesch* **134**:117–145. [5]

———. 2014. The Mammalian Neocortex New Pyramidal Neuron: A New Conception. *Front. Neuroanat.* **7**:51. [5]

Maris, E., P. Fries, and F. van Ede. 2016. Diverse Phase Relations among Neuronal Rhythms and Their Potential Function. *Trends Neurosci.* **39**:86–99. [10]

Markov, N. T., M. Ercsey-Ravasz, C. Lamy, et al. 2013. The Role of Long-Range Connections on the Specificity of the Macaque Interareal Cortical Network. *PNAS* **110**:5187–5192. [7]

Markov, N. T., M. M. Ercsey-Ravasz, A. R. Ribeiro Gomes, et al. 2014. A Weighted and Directed Interareal Connectivity Matrix for Macaque Cerebral Cortex. *Cereb. Cortex* **24**:17–36. [10]

Markram, H., J. Lübke, M. Frotscher, and B. Sakmann. 1997. Regulation of Synaptic Efficacy by Coincidence of Postsynaptic Aps and EPSPs. *Science* **275**:213–215. [10, 13]

Marner, L., J. R. Nyengaard, Y. Tang, and B. Pakkenberg. 2003. Marked Loss of Myelinated Nerve Fibers in the Human Brain with Age. *J. Comp. Neur.* **462**:144–152. [3]

Marr, D. 1970. A Theory for Cerebral Neocortex. *Proc. R. Soc. Lond. B* **176**:161–234. [13]

———. 1982. Vision: A Computational Investigation into the Human Representation and Processing of Visual Information. San Francisco: W. H. Freeman and Co. [9, 17]

Martins, M. D., S. Laaha, E. M. Freiberger, S. Choi, and W. T. Fitch. 2014. How Children Perceive Fractals: Hierarchical Self-Similarity and Cognitive Development. *Cognition* **133**:10–24. [15]

Maruyama, M., C. Pallier, A. Jobert, M. Sigman, and S. Dehaene. 2012. The Cortical Representation of Simple Mathematical Expressions. *Neuroimage* **61**:1444–1460. [15]

Masquelier, T., E. Hugues, G. Deco, and S. J. Thorpe. 2009. Oscillations, Phase-of-Firing Coding, and Spike Timing-Dependent Plasticity: An Efficient Learning Scheme. *Journal of Neuroscience.* **29**:13484–13493. [10]

Matsui, T., T. Murakami, and K. Ohki. 2016. Transient Neuronal Coactivations Embedded in Globally Propagating Waves Underlie Resting-State Functional Connectivity. *PNAS* **113**:6556–6561. [6]

Matsuzaki, F., and A. Shitamukai. 2015. Cell Division Modes and Cleavage Planes of Neural Progenitors during Mammalian Cortical Development. *Cold Spring Harb. Perspect. Biol.* **7**:e015719–015718. [5]

Matus, A. 2009. Dendritic Spine History. In: Encyclopedia of Neuroscience, pp. 453–457. L. R. Squire, series ed. Cambridge, MA: Academic Press. [9]

Mayer, A., C. M. Schwiedrzik, M. Wibral, W. Singer, and L. Melloni. 2016. Expecting to See a Letter: Alpha Oscillations as Carriers of Top-Down Sensory Predictions. *Cereb. Cortex* **26**:3146–3160. [14]

Mayer, C., C. Hafemeister, R. C. Bandler, et al. 2018. Developmental Diversification of Cortical Inhibitory Interneurons. *Nature* **555**:457–462. [5]

Mayer, C., X. H. Jaglin, L. V. Cobbs, et al. 2015. Clonally Related Forebrain Interneurons Disperse Broadly across Both Functional Areas and Structural Boundaries. *Neuron* **87**:989–998. [5]

Mazor, O., and G. Laurent. 2005. Transient Dynamics versus Fixed Points in Odor Representations by Locust Antennal Lobe Projection Neurons. *Neuron* **48**:661–673. [12, 13]

McAdams, C. J., and J. H. Maunsell. 1999. Effects of Attention on Orientation-Tuning Functions of Single Neurons in Macaque Cortical Area V4. *J. Neurosci.* **19**:431–441. [13]

McConnell, S. K., and C. E. Kaznowski. 1991. Cell Cycle Dependence of Laminar Determination in Developing Neocortex. *Science* **254**:282–285. [2, 5]

McCullagh, P., and J. A. Nelder. 1989. Generalized Linear Models. Monographs on Statistics and Applied Probablity 37. London: Chapman and Hall/CRC. [11]

McCurry, C. L., J. D. Shepherd, D. Tropea, et al. 2010. Loss of Arc Renders the Visual Cortex Impervious to the Effects of Sensory Experience or Deprivation. *Nat. Neurosci.* **13**:450–457. [5]

McGinley, M. J., S. V. David, and D. A. McCormick. 2015a. Cortical Membrane Potential Signature of Optimal States for Sensory Signal Detection. *Neuron* **87**:179–192. [13]

McGinley, M. J., M. Vinck, J. Reimer, et al. 2015b. Waking State: Rapid Variations Modulate Neural and Behavioral Responses. *Neuron* **87**:1143–1161. [13]

McGraw, M. B. 1945. The Neuromuscular Maturation of the Human Infant. New York: Columbia Univ. Press. [16]

McKinney, W. T., Jr., and W. E. Bunney, Jr. 1969. Animal Model of Depression. I. Review of Evidence: Implications for Research. *Arch. Gen. Psychiatry* **21**:240–248. [7]

McLean, C. Y., P. L. Reno, A. A. Pollen, et al. 2011. Human-Specific Loss of Regulatory DNA and the Evolution of Human-Specific Traits. *Nature* **471**:216–219. [3, 5]

Medaglia, J. D., M. E. Lynall, and D. S. Bassett. 2015. Cognitive Network Neuroscience. *J. Cogn. Neurosci.* **27**:1471–1491. [7]

Medicus, G. M. 1992. The Inapplicability of the Biogenetic Rule to Behavioral Development. *Human Dev.* **35**:1–8. [16]

Medina, L., I. Legaz, G. Gonzalez, et al. 2004. Expression of Dbx1, Neurogenin 2, Semaphorin 5A, Cadherin 8, and Emx1 Distinguish Ventral and Lateral Pallial Histogenetic Divisions in the Developing Mouse Claustroamygdaloid Complex. *J. Comp. Neur.* **474**:504–523. [2]

Melloni, L., C. M. Schwiedrzik, E. Rodriguez, and W. Singer. 2009. (Micro) Saccades, Corollary Activity and Cortical Oscillations. *Trends Cogn. Sci.* **13**:239–245. [17]

Melozzi, F., M. M. Woodman, V. K. Jirsa, and C. Bernard. 2017. The Virtual Mouse Brain: A Computational Neuroinformatics Platform to Study Whole Mouse Brain Dynamics. *eNeuro* **17**:0111. [7]

Merchant, H., O. Perez, W. Zarco, and J. Gamez. 2013. Interval Tuning in the Primate Medial Premotor Cortex as a General Timing Mechanism. *J. Neurosci.* **33**:9082–9096. [17]

Meshulam, L., J. L. Gauthier, C. D. Brody, D. W. Tank, and W. Bialek. 2017. Collective Behavior of Place and Non-Place Neurons in the Hippocampal Network. *Neuron* **96**:1178–1191. [12]

Messé, A., M.-T. Hütt, and C. C. Hilgetag. 2018. Toward a Theory of Coactivation Patterns in Excitable Neural Networks. *PLoS Comput. Biol.* **14**:e1006084. [11]

Messinger, A., L. R. Squire, S. M. Zola, and T. D. Albright. 2001. Neuronal Representations of Stimulus Associations Develop in the Temporal Lobe during Learning. *PNAS* **98**:12239–12244. [17]

Mesulam, M. M. 1998. From Sensation to Cognition. *Brain* **121**:1013–1052. [14]

Metin, C., C. Alvarez, D. Moudoux, et al. 2007. Conserved Pattern of Tangential Neuronal Migration during Forebrain Development. *Development* **134**:2815–2827. [5]

Meyer, G., J. P. Schaaps, L. Moreau, and A. M. Goffinet. 2000. Embryonic and Early Fetal Development of the Human Neocortex. *J. Neurosci.* **20**:1858–1868. [2, 5]

Meyer, T., and C. R. Olson. 2011. Statistical Learning of Visual Transitions in Monkey Inferotemporal Cortex. *PNAS* **108**:19401–19406. [15]

Mi, D., Z. Li, L. Lim, et al. 2018. Early Emergence of Cortical Interneuron Diversity in the Mouse Embryo. *Science* **360**:81–85. [5]

Micheloyannis, S., E. Pachou, C. J. Stam, et al. 2006. Small-World Networks and Disturbed Functional Connectivity in Schizophrenia. *Schizophr. Res.* **87**:60–66. [7]

Miller, J. A., S.-L. Ding, S. M. Sunkin, et al. 2014. Transcriptional Landscape of the Prenatal Human Brain. *Nature* **508**:199–206. [3, 5]

Miller, R. F., and S. A. Bloomfield. 1983. Electroanatomy of a Unique Amacrine Cell in the Rabbit Retina. *PNAS* **80**:3069–3073. [9]

Miltner, W. H. R., C. Braun, M. Arnold, H. Witte, and E. Taub. 1999. Coherence of Gamma-Band EEG Activity as a Basis for Associative Learning. *Nature* **397**:434–436. [10]

Mirza, M. B., R. A. Adams, C. D. Mathys, and K. Friston. 2016. Scene Construction, Visual Foraging, and Active Inference. *Front. Comput. Neurosci.* **10**:56. [14]

Mishchenko, Y., T. Hu, J. Spacek, et al. 2010. Ultrastructural Analysis of Hippocampal Neuropil from the Connectomics Perspective. *Neuron* **67**:1009–1020. [9]

Mishra-Gorur, K., A. O. Çağlayan, A. E. Schaffer, et al. 2014. Mutations in *KATNB1* Cause Complex Cerebral Malformations by Disrupting Asymmetrically Dividing Neural Progenitors. *Neuron* **84**:1226–1239. [4]

Mitchell, C., and D. L. Silver. 2018. Enhancing Our Brains: Genomic Mechanisms Underlying Cortical Evolution. *Semin. Cell Dev. Biol.* **76**:23–32. [5]

Mitchell, J. F., K. A. Sundberg, and J. H. Reynolds. 2009. Spatial Attention Decorrelates Intrinsic Activity Fluctuations in Macaque Area V4. *Neuron* **63**:879–888. [13]

Mitra, A., A. Kraft, P. Wright, et al. 2018. Spontaneous Infra-Slow Brain Activity Has Unique Spatiotemporal Dynamics and Laminar Structure. *Neuron* **98**:297–305. [6]

Mitra, A., and M. E. Raichle. 2016. How Networks Communicate: Propagation Patterns in Spontaneous Brain Activity. *Phil. Trans. R. Soc. B* **371**:pii: 20150546. [6]

Mitra, A., A. Z. Snyder, T. Blazey, and M. E. Raichle. 2015. Lag Threads Organize the Brain's Intrinsic Activity. *PNAS* **112**:E2235–2244. [6]

Mitra, A., A. Z. Snyder, C. D. Hacker, et al. 2016. Human Cortical-Hippocampal Dialogue in Wake and Slow-Wave Sleep. *PNAS* **113**:E6868–E6876. [6]

Mitra, A., A. Z. Snyder, C. D. Hacker, and M. E. Raichle. 2014. Lag Structure in Resting-State fMRI. *J. Neurophysiol.* **111**:2374–2391. [6]

Miwa, J. M., T. R. Stevens, S. L. King, et al. 2006. The Prototoxin Lynx1 Acts on Nicotinic Acetylcholine Receptors to Balance Neuronal Activity and Survival *in Vivo*. *Neuron* **51**:587–600. [5]

Miyashita-Lin, E. M., R. Hevner, K. M. Wassarman, S. Martinez, and J. L. Rubenstein. 1999. Early Neocortical Regionalization in the Absence of Thalamic Innervation. *Science* **285**:906–909. [5]

Miyata, T., A. Kawaguchi, H. Okano, and M. Ogawa. 2001. Asymmetric Inheritance of Radial Glial Fibers by Cortical Neurons. *Neuron* **31**:727–741. [2]

Mizutani, K., K. Yoon, L. Dang, A. Tokunaga, and N. Gaiano. 2007. Differential Notch Signalling Distinguishes Neural Stem Cells from Intermediate Progenitors. *Nature* **449**:351–355. [2]

Mnih, V., K. Kavukcuoglu, D. Silver, et al. 2015. Human-Level Control through Deep Reinforcement Learning. *Nature* **518**:529–533. [17]

Mochida, G. H., V. S. Ganesh, M. I. de Michelena, et al. 2012. CHMP1A Encodes an Essential Regulator of BMI1-INK4A in Cerebellar Development. *Nat. Genet.* **44**:1260–1264. [4]

Mochida, G. H., and C. A. Walsh. 2001. Molecular Genetics of Human Microcephaly. *Curr. Opin. Neurol.* **14**:151–156. [16]

Molliver, M. E., I. Kostovic, and H. van der Loos. 1973. The Development of Synapses in Cerebral Cortex of the Human Fetus. *Brain Res.* **50**:403–407. [5]

Molnár, Z., and C. Blakemore. 1995. How Do Thalamic Axons Find Their Way to the Cortex? *Trends Neurosci.* **18**:389–397. [5]

Molnár, Z., and G. Clowry. 2012. Cerebral Cortical Development in Rodents and Primates. *Prog. Brain Res.* **195**:45–70. [3]

Molnár, Z., and A. Pollen. 2013. How Unique Is the Human Neocortex? *Development* **141**:11–16. [5]

Montgomery, S. H., and N. I. Mundy. 2014. Microcephaly Genes Evolved Adaptively Throughout the Evolution of Eutherian Mammals. *BMC Evol. Biol.* **14**:120. [4]

Montgomery, S. M., A. Sirota, and G. Buzsáki. 2008. Theta and Gamma Coordination of Hippocampal Networks during Waking and Rapid Eye Movement Sleep. *Journal of Neuroscience.* **28**:6731–6741. [10]

Montiel, J. F., W. Z. Wang, F. M. Oeschger, et al. 2011. Hypothesis on the Dual Origin of the Mammalian Subplate. *Front. Neuroanat.* **5**:25. [5]

Montijn, J. S., G. T. Meijer, C. S. Lansink, and C. M. Pennartz. 2016. Population-Level Neural Codes Are Robust to Single-Neuron Variability from a Multidimensional Coding Perspective. *Cell Rep.* **16**:2486–2498. [12]

Monto, S., S. Palva, J. Voipio, and J. M. Palva. 2008. Very Slow EEG Fluctuations Predict the Dynamics of Stimulus Detection and Oscillation Amplitudes in Humans. *J. Neurosci.* **28**:8268–8272. [6]

Moore, C. I., M. Carlen, U. Knoblich, and J. A. Cardin. 2010. Neocortical Interneurons: From Diversity, Strength. *Cell* **142**:189–193. [10]

Moore, C. I., and R. Cao. 2008. The Hemo-Neural Hypothesis: On the Role of Blood Flow in Information Processing. *J. Neurophysiol.* **99**:2035–2047. [9]

Moore, G. P., D. H. Perkel, and J. P. Segundo. 1966. Statistical Analysis and Functional Interpretation of Neuronal Spike Train Data. *Annu. Rev. Neurosci.* **28**:493–522. [13]

Mora-Bermudez, F., F. Badsha, S. Kanton, et al. 2016. Differences and Similarities between Human and Chimpanzee Neural Progenitors during Cerebral Cortex Development. *eLife* **5**:166. [5]

Moran, J., and R. Desimone. 1985. Selective Attention Gates Visual Processing in the Extrastriate Cortex. *Science* **229**:782–784. [13]

Moreno, A., F. Limousin, S. Dehaene, and C. Pallier. 2018. Brain Correlates of Constituent Structure in Sign Language Comprehension. *Neuroimage* **167**:151–161. [15]

Moreno-Bote, R., J. Beck, I. Kanitscheider, et al. 2014. Information-Limiting Correlations. *Nat. Neurosci.* **17**:1410–1417. [12]

Morishita, H., J. M. Miwa, N. Heintz, and T. K. Hensch. 2010. Lynx1, a Cholinergic Brake, Limits Plasticity in Adult Visual Cortex. *Science* **330**:1238–1240. [5]

Morishita, W., H. Marie, and R. C. Malenka. 2005. Distinct Triggering and Expression Mechanisms Underlie LTD of AMPA and NMDA Synaptic Responses. *Nat. Neurosci.* **8**:1043–1050. [10]

Mortazavi, F., A. L. Oblak, W. Z. Morrison, et al. 2017. Geometric Navigation of Axons in a Cerebral Pathway: Comparing Dmri with Tract Tracing and Immunohistochemistry. *Cereb. Cortex* **28**:1219–1232. [2]

Morton, G. J., D. E. Cummings, D. G. Baskin, G. S. Barsh, and M. W. Schwartz. 2006. Central Nervous System Control of Food Intake and Body Weight. *Nature* **443**:289–295. [9]

Motter, A. E. 2015. Networkcontrology. *Chaos* **25**:097621. [7]

Mountcastle, V. B. 1957. Modality and Topographic Properties of Single Neurons of Cat's Somatic Sensory Cortex. *J. Neurophysiol.* **20**:408–434. [13]

Mountcastle, V. B. 1998. Perceptual Neuroscience: The Cerebral Cortex. Cambridge, MA: Harvard Univ. Press. [13]

Mouse Genome Sequencing Consortium. 2002. Initial Sequencing and Comparative Analysis of the Mouse Genome. *Nature* **420**:520–562. [3]

Mrsic-Flogel, T. D., J. W. H. Schnupp, and A. J. King. 2003. Acoustic Factors Govern Developmental Sharpening of Spatial Tuning in the Auditory Cortex. *Nat. Neurosci.* **6**:981–988. [16]

Muldoon, S. F., F. Pasqualetti, S. Gu, et al. 2016. Stimulation-Based Control of Dynamic Brain Networks. *PLoS Comput. Biol.* **12**:e1005076. [7]

Muller, L., F. Chavane, J. Reynolds, and T. J. Sejnowski. 2018. Cortical Travelling Waves: Mechanisms and Computational Principles. *Nat. Rev. Neurosci.* **19**:255–268. [1, 13]

Muller, L., G. Piantoni, D. Koller, et al. 2016. Rotating Waves during Human Sleep Spindles Organize Global Patterns of Activity That Repeat Precisely through the Night. *eLife* **5**: [13]

Mumford, D. 1992. On the Computational Architecture of the Neocortex II: The Role of Cortico-Cortical Loops. *Biol. Cybern.* **66**:241–251. [14]

Mundinano, I.-C., D. M. Fox, W. C. Kwan, et al. 2018. Transient Visual Pathway Critical for Normal Development of Primate Grasping Behavior. *PNAS* **115**:1364–1369. [9]

Mundinano, I.-C., W. C. Kwan, and J. A. Bourne. 2015. Mapping the Mosaic Sequence of Primate Visual Cortical Development. *Front. Neuroanat.* **9**:132. [9]

Murdock, D. R., G. D. Clark, M. N. Bainbridge, et al. 2011. Whole-Exome Sequencing Identifies Compound Heterozygous Mutations in WDR62 in Siblings with Recurrent Polymicrogyria. *Am. J. Med. Genet. A* **155**:2071–2077. [4]

Murphy, A. C., S. Gu, A. N. Khambhati, et al. 2016. Explicitly Linking Regional Activation and Function Connectivity: Community Structure of Weighted Networks with Continuous Annotation. *arXiv* 1611.07962. [7]

Murphy, R. A., E. Mondragon, and V. A. Murphy. 2008. Rule Learning by Rats. *Science* **319**:1849–1851. [15]

Murray, J. D., A. Bernacchia, D. J. Freedman, et al. 2014. A Hierarchy of Intrinsic Timescales across Primate Cortex. *Nat. Neurosci.* **17**:1661–1663. [14]

Napier, J. R., and P. H. Napier. 1985. The Natural History of the Primates. Cambridge, MA: MIT Press. [16]

Nara, S. 2003. Can Potentially Useful Dynamics to Solve Complex Problems Emerge from Constrained Chaos and/or Chaotic Itinerancy? *Chaos* **13**:1110–1121. [14]

Narayanan, R. T., R. Egger, A. S. Johnson, et al. 2015. Beyond Columnar Organization: Cell Type- and Target Layer-Specific Principles of Horizontal Axon Projection Patterns in Rat Vibrissal Cortex. *Cereb. Cortex* **25**:4450–4468. [8]

Narayanan, R. T., H. Mohan, R. Broersen, et al. 2014. Juxtasomal Biocytin Labeling to Study the Structure-Function Relationship of Individual Cortical Neurons. *J. Vis. Exp.* **25**:e51359. [8]

Narayanan, R. T., D. Udvary, and M. Oberlaender. 2017. Cell Type-Specific Structural Organization of the Six Layers in Rat Barrel Cortex. *Front. Neuroanat.* **11**:91. [9]

Nauta, W. J., and P. A. Gygax. 1954. Silver Impregnation of Degenerating Axons in the Central Nervous System: A Modified Technic. *Stain Technol.* **29**:91–93. [9]

Navlakha, S., A. L. Barth, and Z. Bar-Joseph. 2015. Decreasing-Rate Pruning Optimizes the Construction of Efficient and Robust Distributed Networks. *PLoS Comput. Biol.* **11**:e1004347. [5]

Nayebi, A., D. Bear, J. Kubilius, et al. 2018. Task-Driven Convolutional Recurrent Models of the Visual System. *arXiv* 1807.00053. [17]

Nelson, B. R., R. D. Hodge, F. Bedogni, and R. F. Hevner. 2013. Dynamic Interactions between Intermediate Neurogenic Progenitors and Radial Glia in Embryonic Mouse Neocortex: Potential Role in Dll1-Notch Signaling. *J. Neurosci.* **33**:9122–9139. [2]

Nelson, M. J., I. El Karoui, K. Giber, et al. 2017. Neurophysiological Dynamics of Phrase-Structure Building during Sentence Processing. *PNAS* **114**:E3669–E3678. [15]

Neveu, D., and R. S. Zucker. 1996. Postsynaptic Levels of [Ca2+]I Needed to Trigger LTD and LTP. *Neuron* **16**:619–629. [10]

Newman, M. E. J. 2010. Networks: An Introduction. Boston: MIT Press. [7]

———. 2011. Complex Systems: A Survey. *Am. J. Phys.* **79**:800–810. [7]

Newman, M. E. J., and A. Clauset. 2016. Structure and Inference in Annotated Networks. *Nat. Commun.* **7**:11863. [7]

Ni, A. M., and J. H. R. Maunsell. 2017. Spatially Tuned Normalization Explains Attention Modulation Variance within Neurons. *J. Neurophysiol.* **118**:1903–1913. [13]

Ni, A. M., D. A. Ruff, J. J. Alberts, J. Symmonds, and M. R. Cohen. 2018. Learning and Attention Reveal a General Relationship between Population Activity and Behavior. *Science* **359**:463–465. [12]

Ni, J., C. M. Lewis, T. Wunderle, et al. 2017. Gamma-Band Resonance of Visual Cortex to Optogenetic Stimulation. *bioRxiv* **135467**:1–31. [10]

Nicholas, A. K., M. Khurshid, J. Désir, et al. 2010. WDR62 Is Associated with the Spindle Pole and Is Mutated in Human Microcephaly. *Nat. Genet.* **42**:1010–1014. [4]

Nicholas, A. K., E. A. Swanson, J. J. Cox, et al. 2009. The Molecular Landscape of ASPM Mutations in Primary Microcephaly. *J. Med. Genet.* **46**:249–253. [4]

Nicolis, G., and I. Prigogine. 1977. Self-Organization in Non-Equilibrium Systems. New York: Wiley. [14]

Nicosia, V., P. E. Vértes, W. R. Schafer, V. Latora, and E. T. Bullmore. 2013. Phase Transition in the Economically Modeled Growth of a Cellular Nervous System. *PNAS* **110**:7880–7885. [7]

Niebur, E. N., H. G. Schuster, and D. M. Kammen. 1991. Collective Frequencies and Metastability in Networks of Limit Cycle Oscillators with Time Delay. *Phys. Rev. Lett.* **67**:2753–2756. [10]

Nieder, A., and S. Dehaene. 2009. Representation of Number in the Brain. *Annu. Rev. Neurosci.* **32**:185–208. [15, 17]

Niell, C. M., and M. P. Stryker. 2010. Modulation of Visual Responses by Behavioral State in Mouse Visual Cortex. *Neuron* **65**:472–479. [13]

Nielsen, R., C. D. Bustamante, A. G. Clark, et al. 2005. A Scan for Positively Selected Genes in the Genomes of Humans and Chimpanzees. *PLoS Biol.* **3**:e170. [3]

Nieters, P., J. Leugering, and G. Pipa. 2017. Neuromorphic Computation in Multi-Delay Coupled Models. *IBM J. Res. Dev.* **61**:8:7–8:9. [11]

Nikolic, D., S. Häusler, W. Singer, and W. Maass. 2009. Distributed Fading Memory for Stimulus Properties in the Primary Visual Cortex. *PLoS Biol.* **7**:1–19. [10]

Nimchinsky, E. A., E. Gilissen, J. M. Allman, et al. 1999. A Neuronal Morphologic Type Unique to Humans and Great Apes. *PNAS* **96**:5268–5273. [3, 5]

Nobrega, M. A. 2003. Scanning Human Gene Deserts for Long-Range Enhancers. *Science* **302**:413. [4]

Noctor, S. C., A. C. Flint, T. A. Weissman, R. S. Dammerman, and A. R. Kriegstein. 2001. Neurons Derived from Radial Glial Cells Establish Radial Units in Neocortex. *Nature* **409**:714–720. [2, 5]

Noctor, S. C., A. C. Flint, T. A. Weissman, et al. 2002. Dividing Precursor Cells of the Embryonic Cortical Ventricular Zone Have Morphological and Molecular Characteristics of Radial Glia. *J. Neurosci.* **22**:3161–3173. [2]

Nomura, T., W. Yamashita, H. Gotoh, and K. Ono. 2018. Species-Specific Mechanisms of Neuron Subtype Specification Reveal Evolutionary Plasticity of Amniote Brain Development. *Cell Rep.* **22**:3142–3151. [5]

Nord, A. S., M. J. Blow, C. Attanasio, et al. 2013. Rapid and Pervasive Changes in Genome-Wide Enhancer Usage during Mammalian Development. *Cell* **155**:1521–1531. [5]

Nord, A. S., K. Pattabiraman, A. Visel, and J. L. R. Rubenstein. 2015. Genomic Perspectives of Transcriptional Regulation in Forebrain Development. *Neuron* **85**:27–47. [5]

Noudoost, B., and T. Moore. 2011. The Role of Neuromodulators in Selective Attention. *Trends Cogn. Sci.* **15**:585–591. [9]

Nowak, L., P. Bregestovski, P. Ascher, A. Herbet, and A. Prochiantz. 1984. Magnesium Gates Glutamate-Activated Channels in Mouse Central Neurones. *Nature* **307**:462–465. [10]

Nowakowski, T. J., A. Bhaduri, A. A. Pollen, et al. 2017. Spatiotemporal Gene Expression Trajectories Reveal Developmental Hierarchies of the Human Cortex. *Science* **358**:1318–1323. [3, 5]

Nowakowski, T. J., A. A. Pollen, C. Sandoval-Espinosa, and A. R. Kriegstein. 2016. Transformation of the Radial Glia Scaffold Demarcates Two Stages of Human Cerebral Cortex Development. *Neuron* **91**:1219–1227. [2, 5]

Oberheim, N. A., T. Takano, X. Han, et al. 2009. Uniquely Hominid Features of Adult Human Astrocytes. *J Neurosci* **29**:3276–3287. [5]

Oberlaender, M., C. P. de Kock, R. M. Bruno, et al. 2012. Cell Type-Specific Three-Dimensional Structure of Thalamocortical Circuits in a Column of Rat Vibrissal Cortex. *Cereb. Cortex* **22**:2375–2391. [8]

O'Bleness, M. S., C. M. Dickens, L. J. Dumas, et al. 2012. Evolutionary History and Genome Organization of DUF1220 Protein Domains. *G3 (Bethesda)* **2**:977–986. [3]

O'Connell, C. B., and Y. L. Wang. 2000. Mammalian Spindle Orientation and Position Respond to Changes in Cell Shape in a Dynein-Dependent Fashion. *Mol. Biol. Cell* **11**:1765–1774. [2]

O'Connell, K. F., C. Caron, K. R. Kopish, et al. 2001. The *C. elegans* Zyg-1 Gene Encodes a Regulator of Centrosome Duplication with Distinct Maternal and Paternal Roles in the Embryo. *Cell* **105**:547–558. [4]

Oeschger, F. M., W. Z. Wang, S. Lee, et al. 2012. Gene Expression Analysis of the Embryonic Subplate. *Cereb. Cortex* **22**:1343–1359. [5]

Ogawa, S., T. M. Lee, A. R. Kay, and D. W. Tank. 1990. Brain Magnetic Resonance Imaging with Contrast Dependent on Blood Oxygenation. *PNAS* **87**:9868–9872. [6]

Oh, S. W., J. A. Harris, L. Ng, et al. 2014. A Mesoscale Connectome of the Mouse Brain. *Nature* **508**:207. [7]

Ohno, S. 1970. Evolution by Gene Duplication. Berlin: Springer. [3]

Okamoto, M., T. Miyata, D. Konno, et al. 2016. Cell-Cycle-Dependent Transitions in Temporal Identity of Mammalian Neural Progenitor Cells. *Nat. Commun.* 7:1–16. [5]

O'Keefe, J. 1976. Place Units in the Hippocampus of the Freely Moving Rat. *Exp. Neurol.* **51**:78–109. [17]

O'Keefe, J., and M. L. Recce. 1993. Phase Relationship between Hippocampal Place Units and the EEG Theta Rhythm. *Hippocampus* **3**:317–330. [13, 14]

O'Leary, D. D., S. J. Chou, and S. Sahara. 2007. Area Patterning of the Mammalian Cortex. *Neuron* **56**:252–269. [3]

O'Leary, D. D., A. M. Stocker, and A. Zembrycki. 2013. Area Patterning of the Mammalian Cortex. In: Patterning and Cell Type Specification in the Developing CNS and Pns, ed. J. L. Rubenstein and P. Rakic, vol. 1, pp. 61–85. San Diego: Elsevier. [5]

Olsen, S. R., D. S. Bortone, H. Adesnik, and M. Scanziani. 2012. Gain Control by Layer Six in Cortical Circuits of Vision. *Nature* **483**:47–52. [9]

Olshausen, B. A., and D. J. Field 1996. Emergence of Simple-Cell Receptive Field Properties by Learning a Sparse Code for Natural Images. *Nature* **381**:607–609. [13]

Onat, S., D. Jancke, and P. König. 2013. Cortical Long-Range Interactions Embed Statistical Knowledge of Natural Sensory Input: A Voltage-Sensitive Dye Imaging Study. *F1000Res.* **2**:51. [11]

Onorati, M., Z. Li, F. Liu, et al. 2016. Zika Virus Disrupts Phospho-TBK1 Localization and Mitosis in Human Neuroepithelial Stem Cells and Radial Glia. *Cell Rep.* **16**:2576–2592. [3]

Optican, L., and B. J. Richmond. 1987. Temporal Encoding of Two-Dimensional Patterns by Single Units in Primate Inferior Cortex II: Information Theoretic Analysis. *J. Neurophysiol.* **57**:162–178. [14]

Oram, M. W., P. Foldiak, D. I. Perrett, and F. Sengpiel. 1998. The "Ideal Homunculus": Decoding Neural Population Signals. *Trends Neurosci.* **21**:259–265. [12]

Ørstavik, S., and J. Stark. 1998. Reconstruction and Cross-Prediction in Coupled Map Lattices Using Spatio-Temporal Embedding Techniques. *Phys. Lett. A* **247**:145–160. [14]

Ostojic, S., and N. Brunel. 2011. From Spiking Neuron Models to Linear-Nonlinear Models. *PLoS Comput. Biol.* **7**:e1001056 [11]

Otani, T., M. C. Marchetto, F. H. Gage, B. D. Simons, and F. J. Livesey. 2016. 2d and 3D Stem Cell Models of Primate Cortical Development Identify Species-Specific Differences in Progenitor Behavior Contributing to Brain Size. *Cell Stem Cell* **18**:467–480. [5]

Ouattara, K., A. Lemasson, and K. Zuberbühler. 2009. Campbell's Monkeys Concatenate Vocalizations into Context-Specific Call Sequences. *PNAS* **106**:22026–22031. [15]

Oudeyer, P.-Y., and F. Kaplan. 2007. What Is Intrinsic Motivation? A Typology of Computational Approaches. *Front. Neurorobot.* **1**:6. [14]

Owens, D. F., A. C. Flint, R. S. Dammerman, and A. R. Kriegstein. 2000. Calcium Dynamics of Neocortical Ventricular Zone Cells. *Dev. Neurosci.* **22**:25–33. [2]

Owens, D. F., and A. R. Kriegstein. 1998. Patterns of Intracellular Calcium Fluctuation in Precursor Cells of the Neocortical Ventricular Zone. *J. Neurosci.* **18**:5374–5388. [2]

Padberg, J., E. Disbrow, and L. Krubitzer. 2005. The Organization and Connections of Anterior and Posterior Parietal Cortex in Titi Monkeys: Do New World Monkeys Have an Area 2? *Cereb. Cortex* **15**:1938–1963. [16]

Pagnamenta, A. T., J. E. Murray, G. Yoon, et al. 2012. A Novel Nonsense CDK5RAP2 Mutation in a Somali Child with Primary Microcephaly and Sensorineural Hearing Loss. *Am. J. Med. Genet. A* **158A**:2577–2582. [4]

Pai, A. A., J. T. Bell, J. C. Marioni, J. K. Pritchard, and Y. Gilad. 2011. A Genome-Wide Study of DNA Methylation Patterns and Gene Expression Levels in Multiple Human and Chimpanzee Tissues. *PLoS Genet.* **7**:e1001316. [3]

Pajevic, S., P. J. Basser, and R. D. Fields. 2014. Role of Myelin Plasticity in Oscillations and Synchrony of Neuronal Activity. *Neuroscience* **276**:135–147. [10]

Pakkenberg, B., D. Pelvig, L. Marner, et al. 2003. Aging and the Human Neocortex. *Exp. Gerontol.* **38**:95–99. [3]

Pallier, C., A. D. Devauchelle, and S. Dehaene. 2011. Cortical Representation of the Constituent Structure of Sentences. *PNAS* **108**:2522–2527. [15]

Palm, G., A. Knoblauch, F. Hauser, and A. Schuz. 2014. Cell Assemblies in the Cerebral Cortex. *Biol. Cybern.* **108**:559–572. [12]

Palmer, C. J., A. K. Seth, and J. Hohwy. 2015. The Felt Presence of Other Minds: Predictive Processing, Counterfactual Predictions, and Mentalising in Autism. *Conscious. Cogn.* **36**:376–389. [14]

Palmigiano, A., T. Geisel, F. Wolf, and D. Battaglia. 2017. Flexible Information Routing by Transient Synchrony. *Nat. Neurosci.* **20**:1014–1022. [7, 10]

Palomero-Gallagher, N., and K. Zilles. 2017. Cortical Layers: Cyto-, Myelo-, Receptor- and Synaptic Architecture in Human Cortical Areas. *Neuroimage* **S1053-8119**:30682–30681. [8, 9]

Palva, J. M., and S. Palva. 2012. Infra-Slow Fluctuations in Electrophysiological Recordings, Blood-Oxygenation-Level-Dependent Signals, and Psychophysical Time Series. *Neuroimage* **62**:2201–2211. [6]

Palva, J. M., S. Palva, and K. Kaila. 2005. Phase Synchrony among Neuronal Oscillations in the Human Cortex. *Journal of Neuroscience.* **25**:3962–3972. [10]

Pan, W. J., G. J. Thompson, M. E. Magnuson, D. Jaeger, and S. Keilholz. 2013. Infraslow LFP Correlates to Resting-State fMRI BOLD Signals. *Neuroimage* **74**:288–297. [6]

Papineau, D. 1987. Reality and Representation. Oxford: Blackwell. [7]

Paredes, M. F., D. James, S. Gil-Perotin, et al. 2016. Extensive Migration of Young Neurons into the Infant Human Frontal Lobe. *Science* **354**: 279–287. [5]

Parravicini, A., and T. Pievani. 2018. Continuity and Discontinuity in Human Language Evolution: Putting an Old-Fashioned Debate in Its Historical Perspective. *Topoi* **37**:279–287. [17]

Parthasarathy, S., S. Srivatsa, A. Nityanandam, and V. Tarabykin. 2014. Ntf3 Acts Downstream of Sip1 in Cortical Postmitotic Neurons to Control Progenitor Cell Fate through Feedback Signaling. *Development* **141**:3324–3330. [2]

Pasqualetti, F., S. Zampieri, and F. Bullo. 2014. Controllability Metrics, Limitations and Algorithms for Complex Networks. *IEEE Trans. Control Netw. Syst.* **1**:40–52. [7]

Pastuzyn, E. D., C. E. Day, R. B. Kearns, et al. 2018. The Neuronal Gene Arc Encodes a Repurposed Retrotransposon Gag Protein That Mediates Intercellular RNA Transfer. *Cell* **172**:275–288. [4]

Patel, J., S. Fujisawa, A. Berenyi, S. Royer, and G. Buzsáki. 2012. Traveling Theta Waves Along the Entire Septotemporal Axis of the Hippocampus. *Neuron* **75**:410–417. [12]

Pattabiraman, K., O. Golonzhka, S. Lindtner, et al. 2014. Transcriptional Regulation of Enhancers Active in Protodomains of the Developing Cerebral Cortex. *Neuron* **82**:989–1003. [5]

Pattamadilok, C., S. Dehaene, and C. Pallier. 2015. A Role for Left Inferior Frontal and Posterior Superior Temporal Cortex in Extracting a Syntactic Tree from a Sentence. *Cortex* **75**:44–55. [15]

Patterson, M. L., and J. F. Werker. 2003. Two-Month-Old Infants Match Phonetic Information in Lips and Voice. *Dev. Sci.* **6**:191–196. [16]

Paulignan, Y., C. MacKenzie, R. Marteniuk, and M. Jeannerod. 1990. The Coupling of Arm and Finger Movements during Prehension. *Exp. Brain Res.* **79**:431–435. [17]

Pauling, L., and C. D. Coryell. 1936. The Magnetic Properties and Structure of Hemoglobin, Oxyhemoglobin and Carbonmonoxyhemoglobin. *PNAS* **22**:210–216. [6]

Pecka, M., Y. Han, E. Sader, and T. D. Mrsic-Flogel. 2014. Experience-Dependent Specialization of Receptive Field Surround for Selective Coding of Natural Scenes. *Neuron* **84**:457–469. [10]

Pedraza, M., A. Hoerder-Suabedissen, M. A. Albert-Maestro, Z. Molnár, and J. A. De Carlos. 2014. Extracortical Origin of Some Murine Subplate Cell Populations. *PNAS* **111**:8613–8618. [5]

Penn, D. C., K. J. Holyoak, and D. J. Povinelli. 2008. Darwin's Mistake: Explaining the Discontinuity between Human and Nonhuman Minds. *Behav. Brain Sci.* **31**:109–130. [15]

Pennacchio, L. A., N. Ahituv, A. M. Moses, et al. 2006. *In Vivo* Enhancer Analysis of Human Conserved Non-Coding Sequences. *Nature* **444**:499–502. [4]

Pepperberg, I. M. 2013. Abstract Concepts: Data from a Grey Parrot. *Behav. Processes* **93**:82–90. [15]

Pereira, F., B. Lou, B. Pritchett, et al. 2018. Toward a Universal Decoder of Linguistic Meaning from Brain Activation. *Nat. Commun.* **9**:963. [17]

Pérez, T., G. C. Garcia, V. M. Eguíluz, et al. 2011. Effect of the Topology and Delayed Interactions in Neuronal Networks Synchronization. *PLoS One* **6**:e19900. [10, 11]

Perin, R., T. K. Berger, and H. Markram. 2011. A Synaptic Organizing Principle for Cortical Neuronal Groups. *PNAS* **108**:5419–5424. [12]

Perkel, D. H., G. L. Gerstein, and G. P. Moore. 1967. Neuronal Spike Trains and Stochastic Point Processes. II. Simultaneous Spike Trains. *Biophys. J.* **7**:419–440. [13]

Petanjek, Z., M. Judas, G. Simic, et al. 2011. Extraordinary Neoteny of Synaptic Spines in the Human Prefrontal Cortex. *PNAS* **108**:13281–13286. [5]

Peters, A., and B. R. Payne. 1993. Numerical Relationships between Geniculocortical Afferents and Pyramidal Cell Modules in Cat Primary Visual Cortex. *Cereb. Cortex* **3**:69–78. [12]

Peters, A. J., S. X. Chen, and T. Komiyama. 2014. Emergence of Reproducible Spatiotemporal Activity during Motor Learning. *Nature* **510**:263–267. [12]

Petersen, C. C., A. Grinvald, and B. Sakmann. 2003. Spatiotemporal Dynamics of Sensory Responses in Layer 2/3 of Rat Barrel Cortex Measured *in Vivo* by Voltage-Sensitive Dye Imaging Combined with Whole-Cell Voltage Recordings and Neuron Reconstructions. *J. Neurosci.* **23**:1298–1309. [12]

Petilla Interneuron Nomenclature Group, G. A. Ascoli, L. Alonso-Nanclares, et al. 2008. Petilla Terminology: Nomenclature of Features of GABAergic Interneurons of the Cerebral Cortex. *Nat. Rev. Neurosci.* **9**:557–568. [3, 8]

Petri, G., P. Expert, F. Turkheimer, et al. 2014. Homological Scaffolds of Brain Functional Networks. *J. R. Soc. Interface* **11**:20140873. [7]

Petrides, M. 1995. Impairments on Nonspatial Self-Ordered and Externally Ordered Working Memory Tasks after Lesions of the Mid-Dorsal Part of the Lateral Frontal Cortex in the Monkey. *J. Neurosci.* **15**:359–375. [17]

Petrides, M., G. Cadoret, and S. Mackey. 2005. Orofacial Somatomotor Responses in the Macaque Monkey Homologue of Broca's Area. *Nature* **435**:1235–1238. [16]

Petrides, M., and D. N. Pandya. 2002. Comparative Cytoarchitectonic Analysis of the Human and the Macaque Ventrolateral Prefrontal Cortex and Corticocortical Connection Patterns in the Monkey. *Eur. J. Neurosci.* **16**:291–310. [16]

Pfeffer, C. K., M. Xue, M. He, Z. J. Huang, and M. Scanziani. 2013. Inhibition of Inhibition in Visual Cortex: The Logic of Connections between Molecularly Distinct Interneurons. *Nat. Neurosci.* **16**:1068–1076. [9]

Pfeiffer, M., M. Betizeau, J. Waltispurger, et al. 2016. Unsupervised Lineage-Based Characterization of Primate Precursors Reveals High Proliferative and Morphological Diversity in the OSVZ. *J. Comp. Neur.* **524**:535–563. [2, 5]

Pfurtscheller, G., and F. H. Lopes da Silva. 1999. Event-Related EEG/MEG Synchronization and Desynchronization: Basic Principles. *Clin. Neurophysiol.* **110**:1842–1857. [12]

Pho, G. N., M. J. Goard, J. Woodson, B. Crawford, and M. Sur. 2018. Task-Dependent Representations of Stimulus and Choice in Mouse Parietal Cortex. *Nat. Commun.* **9**:2596. [5]

Piai, V., K. L. Anderson, J. J. Lin, et al. 2016. Direct Brain Recordings Reveal Hippocampal Rhythm Underpinnings of Language Processing. *PNAS* **113**:11366–11371. [17]

Piao, X. 2004. G Protein-Coupled Receptor-Dependent Development of Human Frontal Cortex. *Science* **303**:2033–2036. [4]

Pica, P., C. Lemer, V. Izard, and S. Dehaene. 2004. Exact and Approximate Arithmetic in an Amazonian Indigene Group. *Science* **306**:499–503. [17]

Picco, N., F. Garcia-Moreno, P. K. Maini, T. E. Woolley, and Z. Molnár. 2018. Mathematical Modeling of Cortical Neurogenesis Reveals That the Founder Population Does Not Necessarily Scale with Neurogenic Output. *Cereb. Cortex* **28**:2540–2550. [5]

Pierfelice, T., L. Alberi, and N. Gaiano. 2011. Notch in the Vertebrate Nervous System: An Old Dog with New Tricks. *Neuron* **69**:840–855. [2]

Pikovsky, A. S., and J. Kurths. 1997. Coherence Resonance in a Noise-Driven Excitable System. *Phys. Rev. Lett.* **78**:775–778. [11]

Pilaz, L.-J., J. J. McMahon, E. E. Miller, et al. 2016. Prolonged Mitosis of Neural Progenitors Alters Cell Fate in the Developing Brain. *Neuron* **89**:83–99. [2, 5]

Pilaz, L.-J., D. Patti, G. Marcy, et al. 2009. Forced G1-Phase Reduction Alters Mode of Division, Neuron Number, and Laminar Phenotype in the Cerebral Cortex. *PNAS* **106**:21924–21929. [5]

Pilley, J. W., and A. K. Reid. 2011. Border Collie Comprehends Object Names as Verbal Referents. *Behav. Processes* **86**:184–195. [17]

Pilz, G. A., A. Shitamukai, I. Reillo, et al. 2013. Amplification of Progenitors in the Mammalian Telencephalon Includes a New Radial Glial Cell Type. *Nat. Commun.* **4**:2125. [2]

Pinto, L., D. Drechsel, M. T. Schmid, et al. 2009. AP2gamma Regulates Basal Progenitor Fate in a Region- and Layer-Specific Manner in the Developing Cortex. *Nat. Neurosci.* **12**:1229–1237. [2]

Pinto, L., M. T. Mader, M. Irmler, et al. 2008. Prospective Isolation of Functionally Distinct Radial Glial Subtypes: Lineage and Transcriptome Analysis. *Mol. Cell. Neurosci.* **38**:15–42. [2]

Pletikos, M., A. M. M. Sousa, G. Sedmak, et al. 2014. Temporal Specification and Bilaterality of Human Neocortical Topographic Gene Expression. *Neuron* **81**:321–332. [5]

Pluta, S., A. Naka, J. Veit, et al. 2015. A Direct Translaminar Inhibitory Circuit Tunes Cortical Output. *Nat. Neurosci.* **18**:1631–1640. [9]

Poeppel, D., W. J. Idsardi, and V. van Wassenhove. 2008. Speech Perception at the Interface of Neurobiology and Linguistics. *Phil. Trans. R. Soc. B* **363**:1071–1086. [14]

Pohlmeyer, E. A., S. A. Solla, E. J. Perreault, and L. E. Miller. 2007. Prediction of Upper Limb Muscle Activity from Motor Cortical Discharge during Reaching. *J. Neural Eng.* **4**:369–379. [13]

Poirier, K., N. Lebrun, L. Broix, et al. 2013. Mutations in *TUBG1*, *DYNC1H1*, *KIF5C* and *KIF2A* Cause Malformations of Cortical Development and Microcephaly. *Nat. Genet.* **45**:639–647. [4]

Poirier, K., Y. Saillour, N. Bahi-Buisson, et al. 2010. Mutations in the Neuronal ß-Tubulin Subunit TUBB3 Result in Malformation of Cortical Development and Neuronal Migration Defects. *Hum. Mol. Genet.* **19**:4462–4473. [4]

Pollard, K. S., S. R. Salama, B. King, et al. 2006a. Forces Shaping the Fastest Evolving Regions in the Human Genome. *PLoS Genet.* **2**:e168. [3–5]

Pollard, K. S., S. R. Salama, N. Lambert, et al. 2006b. An RNA Gene Expressed during Cortical Development Evolved Rapidly in Humans. *Nature* **443**:167–172. [3–5]

Pollen, A. A., T. J. Nowakowski, J. Chen, et al. 2015. Molecular Identity of Human Outer Radial Glia during Cortical Development. *Cell* **163**:55–67. [3, 5]

Pollen, A. A., T. J. Nowakowski, J. Shuga, et al. 2014. Low-Coverage Single-Cell mRNA Sequencing Reveals Cellular Heterogeneity and Activated Signaling Pathways in Developing Cerebral Cortex. *Nat. Biotechnol.* **32**:1053–1058. [5]

Polleux, F., K. L. Whitford, P. A. Dijkhuizen, T. Vitalis, and A. Ghosh. 2002. Control of Cortical Interneuron Migration by Neurotrophins and Pi3- Kinase Signaling. *Development* **129**:3147–3160. [2]

Poort, J., and P. R. Roelfsema. 2009. Noise Correlations Have Little Influence on the Coding of Selective Attention in Area V1. *Cereb. Cortex* **19**:543–553. [12]

Popesco, M. C., E. J. Maclaren, J. Hopkins, et al. 2006. Human Lineage-Specific Amplification, Selection, and Neuronal Expression of DUF1220 Domains. *Science* **313**:1304–1307. [3]

Portmann, A. 1990. A Zoologist Looks at Humankind. New York: Columbia Univ. Press. [16]

Poskanzer, K. E., and R. Yuste. 2011. Astrocytic Regulation of Cortical up States. *PNAS* **108**:18453–18458. [6]

Postiglione, M. P., C. Juschke, Y. Xie, et al. 2011. Mouse Inscuteable Induces Apical-Basal Spindle Orientation to Facilitate Intermediate Progenitor Generation in the Developing Neocortex. *Neuron* **72**:269–284. [2]

Power, J. D., K. A. Barnes, A. Z. Snyder, B. L. Schlaggar, and S. E. Petersen. 2012. Spurious but Systematic Correlations in Functional Connectivity MRI Networks Arise from Subject Motion. *Neuroimage* **59**:2142–2154. [6]

Power, J. D., A. L. Cohen, S. M. Nelson, et al. 2011. Functional Network Organization of the Human Brain. *Neuron* **72**:665–678. [6]

Prabhakar, S., J. P. Noonan, S. Pääbo, and E. M. Rubin. 2006. Accelerated Evolution of Conserved Noncoding Sequences in Humans. *Science* **314**:786. [3]

Prechtl, J. C., L. B. Cohen, B. Pesaran, P. P. Mitra, and D. Kleinfeld. 1997. Visual Stimuli Induce Waves of Electrical Activity in Turtle Cortex. *PNAS* **94**:7621–7626. [12]

Prescott, S. L., R. Srinivasan, M. C. Marchetto, et al. 2015. Enhancer Divergence and Cis-Regulatory Evolution in the Human and Chimp Neural Crest. *Cell* **163**:68–83. [3]

Preuss, T. M. 1995. Do Rats Have Prefrontal Cortex? The Rose-Woolsey-Akert Program Reconsidered. *J. Cogn. Neurosci.* **7**:1–24. [5]

———. 2000a. Taking the Measure of Diversity: Comparative Alternatives to the Model-Animal Paradigm in Cortical Neuroscience. *Brain Behav. Evol.* **55**:287–299. [16]

———. 2000b. What's Human About the Human Brain? In: The New Cognitive Neurosciences, 2nd Ed, ed. M. S. Gazzaniga, pp. 1219–1234. Cambridge, MA: MIT Press. [16]

———. 2004. Specializations of the Human Visual System: The Monkey Model Meets Human Reality. In: The Primate Visual System, ed. J. H. Kaas and C. E. Collins, pp. 231–260. Methods and New Frontiers in Neuroscience, S. A. Simon and M. A. L. Nicolelis, series eds. Boca Raton: CRC Press. [17]

———. 2009. Primate Brain Evolution. In: Evolutionary Neuroscience, ed. J. H. Kaas, pp. 793–825. Oxford: Academic Press. [16]

Preuss, T. M., H. Qi, and J. H. Kaas. 1999. Distinctive Compartmental Organization of Human Primary Visual Cortex. *PNAS* **96**:11601–11606. [16]

Pribram, K. H., and P. D. MacLean. 1953. Neuronographic Analysis of Medial and Basal Cerebral Cortex, II: Monkey. *J. Neurophysiol.* **16**:324–340. [9]

Puelles, L., E. Kuwana, E. Puelles, et al. 2000. Pallial and Subpallial Derivatives in the Embryonic Chick and Mouse Telencephalon, Traced by the Expression of the Genes Dlx-2, Emx-1, Nkx-2.1, Pax-6, and Tbr-1. *J. Comp. Neur.* **424**:409–438. [5]

Pulvers, J. N., J. Bryk, J. L. Fish, et al. 2010. Mutations in Mouse Aspm (Abnormal Spindle-Like Microcephaly Associated) Cause Not Only Microcephaly but Also Major Defects in the Germline. *PNAS* **107**:16595–16600. [4, 5]

Purpura, K. P., S. F. Kalik, and N. D. Schiff. 2003. Analysis of Perisaccadic Field Potentials in the Occipitotemporal Pathway during Active Vision. *J. Neurophysiol.* **90**:3455–3478. [14]

Rabinovich, M. I., V. S. Afraimovich, V. Bick, and P. Varona. 2012. Information Flow Dynamics in the Brain. *Phys. Life Rev.* **9**:51–73. [14]

Rabinovich, M. I., R. Huerta, and G. Laurent. 2008. Transient Dynamics for Neural Processing. *Science* **321**:48–50. [14]

Radonjic, N. V., A. E. Ayoub, F. Memi, et al. 2014. Diversity of Cortical Interneurons in Primates: The Role of the Dorsal Proliferative Niche. *Cell Rep.* **9**:2139–2151. [5]

Radons, G., J. D. Becker, B. Dulfer, and J. Kruger. 1994. Analysis, Classification, and Coding of Multielectrode Spike Trains with Hidden Markov Models. *Biol. Cybern.* **71**:359–373. [12]

Raghanti, M. A., C. D. Stimpson, J. L. Marcinkiewicz, et al. 2008. Differences in Cortical Serotonergic Innervation among Humans, Chimpanzees, and Macaque Monkeys: A Comparative Study. *Cereb. Cortex* **18**:584–597. [16]

Raichle, M. E. 2009. A Paradigm Shift in Function Brain Imaging. *J. Neurosci.* **29**:12729–12734. [6]

———. 2011. The Restless Brain. *Brain Connect.* **1**:3–12. [6]

Raichle, M. E., and D. A. Gusnard. 2002. Appraising the Brain's Energy Budget. *PNAS* **99**:10237–10239. [9]

Raichle, M. E., and M. A. Mintun. 2006. Brain Work and Brain Imaging. *Annu. Rev. Neurosci.* **29**:449–476. [6]

Raju C. S., Spatazza J, Stanco A, Larimer P, Sorrells S. F., et. al. 2018 Secretagogin is Expressed by Developing Neocortical GABAergic Neurons in Humans but not Mice and Increases Neurite Arbor Size and Complexity. *Cereb. Cortex* **28**:1946–1958. [5]

Rakic, P. 1972. Mode of Cell Migration to the Superficial Layers of Fetal Monkey Neocortex. *J. Comp. Neur.* **145**:61–83. [5]

———. 1974. Neurons in Rhesus Monkey Visual Cortex: Systematic Relation between Time of Origin and Eventual Disposition. *Science* **183**:425–427. [5]

———. 1975. Timing of Major Ontogenetic Events in the Visual Cortex of the Rhesus Monkey. *UCLA Forum Med. Sci.* **18**:3–40. [2]

———. 1977. Genesis of the Dorsal Lateral Geniculate Nucleus in the Rhesus Monkey: Site and Time of Origin, Kinetics of Proliferation, Routes of Migration and Pattern of Distribution of Neurons. *J. Comp. Neur.* **176**:23–52. [5]

———. 1988a. Intrinsic and Extrinsic Determinants of Neocortical Parcellation: A Radial Unit Model. In: Neurobiology of Neocortex: Report of the Dahlem Workshop on Neurobiology of Neocortex, ed. P. Rakic and W. Singer, pp. 5–27. Chichester: John Wiley and Sons. [2]

———. 1988b. Specification of Cerebral Cortical Areas. *Science* **241**:170–176. [2, 5]

———. 1995. A Small Step for the Cell, a Giant Leap for Mankind: A Hypothesis of Neocortical Expansion during Evolution. *Trends Neurosci.* **18**:383–388. [2]

———. 2003. Elusive Radial Glial Cells: Historical and Evolutionary Perspective. *Glia* **43**:19–32. [5]

———. 2009. Evolution of the Neocortex: A Perspective from Developmental Biology. *Nat. Rev. Neurosci.* **10**:724–735. [4, 5]

Rakic, P., J. P. Bourgeois, M. F. Eckenhoff, N. Zecevic, and P. S. Goldman-Rakic. 1986. Concurrent Overproduction of Synapses in Diverse Regions of the Primate Cerebral Cortex. *Science* **232**:232–235. [5]

Rakic, P., and W. Singer, eds. 1988. Neurobiology of Neocortex: Report of the Dahlem Workshop on Neurobiology of Neocortex. Life Sciences Research Reports, vol. J. R. Lupp, series ed. New York: Wiley. [1, 5, 9, 13, 17]

Rakic, P., I. Suner, and R. W. Williams. 1991. A Novel Cytoarchitectonic Area Induced Experimentally within the Primate Visual Cortex. *PNAS* **88**:2083–2087. [5]

Ramaswamy, S., and H. Markram. 2015. Anatomy and Physiology of the Thick-Tufted Layer 5 Pyramidal Neuron. *Front. Cell. Neurosci.* **9**:233. [8]

Rao, R. P. N., and D. H. Ballard. 1999. Predictive Coding in the Visual Cortex: A Functional Interpretation of Some Extra-Classical Receptive-Field Effects. *Nat. Neurosci.* **2**:79–87. [13, 14]

Rash, B. G., A. Duque, Y. M. Morozov, et al. 2019 Gliogenesis in the Outer Subventricular Zone Promotes Enlargement and Gyrification of the Primate Cerebrum. *PNAS*, in press. [5]

Rasin, M. R., V. R. Gazula, J. J. Breunig, et al. 2007. Numb and Numbl Are Required for Maintenance of Cadherin-Based Adhesion and Polarity of Neural Progenitors. *Nat. Neurosci.* **10**:819–827. [2]

Ray, S., and J. H. R. Maunsell. 2010. Differences in Gamma Frequencies across Visual Cortex Restrict Their Possible Use in Computation. *Neuron* **67**:885–896. [10]

———. 2015. Do Gamma Oscillations Play a Role in Cerebral Cortex? *Trends Cogn. Sci.* **19**:78–85. [10]

Razavi, M. J., T. Zhang, H. Chen, et al. 2017. Radial Structure Scaffolds Convolution Patterns of Developing Cerebral Cortex. *Front. Comput. Neurosci.* **11**:76. [2]

Reddy, D. V. R., A. Sen, and J. L. Johnston. 1998. Time Delay Induced Death in Coupled Limit Cycle Oscillators. *Phys. Rev. Lett.* **80**:5109–5112. [10]

Reid, C. B., S. F. Tavazoie, and C. A. Walsh. 1997. Clonal Dispersion and Evidence for Asymmetric Cell Division in Ferret Cortex. *Development* **124**:2441–2450. [5]

Reillo, I., C. de Juan Romero, M. Á. García-Cabezas, and V. Borrell. 2011. A Role for Intermediate Radial Glia in the Tangential Expansion of the Mammalian Cerebral Cortex. *Cereb. Cortex* **21**:1674–1694. [2–5]

Reilly, S. K., J. Yin, A. E. Ayoub, et al. 2015. Evolutionary Changes in Promoter and Enhancer Activity during Human Corticogenesis. *Science* **347**:1155–1159. [3–5]

Reimann, M. W., M. Nolte, M. Scolamiero, et al. 2017. Cliques of Neurons Bound into Cavities Provide a Missing Link between Structure and Function. *Front. Comput. Neurosci.* **11**:48. [7]

Reimer, J., E. Froudarakis, C. R. Cadwell, et al. 2014. Pupil Fluctuations Track Fast Switching of Cortical States during Quiet Wakefulness. *Neuron* **84**:355–362. [12]

Reimer, J., M. J. McGinley, Y. Liu, et al. 2016. Pupil Fluctuations Track Rapid Changes in Adrenergic and Cholinergic Activity in Cortex. *Nat. Commun.* **7**:13289. [9, 13]

Reina, G. A., D. W. Moran, and A. B. Schwartz. 2001. On the Relationship between Joint Angular Velocity and Motor Cortical Discharge During Reaching. *J. Neurophysiol.* **85**:2576–2589. [13]

Reinagel, P., and R. C. Reid. 2002. Precise Firing Events Are Conserved across Neurons. *Journal of Neuroscience.* **22**:6837–6841. [10]

Reiner, O., R. Carrozzo, Y. Shen, et al. 1993. Isolation of a Miller–Dicker Lissencephaly Gene Containing G Protein ß-Subunit-Like Repeats. *Nature* **364**:717–721. [4]

Renart, A., J. de la Rocha, P. Bartho, et al. 2010. The Asynchronous State in Cortical Circuits. *Science* **327**:587–590. [12]

Rendall, D., and A. Di Fiore. 2007. Homoplasy, Homology, and the Perceived Special Status of Behavior in Evolution. *J. Hum. Evol.* **52**:504–521. [17]

Ress, D., B. T. Backus, and D. J. Heeger. 2000. Activity in Primary Visual Cortex Predicts Performance in a Visual Detection Task. *Nat. Neurosci.* **3**:940–945. [6, 9]

Resulaj, A., S. Ruediger, S. R. Olsen, and M. Scanziani. 2018. First Spikes in Visual Cortex Enable Perceptual Discrimination. *eLife* **7**: [13]

Reyes-Puerta, V., J. J. Sun, S. Kim, W. Kilb, and H. J. Luhmann. 2015. Laminar and Columnar Structure of Sensory-Evoked Multineuronal Spike Sequences in Adult Rat Barrel Cortex *in Vivo*. *Cereb. Cortex* **25**:2001–2021. [12]

Reynolds, J. H., and L. Chelazzi. 2004. Attentional Modulation of Visual Processing. *Annu. Rev. Neurosci.* **27**:611–647. [13]

Reynolds, J. H., and D. J. Heeger. 2009. The Normalization Model of Attention. *Neuron* **61**:168–185. [13]

Richiardi, J., A. Altmann, A. C. Milazzo, et al. 2015. Brain Networks: Correlated Gene Expression Supports Synchronous Activity in Brain Networks. *Science* **348**:1241–1244. [6]

Richman, D. P., R. M. Stewart, J. W. Hutchinson, and V. S. Caviness, Jr. 1975. Mechanical Model of Brain Convolutional Development. *Science* **189**:18–21. [2]

Rickmann, M., B. M. Chronwall, and J. R. Wolff. 1977. On the Development of Non-Pyramidal Neurons and Axons Outside the Cortical Plate: The Early Marginal Zone as a Pallial Anlage. *Anat. Embryol.* **151**:285–307. [5]

Riehle, A., S. Grun, M. Diesmann, and A. Aertsen. 1997. Spike Synchronization and Rate Modulation Differentially Involved in Motor Cortical Function. *Science* **278**:1950–1953. [12]

Rilling, J. K., M. F. Glasser, T. M. Preuss, et al. 2008. The Evolution of the Arcuate Fasciculus Revealed with Comparative DTI. *Nat. Neurosci.* **11**:426–428. [5, 16]

Ringo, J. L., R. W. Doty, S. Demeter, and P. Y. Simard. 1994. Time Is of the Essence: A Conjecture That Hemispheric Specialization Arises from Interhemispheric Conduction Delay. *Cereb. Cortex* **4**:331–343. [11]

Ritter, P., M. Schirner, A. R. McIntosh, and V. K. Jirsa. 2013. The Virtual Brain Integrates Computational Modeling and Multimodal Neuroimaging. *Brain Connect.* **3**:121–145. [7]

Rodenas-Cuadrado, P., J. Ho, and S. C. Vernes. 2014. Shining a Light on CNTNAP2: Complex Functions to Complex Disorders. *Euro. J. Human Gen.* **22**:171–178. [3]

Rodriguez, R., U. Kallenbach, W. Singer, and M. H. J. Munk. 2004. Short- and Long-Term Effects of Cholinergic Modulation on Gamma Oscillations and Response Synchronization in the Visual Cortex. *Journal of Neuroscience.* **24**:10369–10378. [10]

Roelfsema, P. R., A. K. Engel, P. König, and W. Singer. 1997. Visuomotor Integration Is Associated with Zero Time-Lag Synchronization among Cortical Areas. *Nature* **385**:157–161. [10]

Roessler, E., E. Belloni, K. Gaudenz, et al. 1996. Mutations in the Human *Sonic Hedgehog* Gene Cause Holoprosencephaly. *Nat. Genet.* **14**:357–360. [4]

Rojas-Piloni, G., J. M. Guest, R. Egger, et al. 2017. Relationships between Structure, *in Vivo* Function and Long-Range Axonal Target of Cortical Pyramidal Tract Neurons. *Nat. Commun.* **8**:870. [8]

Roland, P. E., A. Hanazawa, C. Undeman, et al. 2006. Cortical Feedback Depolarization Waves: A Mechanism of Top-Down Influence on Early Visual Areas. *PNAS* **103**:12586–12591. [12]

Ronan, L., N. Voets, C. Rua, et al. 2014. Differential Tangential Expansion as a Mechanism for Cortical Gyrification. *Cereb. Cortex* **24**:2219–2228. [2]

Rosenblatt, F. 1958. The Perceptron. A Probabilistic Model for Information Storage and Organization in the Brain. *Psychol. Rev.* **65**:386–408. [10, 11]

Rouaux, C., and P. Arlotta. 2013. Direct Lineage Reprogramming of Post-Mitotic Callosal Neurons into Corticofugal Neurons *in Vivo*. *Nat. Cell Biol.* **15**:214–221. [5]

Roux, L., B. Hu, R. Eichler, E. Stark, and G. Buzsáki. 2017. Sharp Wave Ripples during Learning Stabilize the Hippocampal Spatial Map. *Nat. Neurosci.* **20**:845–853. [13]

Roy, D., R. Sigala, M. Breakspear, et al. 2014. Using the Virtual Brain to Reveal the Role of Oscillations and Plasticity in Shaping Brain's Dynamical Landscape. *Brain Connect.* **4**:791–811. [7]

Rubino, D., K. A. Robbins, and N. G. Hatsopoulos. 2006. Propagating Waves Mediate Information Transfer in the Motor Cortex. *Nat. Neurosci.* **9**:1549–1557. [12]

Rubinov, M., and O. Sporns. 2011. Weight-Conserving Characterization of Complex Functional Brain Networks. *Neuroimage* **56**:2068–2079. [7]

Rudy, B., G. Fishell, S. Lee, and J. Hjerling-Leffler. 2011. Three Groups of Interneurons Account for Nearly 100% of Neocortical GABAergic Neurons. *Dev. Neurobiol.* **71**:45–61. [8]

Ryan, R. M., and E. L. Deci. 1985. Intrinsic Motivation and Self-Determination in Human Behavior. New York: Plenum. [14]

Saalmann, Y. B., M. A. Pinsk, L. Wang, X. Li, and S. Kastner. 2012. The Pulvinar Regulates Information Transmission between Cortical Areas Based on Attention Demands. *Science* **337**:753–756. [10]

Sacher, G. A., and E. F. Staffeldt. 1974. Relation of Gestation Time to Brain Weight for Placental Mammals: Implications for the Theory of Vertebrate Growth. *Am. Natural.* **108**:593–615. [16]

Sadtler, P. T., K. M. Quick, M. D. Golub, et al. 2014. Neural Constraints on Learning. *Nature* **512**:423–426. [12]

Safaai, H., R. Neves, O. Eschenko, N. K. Logothetis, and S. Panzeri. 2015. Modeling the Effect of Locus Coeruleus Firing on Cortical State Dynamics and Single-Trial Sensory Processing. *PNAS* **112**:12834–12839. [7]

Saffary, R., and Z. Xie. 2011. FMRP Regulates the Transition from Radial Glial Cells to Intermediate Progenitor Cells during Neocortical Development. *J. Neurosci.* **31**:1427–1439. [2]

Saffran, J. R., and N. Z. Kirkham. 2018. Infant Statistical Learning. *Annu. Rev. Psychol.* **69**:181–203. [17]

Sahin, M., and M. Sur. 2015. Genes, Circuits, and Precision Therapies for Autism and Related Neurodevelopmental Disorders. *Science* **350**:aab3897. [5]

Sakarya, O., K. A. Armstrong, M. Adamska, et al. 2007. A Post-Synaptic Scaffold at the Origin of the Animal Kingdom. *PLoS One* **2**:e506. [13]

Salakhutdinov, R., J. B. Tenenbaum, and A. Torralba. 2013. Learning with Hierarchical-Deep Models. *IEEE Trans. Pattern Anal. Mach. Intell.* **35**:1958–1971. [14]

Salazar, R. F., N. M. Dotson, S. L. Bressler, and C. M. Gray. 2012. Content-Specific Fronto-Parietal Synchronization during Visual Working Memory. *Science* **338**:1097–1100. [10]

Saleem, A. B., A. Ayaz, K. J. Jeffery, K. D. Harris, and M. Carandini. 2013. Integration of Visual Motion and Locomotion in Mouse Visual Cortex. *Nat. Neurosci.* **16**:1864–1869. [13]

Salinas, E., and L. F. Abbott. 1994. Vector Reconstruction for Firing Rates. *J. Comput. Neurosci.* **1**:89–107. [12]

Salinas, E., and T. J. Sejnowski. 2001. Correlated Neuronal Activity and the Flow of Neural Information. *Nat. Rev. Neurosci.* **2**:539–550. [13]

Sanchez-Alcaniz, J. A., S. Haege, W. Mueller, et al. 2011. CXCR7 Controls Neuronal Migration by Regulating Chemokine Responsiveness. *Neuron* **69**:77–90. [5]

Sautois, B., S. R. Soffe, W. C. Li, and A. Roberts. 2007. Role of Type-Specific Neuron Properties in a Spinal Cord Motor Network. *J. Comput. Neurosci.* **23**:59–77. [7]

Savtchouk, I., and A. Volterra. 2018. Gliotransmission: Beyond Black-and-White. *J. Neurosci.* **38**:14–25. [7]

Scannell, J. W., C. Blakemore, and M. P. Young. 1995. Analysis of Connectivity in the Cat Cerebral Cortex. *J. Neurosci.* **15**:1463–1483. [7]

Schapiro, A. C., E. Gregory, B. Landau, M. McCloskey, and N. B. Turk-Browne. 2014. The Necessity of the Medial Temporal Lobe for Statistical Learning. *J. Cogn. Neurosci.* **26**:1736–1747. [17]

Schapiro, A. C., N. B. Turk-Browne, M. M. Botvinick, and K. A. Norman. 2017. Complementary Learning Systems within the Hippocampus: A Neural Network Modelling Approach to Reconciling Episodic Memory with Statistical Learning. *Phil. Trans. R. Soc. B* **372**:20160049. [17]

Schendan, H. E., M. M. Searl, R. J. Melrose, and C. E. Stern. 2003. An Fmri Study of the Role of the Medial Temporal Lobe in Implicit and Explicit Sequence Learning. *Neuron* **37**:1013–1025. [17]

Schenker, N. M., W. D. Hopkins, M. A. Spocter, et al. 2010. Broca's Area Homologue in Chimpanzees (*Pan troglodytes*): Probabilistic Mapping, Asymmetry, and Comparison to Humans. *Cereb. Cortex* **20**:730–742. [16]

Schillen, T. B., and P. König. 1994. Binding by Temporal Structure in Multiple Feature Domains of an Oscillatory Neuronal Network. *Biol. Cybern.* **70**:397–405. [10]

Schmechel, D. E., and P. Rakic. 1979. A Golgi Study of Radial Glial Cells in Developing Monkey Telencephalon: Morphogenesis and Transformation into Astrocytes. *Anat. Embryol.* **156**:115–152. [5]

Schmidhuber, J. 2006. Developmental Robotics, Optimal Artificial Curiosity, Creativity, Music, and the Fine Arts. *Connect. Sci.* **18**:173–187. [14]

———. 2010. Formal Theory of Creativity, Fun, and Intrinsic Motivation (1990–2010). *IEEE Trans. Auton. Ment. Dev.* **2**:230–247. [14]

Schmitz, T. W., and J. Duncan. 2018. Normalization and the Cholinergic Microcircuit: A Unified Basis for Attention. *Trends Cogn. Sci.* **22**:422–437. [13]

Schneider, E., N. El Hajj, S. Richter, et al. 2014. Widespread Differences in Cortex DNA Methylation of the "Language Gene" CNTNAP2 between Humans and Chimpanzees. *Epigenetics* **9**:533–545. [3]

Schneider, E., S. Mayer, N. El Hajj, et al. 2012. Methylation and Expression Analyses of the 7q Autism Susceptibility Locus Genes Mest , Copg2, and Tsga14 in Human and Anthropoid Primate Cortices. *Cytogenet. Genome Res.* **136**:278–287. [3]

Schneirla, T. C. 1949. Levels in the Psychological Capacities of Animals. In: Philosophy for the Future, ed. R. W. Sellars et al., pp. 243–286. New York: Macmillan. [16]

Schölkopf, B., and A. J. Smola. 2002. Learning with Kernels: Support Vector Machines, Regularization, Optimization, and Beyond. Boston: MIT Press. [11]

Scholvinck, M. L., A. B. Saleem, A. Benucci, K. D. Harris, and M. Carandini. 2015. Cortical State Determines Global Variability and Correlations in Visual Cortex. *J. Neurosci.* **35**:170–178. [13]

Scholz, J., M. C. Klein, T. E. Behrens, and H. Johansen-Berg. 2009. Training Induces Changes in White-Matter Architecture. *Nat. Neurosci.* **12**:1370–1371. [7]

Schroeder, C. E., and P. Lakatos. 2008. Low-Frequency Neuronal Oscillations as Instruments of Sensory Selection. *Trends Neurosci.* **32**:9–18. [6]

Schultz, W. 2015. Neuronal Reward and Decision Signals: From Theories to Data. *Physiol. Rev.* **95**:853–951. [17]

Schumacher, J., R. Haslinger, and G. Pipa. 2012. Statistical Modeling Approach for Detecting Generalized Synchronization. *Phys. Rev. E* **85**:056215. [14]

Schumacher, J., H. Toutounji, and P. Pipa. 2013. An Analytical Approach to Single Node Delay-Coupled Reservoir Computing. In: Artificial Neural Networks and Machine Learning: Icann 2013, ed. V. Mladenov et al., vol. 8131, pp. 26–34. Lecture Notes in Computer Science. Berlin: Springer. [11]

Schumacher, J., T. Wunderle, P. Fries, F. Jäkel, and G. Pipa. 2015. A Statistical Framework to Infer Delay and Direction of Information Flow from Measurements of Complex Systems. *Neural Comput.* **27**:1555–1608. [14]

Schwartz, A. B. 1994. Direct Cortical Representation of Drawing. *Sci. Adv.* **265**:540–542. [13]

Schwartz, A. B., X. T. Cui, D. J. Weber, and D. W. Moran. 2006. Brain-Controlled Interfaces: Movement Restoration with Neural Prosthetics. *Neuron* **52**:205–220. [13]

Schwartz, O., and E. P. Simoncelli. 2001. Natural Signal Statistics and Sensory Gain Control. *Nat. Neurosci.* **4**:819–825. [13]

Schwiedrzik, C. M., S. S. Sudmann, T. Thesen, et al. 2018. Medial Prefrontal Cortex Supports Perceptual Memory. *Curr. Biol.* **28**:R1094–R1095. [17]

Schwiedrzik, C. M., W. Zarco, S. Everling, and W. A. Freiwald. 2015. Face Patch Resting State Networks Link Face Processing to Social Cognition. *PLoS Biol.* **13**:e1002245. [17]

Scott, S. H. 2003. The Role of Primary Motor Cortex in Goal-Directed Movements: Insights from Neurophysiological Studies on Non-Human Primates. *Curr. Opin. Neurobiol.* **13**:671–677. [13]

Seelig, J. D., and V. Jayaraman. 2015. Neural Dynamics for Landmark Orientation and Angular Path Integration. *Nature* **521**:186–191. [13]

Seidemann, E., I. Meilijson, M. Abeles, H. Bergman, and E. Vaadia. 1996. Simultaneously Recorded Single Units in the Frontal Cortex Go through Sequences of Discrete and Stable States in Monkey Performing a Delayed Localization Task. *J. Neurosci.* **16**:752–768. [12]

Seifert, U. 2012. Stochastic Thermodynamics, Fluctuation Theorems and Molecular Machines. *Rep. Prog. Phys.* **75**:126001. [14]

Sejnowski, T. J., and O. Paulsen. 2006. Network Oscillations: Emerging Computational Principles. *J. Neurosci.* **26**:1673–1676. [13]

Semendeferi, K., H. Damasio, R. Frank, and G. W. Van Hoesen. 1997. The Evolution of the Frontal Lobes: A Volumetric Analysis Based on Three-Dimensional Reconstructions of Magnetic Resonance Scans of Human and Ape Brains. *J. Hum. Evol.* **32**:375–388. [16]

Sena, E. S., H. B. van der Worp, P. M. Bath, D. W. Howells, and M. R. Macleod. 2010. Publication Bias in Reports of Animal Stroke Studies Leads to Major Overstatement of Efficacy. *PLoS Biol.* **8**:e1000344. [17]

Sengupta, B., A. Tozzi, G. K. Cooray, P. K. Douglas, and K. Friston. 2016. Towards a Neuronal Gauge Theory. *PLoS Biol.* **14**:e1002400. [14]

Sergent, C., S. Baillet, and S. Dehaene. 2005. Timing of the Brain Events Underlying Access to Consciousness during the Attentional Blink. *Nat. Neurosci.* **8**:1391–1400. [10]

Seth, A. K. 2015. Inference to the Best Prediction. In: Open Mind: Philosophy and the Mind Sciences in the 21st Century, ed. T. K. Metzinger and J. M. Windt. Frankfurt am Main: MIND Group. [14]

Seth, A. K., and K. Friston. 2015. The Cybernetic Bayesian Brain: From Interoceptive Inference to Sensorimotor Contingencies. In: Open Mind: Philosophy and the Mind Sciences in the 21st Century, ed. T. Metzinger and J. M. Windt, vol. 35. Frankfurt am Main: MIND Group. [14]

Seuntjens, E., A. Nityanandam, A. Miquelajauregui, et al. 2009. Sip1 Regulates Sequential Fate Decisions by Feedback Signaling from Postmitotic Neurons to Progenitors. *Nat. Neurosci.* **12**:1373–1380. [2]

Shadlen, M. N., and W. T. Newsome. 1998. The Variable Discharge of Cortical Neurons: Implications for Connectivity, Computation, and Information Coding. *J. Neurosci.* **18**:3870–3896. [13]

Shen, J., W. Eyaid, G. H. Mochida, et al. 2005. ASPM Mutations Identified in Patients with Primary Microcephaly and Seizures. *J. Med. Genet.* **42**:725–729. [4]

Shen, J., E. C. Gilmore, C. A. Marshall, et al. 2010. Mutations in *PNKP* Cause Microcephaly, Seizures and Defects in DNA Repair. *Nat. Genet.* **42**:245–249. [4]

Shenoy, K. V., and J. M. Carmena. 2014. Combining Decoder Design and Neural Adaptation in Brain-Machine Interfaces. *Neuron* **84**:665–680. [12]

Shenoy, K. V., M. Sahani, and M. M. Churchland. 2013. Cortical Control of Arm Movements: A Dynamical Systems Perspective. *Annu. Rev. Neurosci.* **36**:337–359. [12]

Sherman, M. A., S. Lee, R. Law, et al. 2016. Neural Mechanisms of Transient Neocortical Beta Rhythms: Converging Evidence from Humans, Computational Modeling, Monkeys, and Mice. *PNAS* **113**:E4885–E4894. [7]

Sherman, S. M. 2017. Functioning of Circuits Connecting Thalamus and Cortex. *Compr. Physiol.* **7**:713–739. [9]

Sherman, S. M., and R. W. Guillery. 1998. On the Actions That One Nerve Cell Can Have on Another: Distinguishing "Drivers" from "Modulators." *PNAS* **95**:7121–7126. [14]

———. 2011. Distinct Functions for Direct and Transthalamic Corticocortical Connections. *J. Neurophysiol.* **106**:1068–1077. [5, 14]

Sherrington, C. 1906. The Integrative Action of the Nervous System. New Haven: Yale Univ. Press. [13]

Sherwood, C. C., F. Subiaul, and T. W. Zawidzki. 2008. A Natural History of the Human Mind: Tracing Evolutionary Changes in Brain and Cognition. *J. Anat.* **212**:426–454. [16]

Shimamura, K., and J. L. Rubenstein. 1997. Inductive Interactions Direct Early Regionalization of the Mouse Forebrain. *Development* **124**:2709–2718. [5]

Shimogori, T., V. Banuchi, H. Y. Ng, J. B. Strauss, and E. A. Grove. 2004. Embryonic Signaling Centers Expressing BMP, WNT and FGF Proteins Interact to Pattern the Cerebral Cortex. *Development* **131**:5639–5647. [2]

Shin, C. W., and S. Kim. 2006. Self-Organized Criticality and Scale-Free Properties in Emergent Functional Neural Networks. *Phys. Rev. E* **74**:45101. [14]

Shipp, S. 2016. Neural Elements for Predictive Coding. *Front. Psychol.* **7**:1792. [14]

Shitamukai, A., D. Konno, and F. Matsuzaki. 2011. Oblique Radial Glial Divisions in the Developing Mouse Neocortex Induce Self-Renewing Progenitors Outside the Germinal Zone That Resemble Primate Outer Subventricular Zone Progenitors. *J. Neurosci.* **31**:3683–3695. [2, 5]

Shmueli, G. 2010. To Explain or to Predict? *Statis. Sci.* **25**:289–310. [7]

Shneider, A. M. 2009. Four Stages of a Scientific Discipline; Four Types of Scientist. *Trends Biochem. Sci.* **34**:217–223. [7]

Shomrat, T., A. L. Turchetti-Maia, N. Stern-Mentch, J. A. Basil, and B. Hochner. 2015. The Vertical Lobe of Cephalopods: An Attractive Brain Structure for Understanding the Evolution of Advanced Learning and Memory Systems. *J. Comp. Physiol. A* **201**:947–956. [16]

Shuler, M. G., and M. F. Bear. 2006. Reward Timing in the Primary Visual Cortex. *Science* **311**:1606–1609. [13]

Shulha, H. P., J. L. Crisci, D. Reshetov, et al. 2012. Human-Specific Histone Methylation Signatures at Transcription Start Sites in Prefrontal Neurons. *PLoS Biol.* **10**:e1001427. [3]

Shwartz-Ziv, R., and N. Tishby. 2017. Opening the Black Box of Deep Neural Networks via Information. *arXiv* 1703.00810. [11, 13]

Sidman, R. L., and P. Rakic. 1973. Neuronal Migration, with Special Reference to Developing Human Brain: A Review. *Brain Res.* **62**:1–35. [5]

Siegel, M., T. H. Donner, R. Oostenveld, P. Fries, and A. K. Engel. 2008. Neuronal Synchronization Along the Dorsal Visual Pathway Reflects the Focus of Spatial Attention. *Neuron* **60**:709–719. [10]

Siegel, M., M. R. Warden, and E. K. Miller. 2009. Phase-Dependent Neuronal Coding of Objects in Short-Term Memory. *PNAS* **106**:21341–21346. [13]

Siegler, R. S., and J. E. Opfer. 2003. The Development of Numerical Estimation: Evidence for Multiple Representations of Numerical Quantity. *Psychol. Sci.* **14**:237–243. [17]

Sigaard, R. K., M. Kjær, and B. Pakkenberg. 2016. Development of the Cell Population in the Brain White Matter of Young Children. *Cereb. Cortex* **26**:89–95. [3]

Silan, F., M. Yoshioka, K. Kobayashi, et al. 2003. A New Mutation of the Fukutin Gene in a Non-Japanese Patient. *Ann. Neurol.* **53**:392–396. [4]

Silbereis, J. C., S. Pochareddy, Y. Zhu, M. Li, and N. Sestan. 2016. The Cellular and Molecular Landscapes of the Developing Human Central Nervous System. *Neuron* **89**:248–268. [3, 5]

Silver, D., A. Huang, C. J. Maddison, et al. 2016. Mastering the Game of Go with Deep Neural Networks and Tree Search. *Nature* **529**:484–489. [17]

Silver, D., J. Schrittwieser, K. Simonyan, et al. 2017. Mastering the Game of Go without Human Knowledge. *Nature* **550**:354–359. [10]

Simonsen, I., L. Buzna, K. Peters, S. Bornholdt, and D. Helbing. 2008. Transient Dynamics Increasing Network Vulnerability to Cascading Failures. *Phys. Rev. Lett.* **100**:218701. [7]

Singer, W. 1993. Synchronization of Cortical Activity and Its Putative Role in Information Processing and Learning. *Annu. Rev. Physiol.* **55**:349–374. [10]

———. 1999. Neuronal Synchrony: A Versatile Code for the Definition of Relations? *Neuron* **24**:49–65. [10]

———. 2016. The History of Neuroscience in Autobiography, ed. T. D. Albright and L. Squire, vol. 9, pp. 424–483. Washington, D.C.: Society for Neuroscience. [1]

———. 2017. Synchronous Oscillations and Memory Formation. In: Learning Theory and Behavior, vol. 1 of Learning and Memory: A Comprehensive Reference, ed. J. H. Byrne, pp. 591–597. Oxford: Academic Press. [10]

———. 2018a. Neuronal Oscillations: Unavoidable and Useful? *Eur. J. Neurosci.* **48**:2389–2398. [12]

————. 2018b. The Role of Oscillations and Synchrony in the Development of the Nervous System. In: Emergent Brain Dynamics: Prebirth to Adolescence, ed. A. A. Benasich and U. Ribary, vol. 25. Strüngmann Forum Reports, J. Lupp, series ed. Cambridge, MA: MIT Press. [10]

Singer, W., and C. M. Gray. 1995. Visual Feature Integration and the Temporal Correlation Hypothesis. *Annu. Rev. Neurosci.* **18**:555–586. [10, 14]

Singer, W., and F. Tretter. 1976. Unusually Large Receptive Fields in Cats with Restricted Visual Experience. *Exp. Brain Res.* **26**:171–184. [10]

Sir, J.-H., A. R. Barr, A. K. Nicholas, et al. 2011. A Primary Microcephaly Protein Complex Forms a Ring around Parental Centrioles. *Nat. Genet.* **43**:1147–1153. [4]

Sirotin, Y. B., and A. Das. 2009. Anticipatory Haemodynamic Signals in Sensory Cortex Not Predicted by Local Neuronal Activity. *Nature* **457**:475–479. [6, 9]

Sizemore, A. E., and D. S. Bassett. 2018. Dynamic Graph Metrics: Tutorial, Toolbox, and Tale. *Neuroimage* **180**:417–427. [7]

Sizemore, A. E., C. Giusti, A. Kahn, et al. 2017. Cliques and Cavities in the Human Connectome. *J. Comput. Neurosci.* **44**:115–145. [7]

Sizemore, A. E., J. Phillips-Cremins, R. Ghrist, and D. S. Bassett. 2018. The Importance of the Whole: Topological Data Analysis for the Network Neuroscientist. *arXiv* 1806.05167. [7]

Skop, A. R., and J. G. White. 1998. The Dynactin Complex Is Required for Cleavage Plane Specification in Early *Caenorhabditis elegans* Embryos. *Curr. Biol.* **8**:1110–1116. [2]

Sliwa, J., and W. A. Freiwald. 2017. A Dedicated Network for Social Interaction Processing in the Primate Brain. *Science* **356**:745–749. [17]

Smaers, J. B., A. Gómez-Robles, A. N. Parks, and C. C. Sherwood. 2017. Exceptional Evolutionary Expansion of Prefrontal Cortex in Great Apes and Humans. *Curr. Biol.* **27**:714–720. [15]

Smart, I. H., C. Dehay, P. Giroud, M. Berland, and H. Kennedy. 2002. Unique Morphological Features of the Proliferative Zones and Postmitotic Compartments of the Neural Epithelium Giving Rise to Striate and Extrastriate Cortex in the Monkey. *Cereb. Cortex* **12**:37–53. [2, 5]

Smith, G. B., B. Hein, D. E. Whitney, D. Fitzpatrick, and M. Kaschube. 2018. Distributed network interactions and their emergence in developing neocortex. *Nat. Neurosci.* **21**:1600–1608. [13]

Smith, G. B., A. Sederberg, Y. M. Elyada, et al. 2015. The Development of Cortical Circuits for Motion Discrimination. *Nat. Neurosci.* **18**:252–261. [10]

Smith, J. D., M. E. Berg, R. G. Cook, et al. 2012a. Implicit and Explicit Categorization: A Tale of Four Species. *Neurosci. Biobehav. Rev.* **36**:2355–2369. [17]

Smith, J. D., M. J. Crossley, J. Boomer, et al. 2012b. Implicit and Explicit Category Learning by Capuchin Monkeys (Cebus apella). *J. Comp. Psychol.* **126**:294–304. [17]

Smith, S. M., D. Vidaurre, C. F. Beckmann, et al. 2013. Functional Connectomics from Resting-State fMRI. *Trends Cogn. Sci.* **17**:666–682. [9]

Sohal, V. S., F. Zhang, O. Yizhar, and K. Deisseroth. 2009. Parvalbumin Neurons and Gamma Rhythms Enhance Cortical Circuit Performance. *Nature* **459**:698–702. [13]

Sokolov, A. A., R. C. Miall, and R. B. Ivry. 2017. The Cerebellum: Adaptive Prediction for Movement and Cognition. *Trends Cogn. Sci.* **21**:313–332. [5]

Somel, M., H. Franz, Z. Yan, et al. 2009. Transcriptional Neoteny in the Human Brain. *PNAS* **106**:5743–5748. [16]

Somel, M., X. Liu, and P. Khaitovich. 2013. Human Brain Evolution: Transcripts, Metabolites and Their Regulators. *Nat. Rev. Neurosci.* **14**:112–127. [3]

Song, S., P. J. Sjostrom, M. Reigl, S. Nelson, and D. B. Chklovskii. 2005. Highly Nonrandom Features of Synaptic Connectivity in Local Cortical Circuits. *PLoS Biol.* **3**:e68. [12]

Song, W. J., H. Kawaguchi, S. Totoki, et al. 2006. Cortical Intrinsic Circuits Can Support Activity Propagation through an Isofrequency Strip of the Guinea Pig Primary Auditory Cortex. *Cereb. Cortex* **16**:718–729. [12]

Sonnweber, R., A. Ravignani, and W. T. Fitch. 2015. Non-Adjacent Visual Dependency Learning in Chimpanzees. *Anim. Cogn.* **18**:733–745. [15]

Sorensen, S. A., A. Bernard, V. Menon, et al. 2015. Correlated Gene Expression and Target Specificity Demonstrate Excitatory Projection Neuron Diversity. *Cereb. Cortex* **25**:433–449. [8]

Sousa, A. M. M., K. A. Meyer, G. Santpere, F. O. Gulden, and N. Sestan. 2017a. Evolution of the Human Nervous System Function, Structure, and Development. *Cell* **170**:226–247. [3, 5]

Sousa, A. M. M., Y. Zhu, M. A. Raghanti, et al. 2017b. Molecular and Cellular Reorganization of Neural Circuits in the Human Lineage. *Science* **358**:1027–1032. [3]

Spillmann, L. 2014. Receptive Fields of Visual Neurons: The Early Years. *Perception* **43**:1145–1176. [13]

Spitzer, N. C. 2006. Electrical Activity in Early Neuronal Development. *Nature* **444**:707. [13]

Spocter, M. A., W. D. Hopkins, A. R. Garrison, et al. 2010. Wernicke's Area Homologue in Chimpanzees (*Pan troglodytes*) and Its Relation to the Appearance of Modern Human Language. *Proc. R. Soc. Lond. B* **277**:2165–2174. [16]

Sporns, O. 2014. Contributions and Challenges for Network Models in Cognitive Neuroscience. *Nat. Neurosci.* **17**:652–660. [7]

Sporns, O., and R. F. Betzel. 2016. Modular Brain Networks. *Annu. Rev. Psychol.* **67**:613–640. [9]

Sporns, O., G. Tononi, and R. Kötter. 2005. The Human Connectome: A Structural Description of the Human Brain. *PLoS Comput. Biol.* **1**:e42. [7]

Srinivasan, M. V., S. B. Laughlin, and A. Dubs. 1982. Predictive Coding: A Fresh View of Inhibition in the Retina. *Proc R Soc Lond B Biol Sci* **216**:427–459. [14]

Stafford, J. M., B. R. Jarrett, O. Miranda-Dominguez, et al. 2014. Large-Scale Topology and the Default Mode Network in the Mouse Connectome. *PNAS* **111**:18745–18750. [6]

Stahl, R., T. Walcher, C. De Juan Romero, et al. 2013. Trnp1 Regulates Expansion and Folding of the Mammalian Cerebral Cortex by Control of Radial Glial Fate. *Cell* **153**:535–549. [2]

Stam, C. J. 2004. Functional Connectivity Patterns of Human Magnetoencephalographic Recordings: A "Small-World" Network? *Neurosci. Lett.* **355**:25–28. [7]

———. 2014. Modern Network Science of Neurological Disorders. *Nat. Rev. Neurosci.* **15**:683–695. [7, 9]

Stancik, E. K., I. Navarro-Quiroga, R. Sellke, and T. F. Haydar. 2010. Heterogeneity in Ventricular Zone Neural Precursors Contributes to Neuronal Fate Diversity in the Postnatal Neocortex. *J. Neurosci.* **30**:7028–7036. [2, 5]

Stehberg, J., P. T. Dang, and R. D. Frostig. 2014. Unimodal Primary Sensory Cortices Are Directly Connected by Long-Range Horizontal Projections in the Rat Sensory Cortex. *Front. Neuroanat.* **8**:93. [8]

Steinecke, A., C. Gampe, G. Zimmer, J. Rudolph, and J. Bolz. 2014. Epha/ephrin A Reverse Signaling Promotes the Migration of Cortical Interneurons from the Medial Ganglionic Eminence. *Development* **141**:460–471. [5]

Stenzel, D., M. Wilsch-Bräuninger, F. K. Wong, H. Heuer, and W. B. Huttner. 2014. Integrin $Á_v\beta_3$ and Thyroid Hormones Promote Expansion of Progenitors in Embryonic Neocortex. *Development* **141**:795–806. [5]

Steriade, M., A. McCormick, and T. J. Sejnowski. 1993. Thalamocortical Oscillations in the Sleeping and Aroused Brain. *Science* **262**:679–685. [10]

Stettler, D. D., A. Das, J. Bennett, and C. D. Gilbert. 2002. Lateral Connectivity and Contextual Interactions in Macaque Primary Visual Cortex. *Neuron* **36**:739–750. [10]

Still, S., D. A. Sivak, A. J. Bell, and G. E. Crooks. 2012. Thermodynamics of Prediction. *Phys. Rev. Lett.* **109**:120604. [14]

Stiso, J., A. N. Khambhati, T. Menara, et al. 2018. White Matter Network Architecture Guides Direct Electrical Stimulation through Optimal State Transitions. *arXiv* 1805.01260. [7]

Storm, E. E., S. Garel, U. Borello, et al. 2006. Dose-Dependent Functions of FGF8 in Regulating Telencephalic Patterning Centers. *Development* **133**:1831–1844. [2]

Strauss, M., J. D. Sitt, J. R. King, et al. 2015. Disruption of Hierarchical Predictive Coding during Sleep. *PNAS* **112**:E1353–1362. [14]

Stringer, C., M. Pachitariu, N. Steinmetz, et al. 2018. Spontaneous Behaviors Drive Multidimensional, Brain-Wide Population Activity. *bioRxiv*306019. [13]

Stuart, G. J., and M. Häusser. 2001. Dendritic Coincidence Detection of EPSPs and Action Potentials. *Nat. Neurosci.* **4**:63–71. [10]

Suarez, M. 2009. Fictions in Science: Philosophical Essays on Modelling and Idealization. London: Routledge. [7]

Sugihara, I., E. J. Lang, and R. Llinas. 1993. Uniform Olivocerebellar Conduction Time Underlies Purkinje Cell Complex Spike Synchronicity in the Rat Cerebellum. *J. Physiol.* **470**:243–271. [13]

Sultan, K. T., Z. Han, X. J. Zhang, et al. 2016. Clonally Related GABAergic Interneurons Do Not Randomly Disperse but Frequently Form Local Clusters in the Forebrain. *Neuron* **92**:31–44. [5]

Sultan, K. T., W. Shi, and S.-H. Shi. 2014. Clonal Origins of Neocortical Interneurons. *Curr. Opin. Neurobiol.* **26**:125–131. [5]

Sun, H., and B. J. Frost. 1998. Computation of Different Optical Variables of Looming Objects in Pigeon Nucleus Rotundus Neurons. *Nat. Neurosci.* **1**:296–303. [13]

Sur, M., and J. L. R. Rubenstein. 2005. Patterning and Plasticity of the Cerebral Cortex. *Science* **310**:805–810. [2, 3]

Suthers, R. A., and D. Margoliash. 2002. Motor Control of Birdsong. *Curr. Opin. Neurobiol.* **12**:684–690. [10]

Suway, S. B., J. Orellana, A. J. C. McMorland, et al. 2018. Temporally Segmented Directionality in the Motor Cortex. *Cereb. Cortex* **28**:2326-2339. [13]

Swadlow, H. A., I. N. Beloozerova, and M. G. Sirota. 1998. Sharp, Local Synchrony among Putative Feed-Forward Inhibitory Interneurons of Rabbit Somatosensory Cortex. *J. Neurophysiol.* **79**:567–582. [13]

Swanson, L. W. 2000. Cerebral Hemisphere Regulation of Motivated Behavior. *Brain Res.* **886**:113–164. [9]

Takahashi, K., S. Kim, T. P. Coleman, et al. 2015. Large-Scale Spatiotemporal Spike Patterning Consistent with Wave Propagation in Motor Cortex. *Nat. Commun.* **6**:7169. [12]

Takahashi, T., R. S. Nowakowski, and V. S. Caviness, Jr. 1995. The Cell Cycle of the Pseudostratified Ventricular Epithelium of the Embryonic Murine Cerebral Wall. *J. Neurosci.* **15**:6046–6057. [2]

Takesian, A. E., L. J. Bogart, J. W. Lichtman, and T. K. Hensch. 2018. Inhibitory Circuit Gating of Auditory Critical-Period Plasticity. *Nat. Neurosci.* **21**:218–227. [5]

Takesian, A. E., and T. K. Hensch. 2013. Balancing Plasticity/Stability across Brain Development. *Prog. Brain Res.* **207**:3–34. [5]

Tallinen, T., J. Y. Chung, J. S. Biggins, and L. Mahadevan. 2014. Gyrification from Constrained Cortical Expansion. *PNAS* **111**:12667–12672. [5]

Tamamaki, N., K. Nakamura, K. Okamoto, and T. Kaneko. 2001. Radial Glia Is a Progenitor of Neocortical Neurons in the Developing Cerebral Cortex. *Neurosci. Res.* **41**:51–60. [2, 5]

Tang, A., D. Jackson, J. Hobbs, et al. 2008. A Maximum Entropy Model Applied to Spatial and Temporal Correlations from Cortical Networks *in Vitro*. *J. Neurosci.* **28**:505–518. [7]

Tang, E., and D. S. Bassett. 2018. Control of Dynamics in Brain Networks. *Rev. Mod. Phys.* **90**:031003. [7]

Tang, Y., J. R. Nyengaard, D. M. De Groot, and H. J. Gundersen. 2001. Total Regional and Global Number of Synapses in the Human Brain Neocortex. *Synapse* **41**:258–273. [3]

Tani, J. 2003. Learning to Generate Articulated Behavior through the Bottom-up and the Top-Down Interaction Processes. *Neural Netw.* **16**:11–23. [14]

Tani, J., M. Ito, and Y. Sugita. 2004. Self-Organization of Distributedly Represented Multiple Behavior Schemata in a Mirror System: Reviews of Robot Experiments Using Rnnpb. *Neural Netw.* **17**:1273–1289. [14]

Tank, D. W., A. Gelperin, and D. Kleinfeld. 1994. Odors, Oscillations, and Waves: Does It All Compute? *Science* **265**:1819–1820. [13]

Tao, H., L. I. Zhang, G. Bi, and M. Poo. 2000. Selective Presynaptic Propagation of Long-Term Potentiation in Defined Neural Networks. *J. Neurosci.* **20**:3233–3243. [17]

Tasic, B., V. Menon, T. N. Nguyen, et al. 2016. Adult Mouse Cortical Cell Taxonomy Revealed by Single Cell Transcriptomics. *Nat. Neurosci.* **19**:335–346. [5]

Tasic, B., Z. Yao, L. T. Graybuck, et al. 2018. Shared and Distinct Transcriptomic Cell Types across Neocortical Areas. *Nature* **563**:7278. [13]

Taverna, E., M. Götz, and W. B. Huttner. 2014. The Cell Biology of Neurogenesis: Toward an Understanding of the Development and Evolution of the Neocortex. *Annu. Rev. Cell. Dev. Biol.* **30**:465–502. [3, 5]

Taverna, E., and W. B. Huttner. 2010. Neural Progenitor Nuclei in Motion. *Neuron* **67**:906–914. [5]

Taverna, E., F. Mora-Bermudez, P. J. Strzyz, et al. 2016. Non-Canonical Features of the Golgi Apparatus in Bipolar Epithelial Neural Stem Cells. *Sci. Rep.* **6**:21206. [5]

Taylor, D. M., S. I. Tillery, and A. B. Schwartz. 2002. Direct Cortical Control of 3D Neuroprosthetic Devices. *Science* **296**:1829–1832. [12]

Taylor, P. N., J. Thomas, N. Sinha, et al. 2015. Optimal Control Based Seizure Abatement Using Patient Derived Connectivity. *Front. Neurosci.* **9**:202. [7]

Teller, S., C. Granell, M. De Domenico, et al. 2014. Emergence of Assortative Mixing between Clusters of Cultured Neurons. *PLoS Comput. Biol.* **10**:e1003796. [7]

Tenenbaum, J. B., C. Kemp, T. L. Griffiths, and N. D. Goodman. 2011. How to Grow a Mind: Statistics, Structure, and Abstraction. *Science* **331**:1279–1285. [14]

Terrace, H. S., L. A. Petitto, R. J. Sanders, and T. G. Bever. 1979. Can an Ape Create a Sentence? *Science* **206**:891–902. [15, 17]

Tewarie, P., A. Hillebrand, E. van Dellen, et al. 2014. Structural Degree Predicts Functional Network Connectivity: A Multimodal Resting-State fMRI and MEG Study. *Neuroimage* **97**:296–307. [7]

Theesfeld, C. L., J. E. Irazoqui, K. Bloom, and D. J. Lew. 1999. The Role of Actin in Spindle Orientation Changes during the Saccharomyces cerevisiae Cell Cycle. *J. Cell Biol.* **146**:1019–1032. [2]

Thelen, E., D. M. Fisher, and R. Ridley-Johnson. 1984. The Relationship between Physical Growth and a Newborn Reflex. *Infant. Behav. Dev.* **7**:479–493. [16]

Theodoni, P., T. I. Panagiotaropoulos, V. Kapoor, N. K. Logothetis, and G. Deco. 2011. Cortical Microcircuit Dynamics Mediating Binocular Rivalry: The Role of Adaptation in Inhibition. *Front. Hum. Neurosci.* **5**:145. [14]

Thornburgh, C. L., S. Narayana, R. Rezaie, et al. 2017. Concordance of the Resting State Networks in Typically Developing, 6- to 7-Year-Old Children and Healthy Adults. *Front. Hum. Neurosci.* **11**:199. [9]

Thulborn, K. R., J. C. Waterton, P. M. Matthews, and G. K. Radda. 1982. Oxygenation Dependence of the Transverse Relaxation Time of Water Protons in Whole Blood at High Field. *Biochim. Biophys. Acta* **714**:265–270. [6]

Tietje, A., K. N. Maron, Y. Wei, and D. M. Feliciano. 2014. Cerebrospinal Fluid Extracellular Vesicles Undergo Age Dependent Declines and Contain Known and Novel Non-Coding Rnas. *PLoS One* **9**:e113116. [4]

Tischfield, D. J., J. Kim, and S. A. Anderson. 2017. Atypical PKC and Notch Inhibition Differentially Modulate Cortical Interneuron Subclass Fate from Embryonic Stem Cells. *Stem Cell Reports* **8**:1135–1143. [5]

Tissir, F., and A. M. Goffinet. 2003. Reelin and Brain Development. *Nat. Rev. Neurosci.* **4**:496–505. [5]

Todorov, E. 2000. Direct Cortical Control of Muscle Activation in Voluntary Arm Movements: A Model. *Nat. Neurosci.* **3**:391–398. [13]

Tolhurst, D. J., J. A. Movshon, and A. F. Dean. 1983. The Statistical Reliability of Signals in Single Neurons in Cat and Monkey Visual Cortex. *Vision Res.* **23**:775–785. [12]

Tolias, A. S., T. Moore, S. M. Smirnakis, et al. 2001. Eye Movements Modulate Visual Receptive Fields of V4 Neurons. *Neuron* **29**:757–767. [13]

Toma, K., T. Kumamoto, and C. Hanashima. 2014. The Timing of Upper-Layer Neurogenesis Is Conferred by Sequential Derepression and Negative Feedback from Deep-Layer Neurons. *J. Neurosci.* **34**:13259–13276. [2]

Tomé, T. 2006. Entropy Production in Nonequilibrium Systems Described by a Fokker-Planck Equation. *Braz. J. Phys.* **36**:1285–1289. [14]

Tononi, G., and C. Cirelli. 2014. Sleep and the Price of Plasticity: From Synaptic and Cellular Homeostasis to Memory Consolidation and Integration. *Neuron* **81**:12–34. [13]

Tononi, G., O. Sporns, and G. M. Edelman. 1994. A Measure for Brain Complexity: Relating Functional Segregation and Integration in the Nervous System. *PNAS* **91**:5033–5037. [14]

Toon, A. 2012. Models as Make-Believe: Imagination, Fiction, and Scientific Representation. Basingstoke: Palgrave. [7]

Torii, M., K. Hashimoto-Torii, P. Levitt, and P. Rakic. 2009. Integration of Neuronal Clones in the Radial Cortical Columns by Epha and Ephrin-a Signalling. *Nature* **461**:524–528. [5]

Torre, E., P. Quaglio, M. Denker, et al. 2016. Synchronous Spike Patterns in Macaque Motor Cortex during an Instructed-Delay Reach-to-Grasp Task. *J. Neurosci.* **36**:8329–8340. [12]

Tosches, M. A., T. M. Yamawaki, R. K. Naumann, et al. 2018. Evolution of Pallium, Hippocampus, and Cortical Cell Types Revealed by Single-Cell Transcriptomics in Reptiles. *Science* **360**:881–888. [5, 13]

Toutounji, H., and G. Pipa. 2014. Spatiotemporal Computations of an Excitable and Plastic Brain: Neuronal Plasticity Leads to Noise-Robust and Noise-Constructive Computations. *PLoS Comput. Biol.* **10**:e1003512. [11, 13, 14]

Toutounji, H., J. Schumacher, and G. Pipa. 2015. Homeostatic Plasticity for Single Node Delay-Coupled Reservoir Computing. *Neural Comput.* **27**:1159–1185. [11]

Townsend, B., L. Paninski, and R. Lemon. 2006. Linear Encoding of Muscle Activity in Primary Motor Cortex and Cerebellum. *J. Neurophysiol.* **96**:2578–2592. [13]

Tremblay, R., S. Lee, and B. Rudy. 2016. GABAergic Interneurons in the Neocortex: From Cellular Properties to Circuits. *Neuron* **91**:260–292. [8]

Triesch, J. 2007. Synergies between Intrinsic and Synaptic Plasticity Mechanisms. *Neural Comput.* **19**:885–909. [11]

Triplett, J. W., and D. A. Feldheim. 2012. Eph and Ephrin Signaling in the Formation of Topographic Maps. *Semin. Cell Dev. Biol.* **23**:7–15. [5]

Tsao, D. Y., S. Moeller, and W. A. Freiwald. 2008. Comparing Face Patch Systems in Macaques and Humans. *PNAS* **105**:19514–19519. [17]

Tsuda, I. 2001. Toward an Interpretation of Dynamic Neural Activity in Terms of Chaotic Dynamical Systems. *Behav. Brain Sci.* **24**:793–810. [14]

Tsunekawa, Y., R. K. Terhune, I. Fujita, et al. 2016. Developing a de Novo Targeted Knock-in Method Based on *in Utero* Electroporation into the Mammalian Brain. *Development* **143**:3216–3222. [5]

Tufte, E. R. 2001. The Visual Display of Quantitative Information. Cheshire, CT: Graphics Press. [7]

Turchi, J., C. Chang, F. Q. Ye, et al. 2017. The Basal Forebrain Regulates Resting-State fMRI Fluctuations. *Neuron* **97**:940–952. [9]

Turrero Garcia, M., Y. Chang, Y. Arai, and W. B. Huttner. 2016. S-Phase Duration Is the Main Target of Cell Cycle Regulation in Neural Progenitors of Developing Ferret Neocortex. *J. Comp. Neur.* **524**:456–470. [2]

Turrero Garcia, M., and C. C. Harwell. 2017. Radial Glia in the Ventral Telencephalon. *FEBS Lett.* **591**:3942–3959. [5]

Tyler, L. K., W. D. Marslen-Wilson, B. Randall, et al. 2011. Left Inferior Frontal Cortex and Syntax: Function, Structure and Behaviour in Patients with Left Hemisphere Damage. *Brain* **134**:415–431. [15]

Tyler, W. A., and T. F. Haydar. 2010. A New Contribution to Brain Convolution: Progenitor Cell Logistics during Cortex Development. *Nat. Neurosci.* **13**:656–657. [2]

———. 2013. Multiplex Genetic Fate Mapping Reveals a Novel Route of Neocortical Neurogenesis, Which Is Altered in the Ts65Dn Mouse Model of Down Syndrome. *J. Neurosci.* **33**:5106–5119. [2]

Tyler, W. A., M. Medalla, T. Guillamon-Vivancos, J. I. Luebke, and T. F. Haydar. 2015. Neural Precursor Lineages Specify Distinct Neocortical Pyramidal Neuron Types. *J. Neurosci.* **35**:6142–6152. [2, 5]

Tytell, E. D., P. Holmes, and A. H. Cohen. 2011. Spikes Alone Do Not Behavior Make: Why Neuroscience Needs Biomechanics. *Curr. Opin. Neurobiol.* **21**:816–822. [16]

Uhlhaas, P., G. Pipa, B. Lima, et al. 2009. Neural Synchrony in Cortical Networks: History, Concept and Current Status. *Front. Integr. Neurosci.* **3**:17. [10, 11]

Uhrig, L., S. Dehaene, and B. Jarraya. 2014. A Hierarchy of Responses to Auditory Regularities in the Macaque Brain. *J. Neurosci.* **34**:1127–1132. [15]

Ulinski, P. S. 1986. Organization of Corticogeniculate Projections in the Turtle, Pseudemys Scripta. *J. Comp. Neur.* **254**:529–542. [5]

Ungerleider, L. G., and M. Mishkin. 1982. Two Cortical Visual Systems. In: Analysis of Visual Behavior, ed. D. J. Ingle et al., pp. 549–586. Cambridge, MA: MIT Press. [9]

Vaadia, E., I. Haalman, M. Abeles, et al. 1995. Dynamics of Neuronal Interactions in Monkey Cortex in Relation to Behavioural Events. *Nature* **373**:515–518. [12]

van den Heuvel, M. P., E. T. Bullmore, and O. Sporns. 2016. Comparative Connectomics. *Trends Cogn. Sci.* **20**:345–361. [6, 7, 9]

van den Heuvel, M. P., and O. Sporns. 2013. Network Hubs in the Human Brain. *Trends Cogn. Sci.* **17**:683–696. [9]

van der Worp, H. B., D. W. Howells, E. S. Sena, et al. 2010. Can Animal Models of Disease Reliably Inform Human Studies? *PLoS Med.* **7**:e1000245. [17]

van Diessen, E., T. Numan, E. van Dellen, et al. 2015. Opportunities and Methodological Challenges in EEG and MEG Resting State Functional Brain Network Research. *Clin. Neurophysiol.* **126**:1468–1481. [7]

Van Essen, D. C. 1997. A Tension-Based Theory of Morphogenesis and Compact Wiring in the Central Nervous System. *Nature* **385**:313–318. [2]

van Leeuwen, C. 2008. Chaos Breeds Autonomy: Connectionist Design between Bias and Baby-Sitting. *Cogn. Process.* **9**:83–92. [14]

Vanni, M. P., A. W. Chan, M. Balbi, G. Silasi, and T. H. Murphy. 2017. Mesoscale Mapping of Mouse Cortex Reveals Frequency-Dependent Cycling between Distinct Macroscale Functional Modules. *J. Neurosci.* **37**:7513–7533. [6]

Van Rullen, R. 2016. Perceptual Cycles. *Trends Cogn. Sci.* **20**:723–735. [10]

Van Rullen, R., A. Delmore, and S. Thorpe. 2001. Feed-Forward Contour Integration in Primary Visual Cortex Based on Asynchronous Spike Propagation. *Neurocomputing* **38**:1003–1009. [10]

Van Rullen, R., R. Guyonneau, and S. J. Thorpe. 2005. Spike Times Make Sense. *Trends Neurosci.* **28**:1–4. [10]

Van Rullen, R., and S. J. Thorpe. 2001. Rate Coding versus Temporal Order Coding: What the Retinal Ganglion Cells Tell the Visual Cortex. *Neural Comput.* **13**:1255–1283. [13]

Varon, R., C. Vissinga, M. Platzer, et al. 1998. Nibrin, a Novel DNA Double-Strand Break Repair Protein, Is Mutated in Nijmegen Breakage Syndrome. *Cell* **93**:467–476. [4]

Varshney, L. R., B. L. Chen, E. Paniagua, D. H. Hall, and D. B. Chklovskii. 2011. Structural Properties of the Caenorhabditis elegans Neuronal Network. *PLoS Comput. Biol.* **7**:e1001066. [7]

Vasistha, N. A., F. Garcia-Moreno, S. Arora, et al. 2015. Cortical and Clonal Contribution of Tbr2 Expressing Progenitors in the Developing Mouse Brain. *Cereb. Cortex* **25**:3290–3302. [5]

Veit, J., R. Hakim, M. P. Jadi, T. J. Sejnowski, and H. Adesnik. 2017. Cortical Gamma Band Synchronization through Somatostatin Interneurons. *Nat. Neurosci.* **20**:951–959. [10]

Vélez-Fort, M., C. V. Rousseau, C. J. Niedworok, et al. 2014. The Stimulus Selectivity and Connectivity of Layer Six Principal Cells Reveals Cortical Microcircuits Underlying Visual Processing. *Neuron* **83**:1431–1443. [8, 9]

Vermunt, M. W., S. C. Tan, B. Castelijns, et al. 2016. Epigenomic Annotation of Gene Regulatory Alterations during Evolution of the Primate Brain. *Nat. Neurosci.* **19**:494–503. [3]

Vern, B. A., B. J. Leheta, V. C. Juel, et al. 1997. Interhemispheric Synchrony of Slow Oscillations of Cortical Blood Volume and Cytochrome Aa3 Redox State in Unanesthetized Rabbits. *Brain Res.* **775**:233–239. [6]

Vertes, P. E., A. F. Alexander-Bloch, and E. T. Bullmore. 2014. Generative Models of Rich Clubs in Hebbian Neuronal Networks and Large-Scale Human Brain Networks. *Phil. Trans. R. Soc. B* **369**:1653. [7]

Vertes, P. E., A. F. Alexander-Bloch, N. Gogtay, et al. 2012. Simple Models of Human Brain Functional Networks. *PNAS* **109**:5868–5873. [7]

Vicente, R., L. L. Gollo, C. R. Mirasso, I. Fischer, and G. Pipa. 2008. Dynamical Relaying Can Yield Zero Time Lag Neuronal Synchrony Despite Long Conduction Delays. *PNAS* **105**:17157–17162. [10, 11]

Villa, A. E., I. V. Tetko, B. Hyland, and A. Najem. 1999. Spatiotemporal Activity Patterns of Rat Cortical Neurons Predict Responses in a Conditioned Task. *PNAS* **96**:1106–1111. [12]

Vincent, J. L., G. H. Patel, M. D. Fox, et al. 2007. Intrinsic Functional Architecture in the Anaesthetized Monkey Brain. *Nature* **447**:83–86. [6]

Vinck, M., and C. A. Bosman. 2016. More Gamma More Predictions: Gamma-Synchronization as a Key Mechanism for Efficient Integration of Classical Receptive Field Inputs with Surround Predictions. *Front. Syst. Neurosci.* **10**:1–27. [10]

Vinck, M., B. Lima, T. Womelsdorf, et al. 2010. Gamma-Phase Shifting in Awake Monkey Visual Cortex. *J. Neurosci.* **30**:1250–1257. [13]

Visel, A., L. Taher, H. Girgis, et al. 2013. A High-Resolution Enhancer Atlas of the Developing Telencephalon. *Cell* **152**:895–908. [5]

Vogels, R. 1990. Population Coding of Stimulus Orientation by Striate Cortical Cells. *Biol. Cybern.* **64**:25–31. [12]

Vogels, R., W. Spileers, and G. A. Orban. 1989. The Response Variability of Striate Cortical Neurons in the Behaving Monkey. *Exp. Brain Res.* **77**:432–436. [12]

von der Malsburg, C., W. A. Phillips, and W. Singer. 2010. Dynamic Coordination in the Brain: From Neurons to Mind, vol. 5. Strüngmann Forum Reports, J. Lupp, series ed. Cambridge, MA: MIT Press. [10, 14]

von Helmholtz, H. 1896. Handbuch Der Physiologischen Optik, vol. 2. Umgearbeitete Auflage. Hamburg: Leopold Voss. [10]

Vyazovskiy, V. V., and K. D. Harris. 2013. Sleep and the Single Neuron: The Role of Global Slow Oscillations in Individual Cell Rest. *Nat. Rev. Neurosci.* **14**:443–451. [13]

Wahl, M., F. Marzinzik, A. D. Friederici, et al. 2008. The Human Thalamus Processes Syntactic and Semantic Language Violations. *Neuron* **59**:695–707. [17]

Wallace, C. S., and D. L. Dowe. 1999. Minimum Message Length and Kolmogorov Complexity. *Comput. J.* **42**:270–283. [14]

Walsh, C. A. 1999. Genetic Malformations of the Human Cerebral Cortex. *Neuron* **23**:19–29. [4]

Walsh, C. A., and E. C. Engle. 2010. Allelic Diversity in Human Developmental Neurogenetics: Insights into Biology and Disease. *Neuron* **68**:245–253. [4]

Wamsley, B., and G. Fishell. 2017. Genetic and Activity-Dependent Mechanisms Underlying Interneuron Diversity. *Nat. Rev. Neurosci.* **18**:299–309. [3, 5, 9]

Wang, H., H. Yang, C. S. Shivalila, et al. 2013. One-Step Generation of Mice Carrying Mutations in Multiple Genes by CRISPR/Cas-Mediated Genome Engineering. *Cell* **153**:910–918. [5]

Wang, H. Y., H. C. Chien, N. Osada, et al. 2007a. Rate of Evolution in Brain-Expressed Genes in Humans and Other Primates. *PLoS Biol.* **5**:e13. [3]

Wang, L., M. Amalric, W. Fang, et al. 2018. Representation of Spatial Sequences Using Nested Rules in Human Prefrontal Cortex. *Neuroimage* **186**:245–255. [17]

Wang, L., L. Uhrig, B. Jarraya, and S. Dehaene. 2015. Representation of Numerical and Sequential Patterns in Macaque and Human Brains. *Curr. Biol.* **25**:1966–1974. [15]

Wang, W., S. S. Chan, D. A. Heldman, and D. W. Moran. 2007b. Motor Cortical Representation of Position and Velocity during Reaching. *J. Neurophysiol.* **97**:4258–4270. [12]

Wang, X., C. Studholme, P. L. Grigsby, et al. 2017. Folding, but Not Surface Area Expansion, Is Associated with Cellular Morphological Maturation in the Fetal Cerebral Cortex. *J. Neurosci.* **37**:1971–1983. [2]

Wang, X., J.-W. Tsai, B. Lamonica, and A. R. Kriegstein. 2011a. A New Subtype of Progenitor Cell in the Mouse Embryonic Neocortex. *Nat. Neurosci.* **14**:555–561. [5]

Wang, X., J. W. Tsai, J. H. Imai, et al. 2009. Asymmetric Centrosome Inheritance Maintains Neural Progenitors in the Neocortex. *Nature* **461**:947–955. [2]

Wang, X. J. 2010. Neurophysiological and Computational Principles of Cortical Rhythms in Cognition. *Physiol. Rev.* **90**:1195–1268. [17]

Wang, Y., G. Li, A. Stanco, et al. 2011b. CXCR4 and CXCR7 Have Distinct Functions in Regulating Interneuron Migration. *Neuron* **69**:61–76. [5]

Ware, M. L., S. F. Tavazoie, C. B. Reid, and C. A. Walsh. 1999. Coexistence of Widespread Clones and Large Radial Clones in Early Embryonic Ferret Cortex. *Cereb. Cortex* **9**:636–645. [5]

Waterson, R. H., E. S. Lander, and R. K. Wilson. 2005. The Chimpanzee Sequencing Analysis Consortium: Initial Sequence of the Chimpanzee Genome and Comparison with the Human Genome. *Nature* **437**:69–87. [3]

Watts, D. J., and S. H. Strogatz. 1998. Collective Dynamics of "Small-World" Networks. *Nature* **393**:440–442. [7]

Weaver, A. H. 2005. Reciprocal Evolution of the Cerebellum and Neocortex in Fossil Humans. *PNAS* **102**:3576–3580. [5]

Wedeen, V. J., D. L. Rosene, R. Wang, et al. 2012. The Geometric Structure of the Brain Fiber Pathways. *Science* **335**:1628–1634. [2]

Wehr, M., and G. Laurent. 1996. Odour Encoding by Temporal Sequences of Firing in Oscillating Neural Assemblies. *Nature* **384**:162–166. [13]

Weiskrantz, L. 1997. Consciousness Lost and Found: A Neuropsychological Exploration. New York: Oxford Univ. Press. [15]

Weiss, D. S., and A. Keller. 1994. Specific Patterns of Intrinsic Connections between Representation Zones in the Rat Motor Cortex. *Cereb. Cortex* **4**:205–214. [8]

Welagen, J., and S. Anderson. 2011. Origins of Neocortical Interneurons in Mice. *Dev. Neurobiol.* **71**:10–17. [5]

Welker, W. I. 1990. The Significance of Foliation and Fissuration of Cerebellar Cortex. The Cerebellar Folium as a Fundamental Unit of Sensorimotor Integration. *Arch. Ital. Biol.* **128**:87–109. [5]

Werchan, D. M., A. G. Collins, M. J. Frank, and D. Amso. 2015. 8-Month-Old Infants Spontaneously Learn and Generalize Hierarchical Rules. *Psychol. Sci.* **26**:805–815. [17]

Werner, G., and V. B. Mountcastle. 1965. Neural Activity in Mechanoreceptive Cutaneo Afferents: Stimulus-Response Relations, Weber Functions and Information Transmission. *J. Neurophysiol.* **28**: [13]

Wespatat, V., F. Tennigkeit, and S. W. 2004. Phase Sensitivity of Synaptic Modifications in Oscillating Cells of Rat Visual Cortex. *Journal of Neuroscience.* **24**:9067–9075. [10]

West, G. 2017. Scale: The Universal Laws of Growth, Innovation, Sustainability, and the Pace of Life in Organisms, Cities, Economies, and Companies. New York: Penguin. [16]

Whittington, M. A., R. D. Traub, N. Kopell, B. Ermentrout, and E. H. Buhl. 2000. Inhibition-Based Rhythms: Experimental and Mathematical Observations on Network Dynamics. *Int. J. Psychophysiol.* **38**:315–336. [10]

Wildegger, T., F. van Ede, M. Woolrich, C. R. Gillebert, and A. C. Nobre. 2017. Preparatory Alpha-Band Oscillations Reflect Spatial Gating Independently of Predictions Regarding Target Identity. *J. Neurophysiol.* **117**:1385–1394. [14]

Williams, S. E., I. Garcia, A. J. Crowther, et al. 2015. ASPM Sustains Postnatal Cerebellar Neurogenesis and Medulloblastoma Growth in Mice. *Development* **142**:3921–3932. [4]

Willner, P. 1984. The Validity of Animal Models of Depression. *Psychopharmacology* **83**:1–16. [7]

———. 2017. Reliability of the Chronic Mild Stress Model of Depression: A User Survey. *Neurobiol. Stress* **6**:68–77. [7]

Wilsch-Bräuninger, M., M. Florio, and W. B. Huttner. 2016. Neocortex Expansion in Development and Evolution: From Cell Biology to Single Genes. *Curr. Opin. Neurobiol.* **39**:122–132. [5]

Wilson, B., Y. Kikuchi, L. Sun, et al. 2015. Auditory Sequence Processing Reveals Evolutionarily Conserved Regions of Frontal Cortex in Macaques and Humans. *Nat. Commun.* **6**:8901. [17]

Wilson, B., W. D. Marslen-Wilson, and C. I. Petkov. 2017. Conserved Sequence Processing in Primate Frontal Cortex. *Trends Neurosci.* **40**:72–82. [14, 15]

Winfree, A. T. 1967. Biological Rhythms and the Behavior of Populations of Coupled Oscillators. *J. Theor. Biol.* **16**:15–42. [10]

Wise, S. P. 2008. Forward Frontal Fields: Phylogeny and Fundamental Function. *Trends Neurosci.* **31**:599–608. [5]

Witte, R. S., P. J. Rousche, and D. R. Kipke. 2007. Fast Wave Propagation in Auditory Cortex of an Awake Cat Using a Chronic Microelectrode Array. *J. Neural Eng.* **4**:68–78. [12]

Wodlinger, B., J. E. Downey, E. C. Tyler-Kabara, et al. 2015. Ten-Dimensional Anthropomorphic Arm Control in a Human Brain-Machine Interface: Difficulties, Solutions, and Limitations. *J. Neural Eng.* **12**:016011. [12, 13]

Womelsdorf, T., J.-M. Schoffelen, R. Oostenveld, et al. 2007. Modulation of Neuronal Interactions through Neuronal Synchronization. *Science* **316**:1609–1612. [10, 14]

Wonders, C. P., and S. A. Anderson. 2006. The Origin and Specification of Cortical Interneurons. *Nat. Rev. Neurosci.* **7**:687–696. [3, 5]

Woodworth, M. B., K. M. Girskis, and C. A. Walsh. 2017. Building a Lineage from Single Cells: Genetic Techniques for Cell Lineage Tracking. *Nat. Rev. Genet.* **18**:230–244. [5]

Workman, A. D., C. J. Charvet, B. Clancy, R. B. Darlington, and B. L. Finlay. 2013. Modeling Transformations of Neurodevelopmental Sequences across Mammalian Species. *J. Neurosci.* **33**:7368–7383. [9]

Wrobel, M. R., and H. G. Sundararaghavan. 2014. Directed Migration in Neural Tissue Engineering. *Tissue Eng. Part B, Rev.* **20**:93–105. [7]

Wrona, D. 2006. Neural-Immune Interactions: An Integrative View of the Bidirectional Relationship between the Brain and Immune Systems. *J Neuroimmunol.* **172**:38–58. [9]

Xing, D., Y. Shen, S. Burns, et al. 2012. Stochastic Generation of Gamma-Band Activity in Primary Visual Cortex of Awake and Anesthetized Monkeys. *J. Neurosci.* **32**:13873–13880a. [13]

Xu, K., E. E. Schadt, K. S. Pollard, P. Roussos, and J. T. Dudley. 2015. Genomic and Network Patterns of Schizophrenia Genetic Variation in Human Evolutionary Accelerated Regions. *Mol. Biol. Evol.* **32**:1148–1160. [4]

Xu, W., X. Huang, K. Takagaki, and J. Y. Wu. 2007. Compression and Reflection of Visually Evoked Cortical Waves. *Neuron* **55**:119–129. [12]

Yamamoto, J., J. Suh, D. Takeuchi, and S. Tonegawa. 2014. Successful Execution of Working Memory Linked to Synchronized High-Frequency Gamma Oscillations. *Cell* **157**:845–857. [10]

Yamashita, W., M. Takahashi, T. Kikkawa, et al. 2018. Conserved and Divergent Functions of Pax6 Underlie Species-Specific Neurogenic Patterns in the Developing Amniote Brain. *Development* **145**:159764. [5]

Yamins, D. L., H. Hong, C. Cadieu, and J. J. DiCarlo. 2013. Hierarchical Modular Optimization of Convolutional Networks Achieves Representations Similar to Macaque It and Human Ventral Stream. In: Proceedings of the 26th International Conference on Neural Information Processing Systems ed. C. J. C. Burges et al., vol. 2, pp. 3093–3101 Lake Tahoe: NIPS. [13]

Yamins, D. L., H. Hong, C. F. Cadieu, et al. 2014. Performance-Optimized Hierarchical Models Predict Neural Responses in Higher Visual Cortex. *PNAS* **111**:8619–8624. [13, 17]

Yan, G., P. E. Vertes, E. K. Towlson, et al. 2017. Network Control Principles Predict Neuron Function in the Caenorhabditis elegans Connectome. *Nature* **550**:519–523. [7]

Yang, C. 2013. Ontogeny and Phylogeny of Language. *PNAS* **110**:6324–6327. [15, 17]

Yang, M. G., and A. E. West. 2016. Editing the Neuronal Genome: A CRISPR View of Chromatin Regulation in Neuronal Development, Function, and Plasticity. *Yale J. Biol. Med.* **89**:457–470. [5]

Yang, Y. J., A. E. Baltus, R. S. Mathew, et al. 2012. Microcephaly Gene Links Trithorax and Rest/Nrsf to Control Neural Stem Cell Proliferation and Differentiation. *Cell* **151**:1097–1112. [4]

Ye, Z., M. A. Mostajo-Radji, J. R. Brown, et al. 2015. Instructing Perisomatic Inhibition by Direct Lineage Reprogramming of Neocortical Projection Neurons. *Neuron* **88**:475–483. [5]

Yeo, B. T., F. M. Krienen, J. Sepulcre, et al. 2011. The Organization of the Human Cerebral Cortex Estimated by Intrinsic Functional Connectivity. *J. Neurophysiol.* **106**:1125–1165. [6]

Yoon, M. J., B. K. Koo, R. Song, et al. 2008. Mind Bomb-1 Is Essential for Intraembryonic Hematopoiesis in the Aortic Endothelium and the Subaortic Patches. *Mol. Cell. Biol.* **28**:4794–47804. [2]

Young, M. P., J. W. Scannell, G. A. Burns, and C. Blakemore. 1994. Analysis of Connectivity: Neural Systems in the Cerebral Cortex. *Rev. Neurosci.* **5**:227–250. [7]

Yovel, G., and W. A. Freiwald. 2013. Face Recognition Systems in Monkey and Human: Are They the Same Thing? *F1000Prime Rep.* **5**:10. [17]

Yu, A. J., and P. Dayan. 2005. Uncertainty, Neuromodulation and Attention. *Neuron* **46**:681–692. [14]

Yu, B. M., J. P. Cunningham, G. Santhanam, et al. 2009a. Gaussian-Process Factor Analysis for Low-Dimensional Single-Trial Analysis of Neural Population Activity. *J. Neurophysiol.* **102**:614–635. [12]

Yu, Q., L. Wu, D. A. Bridwell, et al. 2016. Building an EEG-fMRI Multi-Modal Brain Graph: A Concurrent EEG-fMRI Study. *Front. Hum. Neurosci.* **10**:476. [7]

Yu, T. W., M. H. Chahrour, M. E. Coulter, et al. 2013. Using Whole-Exome Sequencing to Identify Inherited Causes of Autism. *Neuron* **77**:259–273. [4]

Yu, T. W., G. H. Mochida, D. J. Tischfield, et al. 2010. Mutations in *WDR62*, Encoding a Centrosome-Associated Protein, Cause Microcephaly with Simplified Gyri and Abnormal Cortical Architecture. *Nat. Genet.* **42**:1015–1020. [4]

Yu, Y. C., R. S. Bultje, X. Wang, and S. H. Shi. 2009b. Specific Synapses Develop Preferentially among Sister Excitatory Neurons in the Neocortex. *Nature* **458**:501–504. [2]

Yu, Y. C., S. He, S. Chen, et al. 2012. Preferential Electrical Coupling Regulates Neocortical Lineage-Dependent Microcircuit Assembly. *Nature* **486**:113–117. [5]

Yuille, A., and D. Kersten. 2006. Vision as Bayesian Inference: Analysis by Synthesis? *Trends Cogn. Sci.* **10**:301–308. [14]

Yuste, R., J. N. MacLean, J. Smith, and A. Lansner. 2005. The Cortex as a Central Pattern Generator. *Nat. Rev. Neurosci.* **6**:477–483. [10]

Zeitler, M., P. Fries, and C. C. A. M. Gielen. 2008. Biased Competition through Variations in Amplitude of $\Gamma$-Oscillations. *J. Comput. Neurosci.* **25**:89–107. [10]

Zeng, J., G. Konopka, B. G. Hunt, et al. 2012. Divergent Whole-Genome Methylation Maps of Human and Chimpanzee Brains Reveal Epigenetic Basis of Human Regulatory Evolution. *Am. J. Hum. Genet.* **91**:455–465. [3]

Zhang, H., and J. Jacobs. 2015. Traveling Theta Waves in the Human Hippocampus. *J. Neurosci.* **35**:12477–12487. [12]

Zhang, J. 2003. Evolution of the Human *ASPM* Gene, a Major Determinant of Brain Size. *Genetics* **165**:2063–2070. [4]

Zhang, X., J. Ling, G. Barcia, et al. 2014. Mutations in *QARS*, Encoding Glutaminyl-tRNA Synthetase, Cause Progressive Microcephaly, Cerebral-Cerebellar Atrophy, and Intractable Seizures. *The American Journal of Human Genetics* **94**:547–558. [4]

Zhang, Y. S., and A. A. Ghazanfar. 2018. Vocal Development through Morphological Computation. *PLoS Biol.* **16**:e2003933. [16]

Zhong, S., S. Zhang, X. Fan, et al. 2018. A Single-Cell RNA-Seq Survey of the Developmental Landscape of the Human Prefrontal Cortex. *Nature* **555**:524–528. [3]

Zilles, K., and K. Amunts. 2018. Cytoarchitectonic and Receptorarchitectonic Organization in Broca's Region and Surrounding Cortex. *Curr. Opin. Behav. Sci.* **21**:93–105. [17]

Zilles, K., G. Schlaug, M. Matelli, et al. 1995. Mapping of Human and Macaque Sensorimotor Areas by Integrating Architectonic, Transmitter Receptor, MRI and PET Data. *J. Anat.* **187**:515–537. [16]

Zimmer, F., and S. H. Montgomery. 2015. Phylogenetic Analysis Supports a Link between DUF1220 Domain Number and Primate Brain Expansion. *Genome Biol. Evol.* **7**:2083–2088. [3]

Zohary, E., M. N. Shadlen, and W. T. Newsome. 1994. Correlated Neuronal Discharge Rate and Its Implications for Psychophysical Performance. *Nature* **370**:140–143. [12, 13]

# Subject Index

# Strüngmann Forum Report Series

*Agrobiodiversity: Integrating Knowledge for a Sustainable Future*
Edited by Karl S. Zimmerer and Stef de Haan
ISBN: 9780262038683

*Rethinking Environmentalism: Linking Justice, Sustainability, and Diversity*
Edited by Sharachchandra Lele, Eduardo S. Brondizio, John Byrne,
Georgina M. Mace and Joan Martinez-Alier
ISBN: 9780262038966

*Emergent Brain Dynamics: Prebirth to Adolescence*
Edited by April A. Benasich and Urs Ribary
ISBN: 9780262038638

*The Cultural Nature of Attachment: Contextualizing Relationships and Development*
Edited by Heidi Keller and Kim A. Bard
Hardcover: ISBN: 9780262036900, ebook: ISBN: 9780262342865
Winner of the Ursula Gielen Global Psychology Book Award

*Investors and Exploiters in Ecology and Economics: Principles and Applications*
edited by Luc-Alain Giraldeau, Philipp Heeb and Michael Kosfeld
Hardcover: ISBN: 9780262036122, eBook: ISBN: 9780262339797

*Computational Psychiatry: New Perspectives on Mental Illness*
edited by A. David Redish and Joshua A. Gordon, ISBN: 9780262035422

*Complexity and Evolution: Toward a New Synthesis for Economics*
edited by David S. Wilson and Alan Kirman, ISBN: 9780262035385

*The Pragmatic Turn: Toward Action-Oriented Views in Cognitive Science*
edited by Andreas K. Engel, Karl J. Friston and Danica Kragic
ISBN: 978-0-262-03432-6

*Translational Neuroscience: Toward New Therapies*
edited by Karoly Nikolich and Steven E. Hyman, ISBN: 9780262029865

*Trace Metals and Infectious Diseases*
edited by Jerome O. Nriagu and Eric P. Skaar, ISBN 978-0-262-02919-3

*Rethinking Global Land Use in an Urban Era*
edited by Karen C. Seto and Anette Reenberg, ISBN 978-0-262-02690-1

*Schizophrenia: Evolution and Synthesis*
edited by Steven M. Silverstein, Bita Moghaddam and Til Wykes,
ISBN 978-0-262-01962-0

*Cultural Evolution: Society, Technology, Language, and Religion*
edited by Peter J. Richerson and Morten H. Christiansen,
ISBN 978-0-262-01975-0

Available at https://mitpress.mit.edu/books/series/strungmann-forum-reports